Recent Developments in Materials Science

Recent Developments in Materials Science

Edited by
Heather Dale

Willford Press
www.willfordpress.com

Published by Willford Press,
118-35 Queens Blvd., Suite 400,
Forest Hills, NY 11375, USA

ISBN: 978-1-64728-344-5

Cataloging-in-Publication Data

Recent developments in materials science / edited by Heather Dale.
p. cm.
Includes bibliographical references and index.
ISBN 978-1-64728-344-5
1. Materials science. 2. Materials. I. Dale, Heather.
TA403 .R43 2022
620.11--dc23

For information on all Willford Press publications
visit our website at www.willfordpress.com

Willford Press

Contents

Preface

Every book is initially just a concept; it takes months of research and hard work to give it the final shape in which the readers receive it. In its early stages, this book also went through rigorous reviewing. The notable contributions made by experts from across the globe were first molded into patterned chapters and then arranged in a sensibly sequential manner to bring out the best results.

Materials science includes the design and discovery of new materials. It is an interdisciplinary field which uses the concepts of physics, chemistry and engineering to understand materials and metallurgy. Materials science strives to understand the impacts of the history of a material on its structure, properties and performance. The materials paradigm refers to the understanding of the relationships between processing, structure and properties. It is used for researches in various areas such as nanotechnology, metallurgy and biomaterials. Forensic engineering and failure analysis including investigation of materials, structures and products also utilizes materials science. This book attempts to understand the multiple branches that fall under the discipline of materials science and how such concepts have practical applications. A number of latest researches have been included to keep the readers up-to-date with the global concepts in this area of study.

It has been my immense pleasure to be a part of this project and to contribute my years of learning in such a meaningful form. I would like to take this opportunity to thank all the people who have been associated with the completion of this book at any step.

Editor

1

Quantitative Analysis of Damping Enhancement and Piezoelectric Effect Mechanism of CNTs/PMN/EP Composites

Shiyue Pan,[1] Meijuan Li,[2] Fei Chen,[1] Zhixiong Huang,[3] Qiang Shen,[1] and Lianmeng Zhang[1]

[1]State Key Laboratory of Advanced Technology for Materials Synthesis and Processing, Wuhan University of Technology, Wuhan 430070, China
[2]School of Chemistry, Chemical Engineering and Life Sciences, Wuhan University of Technology, Wuhan 430070, China
[3]School of Materials Science and Engineering, Wuhan University of Technology, Wuhan 430070, China

Correspondence should be addressed to Qiang Shen; sqqf@263.net

Academic Editor: Marino Lavorgna

New types of piezoelectric damping materials, including carbon nanotubes (CNTs)/lead magnesium niobate (PMN)/epoxy (EP) resin, are developed. The tan δ area (TA) analysis method is selected to evaluate the damping properties which obviously clarifies the effect of maximum loss factor (tan δ) and effective temperature range on damping properties. Furthermore, the dominant factor of damping enhancement is quantitatively analyzed via the value of TA. Compared with PMN, the interfacial friction of CNTs acts as the dominant factor for the content less than 0.6 wt.%. The maximum damping percentage of CNTs reaches 29.14%. CNTs form loop circuits gradually with the content of CNTs increasing, and electrical energy generated via piezoelectric effect of PMN is efficiently dissipated through the conductive network. Thus, PMN becomes the dominant factor as the content of CNTs exceeds 0.8 wt.%, and the damping percentage reaches 47.43% at the content of CNTs of 1.0 wt.%.

1. Introduction

In the past few decades, the viscoelastic material with an inherent high loss factor is widely applied in automobiles, aircraft, and military equipment to suppress vibration and impact noises [1]. For example, soft viscoelastic layers in viscoelastic sandwich structures are provided as the damping component in order to improve dynamic response [2]. A new type of damping enhancement with piezoelectric and conductive materials as fillers has been developed by Hagood and von Flotow [3, 4]. Electrical energy, generated by piezoelectric effect of piezoelectric ceramics, can be efficiently dissipated by external electronic circuits [5]. Conductive fillers are introduced into piezoelectric-ceramic/polymer composites as passive electrical network, such as carbon nanotubes (CNTs), carbon black (CB), graphite, and aluminum particles [6–13]. CNTs are widely used due to the high electric current carrying capacity (10^{11}~10^{12} A/cm^2) [14]. It is revealed that the CNTs provide superior conductivity at very low concentrations which can transfer electrical energy into thermal energy efficiently [14]. Tian et al. reported a type of CNTs/lead zirconate titanate (PZT)/EP composites. They found that the generated electrical energy was well dissipated, and the loss factors of composites were improved by the incorporation of PZT and CNT at critical electrical percolation [7, 8]. Carponcin et al. studied the influence of piezoelectric on damping properties of CNTs/PZT/polyamide composites, and it was found that the polarization of piezoelectric ceramic particles improved the loss factor of 20% reaching to 0.052 [9]. PMN as another piezoelectric material with perovskite structure also possess high piezoelectric coefficient and dielectric permittivity. PMN piezoelectric materials were used in 0–3 piezoelectric composites for damping enhancement gradually [10–12]. Li and Du focused on damping properties of nitrile-butadiene rubber/phenolic resin/PMN/CB composite. Mechanical energy was converted efficiently via piezoelectric effect at PMN volume content of 50%, and the maximum loss factor reached 0.813 [10].

Shamir et al. found that adding CB to the composites reduced the peak damping intensity but enhanced damping at higher temperature, and the effective damping temperature range was widened [15]. Chang et al. investigated the damping property of polyurethane composites with different ultraviolet absorbents, and they found that the $\tan\delta$ increased at the temperature range of higher than the glass transition temperature though the $\tan\delta_{max}$ was gradually reduced [16]. Thus, in this study, $\tan\delta$ area (TA) and loss modulus area (LA) values are selected as damping performance parameter of materials which obviously clarify the effect of maximum loss factor and effective temperature range on damping properties [17–20]. Furthermore, TA analysis method can be used for quantitative analysis of damping enhancement.

In this study, conductive phase (CNTs) and piezoelectric phase (PMN) are introduced into epoxy matrix to fabricate piezoelectric damping composites. Electrical energy generated via piezoelectric effect of PMN is efficiently dissipated through conductive network. Thus, damping performance of CNTs/PMN/EP composites is effectively improved. The damping enhancement effect of each component is specifically clarified, and the damping percentage is quantitatively analyzed based on the TA analysis method.

2. Experimental Procedure

2.1. Materials. The lead magnesium niobate (PMN) ceramic powder was supplied by Baoding Hengsheng Acoustics Electron Apparatus Company with the mean particle size of 1–10 μm. The dimensions of CNTs (supplied by XFNANO) were 10–20 nm in diameter, and the purity of CNTs was >95%. Epoxy resin used in experiment was diglycidyl ether of bisphenol-A type with an epoxide equivalent weight of 185–200 g/eq. The curing agent used in experiment was polyethylene polyamine which was supplied by Aladdin. The molecular formula was $C_{2n}H_{5n}N_n$. Epoxy resin was transferred into three-dimensional network structure by catalytic polymerization and addition polymerization reaction between molecular chains when polyethylene polyamine was introduced into epoxy, which resulted in the curing of composites.

2.2. Preparation of CNTs/PMN/EP Composites. CNTs were functionalized by 3 : 1 H_2SO_4 (98%)/HNO_3 (70%) acid treatment to obtain carboxyl and hydroxyl groups. Matrices of epoxy resin (EP) with different contents of acid-treated CNTs were made. Then, PMN ceramic was preprocessed by silane coupling agents KH-560 to improve interfacial contact performance and then added to CNT/EP solution. The solution was stirred at 60°C for 12 h with magnetic stirrer and then placed in vacuum drying oven for gas evaporation. Finally, solution with curing agent was cast into a mold and cured for 5 h at room temperature, with a subsequent step at 105°C for 2 h.

The fracture surface analysis was performed by field emission scanning electron microscope (FE-SEM, Hitachi S-4800). The investigation of viscoelastic behavior was performed by dynamic mechanical thermal analysis (DMA, PYRIS-7e). Resistivity was tested by impedance analyzer (HP4294A).

2.3. TA Analysis Method. The TA analysis method is selected to evaluate the damping properties, which obviously clarifies the effect of maximum loss factor and effective temperature range on damping properties. Fradkin et al. proposed the definition of damping functions [21]:

$$\mathrm{LA} = \int_{T_G}^{T_R} E'' dT \approx \left(E'_G - E'_R\right) \frac{R}{(E_a)_{avg}} \frac{\pi}{2} T_g^2, \tag{1}$$

$$\mathrm{TA} = \int_{T_G}^{T_R} \tan\delta \, dT \approx \left(\ln E'_G - \ln E'_R\right) \frac{R}{(E_a)_{avg}} \frac{\pi}{2} T_g^2, \tag{2}$$

where E'' is the loss modulus, $\tan\delta$ is the loss factor, T_g is the glass transition temperature, E'_G and E'_R are the storage modulus in glassy and rubbery states, T_G and T_R are the initial and final temperature of glass transition, $(E_a)_{avg}$ is the activation energy of the relaxation process, and R is the gas constant.

Materials show obvious damping performances for the loss factor reaching 0.3 around damping peak, which means composites reach effective loss factor. Thus, in this research, we choose effective damping temperature range ($\tan\delta \geq 0.3$) as integral region. T_G and T_R in formulas (1) and (2) are replaced with the initial and final temperature of which $\tan\delta$ reaches 0.3. Then, TA values are obtained through integral calculation of loss factor curves over the temperature range. Toshio Ogawa proposed that loss factor had more practical significance than loss modulus in noise control and damping analysis [17]. Amorphous polymer, semicrystalline polymer, and single damping peak copolymer were used as research objects, and the functional relation between TA values and group numbers was shown as follows:

$$\mathrm{TA} = a_0 + \sum_{i=1}^{n} a_i X_i, \tag{3}$$

where X_i is the number of group i, a_i is the regression coefficient which means the contribution of TA values by each group, and a_0 is a constant.

The group contribution analysis method is used to calculate theoretical TA values and develop a relation between the contribution of each functional group and damping properties. Thus, TA values can be used for quantitative analysis of damping enhancement. The damping percentage of each component is quantitatively analyzed based on following formulas:

$$\mathrm{TA}_i = \mathrm{TA}_c - \mathrm{TA}_{c,0}, \tag{4}$$

$$D_i = \frac{\mathrm{TA}_i}{\mathrm{TA}_c} \times 100\%, \tag{5}$$

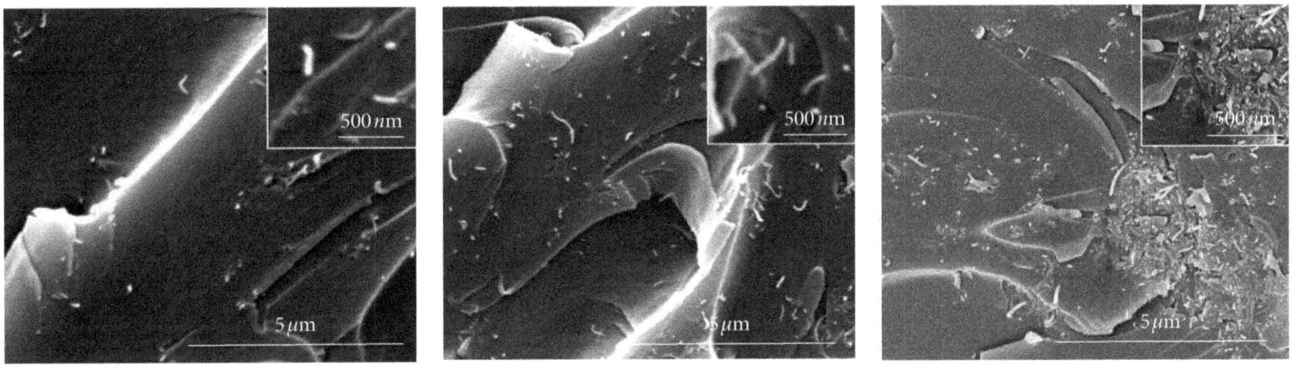

FIGURE 1: Fracture surface of CNTs/EP composites at different contents of CNTs: (a) 0.2 wt.%, (b) 1.0 wt.%, and (c) 2.0 wt.%.

where D_i is the damping percentage of component i, TA_i and TA_c are the TA value of component i and composite, and $TA_{c,0}$ is the TA value of composite at 0 wt.% content of component i.

3. Results and Discussion

As shown in Figure 1, CNTs dispersed well in the chemically functionalized CNTs/EP composites at 0.2 wt.% CNT content, which improves interfacial frictional energy dissipation of CNTs. As we can see from the magnified SEM image, the interparticle distance between carbon nanotubes is very close. When CNT content is up to 1.0 wt.%, no obvious agglomeration is observed in the matrix, and the dispersion state in the whole region still keeps well. With the content of CNTs increasing, some matrix-rich (not containing CNTs) regions appear, and some CNTs entangle with each other as shown in Figure 1(c). As we can see from the magnified SEM image, the agglomeration of CNTs weakens the interfacial contact between CNT fillers and epoxy matrix.

Dynamic mechanical analysis (DMA) of CNTs/EP composites are shown in Figure 2(a). In general, damping materials reach maximum loss factor ($\tan\delta_{max}$) at glass transition temperature (T_g). T_g of CNTs/EP composites is in correspondence with pure epoxy materials. $\tan\delta_{max}$ at T_g increases sharply and reaches optimal value at the CNT content of 0.2 wt.%, due to the interfacial friction when CNTs uniformly dispersed in epoxy matrix. High content of CNTs widens effective temperature range, while $\tan\delta_{max}$ decreases due to agglomeration of fillers and high viscosity of composites. Thus, TA values are selected as damping performance parameter which obviously clarifies the effect of maximum loss factor and wide temperature range on damping properties. Based on loss factor curves of CNTs/EP composites, TA values of composites with different contents of CNTs are obtained as shown in Figure 2(b). The frictional energy dissipation between conductive phase and epoxy matrix rises as the content of CNTs increases. The TA value of CNTs/EP composites increases correspondingly and reaches the maximum value at 0.6 wt.% content of CNTs, for the reason that CNTs widen effective temperature range at 0.6 wt.% content of CNTs compared to 0.2 wt.% CNTs/EP composite. As the CNT content increases gradually, CNTs/EP composites show higher viscosity. The effective temperature range keeps constant when the content of CNTs exceeds 0.8 wt.%. The entanglement of CNTs in epoxy matrix results in the decrease of TA value. However, the effective temperature range is widened significantly when the content of conductive phase exceeds percolation threshold. The TA value of CNTs/EP composites increases at CNT content of 2.0 wt.% consequently. Damping percentage of CNTs and epoxy matrix can be calculated by formulas (4) and (5). The damping percentage of CNTs shows the same trend with TA values and reaches 32.22% at the CNT content of 0.6 wt.%. For the reason that CNTs are uniformly dispersed in epoxy matrix without obvious agglomeration behavior, the interfacial friction is effectively promoted for damping enhancement.

As shown in Figure 3, PMN is uniformly dispersed in epoxy matrix without obvious agglomeration. Composites show strong interfacial compatibility between PMN and epoxy. Though the viscosity of PMN/EP composites increases at 60 wt.% content of PMN, the dispersion state in the whole region can still keep well. The viscosity increases significantly when the content of PMN reaches 80 wt.%. As a result, cohesive force between PMN and epoxy becomes weaker, which causes the agglomeration of PMN ceramic powders and even some holes show up in composites.

As shown in Figure 4(a), T_g of PMN/EP composites is affected obviously by the addition of PMN ceramics. T_g shifts toward high temperature region for the content less than 40 wt.%. T_g of PMN/EP composite is lower than pure epoxy material, and the effective temperature range is efficiently widened. Based on loss factor curves of PMN/EP composites, TA values of composites with different contents of PMN are obtained as shown in Figure 4(b). Piezoelectric effect of PMN ceramic is generated by external force. Bound charge exists on the crystal surface which cannot be transferred into other forms of energy. As a result, the enhancement of damping properties is mainly achieved by the interfacial friction of PMN ceramics. TA values of PMN/EP composites increase with the increase in the content of PMN and reach 14.4 at 40 wt.% content of PMN. On the one hand, the interfacial friction between fillers and polymer chains over

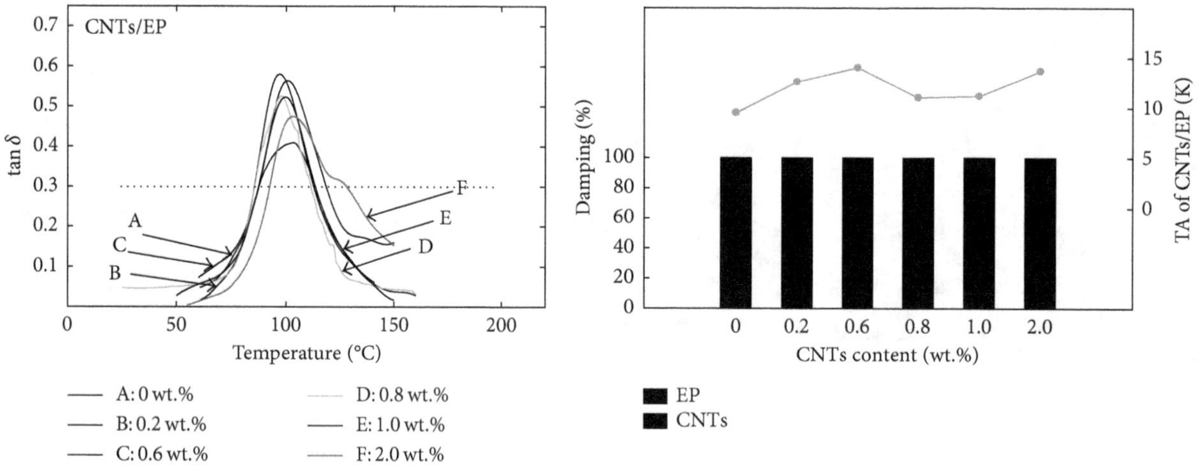

(a) (b)

FIGURE 2: Damping properties of CNTs/EP composites as a function of CNT content: (a) loss factor curves; (b) TA values and damping percentage.

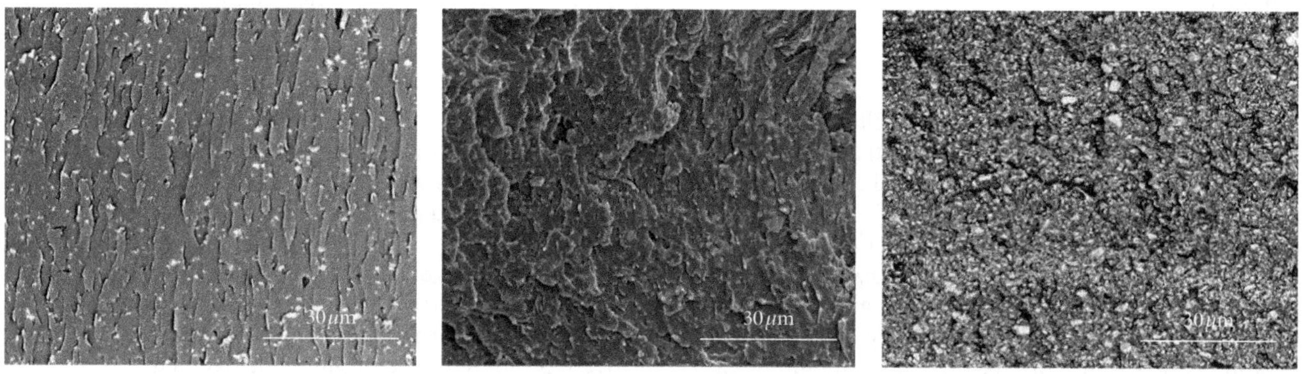

(a) (b) (c)

FIGURE 3: Fracture surface of PMN/EP composites at different contents of PMN: (a) 20 wt.%, (b) 60 wt.%, and (c) 80 wt.%.

the transition range is enhanced with the increase in the content of PMN. On the other hand, fillers reduce the free volume of elastic polymers which results in the decrease of viscoelastic damping. It means that the increase in frictional energy dissipation can be massively offset by the decrease in viscoelastic damping of matrix; thus, TA values of PMN/EP composites decrease correspondingly. Damping percentage of PMN and EP can be calculated by (4) and (5). The damping percentage of PMN shows the similar trend with TA values and reaches 33.95% at PMN content of 40 wt.%. PMN fillers are uniformly dispersed in the matrix and improve damping properties through frictional energy dissipation efficiently while maintaining the inherent viscoelastic damping of epoxy. However, damping properties cannot be improved obviously with the content of PMN exceeding 60 wt.%. The increase in frictional energy dissipation is massively offset by the decrease in viscoelastic damping of epoxy matrix, so the damping percentage of PMN is merely 4.28% at 60 wt.% PMN content.

Loss factor curves of CNTs/60 wt.% PMN/EP composites are shown in Figure 5(a). The increase in frictional energy dissipation can be massively offset by the decrease in free volume of matrix at 60 wt.% content of PMN, and the impact of piezoelectric effect on damping properties can be significantly indicated. As shown in Figure 5(a), $\tan\delta_{max}$ increases sharply at 1.0 wt.% content of CNTs. Combined with TA values of CNTs/EP composites, the damping percentage of each component is shown in Figure 5(b). TA values of CNTs/60 wt.% PMN/EP composites increase gradually with the increasing content of CNTs and reach 21.26 at 1.0 wt.% CNT content. The effect of piezoelectric phase (PMN) and conductive phase (CNTs) on damping properties can be explained from two aspects. On the one hand, the interfacial friction between fillers and epoxy matrix promotes energy dissipation efficiently. On the other hand, piezoelectric effect of PMN results in the enrichment of positive and negative charge on crystal surface. Then, electrical energy converted from mechanical energy is dissipated through continuous conductive network (CNTs). When the content of CNTs is less than 0.6 wt.%, CNT fillers exist as independent units rather than conductive network. Thus, bound charge on the crystal surface cannot be transferred into other forms of energy. The interfacial effect of CNTs is the dominant factor of damping enhancement

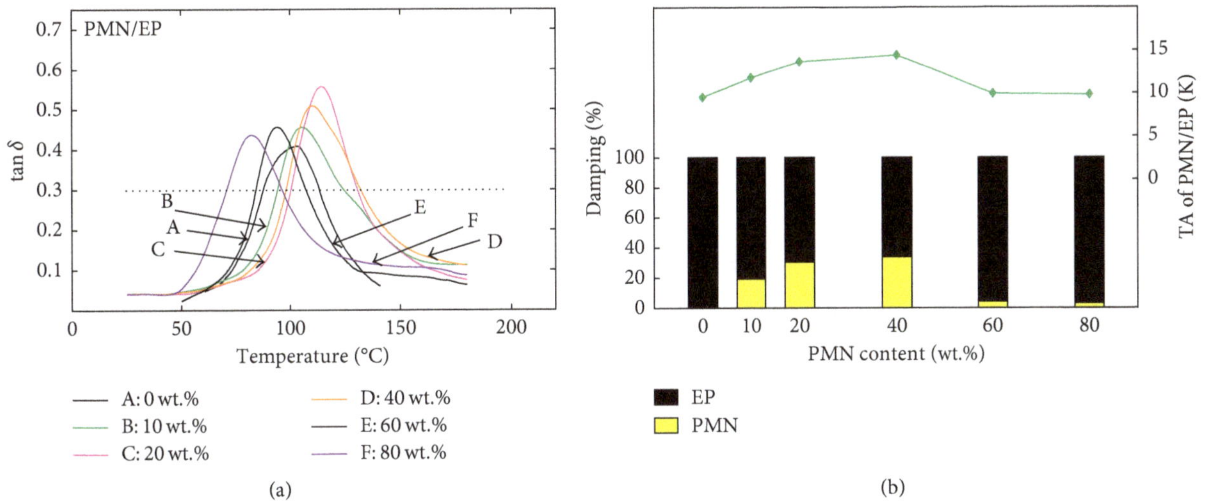

Figure 4: Damping properties of PMN/EP composites as a function of PMN content: (a) loss factor curves; (b) TA values and damping percentage.

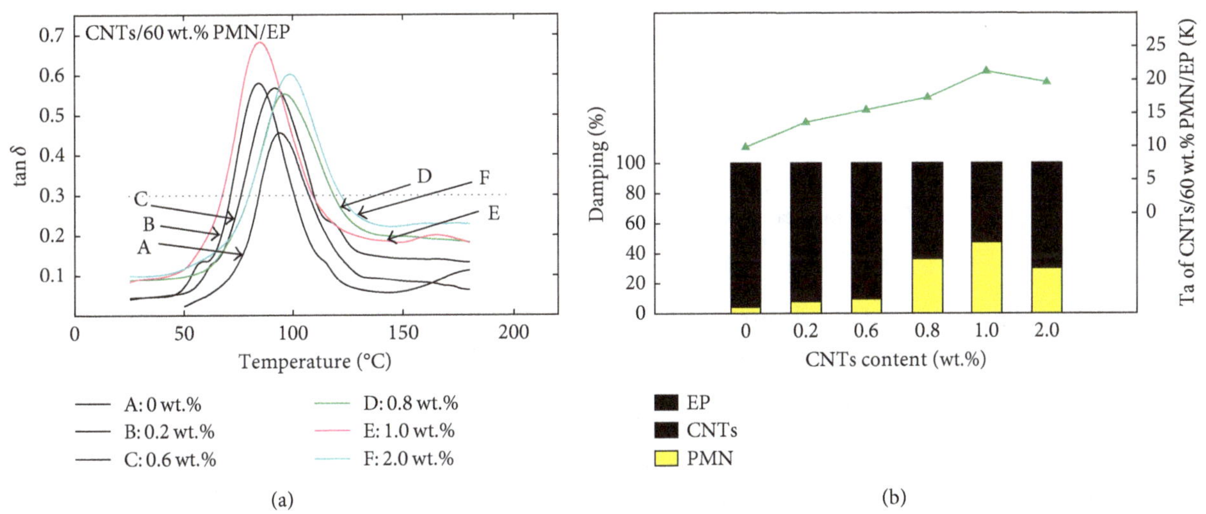

Figure 5: Damping properties of CNTs/PMN/EP composites as a function of CNT content: (a) loss factor curves; (b) TA values and damping percentage.

through frictional energy dissipation. The damping percentage of CNTs reaches 29.14% at CNT content of 0.6 wt.%. The conductive network generates gradually for the content exceeding 0.8 wt.%. Electrical energy generated via piezoelectric effect of PMN is dissipated in the form of thermal energy; thus, piezoelectric phase (PMN) becomes the dominant factor of damping enhancement. Damping percentage of PMN increases obviously and reaches 47.43% at 1.0 wt.% CNT content. When the content of CNTs reaches percolation threshold (1.0 wt.%), the conversion efficiency of electrical energy is the highest. When the CNT content increases to 2.0 wt.%, the high viscosity of composites results in the agglomeration behavior of fillers which reduces functional energy dissipation. Furthermore, high content of CNTs forms short circuit network which decreases energy conversion efficiency. As a result, TA values of CNTs/60 wt.% PMN/EP composites and the damping percentage of PMN decrease. The damping percentage of PMN shows the similar trend with TA values. It is indicated that damping properties of composites are significantly improved by piezoelectric effect of PMN. Thus, the effective method for damping enhancement of piezoelectric damping materials is to realize efficient energy conversion by improving piezoelectric effect.

As shown in Figure 6, the R_v value decreases apparently for the content of CNTs exceeding 0.8 wt.%, which implies that CNT particles gradually contact with each other at 0.8 wt.% CNT content. The R_v value is lower than 10^7 when the CNT content reaches 2.0 wt.%. The formation status of CNT conductive network at different CNT contents is shown in Figure 7. When the content of CNTs is small (Figure 7(a)), less than 0.6 wt.%, the electrical network cannot form. Thus, the piezoelectric effect of PMN has no significant effect on damping enhancement. With the increase in CNT content, carbon nanotubes start to form small loops around PMN particles (Figure 7(b)). Those loop circuits are taken as conversion units to convert electrical energy into thermal

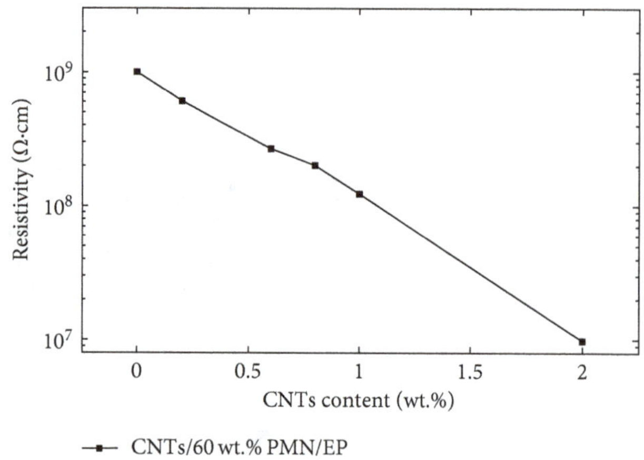

FIGURE 6: Resistivity of composites as a function of CNT content.

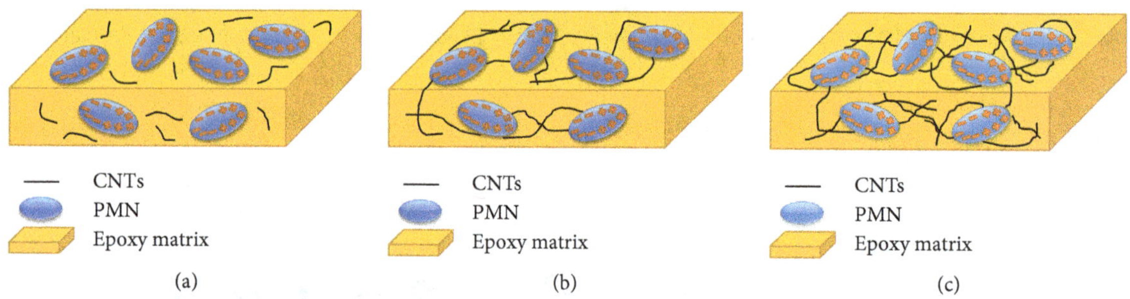

FIGURE 7: Formation status of CNT conductive network at different CNT contents.

energy. As shown in Figure 5(b), the damping percentage of PMN increases sharply at CNT content of 0.8 wt.%, which implies that the electrical energy generated by piezoelectric effect of PMN is dissipated efficiently. The maximum energy conversion rate is obtained when the content of CNTs reaches percolation threshold (1.0 wt.%). However, CNT network forms short circuit when the CNT content reaches 2.0 wt.% (Figure 7(c)), and electrical energy generated via piezoelectric effect of PMN cannot be dissipated efficiently. Thus, the enhancement of damping properties by piezoelectric effect is weakened.

4. Conclusions

The present paper has provided a comprehensive study about quantitative analysis of damping properties based on the TA analysis method. The addition of high content of conductive phase would reduce the damping peak intensity but enhance damping at higher temperatures, and the effective damping temperature range is widened. It is difficult to evaluate the comprehensive damping properties of materials by choosing $\tan\delta_{max}$ as the only performance index. Thus, in this research, the TA value is selected to evaluate damping properties. The TA analysis method obviously clarifies the effect of maximum loss factor ($\tan\delta$) and effective temperature range on damping properties. And synergies of piezoelectric phase and conductive phase are further researched.

(1) The damping percentage of CNTs in CNTs/EP composites reaches 32.22% at CNT content of 0.6 wt.%. The maximum loss factor reaches 0.56, and the effective temperature range is widened.

(2) The damping percentage of PMN ceramics in PMN/EP composites reaches 33.95% at 40 wt.% PMN content. The increase in frictional energy dissipation can be massively offset by the decrease in viscoelastic damping of matrix when the content of PMN exceeds 60 wt.%. So damping properties of PMN/EP composites cannot be improved obviously, and the damping percentage of PMN is merely 4.28%.

(3) The interfacial effect of CNTs acts as the dominant factor of damping enhancement for the content less than 0.6 wt.% in CNTs/PMN/EP composites, and the maximum damping percentage reaches 29.14%. However, PMN becomes the dominant factor of damping enhancement with the content of CNTs increasing, and the damping percentage reaches 47.43% for the content of CNTs reaching percolation threshold. Synergies of piezoelectric phase and conductive phase result in the damping enhancement of CNTs/PMN/EP composites.

Conflicts of Interest

The authors declare that they have no conflicts of interest.

Acknowledgments

This work was financially supported by the National Natural Science Foundation of China (nos. 51472188 and 51521001), Natural Research Funds of Hubei Province (no. 2016CFB583), Fundamental Research Funds for the Central Universities in China, State Key Laboratory of Advanced Electromagnetic Engineering and Technology (Huazhong University of Science and Technology), National Key Research and Development Program of China (nos. 2017YFB0310400 and 2018YFB0905600), and Joint Fund (Grant No. 6141A02022223).

References

[1] Y. B. Liao and H. A. Sodano, "Piezoelectric damping of resistively shunted beams and optimal parameters for maximum damping," *Journal of Vibration and Acoustics*, vol. 132, no. 4, pp. 1014–1020, 2010.

[2] T. H. Cheng, M. Ren, Z. Z. Li, and Y. D. Shen, "Vibration and damping analysis of composite fiber reinforced wind blade with viscoelastic damping control," *Advances in Materials Science and Engineering*, vol. 2015, Article ID 146949, 6 pages, 2015.

[3] R. L. Forward, "Electronic damping of vibrations in optical structures," *Applied Optics*, vol. 18, no. 5, pp. 690–698, 1979.

[4] N. W. Hagood and A. von Flotow, "Damping of structural vibrations with piezoelectric materials and passive electrical network," *Journal of Sound and Vibration*, vol. 146, no. 2, pp. 243–268, 1991.

[5] D. Carponcin, E. Dantras, G. Michon et al., "New hybrid polymer nanocomposites for passive vibration damping by incorporation of carbon nanotubes and lead zirconate titanate particles," *Journal of Non-Crystalline Solids*, vol. 409, pp. 20–26, 2015.

[6] M. B. Bryning, M. F. Islam, J. M. Kikkawa, and A. G. Yodh, "Very low conductivity threshold in bulk isotropic single-walled carbon nanotube-epoxy composites," *Advanced Materials*, vol. 17, no. 9, pp. 1186–1191, 2005.

[7] S. Tian, F. J. Cui, and X. D. Wang, "New type of piezo-damping epoxy-matrix composites with multi-walled carbon nanotubes and lead zirconate titanate," *Materials Letter*, vol. 62, no. 23, pp. 3859–3861, 2008.

[8] S. Tian and X. D. Wang, "Fabrication and performances of epoxy/multi-walled carbon nanotubes/piezoelectric ceramic composites as rigid piezo-damping materials," *Journal of Materials Science*, vol. 43, no. 14, pp. 4979–4987, 2008.

[9] D. Carponcin, E. Dantras, J. Dandurand et al., "Electrical and piezoelectric behavior of polyamide/PZT/CNT multifunctional nanocomposites," *Advanced Engineering Materials*, vol. 16, no. 8, pp. 1018–1025, 2014.

[10] J. Li and M. Du, "Study on the piezoelectric and damping properties of NBR/PF/PMN/CB composites," *Advanced Science Letters*, vol. 4, no. 3, pp. 675–680, 2011.

[11] Y. B. Wang, H. Yan, and Z. X. Huang, "Mechanical, dynamic mechanical and electrical properties of conductive carbon black/piezoelectric ceramic/chlorobutyl rubber composites," *Polymer-Plastics Technology and Engineering*, vol. 51, no. 1, pp. 105–110, 2012.

[12] M. X. Shi, Z. X. Huang, and L. M. Zhang, "The effects of dynamic load on the damping performance of piezoelectric ceramic/conductive carbon/epoxy resin composites," *Polymer-Plastics Technology and Engineering*, vol. 49, no. 10, pp. 979–982, 2010.

[13] H. M. Zhang and X. D. He, "Piezoelectric modal damping performance of 0-3 piezoelectric composite with conducting phase: numerical analysis and experiments," *Polymers and Polymer Composites*, vol. 22, no. 3, pp. 261–268, 2014.

[14] P. Avouris, T. Hertel, R. Martel, T. Schmidt, H. R. Shea, and R. E. Walkup, "Carbon nanotubes: nanomechanics, manipulation and electronic devices," *Applied Surface Science*, vol. 141, no. 3-4, pp. 201–209, 1999.

[15] D. Shamir, A. Siegmann, and M. Narkis, "Vibration damping and electrical conductivity of styrene-butyl acrylate random copolymers filled with carbon black," *Journal of Applied Polymer Science*, vol. 115, no. 4, pp. 1922–1928, 2010.

[16] J. Chang, B. Tian, L. Li, and Y. F. Zheng, "Microstructure and damping property of polyurethane composites hybridized with ultraviolet absorbents," *Advances in Materials Science and Engineering*, vol. 2018, Article ID 9624701, 9 pages, 2018.

[17] T. Ogawa and T. Yamada, "A numerical prediction of peak area in loss factor for polymer," *Journal of Applied Polymer Science*, vol. 53, no. 12, pp. 1663–1668, 1994.

[18] J. J. Fay, C. J. Murphy, D. A. Thomas, and L. H. Sperling, "Effect of morphology, crosslink density, and miscibility on interpenetrating polymer network damping effectiveness," *Polymer Engineering and Science*, vol. 31, no. 24, pp. 1731–1741, 1991.

[19] D. J. Hourston and F.-U. SchäFer, "Poly(ether urethane)/poly (ethyl methacrylate) IPNs with high damping characteristics: the influence of the crosslink density in both networks," *Journal of Applied Polymer Science*, vol. 62, no. 12, pp. 2025–2037, 1996.

[20] K. A. Moly and S. S. Bhagawan, "Correlation between the morphology and dynamic mechanical properties of ethylene vinyl acetate/linear low-density polyethylene blends: effects of the blend ratio and compatibilization," *Journal of Applied Polymer Science*, vol. 100, no. 6, pp. 4526–4538, 2006.

[21] D. G. Franklin, J. N. Foster, and L. H. Sperling, "A quantitative determination of the damping behavior of acrylic-based interpenetrating polymer networks," *Rubber Chemistry and Technology*, vol. 59, no. 2, pp. 255–262, 1986.

2

Advances in Materials Science and Engineering Assessment of Gamma Radiation Hazards Related to Geologic Materials: Comparison of Results by Field Gamma Spectrometry and Laboratory Methods

Miguel Couto,[1] Jorge Sanjurjo-Sánchez,[2] and Carlos Alves[3]

[1]Earth Sciences Department, School of Sciences, University of Minho, 4710-057 Braga, Portugal
[2]Instituto Universitario de Xeoloxía "Isidro Parga Pondal", University of A Coruña, 915001 A Coruña, Spain
[3]Earth Sciences Department, LandS/Lab2PT-Landscape, Heritage and Territory Laboratory (FCT UID/AUR/04509/2013, FEDER COMPETE POCI-01-0145-FEDER-007528), School of Sciences, University of Minho, 4710-057 Braga, Portugal

Correspondence should be addressed to Carlos Alves; casaix@dct.uminho.pt

Academic Editor: Marco Cannas

Geologic materials are a potential source of external gamma radiation in buildings due to the presence of radioisotopes (^{40}K and isotopes from U and Th decay families). The contents of these radioisotopes can be evaluated by laboratory analyses that require the extraction of samples, with associated financial and time constraints, and it is also difficult to perform on materials already applied in existing structures. In this work, portable gamma spectrometry results are compared with laboratory analyses in terms of radioisotopes estimations, and the activity concentration index is calculated from them (which assess the conjoint contribution of the different radioisotopes to the external gamma radiation). The studied objects are rocks and their "in situ" weathering products. The results obtained indicate that despite several problems posed by field measurements with portable gamma spectrometry, namely, in terms of the geometric characteristics of the objects measured, this technique has the potential to identify materials with different hazard levels related to external gamma radiation.

1. Introduction

Natural ionizing radiation includes external and internal sources. While internal sources comprise naturally occurring radioisotopes that are taken into the human body, external ones are mainly cosmic rays and gamma radiation emissions from both terrains and building materials [1]. Isotopes present different activity values (radioactive decay rate, which is expressed in SI units as Becquerel, Bq, with 1 Bq = 1 s^{-1}), and the ratio of the activity of the isotope to the total mass can be considered as an expression of its concentration, referred as activity concentration or specific activity (Bq/kg). The external gamma radiation related to terrestrial sources is mainly due to primordial radionuclides with radioactive decay half-lives that are approximately Earth's age or older [2]. They are mainly ^{40}K and radioisotopes of the radioactive decay series of ^{238}U and ^{232}Th (daughter isotopes of these radioisotopes) present in minerals and rocks used in building materials, being the main cause of indoor radiation exposure, which is about 40% greater than outdoor exposures in middle and high latitudes [3].

Different approaches have been considered to assess the dose that a person will receive from building materials in a given indoor space. The 2013/59/EURATOM directive of the Council of Europe (CEU [4]), which is directed to setting "basic safety standards for protection against the dangers arising from exposure to ionising radiation," indicates the use of an activity concentration index (1) which is based on the sum of factors obtained by dividing the activity concentrations (or specific activities) of isotopes by constants that correspond to certain reference conditions in terms of room characteristics, materials application, and bulk density and exposition time.

$$I = \left(\frac{C_{\text{Ra-226}}}{300}\right) + \left(\frac{C_{\text{Th-232}}}{200}\right) + \left(\frac{C_{\text{K-40}}}{3000}\right), \quad (1)$$

where $C_{\text{Ra-226}}$ is the activity concentration of ^{226}Ra in Bq/kg, $C_{\text{Th-232}}$ is the activity concentration of ^{232}Th in Bq/kg, and $C_{\text{K-40}}$ is the activity concentration of ^{40}K in Bq/kg.

A value of one for this activity concentration index is indicated as a conservative screening tool with the main goal of distinguishing building materials with radioisotope levels that may have a contribution to external gamma radiation in excess of typical outdoor exposure above the reference level. This contribution of the materials will be different for different conditions of those considered to obtain the index, and the question of the typical outdoor exposure might also be a subject of discussion as it can show marked differences depending on the terrain characteristics [5].

There is hence an interest (that might be increased in the future as this directive is implemented) in assessing the concentrations of radioisotopes in building materials. However, direct measurements of isotopes activities are costly and time-consuming and can be considered destructive in the sense that they require sampling, which limits its application to building materials applied in existing structures. This limitation of the extraction of a sample also applies to other, eventually faster, laboratory techniques such as inductively coupled mass spectrometry (ICP-MS) and X-ray fluorescence (XRF) that allow the measuring of the elements amounts (from which specific activities of isotopes can be calculated, e.g., for U, Th, and K by the factors indicated in IAEA [6]).

Portable gamma-ray spectrometry (which is referred here as field gamma spectrometry—FGS) is a classical geophysical technique that has been used for geological exploration as well as for assessing the dose rate in terrains. It can be a fast procedure, and there are variants (such as the one that is considered here) that can be used by a one-person team. When FGS measurements are carried out outdoor on rock outcrops, a $2\pi r$ geometry (plain areas) is needed to avoid over or underestimation of the radioisotopes (K, U, and Th) and gamma radiation [6]. There is, consequently, interest in studying the usability of FGS at several perspectives related to building materials and their contribution to external gamma radiation:

(i) To assess natural variability in quarries (which can be used to avoid portions of the rock mass with undesirable amounts of the radioisotopes)

(ii) To control the products which are going out (again to avoid situations above certain values)

(iii) To assess the potential hazards levels related to a given material applied in existing structures

There are several studies comparing field and laboratory analyses of radioisotopes, some of them regarding building materials, such as dose measurements in spaces and in relation to the applications of materials [7, 8]. In the studies of Mueller et al. [9] and Nuccetelli et al. [10], one can find detailed discussions regarding the issues concerning the comparison of laboratory and field measurements of radioisotopes. However, studies with the perspective that is presented here (assessment in relation to the activity concentration index as a screening tool of building materials) are much scarcer. As far as we know, Stals et al. [11], concerning different types of building materials, and Döse [12], concerning concrete slabs with different types of rock aggregates, are the only works oriented to the assessment of activity concentration index of building materials with portable gamma spectrometry. These two publications report excellent positive linear correlations between portable and laboratory results. Stals et al. [11] found that ^{226}Ra and (especially) ^{40}K results by portable equipment were lower than results from laboratory analyses while ^{232}Th results were higher in the portable equipment results. The final result in terms of activity concentration index (I) was slightly lower for portable measurements (between <0.1 and 3.65) than laboratory analyses (<0.1 to 3.91), getting activity concentration index values between 0.15 and 2.04 for laboratory measurements and between 0.28 and 1.6 for the portable measurements. The results of laboratory analyses were lower for the lower I values and higher by 15%–20% for the higher I range (above 1 in the laboratory results). These studies [11, 12] were both made with laboratory well-defined geometrical specimens, while our study considers outcrops with variable geometrical conditions (hence harder conditions for correlations), which are nearer the field conditions that one can find in quarry mapping. In a field study with a perspective near to this paper (but not in terms of the activity index as screening tool of building materials), Bochiolo et al. [13] found that field measurements gave higher measurements in all the considered radioisotopes (which means, consequently, higher activity concentration index values), in a study performed in an old mine tunnel (therefore in geometrical conditions that are more homogeneous than those considered here).

Our main goal in this paper was to assess the performance of field measurements by portable gamma spectrometry in rock outcrops (including their weathering products) by comparison with laboratory analyses of samples from the same places where the measurements were made.

2. Materials and Methods

2.1. Studied Geologic Objects. The present study considered samples from igneous and metamorphic rocks and their weathering products in outcrops near the towns of A Coruña (NW Spain) and Braga (NW Portugal), which were based on the available geological mapping (IGME [14, 15]; Ferreira et al., [16]), and our field studies can be described as gabbro (3 samples in A Coruña), leucocratic body in gabbro (1 sample in A Coruña), biotite-rich granites (4 samples in Braga), leucocratic granite (3 samples in A Coruña and 5 samples in Braga), and Silurian siliciclastic metamorphic rocks (7 samples in Braga).

2.2. Analytical Methods. The present work considers results from different methods to assess the concentration of radionuclides in such rocks and weathering products to assess their activity index and possible limitations in their use as building materials under the present European regulations established by the directive CEU [4]. The main aim of the work is to compare the assessments of radioisotopes concentrations in these geological materials, namely, in the perspective of the European directive trough the results of

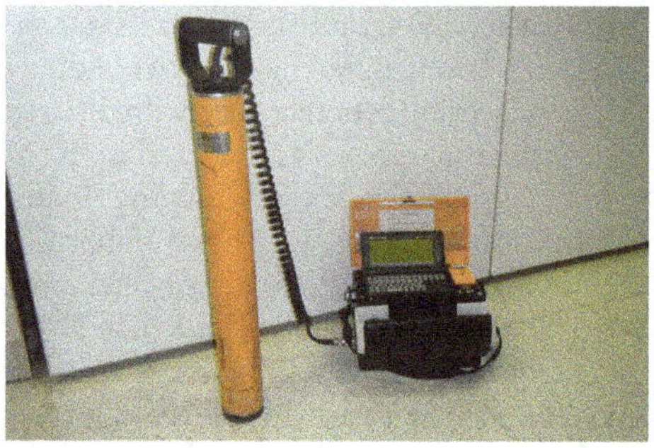

Figure 1: Picture of the portable gamma-ray spectrometer GF Instruments model GRS-2000 used for field measurements in this work. On the left side, the probe with the detector can be seen, and on the right side, the computer system with electric supply can be seen.

the concentration activity index (which can be considered a synthesis parameter of the contribution of the different isotopes in relation to a certain reference state).

2.3. Field Gamma Spectrometry (FGS). Gamma spectra were acquired in situ in the field with a portable gamma spectrometer GF Instruments GRS-2000, equipped with a probe BGO ($Bi_4Ge_3O_{12}$) having a 51 mm × 51 mm detector and a shielded photomultiplier (Figure 1). The equipment allows measuring energies up to 3 MeV. Measurements were made by direct probe contact at outcrops with variable geometry. The recommendations given by the IAEA [6] for static measurements (recording a spectrum in a given point during a certain amount of time) were followed. Each acquired spectrum corresponds to a counting period of 3 minutes. This time period is recommended by the manufacturer to optimize time and precision. According to the manufacturer's specifications for low gamma-ray emissions, statistical uncertainties are estimated to be 6% for K, 30% for U, and 16% for Th by measuring for 180 s. This error is due to background radiation that is basically caused by three main sources: atmospheric radon and its daughters, cosmic rays, and instrument background. Moreover, the geometry of measured surface must be planar ($2\pi r$ geometry) being both possible surface roughness of the measured terrains and any nearby topographic feature possible causes of inaccuracies in the measurements.

From the obtained spectra (an example is presented in Figure 2), we have used some peaks to assess the contents of K (potassium, mass percent), eU (uranium equivalent, in mass parts per million or ppm), and eTh (thorium equivalent, also in ppm). While potassium contents are estimated from the peak of ^{40}K (energy range 1.366–1.564 keV), there are several radioactive isotopes in the uranium and thorium decay series, and hence, "equivalents" of uranium and thorium (eU and eTh, resp.) can be used, assuming secular equilibrium. Estimations of eU and eTh were made from peaks of the spectra corresponding to ^{214}Bi (energy range 1.57–1.959 keV) and ^{208}Tl (energy range 2.42–2.81), respectively, on the decay chain of uranium and thorium. The specific activities of ^{226}Ra (or ^{238}U), ^{232}Th, and ^{40}K were

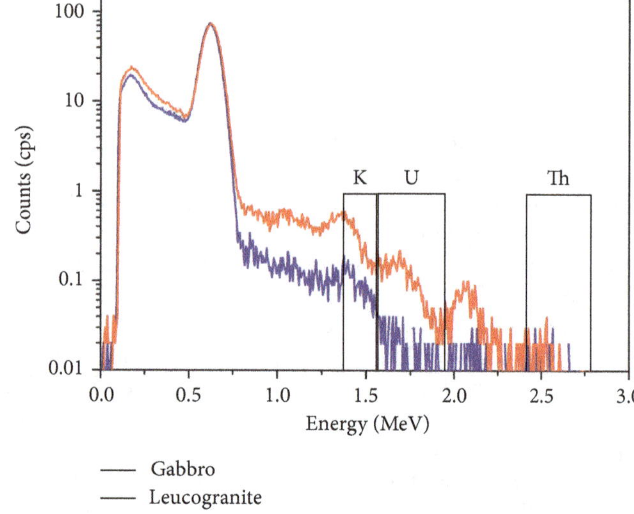

Figure 2: Example of two different field gamma spectrometry (FGS) spectra acquired on outcrops of gabbro (blue) and leucogranite (red) with indication of the energy windows for element estimations.

calculated from eU, eTh, and K with the conversion factors given in IAEA [6].

2.4. Sample Preparation for Laboratory Analyses. For laboratory analyses of the samples, the obtained samples were powdered at a particle size less than 0.5 mm and dried at 100°C to constant weight. Once dried, the powder was fired at 500°C for 20 hours to remove organic matter. A part of the powder was separated for elemental analyses and most of them for low background gamma spectrometry.

2.5. Laboratory Gamma Spectrometry (LGS). For laboratory gamma spectrometry (which is referred here as LGS), a mass of 250 g of each sample was placed in Marinelli polystyrene containers and sealed with a sellotape and left for more than 4 weeks in order to allow for Ra and its short-lived progeny to reach radioactive equilibrium. The radioactivity concentrations were measured in coaxial HPGe Canberra GR6022 of the "Servicio de Radioisotopos (CITIUS)" of the University of Seville (Spain). The detector was shielded with 10 cm thick lead on all sides including the top, to reduce the background contribution from the surroundings. The shielded volume is 180 cm^3; the relative efficiency of the detector is 60%, and it has a resolution of 1.4 keV at 2.3 keV for energies 122 keV and 1332 keV, respectively.

2.6. Element Analyses. The determination of both major and minor elements in the samples was obtained by X-ray fluorescence spectrometry (XRF). Powdered samples were measured in a Bruker-Nonius S4 Pioneer wavelength dispersive fluorescence spectrometer under helium purge at the University of A Coruña. To assess trace elements, inductively coupled plasma-mass spectrometry (ICP-MS) was used.

ICP-MS analyses were carried out in a Thermo Scientific™ ELEMENT XR™ ICP-MS which combines a dual mode secondary electron multiplier with a Faraday detector. Lithium metaborate fusion was used for sample preparation. The powdered samples were mixed with an equal amount of lithium tetraborate flux, placed in a carbon crucible, and fused at 1000°C in a furnace for 30 minutes. Once the melt was cooled, the resultant fusion bead was ground and dissolved in 100 mL of 4% HNO_3/2% HCl_3 solution which was then analysed by ICP-MS. The same factors (given in IAEA [6]) were used to convert U, Th, and K measurements to specific activities of ^{226}Ra (or ^{238}U), ^{232}Th, and ^{40}K.

2.7. Data Treatment. Some notes on the statistical treatment are presented here (additional information can be found in statistical texts such as Dodge [17]). The statistical summary includes diverse measures of points of the value distribution: arithmetic mean (referred simply as mean from now on), harmonic mean, median, geometric mean, minimum, first (lower) quartile, third (upper) quartile, and maximum. While the minimum, mean, and maximum are certainly widely known, some notes are presented in relation to the other measures. The geometric mean is obtained by the n root of the product of the values (or by the arithmetic mean of their logarithms). The harmonic mean is obtained by the quotient between the number of elements of the data set and the sum of the inverse of each element of the data set. The quartiles are the points that divide a set of values into 4 equal parts. The middle quartile divides the data set into two equal parts and is equal to the median. The lower and upper quartiles correspond to the upper limit of the lower 25% and lower limit of the upper 25%. While it is not our goal to characterize the distribution of the sets of values from the studied rocks, dispersion statistics (standard deviation, standard error of the mean, and interquartile range) are presented for the reader to assess the dispersion of the data sets under study, which could be useful for comparison. While we will not discuss the question of the normal distribution of such a geologically heterogeneous set of measurements, one can highlight that, for a normal distribution, the 95% confidence limits of the mean are obtained by the mean more or less the product of the standard deviation by 1.96 (for a higher confidence limit, the error interval is, of course, higher).

As for there are groups of samples from some rock types and since our goal here is to compare techniques and not the geochemical characterization of the geological materials, data treatment is based on robust statistical techniques based on nonparametric approaches (that do not assume any premise concerning the shape of the data distribution or its parameters). The correlation between different measures is assessed by the rank-based Spearman correlation coefficient, which does not require each studied value set that presents a normal distribution, which is not affected by the specific values nor is it affected by arithmetic operations with constants (e.g., dividing by a constant) since it does not depend on the variance of the value sets. One will also present results of the coefficient of determination that assess the fit of the data to a proposed regression equation (it is a measure of the contribution of a given factor to the variance of another under a given statistical model such as a regression straight line). The distribution of the values of the concentration activity index (I) obtained from results of each analytical technique is represented by a boxplot, representing the lower, middle, and upper quartile and raw data. Since the realization of several statistical tests with the same data set can cause an error I type inflation, we limit the use of other statistical tests to the comparison of the calculated values of the concentration activity index. Following the advice of the American Statistical Association (Wasserstein and Lazar [18]), the obtained p values are reported. Comparisons of I values from the different analytical techniques are based on two nonparametric statistical tests for paired statistical samples: the sign test and the Wilcoxon test. The sign test is based on the proportion of the signs of the differences between the paired results. The Wilcoxon test is based on the ranking of the differences between the compared statistical samples.

All statistic tests, the boxplot and the mean and standard error plots were prepared with *Statistica* 11 (*Statsoft*).

3. Results and Discussion

Figure 3 presents a combined boxplot (quartiles) and raw data plot of the logarithms of ratios, in relation to estimations from field measurements, of laboratory estimations of ^{226}Ra, ^{232}Th, and ^{40}K specific activities by laboratory gamma spectrometry and those calculated from uranium, thorium, and potassium quantifications (the first two by MS-ICP and the later by XRF). It can be seen that, in almost all the cases, the ratios of laboratory estimations to field estimations are below one (negative logarithms), indicating that, for the different radioisotopes, estimations from field measurements are higher than laboratory ones.

Bivariate plots of laboratory measurements against the field ones as well as the values of the activity concentration index calculated from the different sets of isotopes estimations (laboratory and field) are plotted in Figure 4. It can be seen that there are clear trends toward correlations between the different sets of measurements, but there is some wider scatter in the case of ^{40}K estimations as shown by the coefficient of determination values, which are lower than those obtained by Stals et al. [11]. The plots also include the line of equality between plotted variables, showing that, in general (with few exceptions), values of radioisotopes estimations from laboratory measurements are less than those from field measurements and that this happens in the comparison of calculated values of the activity concentration index also (in this case, only one value obtained from field estimations is slightly lower).

The p values for the Spearman coefficients calculated for I value estimations from the three sets of analyses are presented in Table 1, showing that they markedly below 0.05 (the highest p value equals to 0.000001). It can also be seen from the Spearman coefficients in Table 1 that the correlation between the estimations from the laboratory analyses

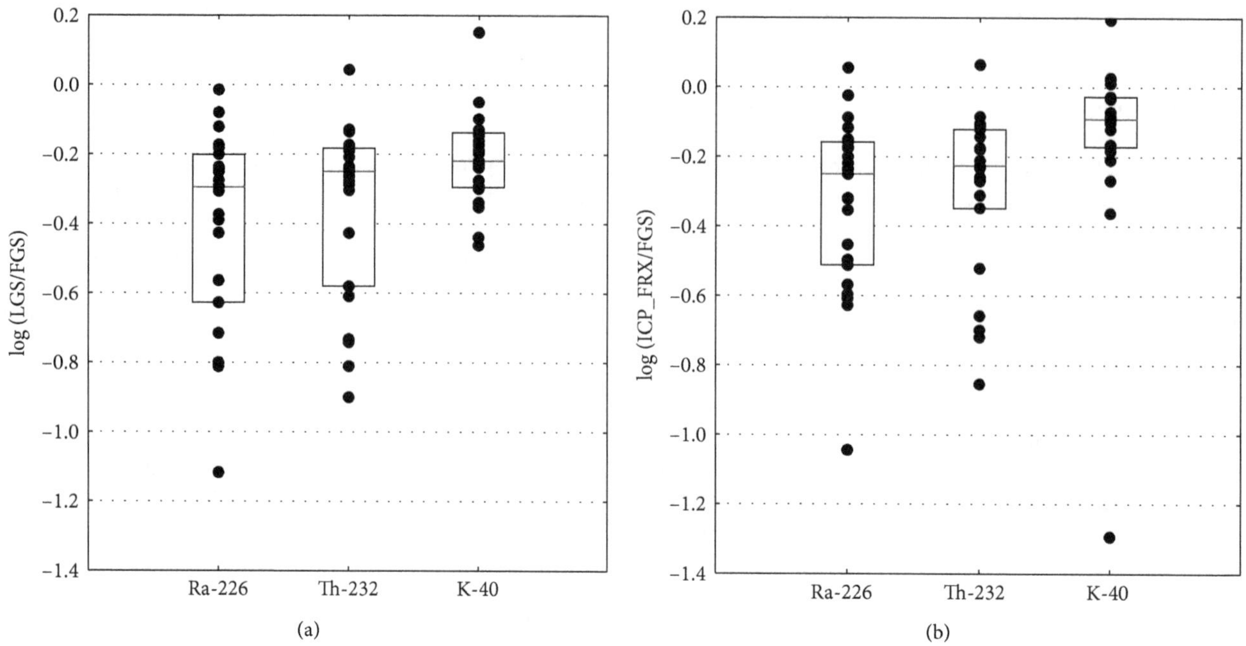

FIGURE 3: Combined boxplots (quartiles) and raw data plots of logarithms of radioisotopes estimations from laboratory analyses to those from field measurements: (a) comparison of laboratory gamma spectrometry (LGS) to field gamma spectrometry (FGS); (b) comparison of estimations from ICP and XRF to FGS.

is better than the correlations of the laboratory analyses with the field measurements.

Table 2 presents the univariate statistics of the activity concentration index (I) values calculated from the three procedures. It can be seen that all location measures are higher in the field set than in the ICP/XRF set that, in turn, are higher than those from the laboratory gamma spectrometry measurements. It can also be seen that I values based on laboratory analyses are always nearer each other than in relation to I values based on field measurements. One can also highlight the relation of the statistics presented in Table 2 in relation to the screening criteria indicated in the European directive ($I = 1$). Most of the average estimations (mean, geometric mean, and median) are above 1.0 for the set of I values from field measurements but does not happen in any of the averages for I values from laboratory measurements. The lower quartile of I values based on field measurements is very near this screening value. These relations are also illustrated in the plot of Figure 5(a) that combines a boxplot and the plotting of raw data. This plot also shows that the number of results that exceeds the $I = 1$ condition is clearly much higher for the field estimations.

The nonparametric statistical tests in the comparison of these paired values (Table 3) show very low p values (the higher one being 0.0005) and being noticeable that the comparison of field values with laboratory values produces lower p values. However, the statistical assessment of higher mean in the field set deserves more caution since for $p = 0.05$, the intervals of mean more or less 1.96 times the standard error of the mean of the different data sets overlap, as can be seen in the plot of Figure 5(b). Statistical nonparametric tests of paired samples also support these relations, with low p values, namely, in the comparison of I values from field and laboratory measurements (with the highest being $p = 0.00003$).

To explain the scatter observed among ^{40}K estimates obtained by both XRF and LGS, two causes can be considered. On the one hand, the ^{40}K and ^{214}Bi peaks (i.e., used to assess the U concentration) are very close in the gamma-ray spectrum. The FGS equipment has low resolution when compared with HPGe used in LGS. Hence, when one of the ^{214}Bi or ^{40}K peaks is high, it can overlap to part of the nearby peak providing an overestimation of the other radioisotope [6]. To check if this effect is a cause of scatter in our results, we have compared eU and eTh estimates from FGS and the ratio of K assessed from FGS to K measured by XRF and LGS (Figure 5). The ratios K (FGS)/K (XRF) and K (FGS)/K (LGS) are near 1 when the estimates of K fit, being higher when the K (FGS) overestimates the K (XRF) or K (LGS) estimates. If the overlap of a high ^{214}Bi peak with the ^{40}K peak is the cause of K (FGS) overestimation, the ratios should be higher and above 1 when higher eU is measured. However, this is not observed when the K/K ratios are plotted versus eU (Figure 6). In fact, when the ratios are plotted versus the ^{208}Tl peak estimates (eTh), the results are similar to those in the former plot, being negligible any peak overlaps on the K and eU estimates by FGS. Anyway, it must not be forgotten that the large uncertainty of the FGS (6%, 30%, and 16% for K, U, and Th, resp.) hinders the assessment of this cause of scatter, and a possible larger data set could be required to get concluding results. Anyway, the disequilibrium in the ^{238}U decay chain should be expected in weathered rocks but not in nonweathered of slightly weathered rocks. In such rocks, secular equilibrium is observed [19]. Moreover, the weathering processes and derived disequilibrium are not necessarily similar in different rocks

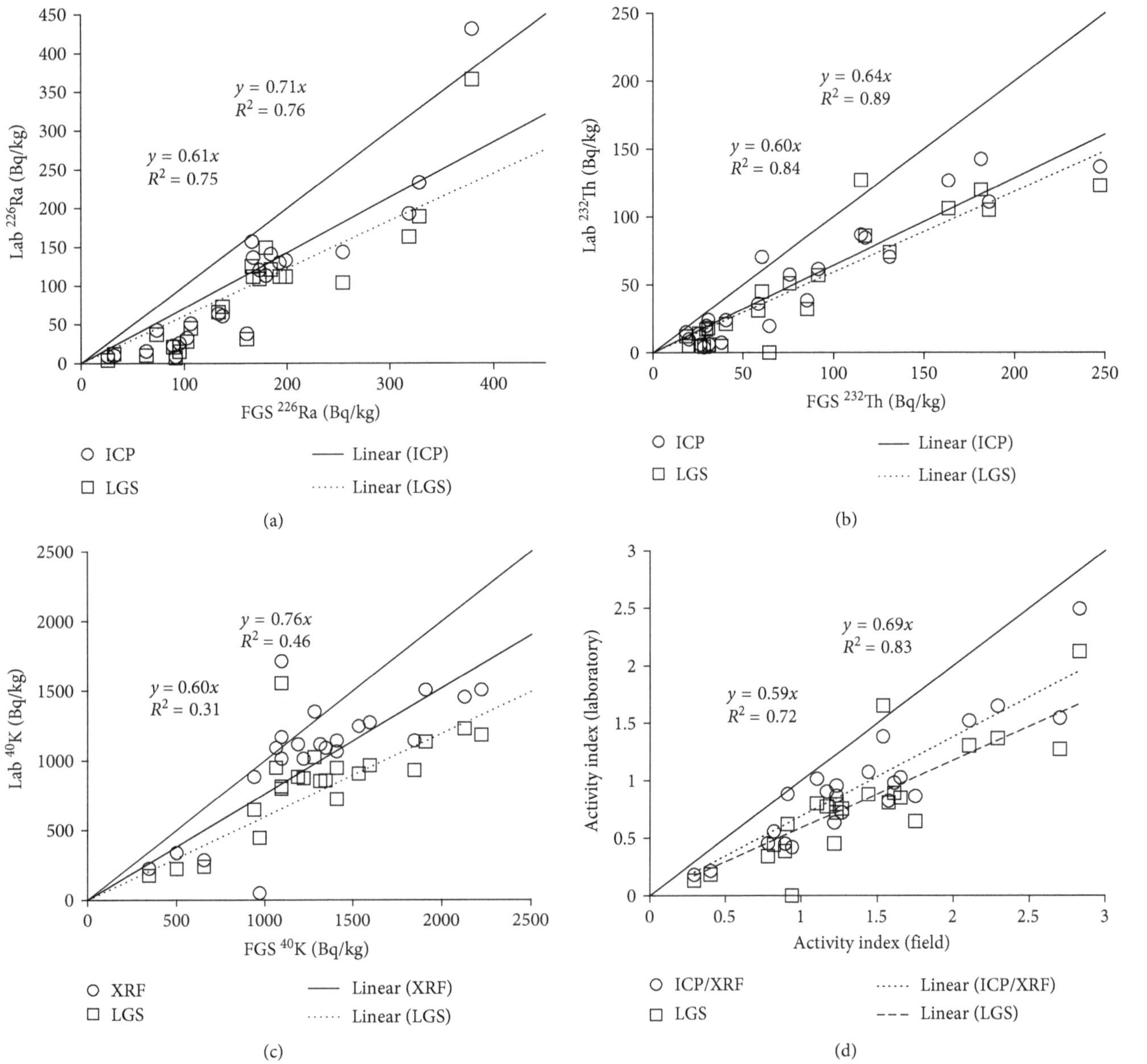

FIGURE 4: Plots of the results from laboratory analyses (ICP, XRF, and LGS: laboratory gamma spectrometry) against field estimations: (a) ^{226}Ra activity concentration; (b) ^{232}Th activity concentration; (c) ^{40}K activity concentration; (d) activity concentration index calculated from the isotopes estimations.

TABLE 1: Spearman correlation coefficients between the calculated values of the concentration activity index (I) from different procedures for estimation of isotopes specific activity and related p values.

	N	R	p value
I (field) and I (ICP/XRF)	23	0.83	1.08×10^{-6}
I (field) and I (LGS)	23	0.87	5.71×10^{-8}
I (ICP/XRF) and I (LGS)	23	0.95	3×10^{-12}

LGS: laboratory gamma spectrometry.

(e.g., gabbro and granite), being necessary to consider any rock type to assess this factor as a possible cause of the scatter results. Other sources of inaccuracy must be taken into account in the estimates, as it is not possible to assess its effect on the results. A flat geometry is required for FGS measurements ($2\pi r$). Although planar surfaces have been chosen for the measurements, the work areas are not completely flat, being surface roughness and topographic features a possible cause of such inaccuracies and scattering.

4. Conclusions

Field portable gamma spectrometry measurements in igneous and metamorphic rocks and their weathering products showed good ordinal correlations with two laboratory techniques of assessment of radioisotopes specific activity, suggesting that it can be used to rank materials in terms of radiological hazard concerning the contribution of materials to the external gamma radiation dose. The results tend to be

Table 2: Statistic measures of the activity concentration index (I) sets from different procedures for estimation of isotopes specific activity and related p values.

	Minimum	Maximum	Mean	G.M.	H.M.	Med.	S.D.	S.E.	L.Q.	U.Q.	I.Q.R.
I (field)	0.29	2.83	1.38	1.22	1.04	1.23	0.65	0.13	0.91	1.65	0.74
I (ICP/XRF)	0.18	2.49	0.94	0.80	0.65	0.88	0.52	0.11	0.56	1.07	0.52
I (LGS)	0.13	2.12	0.80	0.69	0.52	0.77	0.48	0.10	0.44	0.89	0.45

G.M.: geometric mean; H.M.: harmonic mean; Med.: median; S.D.: standard deviation; S.E.: standard error of the mean; L.Q.: lower quartile; U.Q.: upper quartile; I.Q.R.: interquartile range.

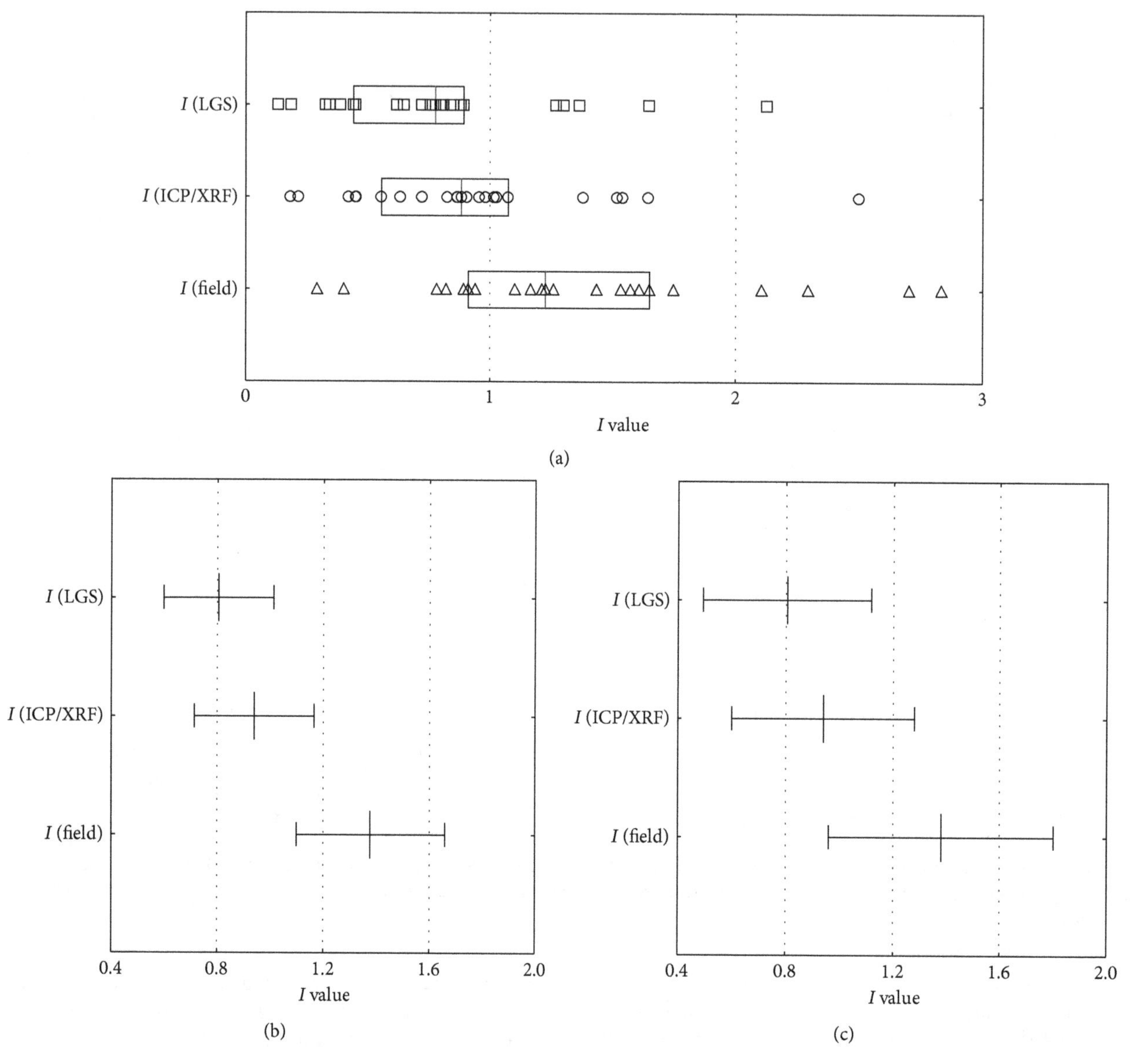

Figure 5: Plots for comparison of the activity concentration index values (I) calculated from the laboratory analyses (ICP, XRF, and LGS: laboratory gamma spectrometry) and field estimations: (a) boxplot (quartiles) and raw data plot; (b) mean and standard error plot for 95% confidence; (c) mean and standard error plot for 99.5% confidence.

Table 3: Results (p values) of statistical tests for paired comparisons of estimations of the activity concentration index (I) values.

	N	Sign test p value	Wilcoxon p value
I (field) and I (ICP)	23	0.000004	0.00003
I (field) and I (LGS)	23	0.000030	0.00003
I (ICP) and I (LGS)	23	0.000175	0.00047

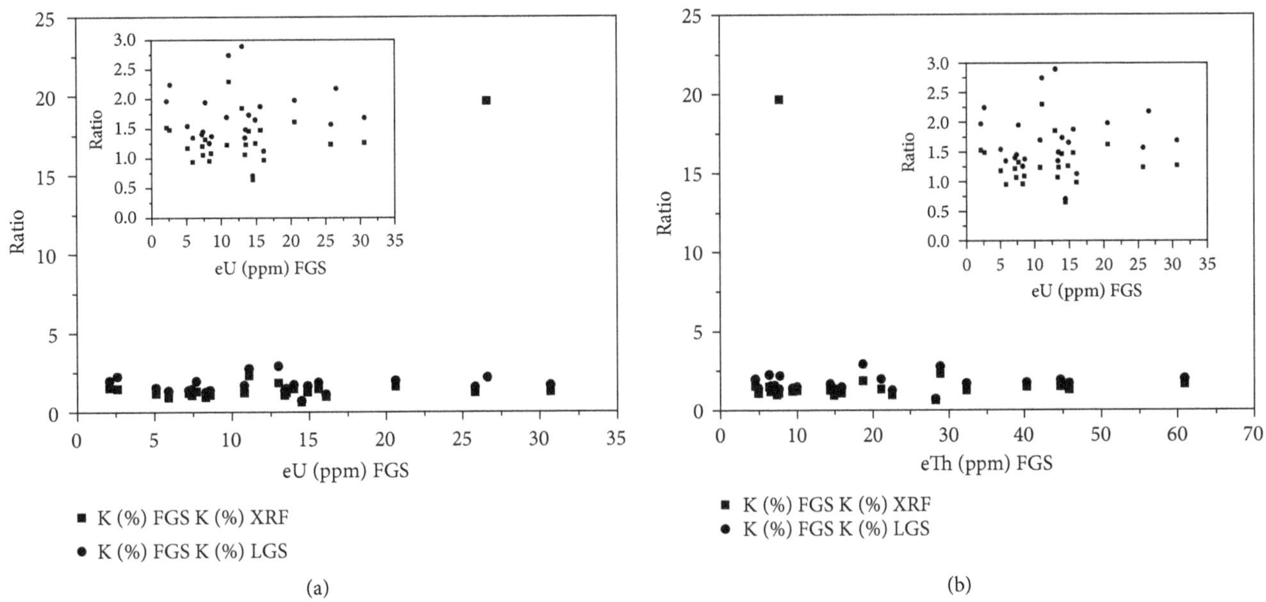

FIGURE 6: Plots for the comparison of estimates of eU (a) and eTh (b) versus K (%) ratios obtained by FGS, XRF, and LGS. The K ratios correspond to K (FGS)/K (XRF) and K (FGS)/K (LGS). The insets show a detail of the whole plots in the range 0–3 of the ratio and 0 to 35 ppm for eU and eTh. No correlation is observed between the data.

higher in the case of field measurements, which is reflected in a higher set of results above the guideline criteria $I = 1$. While one should read carefully in this matter, namely, given the limited sample set considered, these results strongly suggest that portable gamma spectrometry could be a conservative tool in assessing external gamma radiation posed by materials and, of this is confirmed, be used as a screening tool to separated samples to be sent for laboratory analyses.

It is possible that these outcrop measurements could work in a conservative way given the great amount of the geological substance that is present in the measuring point. But it is admitted that portable measurements might not be as efficient in the case of measuring specimens of materials as the effect of the surroundings could behave a higher impact (due to the smaller amount of the measured object).

Conflicts of Interest

The authors declare that there are no conflicts of interest regarding the publication of this paper.

Acknowledgments

The Lab2PT-Landscapes, Heritage and Territory laboratory-AUR/04509 is supported by the Portuguese “Fundação para a Ciência e a Tecnologia” (Portuguese funds and where applicable the FEDER co-financing, in the aim of the new partnership agreement PT2020 and COMPETE2020-POCI 01 0145 FEDER 007528). The University Institute of Geology of the University of A Coruña (Spain) received support from Xunta de Galicia with funds from “Consolidación y estructuración de unidades de investigación competitivas—Grupo de potencial de crecimiento” (GPC2015/024). The authors would like to thank A. Pinto, F. Dias, M. Cunha, and P. Pereira for help during the field work.

References

[1] UNSCEAR, *Ionizing Radiation: Sources and Biological Effects. United Nations Scientific Committee on the Effects of Atomic Radiation. 1982 Report to the General Assembly, with Annexes*, United Nations, New York, NY, USA, 1982.

[2] J. S. Gaffney and N. A. Marley, “Radionuclide sources,” in *Radionuclide Concentrations in Food and the Environment*, M. Pöschl and L. M. L. Nollet, Eds., pp. 23–36, Taylor & Francis, Boca Raton, FL, USA, 2007.

[3] UNSCEAR, *Ionizing Radiation: Sources and Biological Effects. United Nations Scientific Committee on the Effects of Atomic Radiation UNSCEAR 2000 Report to the General Assembly, with Annexes*, United Nations, New York, NY, USA, 2000.

[4] CEU-Council of the European Union, “Council Directive 2013/59/EURATOM of 5 December 2013 laying down basic safety standards for protection against the dangers arising from exposure to ionising radiation, and repealing directives 89/618/Euratom, 90/641/Euratom, 96/29/Euratom, 97/43/Euratom and 2003/122/Euratom,” *Official Journal of the European Union*, vol. 2013, no. 13, 2014.

[5] J. Sanjurjo-Sánchez and C. Alves, “Geologic materials and gamma radiation in the built environment,” *Environmental Chemistry Letters*, vol. 15, no. 4, pp. 561–589, 2017.

[6] IAEA-International Atomic Energy Agency, *Guidelines for Radioelement Mapping Using Gamma Ray Spectrometry Data. IAEA-TECDOC-1363*, International Atomic Energy Agency, Vienna, Austria, 2003.

[7] R. M. Anjos, J. J. Ayub, A. S. Cid, R. Cardoso, and T. Lacerda, “External gamma-ray dose rate and radon concentration in indoor environments covered with Brazilian granites,” *Journal of Environmental Radioactivity*, vol. 102, no. 11, pp. 1055–1061, 2011.

[8] C.-J. Chen and Y.-M. Lin, "Assessment of building materials for compliance with regulations of ROC," *Environment International*, vol. 22, pp. 221–226, 1996.

[9] W. F. Mueller, G. Ilie, H.-J. Lange, M. Rotty, and W. R. Russ, "In-situ measurements and analysis of naturally occurring radioactive materials," in *Proceedings of the 2013 3rd International Conference on Advancements in Nuclear Instrumentation, Measurement Methods and their Applications (ANIMMA)*, pp. 1–8, Marseille, France, June 2013.

[10] C. Nuccetelli, R. Trevisi, F. Leonardi et al., "Radiological characterization of the ancient Roman tuff-pozzolana underground quarry in Orvieto (Italy): a natural laboratory to revisit the interactions between radionuclides and aerosols," *Journal of Environmental Radioactivity*, vol. 168, pp. 54–60, 2017.

[11] M. Stals, S. Verhoeven, M. Bruggeman, V. Pellens, W. Schroeyers, and S. Schreurs, "The use of portable equipment for the activity concentration index determination of building materials: method validation and survey of building materials on the Belgian market," *Journal of Environmental Radioactivity*, vol. 127, pp. 56–63, 2014.

[12] M. Döse, "Risk assessment of Swedish concrete as a construction material in relation to naturally occurring radiation from different aggregates," in *Engineering Geology for Society and Territory-Volume 5*, G. Lollino, A. Manconi, F. Guzzetti, M. Culshaw, P. Bobrowsky, and F. Luino, Eds., pp. 101–105, Springer International Publishing, Cham, Switzerland, 2015.

[13] M. Bochiolo, M. Verdoya, P. Chiozzi, and V. Pasquale, "Radiometric surveying for the assessment of radiation dose and radon specific exhalation in underground environment," *Journal of Applied Geophysics*, vol. 83, pp. 100–106, 2012.

[14] IGME (Instituto Geologico y Minero de España), *Mapa Geológico de España, Escala 1:50.000, Hoja 21 La Coruña*, IGME, Madrid, Spain, 1973.

[15] IGME (Instituto Geologico y Minero de España), *Mapa Geológico de España, Escala 1:50.000, Hoja 20-44 Sisargas-Carballo*, IGME, Madrid, Spain, 1978.

[16] N. Ferreira, G. Dias, C. Meireles, and M. A. Sequeira Braga, *Carta Geológica de Portugal, na Escala de 1:50 000. Folha 05-D (Braga). Notícia Explicativa da Folha 5-D (Braga)*, IGM, Aveiro, Portugal, 2000, in Portuguese.

[17] Y. Dodge, *The Concise Encyclopedia of Statistics*, Springer, New York, NY, USA, 2008.

[18] R. L. Wasserstein and N. A. Lazar, "The ASA's statement on p-values: context, process, and purpose," *The American Statistician*, vol. 70, no. 2, pp. 129–133, 2016.

[19] M. Gascoyne, "Geochemistry of the actinides and their daughters," in *Uranium-Series Disequilibrium: Applications to Earth, Marine, and Environmental Sciences*, M. Ivanovich and R. S. Harmon, Eds., pp. 34–61, Clarendon Press, Oxford, UK, 1992.

3

Real-Time and Quantitative Measurement of Crack-Tip Stress Intensity Factors Using Digital Holographic Interferometry

Haiting Xia[1,2] **Rongxin Guo**[1,2] **Feng Yan,**[1,2] **and Heming Cheng**[1]

[1]*Faculty of Civil Engineering and Mechanics, Kunming University of Science and Technology, Kunming, Yunnan 650500, China*
[2]*Key Laboratory of Yunnan Province for Disaster Reduction in Civil Engineering, Kunming University of Science and Technology, Kunming, Yunnan 650500, China*

Correspondence should be addressed to Rongxin Guo; guorx@kmust.edu.cn

Academic Editor: Hiroshi Noguchi

Detection of the crack in an object is a critical problem for the health monitoring of a transparent object. The real-time and quantitative measurement of the crack-tip stress intensity factor (SIF) remains an open issue. In this paper, an approach for real-time and quantitative measurement for the SIFs of a Mode I crack is presented based on digital holographic interferometry (DHI). A transmission digital holographic system is established to measure the phase difference of an object wave during loading. The expression to achieve the SIF from the phase difference is formulated. To enhance the accuracy of measurement, calibrated phase unwrapping based on least-squares and iteration and median filtering is applied to retrieve the actual phase from the noisy wrapped one. The SIFs of the Mode I crack in a transparent polymethyl methacrylate (PMMA) specimen are measured by this approach. The results are compared with the theoretical ones to demonstrate the feasibility of the proposed approach.

1. Introduction

Transparent materials are widely used in many engineering applications such as conductive materials [1–3], electrode materials [4–7], thermoelectric materials [8], encapsulation materials [9], and bulletproof materials [10, 11]. The fracture and fatigue problems are critical for many applications of the transparent materials. The crack is the key reason of materials failure. Usually the stress intensity factor (SIF) is the main quantity to evaluate the influence of the crack in the materials in elasticity. In the last decades, many measurement approaches were used to detect the SIF near the crack in transparent materials including photoelasticity [12, 13], coherent gradient sensing (CGS) [14–16], and digital coherent sensing (DGS) [17–19]. However, these approaches need birefringent materials, identification of the fringes, phase shifting, or fabrication of speckles which affect the application and accuracy of the measurement of the SIF.

As a noninvasive and full-field measurement method, digital holographic interferometry (DHI) has been widely used in many measurements including deformation [20–22], contour [23, 24], the field of fluid [25, 26], vibration [27, 28], and microscopy [29, 30]. By applying DHI in the measurement of transparent materials, one can get the stress-optic constant, stress field, and strain field [31, 32]. Especially, DHI can detect the damage and cracks in transparent materials which means it can be used in the health monitoring of a transparent structure and experimental study on the fracture and damage mechanics [33]. Usually, DHI obtains physical quantities through the phase difference of an object wave. As it is obtained through numerical operations based on the arctangent function, the phase is wrapped into the interval $[-\pi, \pi]$ and needed to be unwrapped to retrieve the actual nonwrapped phase [34]. However, because stress near the crack tip is concentrated, the phase fringes are very close which results high speckle noise in the phase maps. The existence of high noise makes phase unwrapping very difficult and even to fail. Spatial filtering can be applied so as to reduce noise before phase unwrapping [35]. However, filtering will smear the phase jumps which will make unwrapping to fail. So, detection of the crack by DHI remains a qualitative analysis and does not realize the quantitative measurement which limits the application of DHI in the measurement of transparent materials.

In this paper, we aim to realize unwrapping the phase near the crack tip by using the calibrated phase unwrapping method based on least-squares and iterations [36]. Moreover, we present the scheme for the measurement of SIF near the crack tip in a transparent plate by DHI. In addition, the measured results are compared with the theoretical ones to validate the feasibility of the proposed approach. This paper is organized as follows: Section 2 first presents the principles of the proposed approach; Section 3 introduces the experimental scheme of the measurement for K_I in transparent materials by the proposed approach; in Section 4, the measured results are given and validated by the comparison to the theoretical ones; and some conclusions are given in Section 5 finally.

2. Principles of the Proposed Approach

2.1. Digital Holographic Interferometry. The principle diagram of transmission digital holographic interferometry [31] is shown in Figure 1. The input laser is divided into the object wave and reference wave by a beam splitter. The object wave and reference wave are both expanded and collimated by a spatial filter and a lens, respectively. The transparent object is placed in the collimated object wave and is imaged to the CCD target plane by an equivalent 4f system. Here, the negative lens in the system is used to adjust the field of view of interest. The collimated reference wave is refracted by a beam splitter and interfered with the object wave at the CCD plane. The principle axis of the reference wave forms a certain specified angle with that of the object wave which is normal to the CCD plane so that this system belongs to the image-plane off-axis holography. The CCD camera captures holograms at different loading states. The phase difference

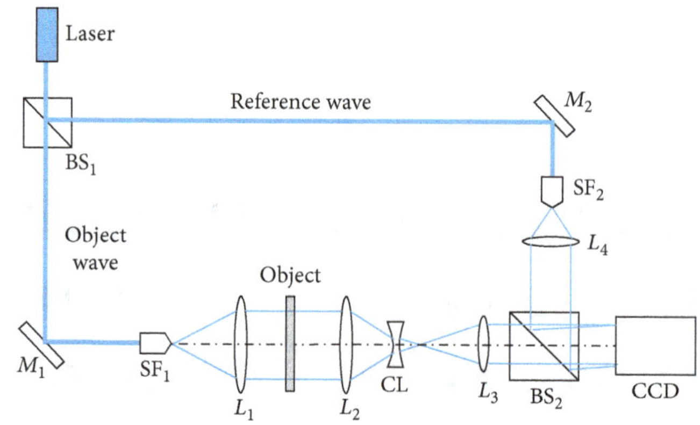

FIGURE 1: Principle diagram of the digital holographic system. BS_1 and BS_2 are the beam splitters; M_1 and M_2 are the mirrors; L_1, L_2, L_3, and L_4 are the positive lenses; CL is the negative lens; SF_1 and SF_2 are the spatial filters.

related to the stress in the specimen is reconstructed according to the following principles shown in Figure 1.

Establishing x and y coordinates in the CCD plane and z coordinate normal to the plane, the object wave $O(x, y)$ and the reference wave $R(x, y)$ can be expressed as

$$\begin{aligned} O(x, y) &= o(x, y)\exp[jk\varphi(x, y)], \\ R(x, y) &= r(x, y)\exp\left[jk\left(\theta_x x + \theta_y y\right)\right], \end{aligned} \tag{1}$$

where $j = \sqrt{-1}$, $k = 2\pi/\lambda$, λ is the wavelength of the laser, θ_x and θ_y are the angles between z axis and the projection of the reference wave vector on the xz plane and yz plane, respectively.

The hologram captured by the CCD is described as

$$\begin{aligned} I(x, y) = {} & o^2(x, y) + r^2(x, y) + o(x, y)\exp[jk\varphi(x, y)]r(x, y)\exp\left[-jk\left(\theta_x x + \theta_y y\right)\right] \\ & + o(x, y)\exp[-jk\varphi(x, y)]r(x, y)\exp\left[jk\left(\theta_x x + \theta_y y\right)\right]. \end{aligned} \tag{2}$$

Performing Fourier transform on both sides of the above equation, we can get the spectrum of the hologram:

$$\begin{aligned} \mathfrak{F}\{I(x, y)\} = {} & G_0\left(f_x, f_y\right) + G\left(f_x + \frac{\theta_x}{\lambda}, f_y + \frac{\theta_y}{\lambda}\right) \\ & + G^*\left(f_x - \frac{\theta_x}{\lambda}, f_y - \frac{\theta_y}{\lambda}\right). \end{aligned} \tag{3}$$

On the right side of this equation, the first item is the spectrum of the zero-order diffracted wave, the second is the spectrum of the object wave which is modulated by the reference wave and translated to the point $(-\theta_x/\lambda, -\theta_y/\lambda)$ relative to the center of the spectrum plane, and the third is the spectrum of the conjugate object wave. So, by selecting the appropriate angles of θ_x and θ_y, we can separate the spectrum of the object wave $G(f_x, f_y)$ from the other spectrums. A rectangular window filter is used to extract $G(f_x, f_y)$. Then, the inverse Fourier transform of $G(f_x, f_y)$ is carried out to reconstruct the complex modulated object wave $f(x, y)$ as follows:

$$f(x, y) = \mathfrak{F}^{-1}\left\{G\left(f_x, f_y\right)\right\}. \tag{4}$$

The phase difference of object waves at two loading states, due to the specimen's deformation, is expressed as

$$\psi(x, y) = \text{angle}\left[\frac{f_1(x, y)}{f_2(x, y)}\right], \tag{5}$$

where $\psi(x, y)$ is the wrapped phase difference, $f_1(x, y)$ and $f_2(x, y)$ are the complex object waves at different loading states, and angle (.) is the function to calculate the phase angle of a complex value.

2.2. Phase Recovering Approach. In this paper, we use calibrated phase unwrapping algorithm based on the least-squares and iterations (CPULSI) [36] and median filtering to recover the true phase from the noisy wrapped phase. Here, φ_{ij} represents the true phase and ψ_{ij} ($\in [-\pi, \pi]$, i.e., modulo 2π) represents the wrapped phase at the grid point (i, j) of a phase map. Their relationship can be expressed by the wrapping operator $W(.)$ which is defined as

$$W(\varphi_{ij}) = \psi_{ij} = \varphi_{ij} + 2\pi k_{ij} \quad (i = 0, 1, \ldots, M-1;\ j = 0, 1, \ldots, N-1), \tag{6}$$

where $-\pi \le \psi_{ij} \le \pi$, k_{ij} is an integer, M and N are, respectively, the number of grid points with respect to the i and j indexes. The 1st order spatial wrapped phase derivatives are defined as

$$\begin{aligned}
\Delta_{ij}^{x} &= W(\psi_{(i+1)j} - \psi_{ij}) \quad (i = 0, 1, \ldots, M-2;\ j = 0, 1, \ldots, N-1),\\
\Delta_{ij}^{x} &= 0, \quad \text{otherwise},\\
\Delta_{ij}^{y} &= W(\psi_{i(j+1)} - \psi_{ij}) \quad (i = 0, 1, \ldots, M-1;\ j = 0, 1, \ldots, N-2),\\
\Delta_{ij}^{y} &= 0, \quad \text{otherwise},
\end{aligned} \tag{7}$$

where Δ_{ij}^{x} and Δ_{ij}^{y} are, respectively, the difference with respect to the i and j indexes.

In practice, the presence of noise will generate errors between noisy and noise-free phase derivatives and make unwrapping difficult, even to fail. Here, we proposed a calibration approach in [36] to calibrate the phase derivatives exhibiting large errors:

$$\begin{aligned}
\Delta_{ij}^{x} &= \operatorname{sgn}(\Delta_{ij}^{x})|G_x| \quad &&\text{if } |\Delta_{ij}^{x}| \ge T_x,\\
\Delta_{ij}^{x} &= \Delta_{ij}^{x}, &&\text{otherwise},\\
\Delta_{ij}^{y} &= \operatorname{sgn}(\Delta_{ij}^{y})|G_y| \quad &&\text{if } |\Delta_{ij}^{y}| \ge T_y,\\
\Delta_{ij}^{y} &= \Delta_{ij}^{y}, &&\text{otherwise},
\end{aligned} \tag{8}$$

where $\operatorname{sgn}(\ldots)$ is the signum function, T_x and T_y are the thresholds, and G_x and G_y are the calibrated phase derivatives. These parameters are defined as ($E[\ldots]$ means the statistical average)

$$\begin{aligned}
T_x &= \sqrt{E\left[(\Delta_{ij}^{x})^2\right] - \left(E[\Delta_{ij}^{x}]\right)^2},\\
T_y &= \sqrt{E\left[(\Delta_{ij}^{y})^2\right] - \left(E[\Delta_{ij}^{y}]\right)^2},\\
G_x &= \frac{1}{MN}\sum_{i=0}^{i=M-1}\sum_{j=0}^{j=N-1}\Delta_{ij}^{x},\\
G_y &= \frac{1}{MN}\sum_{i=0}^{i=M-1}\sum_{j=0}^{j=N-1}\Delta_{ij}^{y}.
\end{aligned} \tag{9}$$

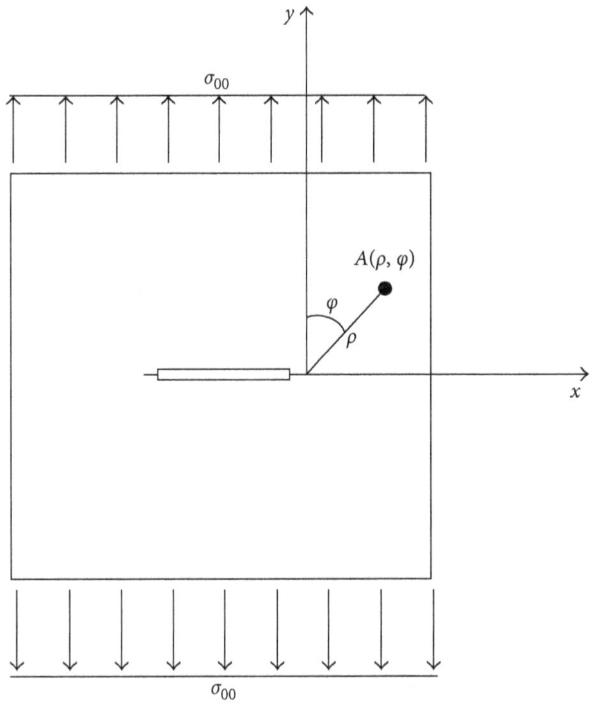

Figure 2: Specimen shape and loading way.

This calibration approach means that the phase derivatives whose values are larger than the standard deviation of phase derivatives are replaced by the average value of phase derivatives.

In the least-squares sense, the optimal solution φ_{ij} can be obtained from the discrete Poisson equation with the Neumann boundary conditions:

$$\left(\varphi_{(i+1)j} - 2\varphi_{ij} + \varphi_{(i-1)j}\right) + \left(\varphi_{i(j+1)} - 2\varphi_{ij} + \varphi_{i(j-1)}\right) = \rho_{ij}, \tag{10}$$

where

$$\rho_{ij} = \left(\Delta_{ij}^{x} - \Delta_{(i-1)j}^{x}\right) + \left(\Delta_{ij}^{y} - \Delta_{i(j-1)}^{y}\right). \tag{11}$$

The discrete Poisson equation can be solved by many methods such as fast Fourier transform (FFT), discrete cosine transform (DCT), or the multigrid method. In this algorithm, the DCT method is selected to solve the least-squares phase unwrapping problem.

Theoretically, the solution obtained from (10) is the exact one. However, there exist errors between the unwrapped phase and the true phase due to noise and to the smoothing performance of the least-squares method. In the proposed approach, iterations of unwrapped phase errors are utilized to seek more accurate results [36].

After unwrapping, median filtering is used to reduce the noise in the phase map. Median filtering is one of the most commonly used spatial filtering methods. It uses the median value in the kernel to substitute the value of center of the kernel [37]. So, it is easy to implement and efficient for impulsive noise and speckle noise. In addition, the phase

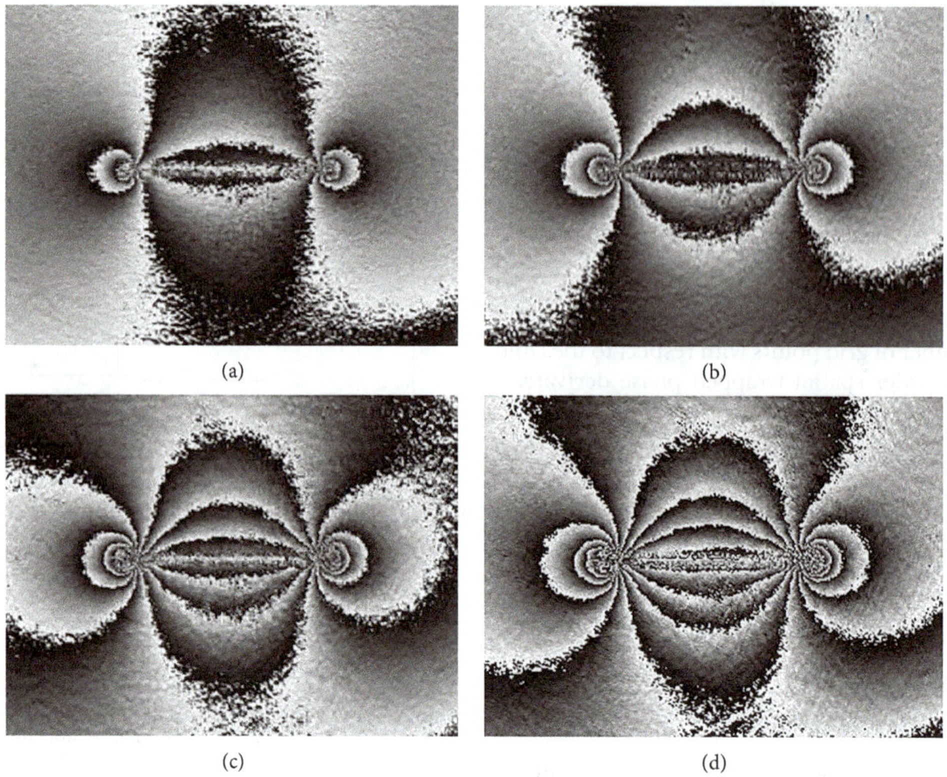

FIGURE 3: The wrapped phase of the object wave through the PMMA specimen with a crack. (a) $\sigma_{00} = 1\,\mathrm{MPa}$, (b) $\sigma_{00} = 1.5\,\mathrm{MPa}$, (c) $\sigma_{00} = 2\,\mathrm{MPa}$, and (d) $\sigma_{00} = 2.5\,\mathrm{MPa}$.

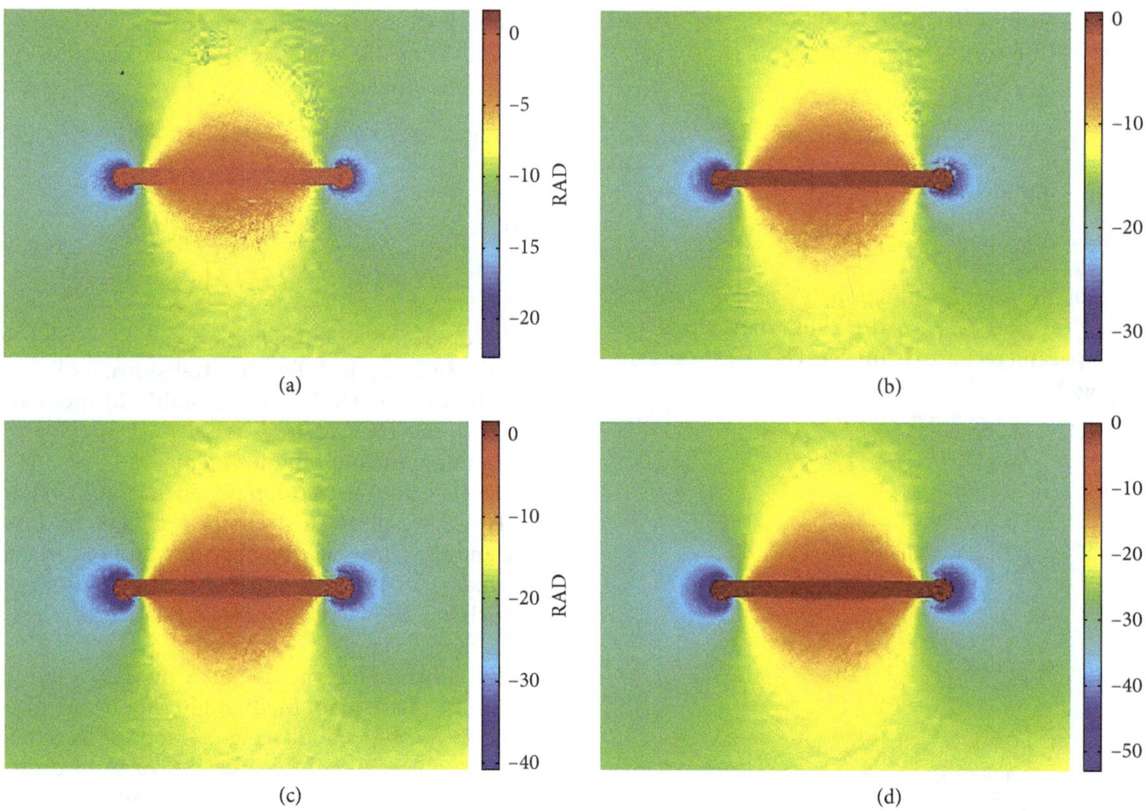

FIGURE 4: The unwrapped phase of the object wave through the PMMA specimen with a crack. (a) $\sigma_{00} = 1\,\mathrm{MPa}$, (b) $\sigma_{00} = 1.5\,\mathrm{MPa}$, (c) $\sigma_{00} = 2\,\mathrm{MPa}$, and (d) $\sigma_{00} = 2.5\,\mathrm{MPa}$.

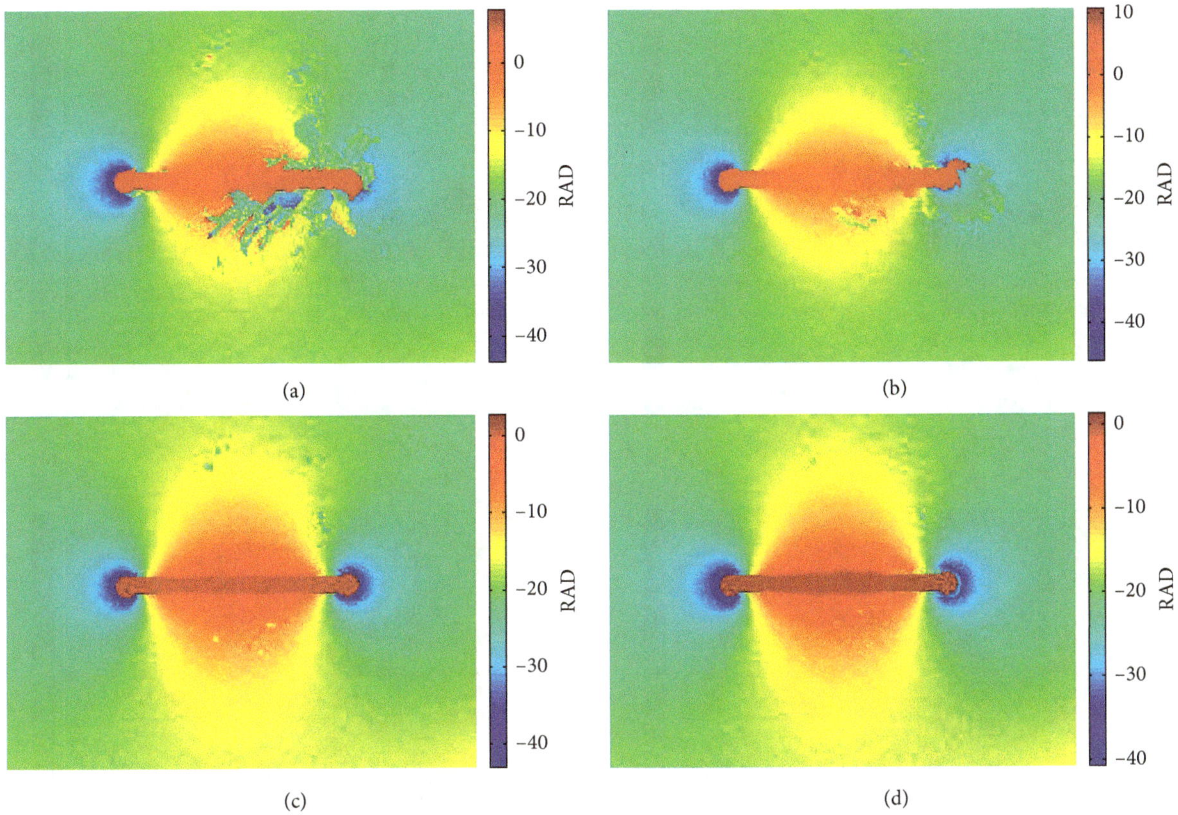

FIGURE 5: The unwrapped phase of the object wave as $\sigma_{00} = 2$ MPa obtained from (a) Gold, (b) Quality, (c) Flynn, and (d) Lp.

map must be padded out of the frontiers with the length of the kernels symmetrically.

2.3. Calculation of the Stress Intensity Factor. For the uniform transparent plate specimen normal to the object wave principle axis, the relationship between the stress and phase difference of the object wave in plane stress state can be expressed as [31]

$$\left(\sigma_x + \sigma_y\right) = \frac{\lambda}{2\pi hc} \Delta\varphi, \tag{12}$$

where σ_x and σ_y are the normal stresses in the plane, h is the thickness of the plate, c is the stress-optic constant of materials, and $\Delta\varphi$ is the unwrapped denoised phase.

Taking into account the specimen with a Mode I through crack as shown in Figure 2, the stress near the crack tip can be expressed as [38]

$$\sigma_x = \frac{K_{\mathrm{I}}}{\sqrt{2\pi r}} \cos\frac{\theta}{2}\left(1 - \sin\frac{\theta}{2}\sin\frac{3\theta}{2}\right), \tag{13}$$

$$\sigma_y = \frac{K_{\mathrm{I}}}{\sqrt{2\pi r}} \cos\frac{\theta}{2}\left(1 + \sin\frac{\theta}{2}\sin\frac{3\theta}{2}\right), \tag{14}$$

where K_{I} is the stress intensity factor (SIF) of the Mode I crack and r and θ are the polar coordinates relative to the crack tip.

By adding (13) and (14), we can get the equation to calculate K_{I} from the stresses as

$$K_{\mathrm{I}} = \frac{\left(\sigma_x + \sigma_y\right)\sqrt{2\pi r}}{2\cos(\theta/2)}. \tag{15}$$

So, from (12) and (15), one can calculate the stress sum and SIF near the crack tip.

3. Experiment

In order to validate its feasibility, the proposed approach is used to measure the SIF near the crack tip in a transparent specimen. The specimen is fabricated with polymethyl methacrylate (PMMA). The shape and load of the specimen are shown in Figure 2. The size of the specimen is 400 mm × 100 mm × 4 mm. A through sharp crack is fabricated in the center of the specimen by the fatigue testing machine, and its length is 22 mm. The specimen is subjected to the vertical uniform tensile load σ_{00} increasing from 0 to 3 MPa which has been estimated to ensure linear elasticity. The elastic modulus of PMMA is $E = 3.24$ GPa, and Poisson's ratio is $\nu = 0.35$. The wavelength of the laser is $\lambda = 532$ nm. A 1024 × 768 pixel CCD camera with a pixel size of 6.4 μm is used to record digital holograms.

4. Experimental Results and Validation

4.1. Experimental Results of Phase Difference. By using the image-plane reconstruction algorithm given in Section 2.1, we can obtain the phase difference of any applied load relative to the initial load ($\sigma_{00} = 0$). The wrapped phase

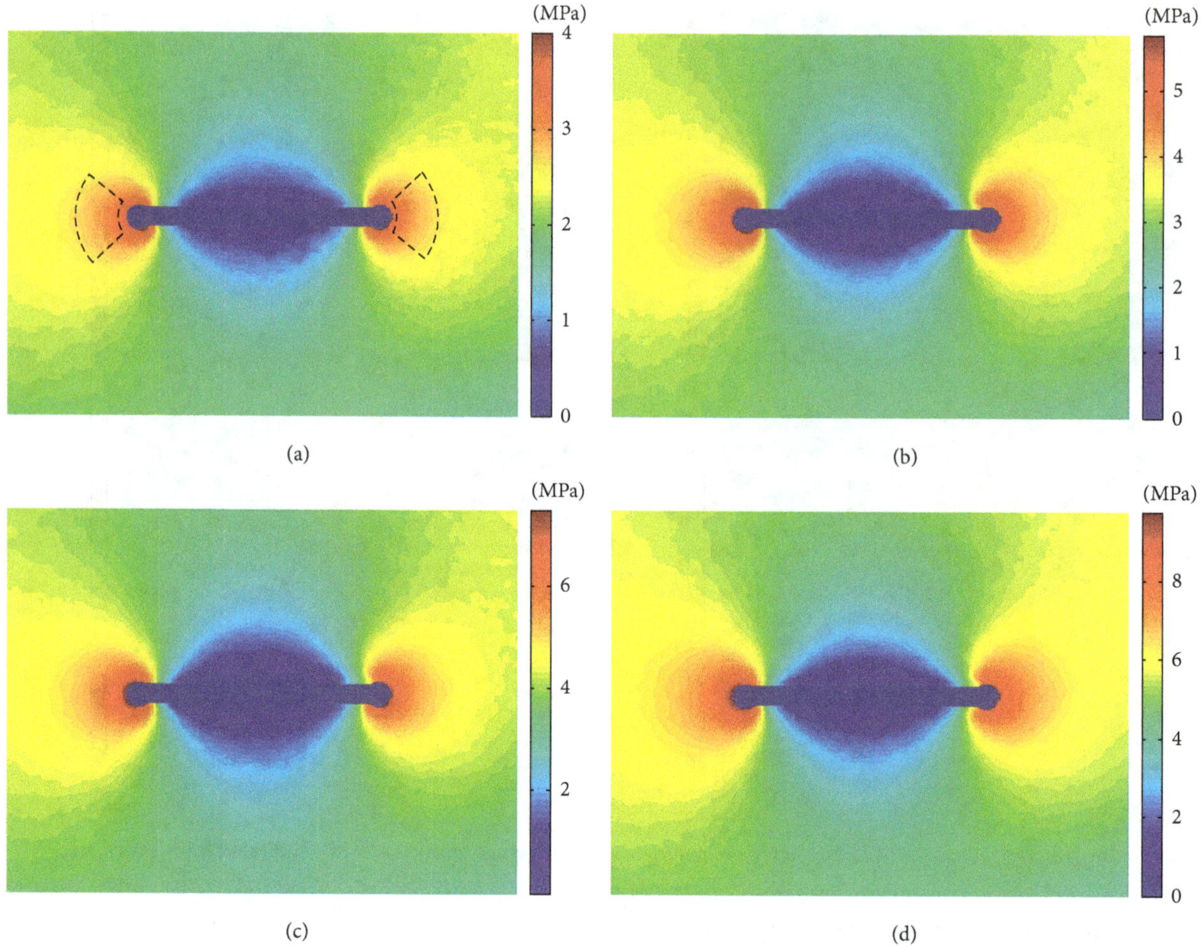

FIGURE 6: The sum of stress ($\sigma_x + \sigma_y$) in the PMMA specimen with a crack. (a) $\sigma_{00} = 1$ MPa, (b) $\sigma_{00} = 1.5$ MPa, (c) $\sigma_{00} = 2$ MPa, and (d) $\sigma_{00} = 2.5$ MPa.

differences at different load are shown in Figure 3. From these wrapped phase maps, it can be seen that there exists high speckle noise, the fringes near the crack tip are very close, and the phase data inside the crack region is invalid. So, these phase maps are very difficult to unwrap.

In order to unwrap the phase data in Figure 3, we use a mask to shield the crack region; then, CPULSI is used to unwrap the masked phase maps. The unwrapped phase maps are shown in Figure 4, and we can see that all the phase maps are unwrapped successfully.

In order to illustrate the necessity of CPULSI used in this paper, four other different classical phase unwrapping algorithms were selected: Goldstein branch cut algorithm (namely, "Gold"), quality-guided path following algorithm ("Quality"), Flynn's minimum-weight-discontinuity algorithm ("Flynn"), and minimum Lp norm algorithm ("Lp") [34]. Figure 5 shows the unwrapped phase maps of the object wave as $\sigma_{00} = 2$ MPa obtained from these four algorithms. Comparing Figure 4(c), we can see that the unwrapped result from CPULSI is better than those from the other algorithms obviously. In fact, the other algorithms cannot successfully unwrap the phase data in this study because there all exist discontinuities in the maps of Figure 5. So, it is necessary to use CPULSI in this study.

4.2. Calculation of the Stress Field and SIF near the Crack Tip. In order to reduce the noise in the phase maps, median filtering with the kernel size of 21×21 pixels is applied on the unwrapped phase maps. The denoised unwrapped phase is used to calculate stress sum near the crack tip. Figure 6 shows the stress sum as $\sigma_{00} = 1$ MPa, 1.5 MPa, 2 MPa, and 2.5 MPa. From these results, it can be seen that the maximum stress sums exist in the zone near the crack tip and those near the crack plane close to zero which is in accordance with the fracture theory.

According to (15), we can calculate SIFs from stress sums at different loads. Because (13) and (14) only express the stresses infinitely close to the crack tip in the range of linear elastic, and the K-dominate zone must be determined before calculation. Here, we select the dashed sectorial domain in Figure 6(a) which is $1.5\,\text{mm} \le r \le 2.5\,\text{mm}$ and $-45^\circ \le \theta \le 45^\circ$ as the K-dominate zone is referring to some literatures [14, 15, 18, 19]. The average value of the SIFs in the

TABLE 1: The comparison between the measured values and theoretical values of K_I.

Tensile load (MPa)	Measured K_I (MPa·m$^{0.5}$)	Theoretical K_I (MPa·m$^{0.5}$)	Relative error (%)
1.0	0.193	0.191	1.0
1.5	0.291	0.287	1.4
2.0	0.379	0.383	−1.0
2.5	0.503	0.479	5.0

K-dominate zone is set as the measured value. The measured SIFs at different loads are shown in Table 1.

In order to validate the accuracy of the measurement, the theoretical values of the SIFs are calculated by using the method in [39]. For the crack in Figure 2, K_I can be calculated as

$$K_I = F\sigma_{00}\sqrt{\pi a}, \tag{16}$$

where σ_{00} is the vertical uniform tensile load, a is the half length of the crack, and F is the calibration coefficient for the size of the specimen which can be expressed as

$$F = \left[1 - 0.025\left(\frac{a}{b}\right)^2 + 0.06\left(\frac{a}{b}\right)^4\right]\sqrt{\sec\left(\frac{\pi a}{2b}\right)}, \tag{17}$$

where b is the half width of the specimen. The calculated theoretical values of SIFs at different loads are also shown in Table 1. We can obtain the relative errors by comparing the measured values and theoretical ones. From Table 1, it can be seen that the error of the measured SIF does not exceed 5% which means the results of the proposed approach are accurate.

5. Conclusions

In this paper, we address the real-time and quantitative measurement for the SIFs of the Mode I crack in the transparent plate by DHI. For the phase processing near the crack tip which is affected by the close fringes and high speckle noise, we use phase unwrapping algorithm CPULSI and median filtering to recover the phase maps. The calculation method of K_I from the phase difference is formulated based on linear elastic fracture mechanics. The proposed approach is used to measure K_I of the crack in the PMMA specimen. The measured results are validated in comparison with the theoretical ones. The proposed approach can be used for the real-time and quantitative test of transparent materials.

Conflicts of Interest

The authors declare that they have no conflicts of interest.

Acknowledgments

This work was supported by the Natural Science Foundation of China (11462009 and 11362007).

References

[1] F.-G. Zhao, Y.-T. Kong, Z.-W. Xu, X. Yao, B. Zuo, and W.-S. Li, "High-performance flexible transparent conductive films achieved by cooperation between 1D copper nanowires and 2D graphene materials," *Journal of Materials Chemistry C*, vol. 5, pp. 5509–5516, 2017.

[2] B. Yang, C. Yao, Y. Yu, Z. Li, and X. Wang, "Nature degradable, flexible, and transparent conductive substrates from green and earth-abundant materials," *Scientific Reports*, vol. 7, no. 1, 2017.

[3] Q. Tang, H. Shen, H. Yao, Y. Jiang, C. Zheng, and K. Gao, "Preparation of silver nanowire/AZO composite film as a transparent conductive material," *Ceramics International*, vol. 43, no. 1, pp. 1106–1113, 2017.

[4] M. Wu, D. Sun, C. Tan, X. Tian, and Y. Huang, "Al-doped ZnO monolayer as a promising transparent electrode material: a first-principles study," *Materials*, vol. 10, no. 4, p. 359, 2017.

[5] G. Wroblewski, B. Swatowska, W. Powroznik, M. Jakubowska, and T. Stapinski, "Structural and optical properties of spray coated carbon hybrid materials applied to transparent and flexible electrodes," *Journal of Nanomaterials*, vol. 2017, Article ID 3424672, 7 pages, 2017.

[6] M. D. S. L. Wimalananda, J.-K. Kim, and J.-M. Lee, "Toward the ultra-transparent electrode by using patterned silver nanowire and graphene layered material," *Carbon*, vol. 125, pp. 9–19, 2017.

[7] T. Q. Trung and N.-E. Lee, "Materials and devices for transparent stretchable electronics," *Journal of Materials Chemistry C*, vol. 5, no. 9, pp. 2202–2222, 2017.

[8] C. Yang, D. Souchay, M. Kneiβ et al., "Transparent flexible thermoelectric material based on non-toxic earth-abundant p-type copper iodide thin film," *Nature Communications*, vol. 8, p. 16076, 2017.

[9] D. J. Kang, G. U. Park, H. Y. Park, and H.-G. Im, "A robust transparent encapsulation material: Silica nanoparticle-embedded epoxy hybrid nanocomposite," *Composites Science and Technology*, vol. 144, pp. 107–113, 2017.

[10] T. Benitez, S. Y. Gómez, A. P. N. de Oliveira, N. Travitzky, and D. Hotza, "Transparent ceramic and glass-ceramic materials for armor applications," *Ceramics International*, vol. 43, no. 16, pp. 13031–13046, 2017.

[11] G.-I. Shim, S.-H. Kim, D.-L. Ahn et al., "Experimental and numerical evaluation of transparent bulletproof material for enhanced impact-energy absorption using strengthened glass/polymer composite," *Composites Part B*, vol. 97, pp. 150–161, 2016.

[12] S. Khaleghian, A. Emami, M. Tehrani, and N. Soltani, "Analysis of effective parameters for stress intensity factors in the contact problem between an asymmetric wedge and a half-plane using an experimental method of photoelasticity," *Materials and Design*, vol. 43, pp. 447–453, 2013.

[13] M. D. Ellingsen and S. K. Khanna, "Experimental investigation of static interfacial fracture in orthotropic polymer composite bimaterials using photoelasticity," *Journal of Engineering Materials and Technology*, vol. 132, no. 2, p. 021007, 2010.

[14] K. Ma and H. Xie, "Mixed-mode fracture investigation of PMMA with initial single/double crack(s) interference using phase-shifted coherent gradient sensing method," *Polymer Testing*, vol. 59, pp. 296–307, 2017.

[15] K. Ma and H. Xie, "A novel phase shifting technique of coherent gradient sensing method for measuring crack-tip

K-dominance," *Experimental Mechanics*, vol. 57, no. 8, pp. 1239–1248, 2017.

[16] X. F. Yao, H. Y. Yeh, and W. Xu, "Fracture investigation at V-notch tip using coherent gradient sensing (CGS)," *International Journal of Solids and Structures*, vol. 43, no. 5, pp. 1189–1200, 2006.

[17] B. M. Sundaram and H. V. Tippur, "Dynamic mixed-mode fracture behaviors of PMMA and polycarbonate," *Engineering Fracture Mechanics*, vol. 176, pp. 186–212, 2017.

[18] B. M. Sundaram and H. V. Tippur, "Dynamics of crack penetration vs. branching at a weak interface: an experimental study," *Journal of the Mechanics and Physics of Solids*, vol. 96, pp. 312–332, 2016.

[19] C. Periasamy and H. V. Tippur, "Measurement of crack-tip and punch-tip transient deformations and stress intensity factors using digital gradient sensing technique," *Engineering Fracture Mechanics*, vol. 98, pp. 185–199, 2013.

[20] S. Montrésor, P. Picart, O. Sakharuk, and L. Muravsky, "Error analysis for noise reduction in 3D deformation measurement with digital color holography," *Journal of the Optical Society of America B*, vol. 34, pp. B9–B15, 2017.

[21] D. Khodadad, A. K. Singh, G. Pedrini, and M. Sjödahl, "Full-field 3D deformation measurement: comparison between speckle phase and displacement evaluation," *Applied Optics*, vol. 55, no. 27, pp. 7735–7743, 2016.

[22] G. N. Oliveira, D. M. C. Rodrigues, L. C. S. Nunes, and P. A. M. dos Santos, "Digital Fourier transform holography applied to investigate mechanical deformation in polymers," *Optics and Lasers in Engineering*, vol. 50, no. 12, pp. 1798–1803, 2012.

[23] V. Lédl, P. Psota, F. Kaván, O. Matoušek, and P. Mokrý, "Surface topography measurement by frequency sweeping digital holography," *Applied Optics*, vol. 56, pp. 7808–7814, 2017.

[24] M. Dekiff, P. Berssenbrügge, B. Kemper, C. Denz, and D. Dirksen, "Simultaneous acquisition of 3D shape and deformation by combination of interferometric and correlation-based laser speckle metrology," *Biomedical Optics Express*, vol. 6, pp. 4825–4840, 2015.

[25] C. Hesseling, T. Homeyer, J. Peinke, and G. Gülker, "Particle depth position detection by 2D correlation in digital in-line holography," *Optics Letters*, vol. 41, no. 21, pp. 4947–4950, 2016.

[26] J.-M. Desse, P. Picart, and F. Olchewsky, "Quantitative phase imaging in flows with high resolution holographic diffraction grating," *Optics Express*, vol. 23, no. 18, pp. 23726–23737, 2015.

[27] K. A. Stetson, "Effect of changing speckles in digital holography on measurements of static and vibratory displacements," *Applied Optics*, vol. 55, no. 16, pp. 4485–4489, 2016.

[28] J. Poittevin, P. Picart, F. Gautier, and C. Pezerat, "Quality assessment of combined quantization-shot-noise-induced decorrelation noise in high-speed digital holographic metrology," *Optics Express*, vol. 23, no. 24, pp. 30917–30932, 2015.

[29] W. J. Eldridge, A. Sheinfeld, M. T. Rinehart, and A. Wax, "Imaging deformation of adherent cells due to shear stress using quantitative phase imaging," *Optics Letters*, vol. 41, no. 2, pp. 352–355, 2016.

[30] M. Malek, H. Khelfa, P. Picart, D. Mounier, and C. Poilâne, "Microtomography imaging of an isolated plant fiber: a digital holographic approach," *Applied Optics*, vol. 55, no. 3, pp. A111–A121, 2016.

[31] H. Xia, R. Guo, F. Yan et al., "Simultaneous measurement of stress-optic constant and stress field of transparent plate by digital holographic interferometry," *Optik-International Journal for Light and Electron Optics*, vol. 127, no. 24, pp. 11974–11981, 2016.

[32] H. Xia, R. Guo, Z. Fan, H. Cheng, and B. Yang, "Non-invasive mechanical measurement for transparent objects by digital holographic interferometry based on iterative least-squares phase unwrapping," *Experimental Mechanics*, vol. 52, no. 4, pp. 439–445, 2012.

[33] R. Guo, H. Xia, Z. Fan, B. Yang, and T. He, "Research on the interaction of collective cracks with digital holography," *Laser Journal*, vol. 30, pp. 38-39, 2009, in Chinese.

[34] D. C. Ghiglia and M. D. Pritt, *Two-Dimensional Phase Unwrapping: Theory, Algorithms, and Software*, Wiley, Hoboken, NJ, USA, 1998.

[35] S. Montresor and P. Picart, "Quantitative appraisal for noise reduction in digital holographic phase imaging," *Optics Express*, vol. 24, no. 13, pp. 14322–14343, 2016.

[36] H. Xia, S. Montresor, R. Guo et al., "Phase calibration unwrapping algorithm for phase data corrupted by strong decorrelation speckle noise," *Optics Express*, vol. 24, no. 25, pp. 28713–28730, 2016.

[37] S. A. Villar, S. Torcida, and G. G. Acosta, "Median filtering: a new insight," *Journal of Mathematical Imaging and Vision*, vol. 58, no. 1, pp. 130–146, 2017.

[38] Z. N. Li, *Applied Fracture Mechanics*, Behang University Press, Beijing, China, 2012, in Chinese.

[39] Chinese Aeronautical Establishment, *Handbook of Stress-Intensity Factors-Revised Edition*, Science Press, Beijing, China, 1993, in Chinese.

4

Investigation on Mechanical Properties and Reaction Characteristics of Al-PTFE Composites with Different Al Particle Size

Jia-xiang Wu, Xiang Fang, Zhen-ru Gao, Huai-xi Wang, Jun-yi Huang, Shuang-zhang Wu, and Yu-chun Li

College of Field Engineering, PLA Army Engineering University, Hou Biaoying Road No. 88, Nanjing, Jiangsu 210007, China

Correspondence should be addressed to Yu-chun Li; liyuchunmail@sina.com

Academic Editor: Zhiping Luo

Al-PTFE (aluminum-polytetrafluoroethylene) serves as one among the most promising reactive materials (RMs). In this work, six types of Al-PTFE composites with different Al particle sizes (i.e., 50 nm, 1~2 μm, 6~7 μm, 12~14 μm, 22~24 μm, and 32~34 μm) were prepared, and quasistatic compression and drop weight tests were conducted to characterize the mechanical properties and reaction characteristics of Al-PTFE composites. The reaction phenomenon and stress-strain curves were recorded by a high-speed camera and universal testing machine. The microstructure of selected specimens was anatomized through adopting a scanning electron microscope (SEM) to correlate the mesoscale structural characteristics to their macroproperties. As the results indicated, in the case of quasistatic compression, the strength of the composites was decreased (the yield strength falling from 22.7 MPa to 13.6 MPa and the hardening modulus declining from 33.3 MPa to 25 MPa) with the increase of the Al particle size. The toughness rose firstly and subsequently decreased and peaked as 116.42 MJ/m^3 at 6~7 μm. The reaction phenomenon occurred only in composites with the Al particle size less than 10 μm. In drop weight tests, six types of specimens were overall reacted. As the Al particle size rose, the ignition energy of the composites enhanced and the composites turned out to be more insensitive to reaction. In a lower strain rate range ($10^{-2}\cdot s^{-1}$~$10^{2}\cdot s^{-1}$), Al-PTFE specimens take on different mechanical properties and reaction characteristics in the case of different strain rates. The formation of circumferential open cracks is deemed as a prerequisite for Al-PTFE specimens to go through a reaction.

1. Introduction

Reactive materials (RMs), comprised of two or more non-explosive solid components and also known as impact-initiated energetic material, have aroused great attentions in recent years. Polytetrafluoroethylene filled by aluminum particles (Al-PTFE) is one of the most promising RMs and has been studied extensively, owing to its easiness to deform and higher strength, stability, and energy density compared with traditional explosive [1–3]. When Al-PTFE is subjected to sufficiently strong mechanical stimulus, violent chemical reaction will occur with new substance produced and large amount of heat released. In this regard, Al-PTFE takes on numerous prominent applications in military applications. For instance, it can be made into the warhead shell functionalized to cause secondary damage to evidently increase the damage effect. Furthermore, it can also be adopted in the systems of air defense, antimissile, obstacle breaking, and rocket propulsion [4].

Al-PTFE pertains to both composite materials and energetic materials; thus, this paper firstly focuses on researching the mechanical properties taken on by materials, describing the deformation, yield, and failure processes of materials in the case of different strain rates to determine the constitutive relation, fracture characteristics, and fatigue resistance in different applications [5–8]. Secondly, this paper shall anatomize the reaction characteristics of materials, calculate the antioverload capacity, reaction threshold, impact sensitivity, and energy release rate through characterizing the reaction process, and establish a reaction mechanism model [9–11].

Currently, the experimental researches performed worldwide primarily focus on the engineering strain of Al-PTFE in the case of different influence factors, and the performance shall be varied with the addition of different fillers. Zheng et al. [12] studied the shock initiation process of nano-Al-PTFE with time- and wavelength-resolved emission spectroscopy. The results showed that the reaction was associated with the arrival of the compressive release wave and shear release waves from the flyer plate edges. Eventually, a mechanism was proposed in which tensile stresses resulting from compressive and shear waves cause the Al shells to crack open, consequently allowing hot PTFE to react with Al. Feng et al. [13, 14] discovered that Al-PTFE mixtures through special heat treatment could go through severe reaction under quasistatic compression. The influence of sintering temperature and equivalence ratio on reaction of Al-PTFE was ascertained. On that basis, a crack-induced initiation mechanism was put forward. Ge et al. [15] prepared three kinds of PTFE-Al-W samples with different component mass ratios which were fabricated under a much lower sintering temperature and for a much shorter duration. They evaluated the mechanical and impact-induced initiation characteristics of the brittle PTFE-Al-W materials through carrying out quasistatic compression tests, dynamic compression tests, and drop weight tests at room and elevated temperatures. Wang et al. [16] obtained the compress stress-strain curves for a pressed and sintered PTFE-Al-Ni mixtures under uniaxial loading at strain rates ranging from 10^{-2} to $3 \times 10^{3} \cdot s^{-1}$ using an electrohydraulic press and SHPB system. The data measured were adapted to the Johnson-Cook model and showed that this kind of material took on evident strain hardening and strain rate hardening. Herbold et al. [17] studied the impact exerted by the particle size on strength, failure, and shock behavior in PTFE-Al-W granular composites. The result revealed that dynamic mechanical properties of high-density PTFE-Al-W mixtures could be tailored by changing the particle size and the mixture porosity. The reason for this phenomenon was that the force chains between fine metallic particles were formed under dynamic loading.

As the foregoing information indicates, the slight differences of raw material exert different effects on the mechanical properties and reaction characteristics of Al-PTFE as composite materials. For the Al particle size serving as the basic characteristic, the effect of the Al particle size on the mechanical properties and reaction characteristics of Al-PTFE materials under quasistatic compression condition are needed to be considered when assessing the safety of the Al-PTFE materials. Besides, it is also extremely significant to study the impact sensitivity of materials, considering materials are vulnerable to impact in the process of production, machining, transportation, and storage. In this paper, six types of Al-PTFE specimens with different Al particle sizes were prepared, and their mechanical properties and reaction characteristics under quasistatic compression and dynamic impact were investigated. And the failure and ignition mechanism were analyzed associated with the reaction phenomena.

2. Experimental Setup

2.1. Specimen Preparation. The initial powders adopted to fabricate samples have the following average size: PTFE: 25 μm (purity > 99.0%, from 3M, Shanghai) and Al: 50 nm, 1~2 μm, 6~7 μm, 12~14 μm, 22~24 μm, and 32~34 μm (purity > 99.0%, from JT, Hunan). In this study, six types of Al-PTFE specimens which graded from #1 to #6, respectively, in accordance with the order of the Al particle size from small to large were fabricated. And the mass fraction of Al and PTFE was proportional to the chemical equilibrium ratio (26%/74%) to ensure the complete reaction of the material.

The preparation process of specimens involved mixing, cold isostatic pressing, and vacuum sintering on the basis of the patent of Nielson's [18]. Procedure was described as follows: the Al-PTFE mixture was stirred by a motor-driven blender for 20 min in an ethanol solution, then put into a vacuum drying oven for 48 h at 60°C, and cold pressed into cylinders with sizes of Φ10 mm × 10 mm for the quasistatic compression test and Φ10 mm × 3 mm for the drop weight test under a compressive pressure of 300 MPa. Lastly, the pressed specimens were heated in a vacuum oven at the sintering temperature of 360°C for 5 h in the cases of the heating rate of $90°C \cdot h^{-1}$ and the cooling rate of $50°C \cdot h^{-1}$. The temperature history of the sintering cycle is depicted in Figure 1.

2.2. Experimental Procedures. Quasistatic compression tests of the six types of specimens were conducted by using a CMT5105 electrohydraulic press with a maximum loading capacity of 100 kN at 26°C of the ambient temperature. The load was applied at the speed of $6\ mm \cdot min^{-1}$ corresponding to the nominal strain rate of $0.01\ s^{-1}$. Prior to tests, all contact surfaces of the specimens were lubricated with petroleum jelly to alleviate the effect of friction. Triplicate experiments were carried out for each type of specimens. In total, eighteen specimens—three for each of the six types of specimens—were tested to ensure the reliability of obtained results.

A fall hammer impact sensitivity test was performed to investigate the sensitivity and impact-initiation characteristics of Al-PTFE materials in the ambient temperature of 26°C. The instrument has a drop mass of 10 kg which can be released from a maximum height of 156 cm, so the maximum strain rate of nearly $500\ s^{-1}$ and the maximum output energy of 152.9 J can be achieved. The specimens were placed on the anvil of the tester without constraint and impacted directly by the drop hammer falling free. The impact sensitivities of the six types of materials were calculated by the characteristic drop height (H_{50}), at which Al-PTFE specimens can react with a 50% possibility. To observe the impact process more accurately, time sequences of the drop weight tests were recorded by a high-speed camera with the frame rate up to 20,000 frames/s.

To investigate the relation between the microstructure of the materials and the reaction mechanisms, Hitachi S-3400N II scanning electron microscope (SEM) was used to observe the interior microstructures of the materials.

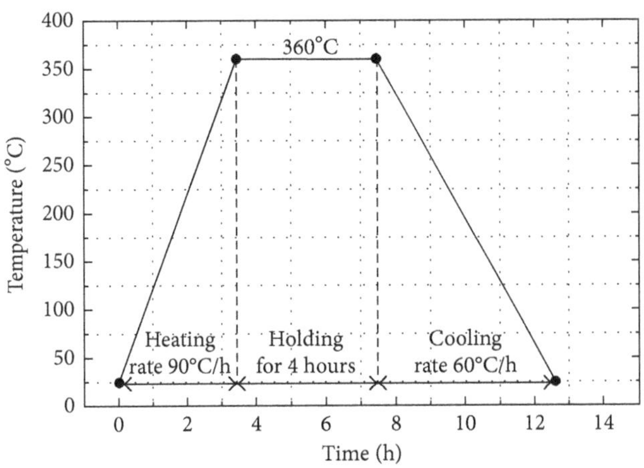

FIGURE 1: The temperature history of the sintering cycle.

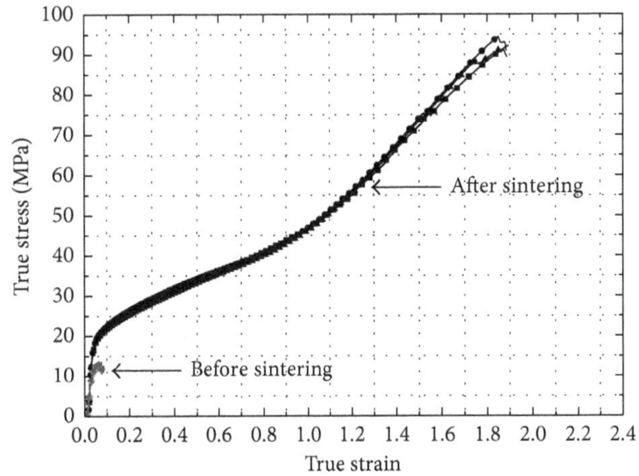

FIGURE 2: True stress-strain curves of specimen #2 before sintering and after sintering.

3. Results and Discussion

3.1. Mechanical Properties under Quasistatic Compression. Figure 2 shows the true stress-strain curves obtained by triplicate experiments of specimen #2 before sintering and after sintering. It can be found that specimen #2 after sintering had much higher strength than that of before sintering. The difference of the strength was mainly attributed to the recrystallization of PTFE, which can be confirmed from the scanning electron microscope (SEM) photographs in Figures 3(a) and 3(b). As can be seen from Figures 3(a) and 3(b), before sintering, many of voids and pores existed between Al particles and PTFE. After sintering, due to the recrystallization of PTFE, Al particles were embedded in the PTFE matrix very tightly without voids, resulting in which the strength of the specimen was enhanced greatly. Besides, it can be observed that the curves take on a high degree of agreement, meaning that the experimental results are in good consistency.

The true stress-strain curves of the six types of Al-PTFE specimens under quasistatic compression are shown in Figure 4. The overall shape of the compressive stress-strain curves of the materials is greatly affected by the Al particle size, and the effect is primarily reflected in the strain hardening stage. The mechanical property parameters of the six types of specimens calculated based on the stress-strain data are presented in Table 1, and it can be found that the yield strength and strain hardening modulus which represent the strength of Al-PTFE materials decreased as the Al particle size rose. When the Al particle size increased from 0.05 μm to 32 μm, the yield strength of Al-PTFE specimens decreased from 22.7 MPa to 13.6 MPa, and the strain hardening modulus (the slope of the tangent line of the strain hardening region of the stress-strain curve) decreased from 33.3 MPa to 25.0 MPa. The toughness indicates the energy absorbed by the unit volume before the failure, which characterizes the ability to resist fracture of the specimens. Table 1 presents that the toughness (the area under the stress-strain curve) firstly increased and decreased subsequently, reaching the maximum of 116.42 MJ/m^3 at 6~7 μm.

To anatomize the effect of the Al particle size on microstructure of the Al-PTFE specimen, the sintered specimens were photographed by SEM, and Figures 3(b) and 3(d) compare the internal microstructures of specimens #2, #1, and #4. As can be seen, the distribution of Al particles of 50 nm in the PTFE matrix was best uniform in all of specimens; this uneven distribution resulted in specimen #4 tending to form the stress concentration; thus, specimen #4 was more prone to local deformation and even failure compared with specimens #1 and #2. In the perspective of fracture mechanics, separation of metal particles from the matrix and the fracture of matrix were two major mesoscale mechanisms for failure [19]. Metal particles with a smaller size were more difficult to peel off from the matrix because particles were bonded to the matrix more tightly and the gaps between each other were smaller (Figure 3(c)); consequently, specimen#1 was more difficult to be damaged when compressed and presented a higher strength. Herbold et al. [17] found that metal particles could form force chains that can transmit and bear the stress when the material was compressed. The characteristics of the force chains had a significant influence on the macroscopic response behavior of the specimen. During compression, the force chains went through three processes of generation, destruction, and regeneration. The smaller the particle size, the stronger the regeneration ability of the force chains was, and therefore, the strength of specimen #1 was highest.

3.2. Reaction Characteristics under Quasistatic Compression. Quasistatic compression tests were carried out for six types of Al-PTFE specimens. It can be observed that violent reaction happened to specimens #1, #2, and #3, while no reaction occurred in specimens #4, #5 and #6. The reaction phase of specimens #1, #2, and #3 recorded by high-speed camera are shown in Figure 5. As can be seen, the reaction of specimen #1 was the most violent, and the ignition phenomenon was the most evident. By comparison, the reaction intensity of specimen #3 was far less than that of specimens #1 and #2, and the ignition was comparatively weaker.

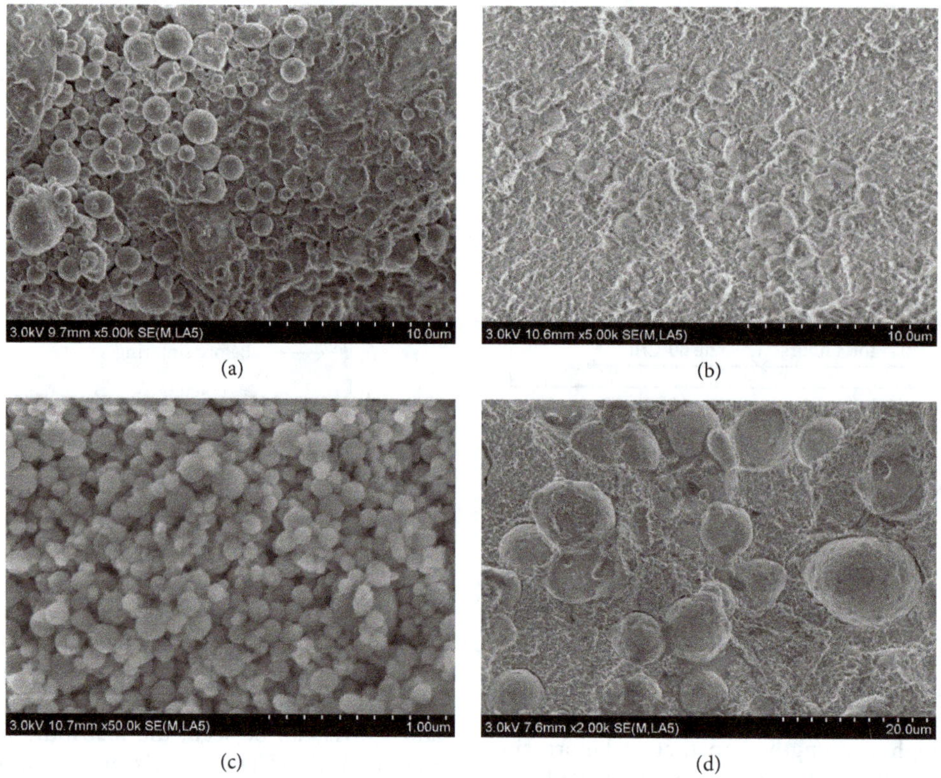

FIGURE 3: The internal microstructures of specimens. (a) Specimen #2 before sintering. (b) Specimen #2 after sintering. (c) Specimen #1 after sintering. (d) Specimen #4 after sintering.

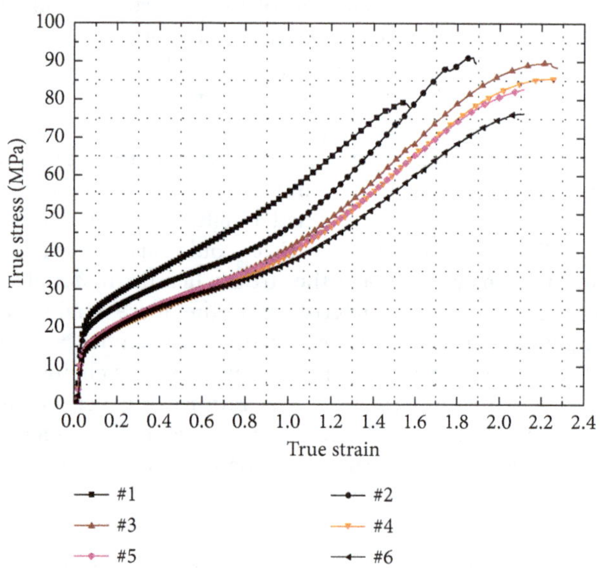

FIGURE 4: True stress-strain curves of specimens #1–#6.

Figure 6 presents the state of six types of Al-PTFE specimens after quasistatic compression. It can be found that specimen #1 reacted completely while specimens #2 and #3 reacted partly, and the reaction was initiated by the circumferential open crack which propagated outward from the radial direction. The black solid residuals can be judged as C (carbon black) based on the following reaction equation:

$$4\mathrm{Al} + 3(-\mathrm{C_2F_4}-) \rightarrow 4\mathrm{AlF_3} + 6\mathrm{C} \tag{1}$$

On the contrary, specimens #4, #5, and #6 failed arising from the formation of internal developed crack, and no circumferential open crack was observed. In accordance with the frame rate and the number of photos captured by the high-speed camera, the duration of the reaction can be approximately calculated through defining the first observation of the ignition phenomenon as the $t = 0$ moment. Table 2 illustrates the reaction duration of specimens #1, #2, and #3 under quasistatic compression. As can be seen from Table 2, the reaction duration of specimen #1 was the shortest (100~150 ms), and the reaction duration of specimen #3 was the longest (1400~1800 ms). The conclusion can be reached that the reaction speed of Al-PTFE specimens takes on a downward trend with the increase of the Al particle size.

The foregoing phenomena can be explicated as follows. For one thing, the specific surface area of particles decreases with the increase of the particle size, that is, surface effect [20]. The Al particles in specimen #1 have the largest specific surface area compared with the others (Table 2), and the contact area between particles is the largest among those under the identical mass condition, thereby enlarging the friction area between the particles and releasing more heat in the process of quasistatic compression, which impel specimen #1 to react more easily and rapidly. For another, during studying the effect of the Al particle size on the thermal degradation of Al-PTFE mixtures, Osborne and Pantoya [21]

TABLE 1: Mechanical properties of specimens #1–#6 under quasistatic compression.

Number	Al particle size (μm)	Elastic modulus (MPa)	Yield strength (MPa)	Hardening modulus (MPa)	Failure stress (MPa)	Toughness (MJ/m^3)
1	0.05	410.6	22.7	33.3	79.3	74.37
2	1~2	400.7	19.8	29.5	92.8	106.38
3	6~7	352.6	14.8	26.4	89.7	116.42
4	12~14	344.9	14.1	25.3	85.6	108.98
5	22~24	340.2	14.3	25.2	82.9	97.18
6	32~34	322.4	13.6	25.0	76.4	90.47

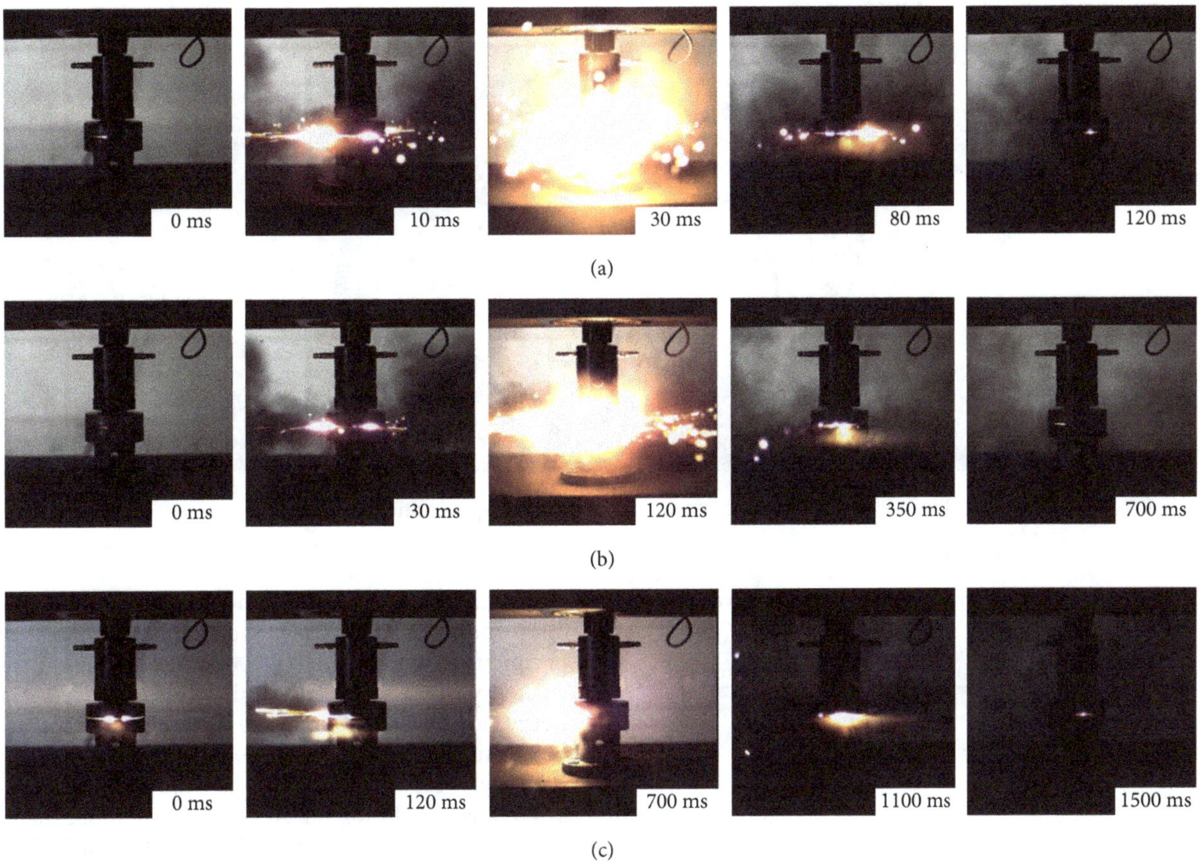

FIGURE 5: Reaction phase of specimens #1, #2, and #3 under quasistatic compression. (a) Specimen #1. (b) Specimen #2. (c) Specimen #3.

found that nano-Al-PTFE mixtures had obvious preignition reaction (PIR) phenomenon when heated over 400°C, which contributes to accelerating the reaction of Al-PTFE.

Through comparing the change of mechanical properties of different specimens and analyzing the reaction phenomena in the quasistatic compression test, the following conclusions can be drawn. On the one hand, for the specimen #1 which can react under quasistatic compression, it was easier to form circumferential open crack by virtue of its highest strength though it took on the worst toughness. The local hot spots were generated at the tip of the crack to trigger the specimen reaction. There was a similar case for specimens #2 and #3 in which activation reaction showed up as circumferential open crack was formed. On the other hand, for the specimens #4, #5, and #6 not capable of reacting under quasistatic compression, although they took on toughness outstripping that of specimen #1, the internal developed crack appeared in the specimen before the absorbed energy was inadequate to trigger the materials by virtue of their low strength. And the crack could not generate the local hot spots of the activation reaction, so the specimen failed to react.

3.3. Reaction Characteristics under Drop Weight Test. The impact sensitivities of the six types of Al-PTFE specimens were calculated by the characteristic drop height (H_{50}), at which the specimen can react with a 50% possibility. The experimental procedure by which the 50% point is obtained is an application of the well-known "up-and-down technique" [22]. In this study, fifteen specimens for each of the types of materials were tested, and the experimental data are recorded in Figure 7 based on "up-and-down technique."

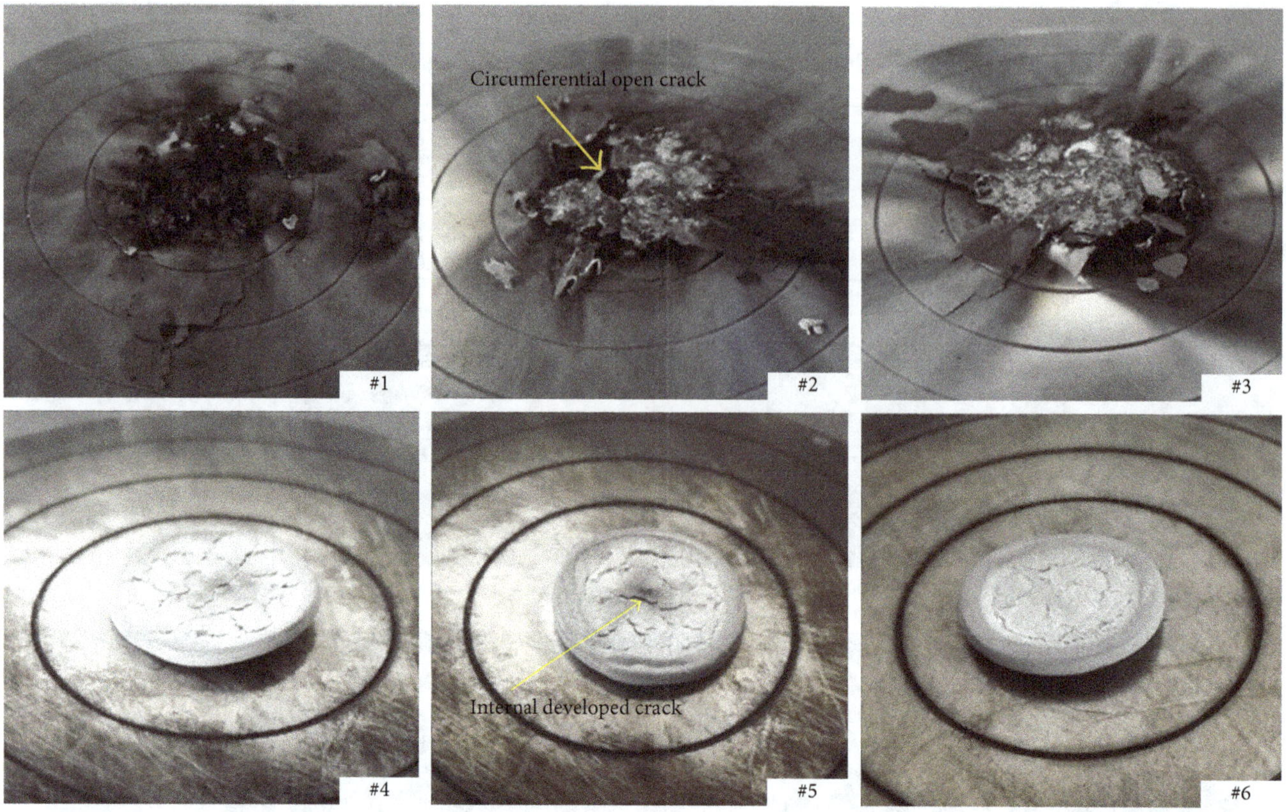

FIGURE 6: The states of specimens #1–#6 after quasistatic compression.

TABLE 2: The reaction duration and specific surface area of specimens #1–#6.

Number	Reaction duration (ms)	Specific surface area (m^2/g)
1	100~150	44.44
2	600~800	2.22
3	1400~1800	0.37
4	—	0.19
5	—	0.10
6	—	0.07

Because the absorbed energy of the specimen is similar to the potential energy of the drop mass, the characteristic drop height (H_{50}) can be used to estimate the ignition energy. The computational formulas of H_{50} and ignition energy are as follows:

$$H_{50} = \left[A + B\left[\frac{\sum iN_i}{N} - \frac{1}{2}\right]\right], \qquad (2)$$

$$E_i \approx E_p = \mathrm{mg}H_{50},$$

where A is the lowest height in the test, B is the increment of height, i is the order of the change of height beginning from 0, N_i is the number of reaction events under certain height which corresponds to i, N is the number of reaction events in all tests, E_i is the ignition energy of the specimen, E_p is the potential energy of the drop mass, and m is the mass of the drop hammer. The calculated results are listed in Table 3. As can be seen from Table 3, all six types of Al-PTFE specimens could react under drop hammer impact, and H_{50} and ignition energy of specimen #1 were highest, whereas those of specimen #6 were lowest. It can be concluded that the impact sensitivity of Al-PTFE material decreases and the ignition energy raises as the Al particle size rises. Furthermore, the change turned out to be comparatively evident for the Al particle size less than 10 μm, while the change tended to be progressively mild for the Al particle size more than 10 μm. In the meantime, the conclusion can be reached through comparing the unreacted phenomena of specimens #4, #5, and #6 under quasistatic compression that the material takes on different mechanical properties and reaction characteristics under different strain rates, which demonstrates that the strain rate effect exists.

The high-speed video sequences of six types of Al-PTFE specimens are presented in Figure 8 under drop hammer impact, and the drop hammer fell from H_{50} for each type of the specimen. The states of six types of specimens after impact are shown in Figure 9. As Figures 8 and 9 bespeak, a violent exothermic reaction occurred for specimens #1, #2, and #3 under drop hammer impact, and the impact state indicates that the specimens reacted completely. Following the chemical equation between Al and PTFE, the black solid residuals remaining on the base of drop hammer tester can be proved to be C. In contrast, the reaction phenomena of specimens #4, #5, and #6 after impact were not obvious as a result of partial reaction, merely a faint flame occurred in specimen #6, and then the reaction was terminated suddenly (Figure 8(f)). With the increase of the Al particle size, the

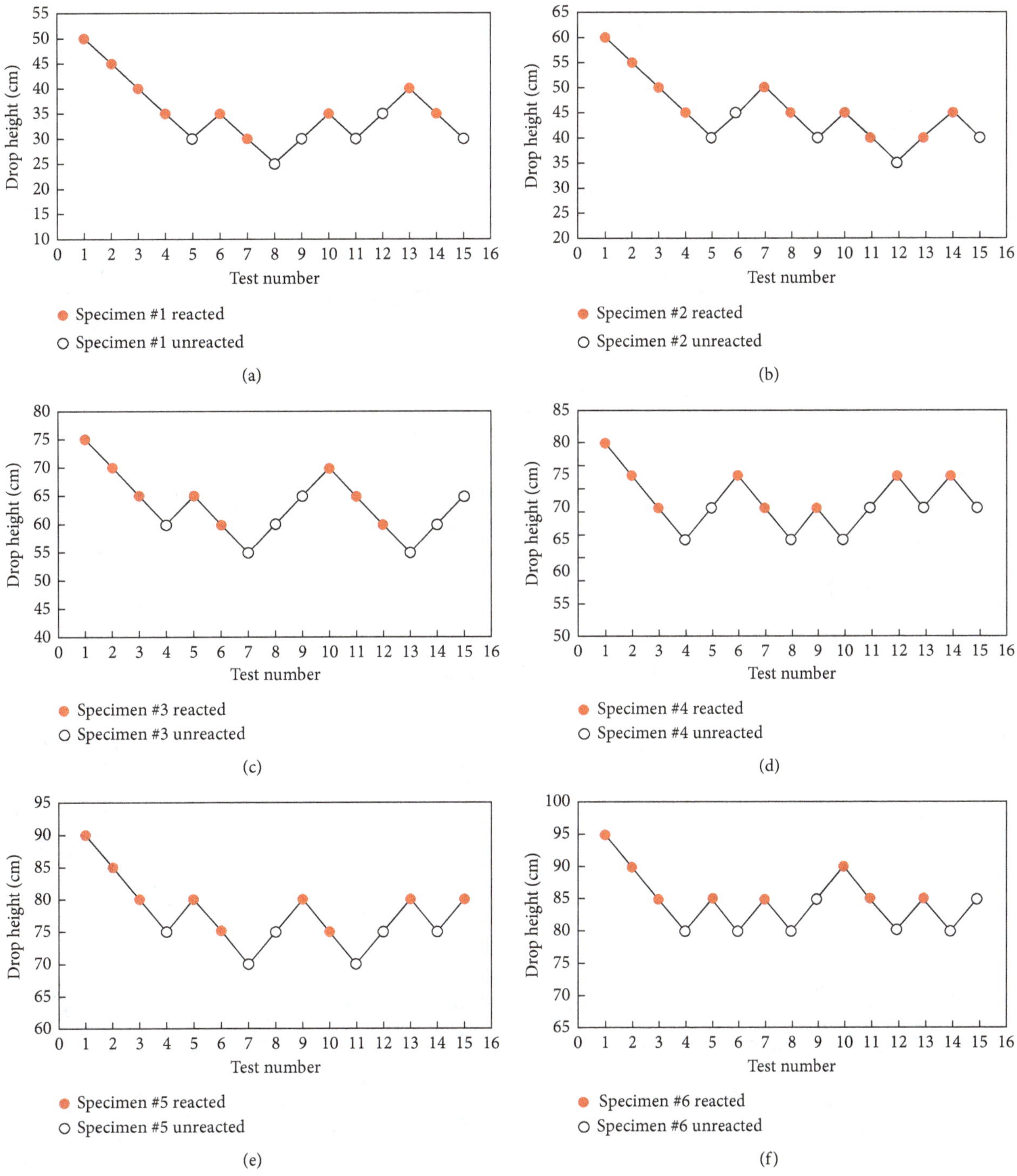

FIGURE 7: The drop weight test data points for specimens #1–#6.

TABLE 3: The characteristic drop height and ignition energy for specimens #1–#6.

Number	H_{50} (cm)	E_i (J)
1	31.07	30.45
2	44.50	43.61
3	63.75	62.48
4	71.25	69.83
5	78.06	76.50
6	85.00	83.30

component of specimens participating in the reaction decreased accordingly, which resulted in the reaction intensity showing a downward trend. It is found through observing the morphology of the impacted specimens that the initial reaction of the specimens occurred in the circumferential open crack, which was consistent with the reaction phenomena manifested in the quasistatic compression test.

As the analysis of experimental results indicates, the impact sensitivity of the specimen was sensitive to the Al particle size, bespeaking that the reaction mechanism took

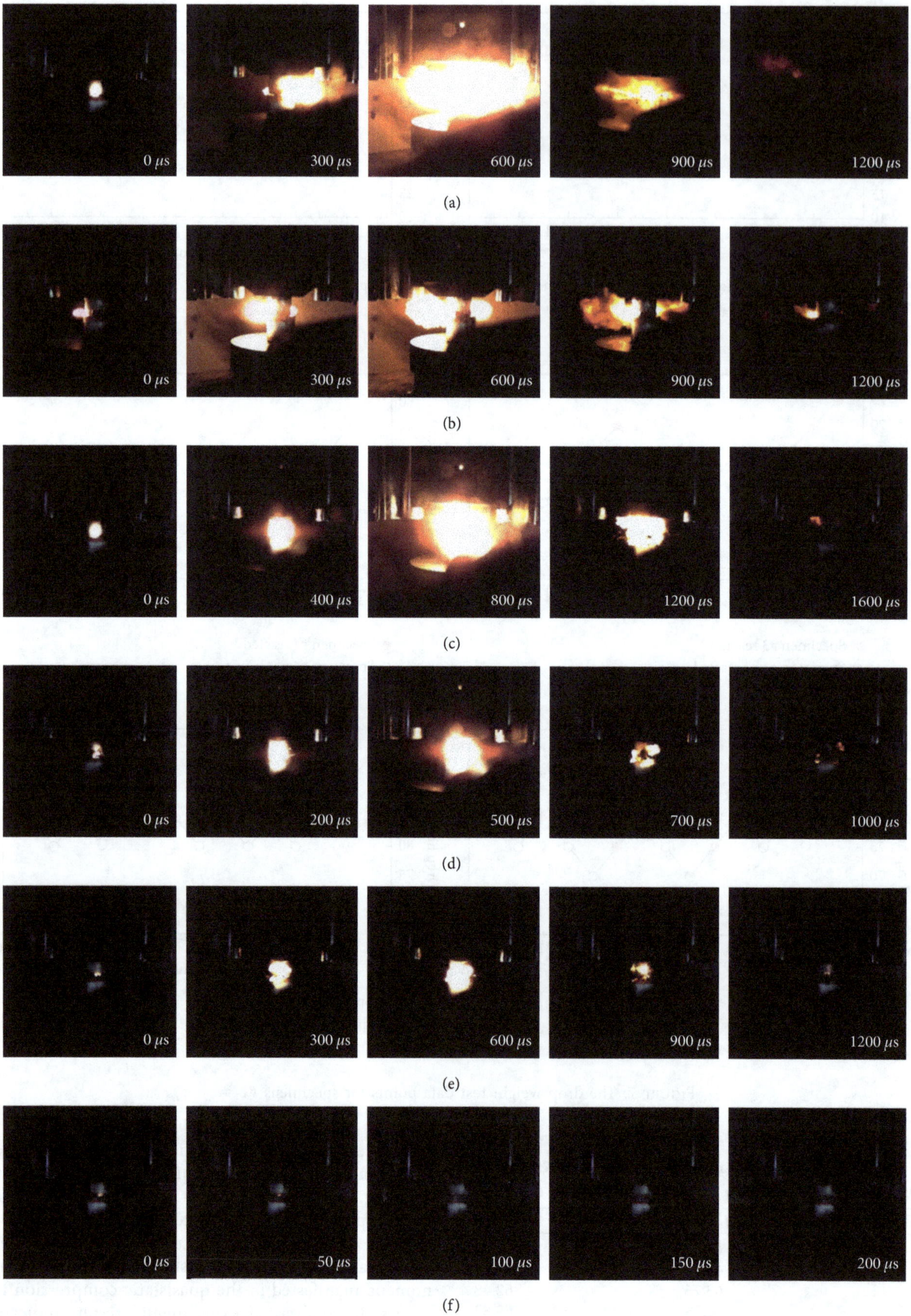

FIGURE 8: The reaction process of specimens #1–#6 under drop hammer impact. (a) Specimen #1. (b) Specimen #2. (c) Specimen #3. (d) Specimen #4. (e) Specimen #5. (f) Specimen #6.

FIGURE 9: The states of specimens #1–#6 after drop hammer impact.

on obvious thermal diffusion reaction characteristics; that is, the oxidant and reductant propagate from the reaction interface to the periphery through the diffusion effect [23, 24]. The specific surface area of Al particles decreased with the increase of the particle size, leading to the decline of diffusion distance between particles and the enlargement of friction area, which promoted the reaction activity of the material. Besides, when the Al particle size was in the level of micron, H_{50} of the specimens changed linearly with the Al particle size; however, when the Al particle size was in the level of nano, it deviated from the trend obviously. The phenomenon was also arising from the acceleration of PIR reaction to the Al-PTFE reaction, and the excitation energy required for the reaction was reduced.

4. Conclusions

In this study, six types of Al-PTFE specimens with different Al particle sizes were prepared, and their mechanical properties and reaction characteristics under quasistatic compression and dynamic impact were investigated. The main conclusions are as follows:

(1) The stress-strain data under quasistatic compression show that the strength of the Al-PTFE specimen decreased with the increase of the Al particle size; the yield strength decreased from 22.7 MPa to 13.6 MPa; and the hardening modulus decreased from 33.3 MPa to 25 MPa. And the toughness rose firstly and subsequently decreased, peaked as 116.42 MJ/m^3 at 6~7 μm.

(2) In the quasistatic compression test, no reaction was occurred for specimens with the Al particle size of 12~14 μm, 22~24 μm, and 32~34 μm, while an exothermic reaction was occurred for specimens with the Al particle size of 50 nm, 1~2 μm, and 6~7 μm. Arising from the function of surface effect and PIR, the reaction intensity and reaction speed showed a downward trend as the Al particle size rose.

(3) In the drop weight test, six types of specimens were overall reacted, and the characteristic drop height and ignition energy increased with the increase of the Al particle size, bespeaking that the material was more sensitive to reaction as the Al particle size rose. The change turned out to be comparatively evident for the Al particle size less than 10 μm, while the change tended to be progressively mild for the Al particle size more than 10 μm.

(4) In a lower strain rate range ($10^{-2}{\cdot}s^{-1}$~$10^{2}{\cdot}s^{-1}$), Al-PTFE specimens takes on different mechanical properties and reaction characteristics under different strain rates. The formation of circumferential open cracks is deemed as a prerequisite for Al-PTFE specimens to go through a reaction. With the increase of the Al particle size, the specimen is more difficult to form circumferential open crack and the reaction is more difficult to be triggered.

Conflicts of Interest

The authors declare that they have no conflicts of interest regarding the publication of this paper.

Acknowledgments

The financial support from the National Natural Science Foundation of China (General Program. Grant no. 51673213) is gratefully acknowledged.

References

[1] D. B. Nielson, R. M. Truitt, and B. N. Ashcroft, "Reactive material enhanced projectiles and related methods," US Patent US7603951, 2009.

[2] W. Mock Jr. and W. H. Holt, "Impact initiation of rods of pressed polytetrafluoroethylene (PTFE) and aluminum powders," *American Institute of Physics*, vol. 845, pp. 1097–1100, 2006.

[3] E. M. Hunt, S. Malcolm, M. L. Pantoya, and F. Davis, "Impact ignition of nano and micron composite energetic materials," *International Journal of Impact Engineering*, vol. 36, no. 6, pp. 842–846, 2009.

[4] E. M. Hunt and M. L. Pantoya, "Impact sensitivity of intermetallic nanocomposites: a study on compositional and bulk density," *Intermetallics*, vol. 18, no. 8, pp. 1612–1616, 2010.

[5] S. Xu, S. Yang, and W. Zhang, "The mechanical behaviors of polytetrafluorethylene/Al/W energetic composites," *Journal of Physics Condensed Matter*, vol. 21, no. 28, p. 285401, 2009.

[6] J. Cai, S. M. Walley, R. J. A. Hunt, W. G. Proud, V. F. Nesterenko, and M. A. Meyers, "High-strain, high-strain-rate flow and failure in PTFE/Al/W granular composites," *Materials Science & Engineering A*, vol. 472, no. 1-2, pp. 308–315, 2008.

[7] C. Ge, Y. Dong, and W. Maimaitituersun, "Microscale simulation on mechanical properties of Al/PTFE composite based on real microstructures," *Materials*, vol. 9, no. 12, p. 590, 2016.

[8] X. F. Zhang, J. Zhang, L. Qiao et al., "Experimental study of the compression properties of Al/W/PTFE granular composites under elevated strain rates," *Materials Science & Engineering A*, vol. 581, no. 10, pp. 48–55, 2013.

[9] H. Wang, Y. Zheng, Q. Yu, and W. Yu, "Impact-induced initiation and energy release behavior of reactive materials," *Journal of Applied Physics*, vol. 110, no. 7, pp. 239–H03, 2011.

[10] L. Wang, J. Liu, S. Li, and X. Zhang, "Investigation on reaction energy, mechanical behavior and impact insensitivity of W–PTFE–Al composites with different W percentage," *Materials & Design*, vol. 92, no. 5, pp. 397–404, 2016.

[11] L. Wang, J. Liu, S. Li, and X. Zhang, "Insensitive high-energy energetic structural material of tungsten-polytetrafluoroethylene-aluminum composites," *AIP Advances*, vol. 5, no. 11, pp. 169–174, 2015.

[12] X. Zheng, A. D. Curtis, W. L. Shaw, and D. D. Dlott, "Shock initiation of nano-Al + Teflon: time-resolved emission studies," *Journal of Physical Chemistry C*, vol. 117, no. 9, pp. 4866–4875, 2013.

[13] B. Feng, Y. C. Li, S. Z. Wu, H.-X. Wang, Z.-M. Tao, and X. Fang, "A crack-induced initiation mechanism of Al-PTFE under quasi-static compression and the investigation of influencing factors," *Materials & Design*, vol. 108, pp. 411–417, 2016.

[14] B. Feng, X. Fang, Y. C. Li, S.-Z. Wu, Y.-M. Mao, and H.-X. Wang, "Reactions of Al-PTFE under impact and quasi-static compression," *Advances in Materials Science & Engineering*, vol. 2015, Article ID 582320, 6 pages, 2015.

[15] C. Ge, W. Maimaitituersun, Y. Dong, and C. Tian, "A study on the mechanical properties and impact-induced initiation characteristics of brittle PTFE/Al/W reactive materials," *Materials*, vol. 10, no. 5, p. 452, 2017.

[16] H. X. Wang, Y. C. Li, B. Feng, J.-Y. Huang, S. Zhang, and X. Fang, "Compressive properties of PTFE/Al/Ni composite under uniaxial loading," *Journal of Materials Engineering & Performance*, vol. 26, no. 5, pp. 2331–2336, 2017.

[17] E. B. Herbold, V. F. Nesterenko, D. J. Benson et al., "Particle size effect on strength, failure, and shock behavior in polytetrafluoroethylene-Al-W granular composite materials," *Journal of Applied Physics*, vol. 104, no. 10, p. 103903, 2008.

[18] D. B. Nielson, R. L. Tanner, and G. K. Lund, "High strength reactive materials," US Patent US6593410, 2003.

[19] J. Cai, V. F. Nesterenko, K. S. Vecchio et al., "The influence of metallic particle size on the mechanical properties of polytetraflouroethylene-Al–W powder composites," *Applied Physics Letters*, vol. 92, no. 3, p. 1007, 2008.

[20] H. Altenbach, V. A. Eremeyev, and N. F. Morozov, "Mechanical properties of materials considering surface effects," *Iutam Bookseries*, vol. 31, pp. 105–115, 2013.

[21] D. T. Osborne and M. L. Pantoya, "Effect of Al particle size on the thermal degradation of Al/Teflon mixtures," *Combustion Science & Technology*, vol. 179, no. 8, pp. 1467–1480, 2007.

[22] W. R. Blackstone, B. B. Baber, and P. M. Ku, "New test techniques for evaluating the compatibility of materials with liquid oxygen under impact," *Tribology Transactions*, vol. 11, no. 3, pp. 216–227, 1968.

[23] S. Chowdhury, K. Sullivan, N. Piekiel, L. Zhou, and M. R. Zachariah, "Diffusive versus explosive reaction at the nanoscale," *Journal of Physical Chemistry C*, vol. 114, no. 20, pp. 9191–9195, 2010.

[24] A. Rai, K. Park, L. Zhou, and M. R. Zachariah, "Understanding the mechanism of aluminium nanoparticle oxidation," *Combustion Theory & Modelling*, vol. 10, no. 5, pp. 843–859, 2006.

5

The Mechanical and Reaction Behavior of PTFE/Al/Fe_2O_3 under Impact and Quasi-Static Compression

Jun-yi Huang, Xiang Fang, Yu-chun Li, Bin Feng, Huai-xi Wang, and Kai Du

College of Field Engineering, PLA University of Science and Technology, Hou Biaoying Road No. 88, Nanjing, Jiangsu 210007, China

Correspondence should be addressed to Xiang Fang; fangxiang3579@163.com

Academic Editor: Jun Liu

Quasi-static compression and drop-weight test were used to characterize the mechanical and reaction behavior of PTFE/Al/Fe_2O_3 composites. Two kinds of PTFE/Al/Fe_2O_3 composites were prepared with different mass of PTFE, and the reaction phenomenon and stress-strain curves were recorded; the residuals after reaction were analyzed by X-ray diffraction (XRD). The results showed that, under quasi-static compression condition, the strength of the materials is increased (from 37.1 Mpa to 77.2 Mpa) with the increase of PTFE, and the reaction phenomenon occurred only in materials with high PTFE content. XRD analysis showed that the reaction between Al and Fe_2O_3 was not triggered with identical experimental conditions. In drop-weight tests, PTFE/Al/Fe_2O_3 specimens with low PTFE content were found to be more insensitive by high-speed photography, and a High Temperature Metal Slag Spray (HTMSS) phenomenon was observed in both kinds of PTFE/Al/Fe_2O_3 composites, indicating the existence of thermite reaction, which was confirmed by XRD. In PTFE/Al/Fe_2O_3 system, the reaction between PTFE and Al precedes the reaction between Al and Fe_2O_3.

1. Introduction

Reactive materials (RMs) or impact-initiated materials and their applications have attached great interest in recent years due to their uniquely mechanical and chemical properties. The mixture of polytetrafluoroethylene (PTFE) and aluminum (Al) is one of the typical RMs and has been extensively studied, because of its high energy density and a favorable combination of properties of PTFE such as low friction coefficient, high thermal stability, high electrical resistance, high chemical inertness, high melting point (327°C), high thermal energy release when decomposed, and the easiness to deform [1]. A lot of experiments have been conducted to probe the mechanical and reaction behavior of PTFE/Al composites by different means. In order to improve the density and strength of the materials, many researchers have tried to add tungsten (W), the most common "heavy" metal, into the PTFE/Al composites. Dynamic compression experiments were performed by Cai et al. on a pressed PTFE/Al/W mixture to understand the mechanical response and failure at high-strain and high-strain rate [1]. The results showed that failure was proceeded by extensive plastic deformation concentrated mainly in the PTFE matrix, and the initiation as well as the propagation of cracks was caused by the separation of the W particle-PTFE interface. Herbold et al. investigated the particle size effect on the strength, failure, shock behavior, and dynamic properties of PTFE/Al/W composites combining simulation and experiments [2, 3]. They found that the increased strength of the sample with fine metallic particles is due to the formation of force chains under dynamic loading. Numerical modeling of shock loading of this granular composite material demonstrated that the internal energy, specifically thermal energy, of the soft PTFE matrix can be tailored by the W particle size distribution. Xu et al. tested four kinds of PTFE/Al/W with different contents and sizes of W particles [4]. The tensile and impact properties of PTFE/Al/W reactive energetic composites were probed using a universal materials testing machine and an improved pendulum impact tester at room temperature. The results revealed that the failure behavior of PTFE/Al/W included deformation, fracture, disorganization, and reaction. They also drew the conclusion that the addition of 30 wt.% of coarse W particles improved the impact strength of the material but led to the increase of reactive activity and

(a) (b)

FIGURE 1: Samples of PTFE/Al/Fe_2O_3 composites: (a) for quasi-static compression and (b) for drop-weight tests.

the decrease of perfectibility of the reaction. Other scholars also carried out in-depth study of PTFE/Al/W from other perspectives using different techniques including quasi-static and dynamic compression (both drop-weight test and Split-Hopkinson Pressure Bar), scanning electron microscopy (SEM), X-ray diffraction (XRD), and differential scanning calorimetry (DSC) [5–9].

As can be seen above, much attention has been put to the addition of W particles to increase the strength and density of PTFE/Al, while literatures about the addition of metal oxides, such as CuO and Fe_2O_3, are scarce. In the above experiments about PTFE/Al/W, no reaction between Al and W was observed; only a small amount of WC and W_2C formed, which were activated by the high heat as a result of fluorination of Al [7]. But the heat generated by the reaction between Al and metal oxides (thermite) cannot be neglected. Zhou et al. studied the reactions of Al/CuO, Al/Fe_2O_3, and Al/ZnO systems using T-Jump/time-of-flight mass spectrometer [10]. They found the temperatures of the reactions can reach 2837 K, 3135 K, and 1822 K, respectively. And the enthalpy of combustion is −604.05 kJ/mol, −425.1 KJ/mol, and −312.1 KJ/mol (per mol of Al) correspondingly. Also, Wang et al. revealed that the heat generated from the reaction between Al and Fe_2O_3 can raise the material to a temperature of ca. 3000°C, higher than the melting point of Fe (1535°C) and Al_2O_3 (2059°C) [11]. However, because of the low energy release rate and the difficulty to shape, the application of the thermites is limited. But experiments confirmed that the addition of 10 wt.% PTFE into Fe_2O_3/Al mixture can reduce the time to reach the exothermic peak by two orders of magnitude [12].

The PTFE/Al/Fe_2O_3 composites are promising reactive materials, which have the dual properties of PTFE/Al and Al/Fe_2O_3 and can greatly expand the application fields of the PTFE/Al composites if the Al/Fe_2O_3 reaction can be triggered. But to the best of our knowledge, few researchers investigated the mechanical and reaction behavior of PTFE/Al/Fe_2O_3 systematically. In this paper, the mechanical and reaction behavior of PTFE/Al/Fe_2O_3 under impact and quasi-static compression were investigated, and the failure and ignition mechanism were analyzed. The insensitivity of PTFE/Al/Fe_2O_3 to impact was measured quantitatively, which is of great importance for applications of reactive materials.

2. Experimental Setup

2.1. Specimens Preparation. In this study, two kinds of composites, 50PTFE-28Al-22Fe_2O_3 (wt.%) (type A specimen) and 75PTFE-14Al-11Fe_2O_3 (wt.%) (type B specimen), were fabricated by powder mixture, cold isostatic pressing (CIPing), and vacuum sintering based on the patent of Nielson's [13]. The initial powers have the following average sizes: PTFE: 25 um (3 M, Shanghai); Al: 1-2 um (JT, Hunan); Fe_2O_3: 3–5 um (NAO, Shanghai). Firstly, the PTFE/Al/Fe_2O_3 powders were suspended in an ethanol solution and mixed by a motor-driven blender for 20 minutes and then dried in a vacuum oven for 48 hours at 60°C, followed by a screening process to break up agglomerates. Next, the dried mixture was cold pressed into cylinders (for quasi-static compression and for drop-weight test, Figure 1) under the pressure of 60 Mpa. Finally, specimens were sintered in a vacuum oven at the temperature of 370°C for 4 h with a cooling rate of 50°C/h.

2.2. Methods. Quasi-static compression tests were performed using a SFLS-30T electrohydraulic press with a maximum loading of 30 kN. The sample was placed between two hardened steel anvils and was compressed at the speed of 9 mm/min, corresponding to a normal strain rate of 0.01/s. The load was terminated when the sample was fractured or the maximum load was reached. All tests were conducted at room temperature and the stress-strain curves of all the samples were recorded by the hydraulic press automatically. To ensure repeatability, at least five specimens of each kind were tested.

Drop-weight tests were carried out with a fall hammer impact sensitivity tester with a drop mass of 10 kg which can be dropped from a maximum height of 1.5 m. So the maximum speed of 5.5 m/s of the drop hammer and the strain rate of almost 1000/s can be achieved. In order to observe the reaction process, high-speed photography with the frame rate up to 20000 frames/s was adopted.

Hitachi S-4800 scanning electron microscope (SEM) was used to observe the microstructure of the materials and to analyze the reaction and fracture mechanisms.

The residues of the specimen after quasi-static compression and drop-weight test were collected and characterized by X-ray diffraction (Bruker D8 ADVANCE, operating at 40 kV and 40 mA with unfiltered Cu Kα radiation).

(a) (b)

FIGURE 2: Samples before and after quasi-static compression. (a) Type A specimen and (b) type B specimen.

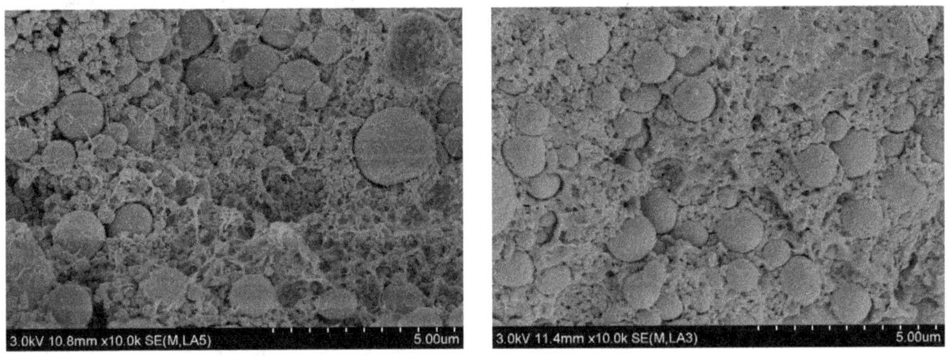

(a) (b)

FIGURE 3: The microstructures of PTFE/Al/Fe_2O_3 composites. (a) Type A specimen and (b) type B specimen.

3. Results and Discussion

3.1. Mechanical Behavior under Quasi-Static Compression. Typical photographs of uncompressed and compressed PTFE/Al/Fe_2O_3 reactive materials of the two kinds are shown in Figure 2. It can be found that there are quite differences between the two materials. For type A specimens, cracks appeared soon after the compression load was added and the specimens were fractured at a relatively low strain level. But type B specimens stayed intact even when they are pressed to about 0.2 of their original length, indicating a greater plastic deformation. The difference of the failure behavior of the two materials is mainly attributed to the different content of PTFE. More PTFE ensures better matrix continuity and thus leads to a great plasticity, which can be confirmed from the SEM photographs in Figure 3. As can be seen from Figure 3, the combination between Al, Fe_2O_3, and PTFE matrix in type B specimens is better than that in type A specimens, indicating more perfect mechanical properties for type B specimens. The stress-strain curves of the two materials are given in Figure 4, and their corresponding parameters are listed in Table 1. When compressed, both types of specimen underwent the process of elastic and plastic deformation, crack propagation, and fracture. As can be seen, the two specimens have almost identical elastic deformation and possess similar yield stresses. After the pressure exceeded the yield stress, the strain hardening phenomenon occurred for both types of PTFE/Al/Fe_2O_3. However, the strain hardening modulus of type B specimen (34.8 Mpa, averaged from strain 0.1 to 1.72) is 2.88 times higher than that of type A specimen (12.1 Mpa, averaged from strain 0.1 to 1.45), which causes the toughness (the area under the stress-strain curve) of type B specimen to be higher than that of type A specimen. That is, in the case of quasi-static compression, type B specimens can absorb more energy than type A specimens before fracture. So type B specimens are more likely to react in the same condition and the experiments had confirmed this hypothesis, which will be discussed later. As for the failure process, type B specimens failed more suddenly than type A specimens. With the increase of compression load, the axial cracks and shear cracks in type A specimens developed gradually; when the specimens cannot resist the pressure, they fractured. But for type B specimens, no obvious visible macrocracks generated until the maximum strength was reached. However, once the fracture condition was met, type B specimens fractured catastrophically, accompanied with a sharp drop in stress and intensive reaction phenomenon. The ignition point was marked in Figure 4.

3.2. Reaction Behavior under Quasi-Static Compression. Figure 5 shows the initiation phenomenon of a type B specimen under quasi-static compression. When the true stress reached about 75 Mpa, the initiation started from the vicinity of the outer surface of the cylinder where the shear force is the strongest, accompanied by explosion sound and bright lights. Then the reaction propagated from the edge to the interior of the specimen and quenched quickly; the whole ignition process lasted about 1 s, with about 90% of the

TABLE 1: Parameters of PTFE/Al/Fe_2O_3 reactive material.

Type	PTFE : Al : Fe_2O_3 (wt.%)	Density (g/cm^3)	Yield stress (Mpa)	Average strain hardening modulus (Mpa)	Ultimate strength (Mpa)	Ignition or not
A	50 : 28 : 22	2.23	20.1	12.1	37.1	No
B	75 : 14 : 11	2.33	19.9	38.4	77.2	Yes

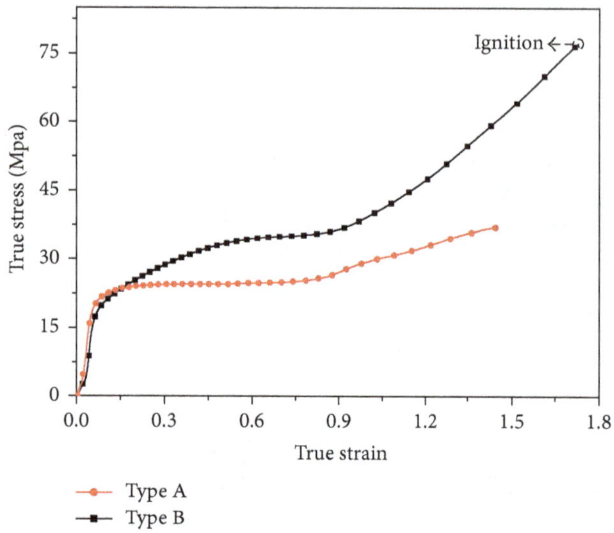

FIGURE 4: Stress-strain curves of two kinds of PTFE/Al/Fe_2O_3 composites.

specimen unreacted. Feng et al. also observed the initiation phenomenon with PTFE/Al composites under similar experimental condition [14], in which PTFE and Al had reacted with each other to form AlF_3 and C (carbon black). However, no initiation phenomenon was observed for any type A specimen, which could be attributed to the low toughness as mentioned above.

The specimen of type B after reaction and the microstructures of the reacted and unreacted regions are shown in Figure 6. As can be seen from Figure 6(a), the black parts correspond to the reacted regions, which is marked with "reacted region." And there is a clear boundary between the unreacted and reacted region, which can be seen more clearly in Figure 6(b). The microstructures of the unreacted and reacted regions are given in Figures 6(c) and 6(d). In unreacted region (Figure 6(c)), PTFE has a distinct orientation and forms a dense nanofiber network, with Al and Fe_2O_3 particles embedded in the PTFE matrix without significant deformation. In reacted region (Figure 6(d)), the crack section is covered by smooth melted PTFE and carbon black, and PTFE nanofiber network completely disappeared. The specimen after reaction is displayed in Figure 6(a), the black part corresponds to the reacted region, which is marked with "initiation." Moreover, the opening crack and shear plane caused by compression are clearly visible, which are closely related to the ignition phenomenon [15]. Although there has been much controversy about the mechanical initiation of explosions over these years, many scholars believe that the initiation is due to the existence of "hot spots." Swallowe and Field found that the hot spots formed at the polymer crack tip and shear band can activate the explosive [16]. In the PTFE/Al/Fe_2O_3 system, PTFE is the predominant pressure-bearing component; when subjected to compression, PTFE matrix was deformed heavily, causing the crack generation and spread. The crack propagation speed can reach 200 m/s to 800 m/s and the temperature of the crack tip can reach 300°C to 1000°C [17] and the high temperature can greatly promote the reaction. Feng et al. proposed a crack-induced initiation mechanism of Al-PTFE under quasi-static compression, which can also explain the reaction mechanism of PTFE/Al/Fe_2O_3 system [15]. As can be seen from the XRD analysis in Figure 7, the reaction products showed that Fe_2O_3 was not involved in the reaction; the initiation phenomenon was actually caused by PTFE and Al.

The XRD results of the residuals of both type A and type B specimens after compression are shown in Figure 7. No other species were found besides the original three components (PTFE, Al, and Fe_2O_3) in the compressed specimen of type A specimens, which means that there was no reaction between PTFE, Al, and Fe_2O_3. This is due to the brittle nature of this material and the lower strength and toughness. When compressed, type A specimens fractured quickly before the formation of high temperature of the crack tip, and thus there was no enough energy to induce the reaction. However, AlF_3 was found in the compressed specimen of type B materials, and the black powder remaining on the anvil proved to be carbon black, indicating the following reaction had taken place:

$$Al + (\text{-}C_2F_4\text{-}) \longrightarrow AlF_3 + C \quad (1)$$

AlF3 was formed from the Al oxidized by fluorine from the PTFE. But besides iron oxide (Fe_2O_3), no other iron compounds or iron (Fe) was found in the reaction products, illustrating that Fe_2O_3 did not involve in the reaction. That is, the thermite reaction between Al and Fe_2O_3 was not triggered under quasi-static compression. Firstly, the energy or temperature required to initiate the thermite reaction is relatively high. Williams et al. found the ignition temperature of Al and Fe_2O_3 was 800 K–950 K by experiment [18], but Zhou et al. concluded that the ignition temperature of Al and Fe_2O_3 was 1100 K using T-Jump/time-of-flight mass spectrometer [10], while Fan et al. discovered that the exothermic peak of DSC curves of Al and Fe_2O_3 powder mixture appeared at 853.5°C–949.7°C, and the exothermic peak represented the thermite reaction [19]. The differences of ignition temperature in the three studies are attributed to the diversity of materials and the difference of heating rate, but all three studies showed that thermite reaction

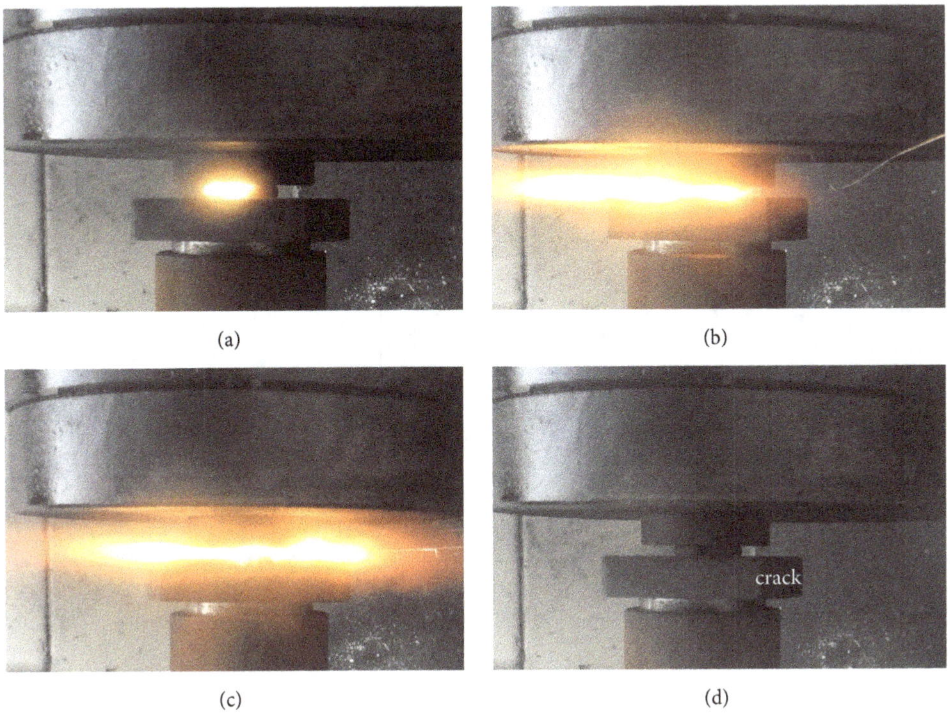

FIGURE 5: The initiation process of a type B specimen. (a) First sign of initiation, (b) reaction propagated from the edge to the interior of the specimen, (c) violent reaction, and (d) after initiation and the formed crack.

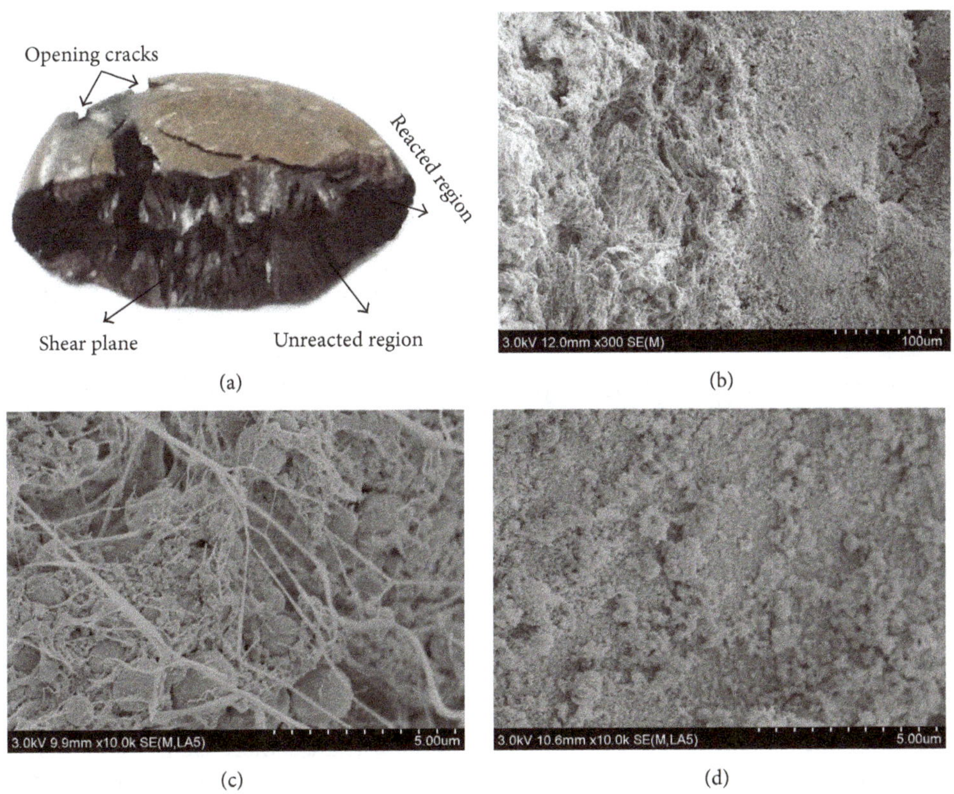

FIGURE 6: SEM photographs of the initiation region and specimen after reaction. (a) Specimen after reaction, (b) boundary of reacted region (right) and unreacted region (left), (c) unreacted region, and (d) reacted region.

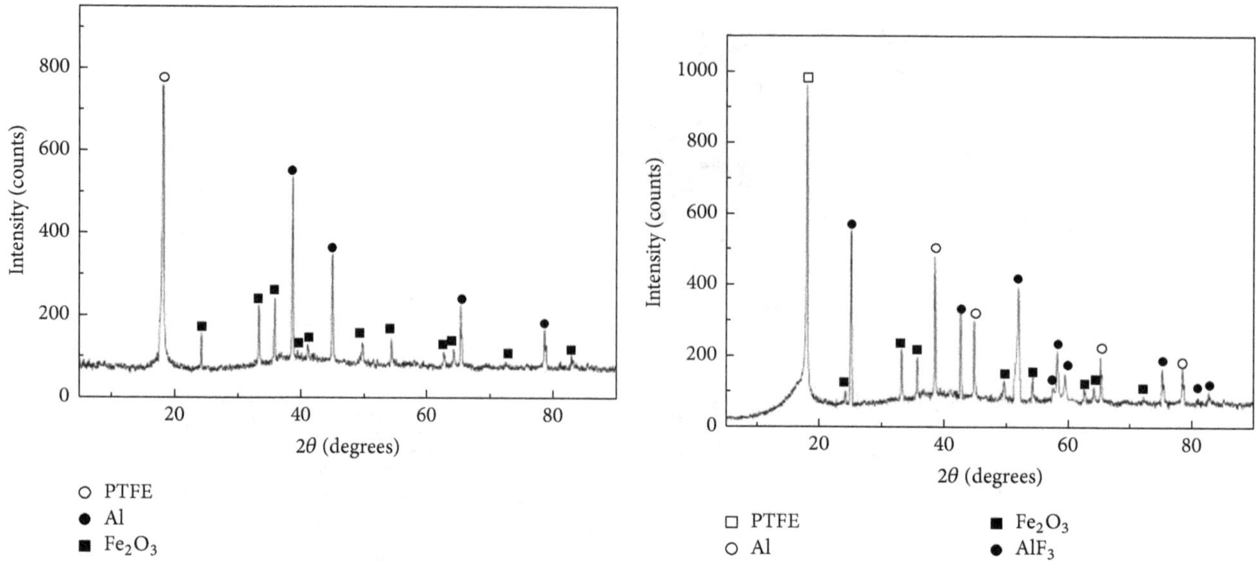

(a) (b)

FIGURE 7: The XRD results of specimens after quasi-static compression. (a) Type A specimen and (b) type B specimen.

needs high temperature (hundreds of degrees centigrade) to trigger. However, the temperature rise caused by quasi-static compression is low relatively (tens of degrees centigrade) according to the simulation result of Feng et al. [14]; therefore it is not enough to trigger the reaction of Al and Fe_2O_3. Furthermore, despite the intense reaction of PTFE with Al, the duration may be too short to induce thermite reaction between Al and Fe_2O_3. Besides, the content of Al and Fe_2O_3 was small and they were dispersed in the PTFE matrix; the less contact of the two substances resulted in the difficulty to react. Lastly, compared with other active oxides, such as CuO and Cu_2O, the activity of Fe_2O_3 is relatively lower because of its much less oxygen release, since the analysis by Wang et al. suggested that the initial steps of the thermite reaction are the decomposition of the oxidizer particles to form oxygen as well as Al diffusion through the Al_2O_3 shell [11].

3.3. Reaction Behavior under Drop-Weight Test. The impact sensitivity of both type A and type B materials was investigated using an impact sensitivity tester and a high-speed photography, and the sensitivity of the materials is measured by the lowest drop height from where the drop mass is released to produce greater than 50% reaction, which is on the basis of an abbreviated Bruceton method [20], and the characteristic drop height is represented by H_{50} in Table 2. Characteristic drop height (H_{50}) can be used to calculate ignition energy based on potential energy as in

$$E_p = mgh, \quad (2)$$

where E_p is the potential energy of the drop mass, m is the mass of the drop mass, and h is the optimal height for reaction. To compare the absorbed energy of the materials under quasi-static compression and drop-weight test, the estimated ignition energy under quasi-static compression is

TABLE 2: Results of ignition energy under static and dynamic load.

Type	Drop height (H_{50})/cm	Ignition energy (static)/J	Ignition energy (impact)/J
A	80	—	78
B	51	71	50

also calculated by the work done by the machine along the compress direction, which can be expressed as follows:

$$W = \sum F_t \Delta s, \quad (3)$$

where W is the work done by the machine, F_t is the transient force, and Δs is infinitesimal increases of displacement; both F_t and Δs can be retrieved from the testing machine directly. The information of the two materials under the two test conditions is listed in Table 2.

As can be seen from Table 2, the ignition energy of type B materials under quasi-static compression (71 J) is higher than that under drop-weight impact (50 J). But if taking consideration of the energy dissipated into the surrounding environment during the quasi-static compression process, they may be similar, because more heat will be lost during quasi-static compression, while dynamic impact can be considered as an adiabatic process. Comparing the behavior of the two materials under dynamic impact, we can conclude that type B specimens are easier to ignite than type A specimens. More energy or higher drop height is needed for type A specimens to react, which is consistent with the results of quasi-static compression.

Figures 8 and 9 show a sequence of images taken from high-speed camera of the two kinds of PTFE/Al/Fe_2O_3 composites under impact loading, and the drop mass was dropped from the same height of 1.5 m. Obviously, type B specimens reacted more violently than type A specimens.

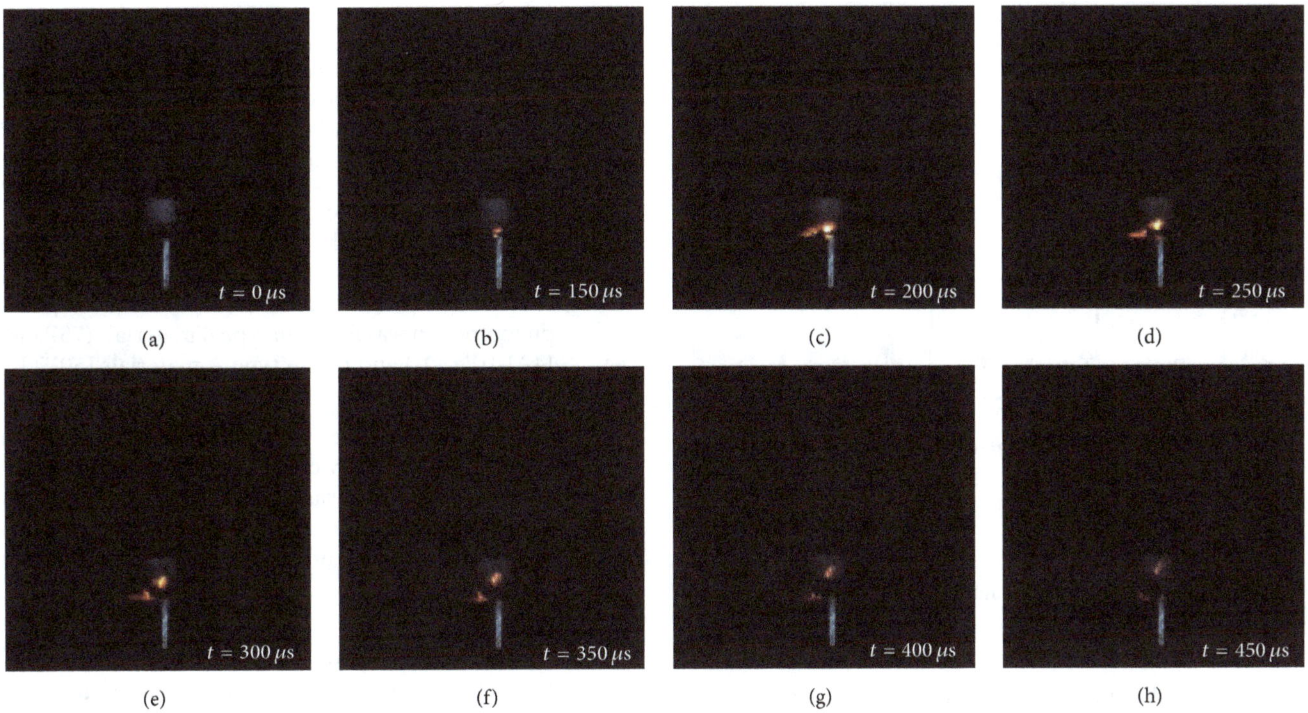

FIGURE 8: Selected images for a type A specimen reaction recorded by a high-speed camera (from the height of 1.5 m).

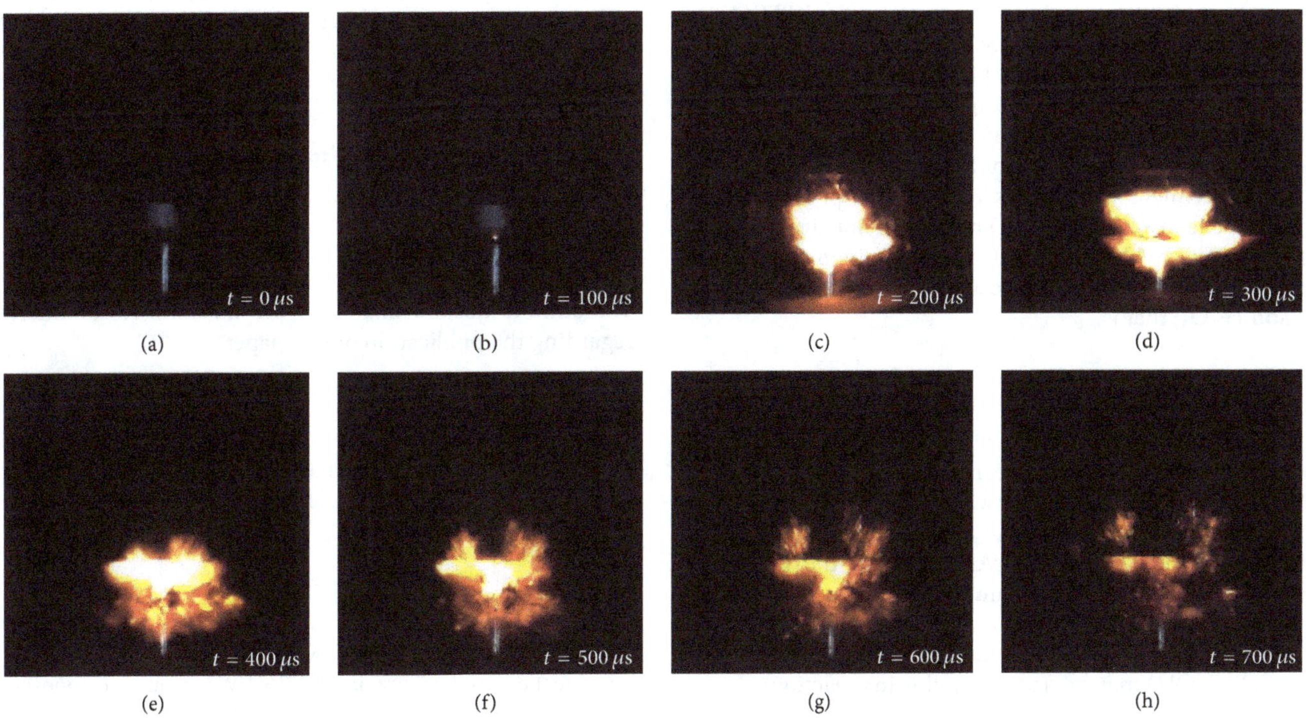

FIGURE 9: Selected images for a type B specimen reaction recorded by a high-speed camera (from the height of 1.5 m).

For type A specimens, when impacted, only a weak fire was produced and the duration of reaction was only 300 μs. But for type B specimens, intensive light was captured by high-speed photography, accompanied with explosion noise. The reaction time is more than 600 μs, which is much longer than that of type A specimens. The difference in reaction time is mainly due to the difference in completeness of the reaction. For type A, only a small part of the specimen was involved in the reaction, while most of type B specimen reacted. Furthermore, the ignition time of type A specimens

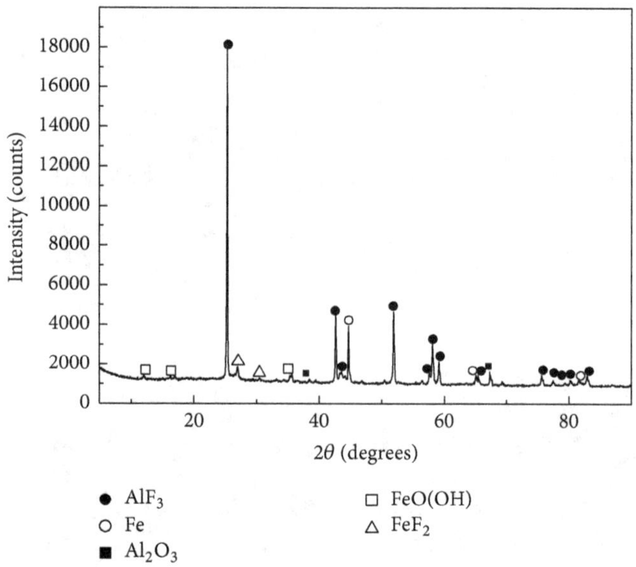

FIGURE 10: XRD results of type B materials after drop-weight test.

is 150 μs, while the ignition time of type B specimens is 100 μs, indicating that type A specimens are more insensitive.

Interestingly, we observed a High Temperature Metal Slag Spray (HTMSS) phenomenon in both kinds of materials upon dynamic compression. However, when PTFE/Al or PTFE/Al/W was impacted, only bright white light appeared with no HTMSS phenomenon [4, 7, 21]. Therefore, Fe_2O_3 is definitely involved in the reaction and local thermite reaction may be triggered, and the XRD of the reaction products of type B materials (Figure 10) confirmed this. The reaction of type A specimens is too weak to allow the products to be analyzed effectively, so the XRD results are not given here. As shown in Figure 10, Fe and Al_2O_3 were found in the reaction products, demonstrating the occurrence of reaction between Al and Fe_2O_3; that is,

$$Al + Fe_2O_3 = Fe + Al_2O_3 \quad (4)$$

According to the results of quasi-static compression, we believe that the reaction between PTFE and Al occurs preferentially, and the reaction between Al and Fe_2O_3 followed, which is triggered by the huge energy generated from the reaction between PTFE and Al and from the impact of drop mass. However, a small amount of FeF_2 and FeO(OH) was also found in the products of type B specimens. FeF_2 is probably from the complex reaction between Fe_2O_3 and PTFE. FeO(OH) may be from the following reaction:

$$4Fe^{2+} + O_2 + 6H_2O \longrightarrow 4FeO(OH) + 8H^+ \quad (5)$$

Fe^{2+} is probably derived from FeF_2, when contacting with O_2 and H_2O in the air; Fe^{2+} will be converted to FeO(OH) under some conditions. However, FeF_2 and FeO(OH) are all traceable, which have little effect on the energy release of the material, so they both can be regarded as impurity and can be ignored temporarily; the specific details will be studied later.

4. Conclusions

In this study, two kinds of PTFE/Al/Fe_2O_3 composites were prepared and their mechanical and reaction behavior under quasi-static compression and dynamic impact were investigated. The main conclusions are shown below.

(1) Under quasi-static compression, with the increase of PTFE content, the strength of the materials is increased (from 37.1 Mpa to 77.2 Mpa), and the fracture mode changes from brittle to ductile. Ignition phenomenon was found in type B materials (75PTFE-14Al-11Fe_2O_3) and not in type A materials (50PTFE-28Al-22Fe_2O_3).

(2) The residuals of the reacted specimen after quasi-static compression were analyzed and no reaction of Al and Fe_2O_3 was found. This is due to the high temperature threshold of thermite reaction, the insufficient contact of Al and Fe_2O_3, and the low reactivity of Fe_2O_3.

(3) In the case of drop-weight test, more energy is needed for type A specimens to ignition than type B specimens. And it takes less time to ignite for the former (100 μs) than the latter (150 μs), indicating that type A specimens are more insensitive.

(4) A High Temperature Metal Slag Spray (HTMSS) phenomenon had been observed in PTFE/Al/Fe_2O_3 composites in the drop-weight test, which indicates that the reaction between Al and Fe_2O_3 was triggered.

(5) In PTFE/Al/Fe_2O_3 system, the reaction between PTFE and Al precedes the reaction between Al and Fe_2O_3.

Conflicts of Interest

The authors declare that there are no conflicts of interest regarding the publication of this paper.

Acknowledgments

The financial support from the National Natural Science Foundation of China (General Program Grant no. 51673213) is gratefully acknowledged.

References

[1] J. Cai, S. M. Walley, and R. J. A. Hunt, "High-strain, high-strain-rate flow and failure in PTFE/Al/W granular composites," *Materials Science & Engineering A*, vol. 472, no. 1-2, pp. 308–315, 2008.

[2] E. B. Herbold, V. F. Nesterenko, D. J. Benson et al., "Particle size effect on strength, failure, and shock behavior in polytetrafluoroethylene-Al-W granular composite materials," *Journal of Applied Physics*, vol. 104, no. 10, Article ID 103903, 2008.

[3] E. B. Herbold, J. Cai, D. J. Benson, and V. F. Nesterenko, "Simulation of particle size effect on dynamic properties and fracture of PTFE-W-Al composites," in *Proceedings of the 15th Biennial International Conference of the APS Topical Group on*

Shock Compression of Condensed Matter (SCCM '07), vol. 955, pp. 785–788, American Institute of Physics, June 2007.

[4] S. Xu, S. Yang, and W. Zhang, "The mechanical behaviors of polytetrafluorethylene/Al/W energetic composites," *Journal of Physics Condensed Matter*, vol. 21, no. 28, Article ID 285401, 2009.

[5] X. F. Zhang, J. Zhang, and L. Qiao, "Experimental study of the compression properties of Al/W/PTFE granular composites under elevated strain rates," *Materials Science & Engineering A*, vol. 581, no. 10, pp. 48–55, 2013.

[6] F. Y. Xu, S. B. Liu, Y. F. Zheng, Q. B. Yu, and H. F. Wang, "Quasi-static compression properties and failure of PTFE/Al/W reactive materials," *Advanced Engineering Materials*, vol. 19, no. 1, Article ID 1600350, 2017.

[7] L. Wang, J. Liu, and S. Li, "Investigation on reaction energy, mechanical behavior and impact insensitivity of W–PTFE–Al composites with different W percentage," *Materials & Design*, vol. 92, no. 5, pp. 397–404, 2016.

[8] J. Addiss, J. Cai, S. Walley, W. Proud, and V. Nesterenko, "High strain and strain-rate behaviour of ptfe/aluminuim/tungsten mixtures," in *Proceedings of the 15th Biennial International Conference of the APS Topical Group on Shock Compression of Condensed Matter (SCCM '07)*, pp. 773–776, June 2007.

[9] J. Cai, F. Jiang, K. S. Vecchio, M. A. Meyers, and V. F. Nesterenko, "Mechanical and microstructural properties of PTFE/Al/W system," in *Proceedings of the 15th Biennial International Conference of the APS Topical Group on Shock Compression of Condensed Matter (SCCM '07)*, pp. 723–726, June 2007.

[10] L. Zhou, N. Piekiel, S. Chowdhury, and M. R. Zachariah, "Time-resolved mass spectrometry of the exothermic reaction between nanoaluminum and metal oxides: the role of oxygen release," *Journal of Physical Chemistry C*, vol. 114, no. 33, pp. 14269–14275, 2010.

[11] L. L. Wang, Z. A. Munir, and Y. M. Maximov, "Thermite reactions: their utilization in the synthesis and processing of materials," *Journal of Materials Science*, vol. 28, no. 14, pp. 3693–3708, 1993.

[12] St Petersberg, *Shock Induced Chemical Processing*, US army research office, 1997.

[13] D. B. Nielson, R. L. Tanner, and G. K. Lund, "High strength reactive materials: US, US6593410," 2003.

[14] B. Feng, X. Fang, Y.-C. Li, H.-X. Wang, Y.-M. Mao, and S.-Z. Wu, "An initiation phenomenon of Al-PTFE under quasi-static compression," *Chemical Physics Letters*, vol. 637, Article ID 33187, pp. 38–41, 2015.

[15] B. Feng, Y. C. Li, and S. Z. Wu, "A crack-induced initiation mechanism of Al-PTFE under quasi-static compression and the investigation of influencing factors," *Materials & Design*, vol. 108, pp. 411–417, 2016.

[16] G. M. Swallowe and J. E. Field, "Ignition of a thin layer of explosive by impact; the effect of polymer particles," *Proceedings of The Royal Society of London, Series A: Mathematical and Physical Sciences*, vol. 379, no. 1777, pp. 389–408, 1982.

[17] K. N. Fuller, P. G. Fox, and J. E. Field, "The temperature rise at the tip of fast-moving cracks in glassy polymers," *Proceedings of the Royal Society A: Mathematical, Physical and Engineering Sciences*, vol. 341, no. 1627, pp. 537–557, 1975.

[18] R. A. Williams, J. V. Patel, A. Ermoline, M. Schoenitz, and E. L. Dreizin, "Correlation of optical emission and pressure generated upon ignition of fully-dense nanocomposite thermite powders," *Combustion and Flame*, vol. 160, no. 3, pp. 734–741, 2013.

[19] R.-H. Fan, H.-L. Lü, K.-N. Sun, W.-X. Wang, and X.-B. Yi, "Kinetics of thermite reaction in Al-Fe_2O_3 system," *Thermochimica Acta*, vol. 440, no. 2, pp. 129–131, 2006.

[20] United Nations, *Recommendations on the Transport of Dangerous Goods Manual of Tests and Criteria*, New York, NY, USA, 3rd edition, 1999, ST/SG/AC.10/11/Rev. 3.

[21] W. Mock Jr. and W. H. Holt, "Impact initiation of rods of pressed polytetrafluoroethylene (PTFE) and aluminum powders," in *Proceedings of the AIP Conference*, vol. 845, pp. 1097–1100, August 2006.

6

Fresh-State Properties of Mortars with Recycled Glass Aggregates: Global Unification of Behavior

José Manuel Gómez-Soberón,[1] Francisca Guadalupe Cabrera-Covarrubias,[2] Jorge Luis Almaral-Sánchez,[2] and María Consolación Gómez-Soberón[3]

[1]*Barcelona School of Building Construction (EPSEB), Polytechnic University of Catalonia, Barcelona 08028, Spain*
[2]*Faculty of Engineering Mochis, Autonomous University of Sinaloa, Los Mochis 81210, Mexico*
[3]*Civil Engineering School, Metropolitan Autonomous University, Mexico City, Mexico*

Correspondence should be addressed to José Manuel Gómez-Soberón; josemanuel.gomez@upc.edu

Academic Editor: Jun Liu

Due to the current problems related to the generation of diverse wastes and the extraction of nonrenewable materials to be used in the construction sector, the alternative use of waste glass could be a sustainable option with environmental and economic benefits, in case of being feasible its use as a replacement of the usual aggregates to manufacture recycled mortars. In this research, one presents a study of the fresh-state properties of the mortars containing 15, 30, 60, and 100% recycled glass aggregates as a replacement for the usual aggregate, providing the experimental results of consistency, density, and air content. Using the experimental results, and by means of a numerical and statistical analysis of these, a diagram of triple interaction that allows us to unify the behavior of the studied properties is constituted; making feasible with this, the prediction of the behavior of these properties with respect to variables as their ratio water/cement, aggregate/cement, and different percentages of replacement of aggregates.

1. Introduction

The application of materials derived from recycling is an alternative that should be promoted nowadays, as they may be capable of providing suitable properties to the elements to which they are added, even being favorably comparable to the traditional "virgin" materials that they replace [1]. A large amount of the natural materials currently extracted (in 2002, 55 billion tons) are used—directly or indirectly—in construction (42% of the total material extracted) [2].

As a result, there has emerged the requirement and a need not to waste what have until now been considered "waste" materials to prevent the contamination that their dumping (in a landfill) produces and ultimately to slow the extraction of nonrenewable materials for use in construction. Therefore, the advances in the feasibility of using recycled materials applied to construction may lead to a rational use of the available resources, while also safeguarding natural resources for future generations [3]. Glass is one of the different types of recyclable materials available which may help us to satisfy the demand for construction materials [4], and has been categorized as a solid residue which currently shows a gradual increase in its use; when added to the fact that it is nonbiodegradable, it may cause serious environmental problems in landfills [4–9]. On the contrary, the alternative of reuse is unfeasible in many cases due to problems in collection, the mixture of different qualities, or the economic cost involved [10]; therefore the alternative of recycling as a building material seems feasible as a mitigation of its own condition [7, 11, 12].

Glass is quite easily segregated from other residues, and the study of its physical and chemical characteristics has led to experimentation in its ability to replace "natural" sands in mortars and concrete [13]. This type of praxis is considered ecofriendly, of particular interest in certain construction applications (waterproofing in the structure, resistance properties, durability, and aesthetics), and even on occasion for its profitability [4, 14]. However, for the residue to bond effectively with the matrix of these composites, its properties and resulting products must be established, in order to ensure the suitability of the new recycled composite for use in construction [13].

Previous studies into the behavior of the properties of recycled mortars that incorporate recycled glass aggregate (RGA) to replace the usual aggregate (UA) have dealt with different areas, such as the effect of the quantity of replacements (substitution percentages of UA by RGA) [7, 8, 15], the variations attributable to the color of the glass [10, 16], the effect of the RGA particle size [17–20], and the side effects caused by the type of RGA and its previous use (e.g., proceeding from glass bottles and cathode ray tubes) [15, 21]. Similarly, there are also precedents in the use of glass with dust-sized particles as a cement substitute, which has permitted more studies to be included, such as the pozzolanic properties that they provide [1, 22, 23] and the effect of the size of the dust particles themselves [9, 24].

Previous studies have focused on explaining the behavior of the mechanical properties of the mortars in their hardened state; few have been aimed specifically at the study of their behavior in their fresh state. Although in part this is justified, as the former are the desired properties and the latter should not be ignored, as the final mechanical behavior is partly linked to the behavior in their fresh state.

Regarding the consistency of mortars with recycled glass (RGA substituting UA) (RGM) and in comparison with the usual mortars (UM) of reference, it has been established that higher substitution percentages lead to increases in consistency, which has been shown by the increase of the diameter reported in the flow table test [15, 25]. As a direct implication of the former, it has been deduced that the water needs of the RGM mixtures can be reduced in order to obtain a similar flow to that of the reference UM and, consequently, a lower is possible water/cement (w/c) ratio without causing later side effects [26]. Similar conclusions were obtained regarding the RGM in which the RGA came from different types of glass (drinks bottles and TV cathode ray tubes); however the latter have greater flow values caused by an increase in their particle size (greater fineness modulus and therefore smaller specific surface) which leads to less need for water to cover the particles [21]. The use of fly ash (FA) or ground-granulated blast-furnace slag (GGBFS) to replace up to 25% of the cement has also been tested, as well as different RGA contents, showing in this study that the mineral composition of the materials for substitution of the cement and the percentage of RGA used may cause variations in the consistency of the RGM; the previous behavior (exclusive use of RGA in RGM) may even be reversed when 75% of RGA and 25% of GGBFS are used [27].

Other studies have managed to stabilize the flow of RGM with different RGA contents by means of superplasticizer additives; contents of between 0.5 and 1.5% of superplasticizer per weight of cement manage to establish as a constant the quantity of water needed for flows of similar consistency (between 243 and 259 mm) with up to 75% of RGA [7, 12]. In a similar study (the use of superplasticizer and RGA), which also made changes in the size of the particles, the previous behavior remained in force, with the exception that the increase in RGA particle size permitted a reduction in the additive contents to be used, i.e., the increase in the RGA particles causes an increase in flow, due to the RGA showing a smoother surface and an absorption coefficient close to zero [19].

Another similar case is presented, of RGM with 100% of glass from drink bottles, and RGM with 100% of cathode ray tube glass (CRT) (particle size < 5 mm) has shown an increase in the fluidity due to the smoother surface of the glass and a lower water absorption capacity. However, in this very same study, RGM with smaller particles caused a reduction in fluidity, most notably in the mixture with the finest particles (<600 μm), this behavior being attributed to the fineness of the aggregate with the greatest total surface per unit of volume [20]. The study presented by Öznur Öz et al. [28] reported similar reduction behavior; this as a consequence of the use of RGA in powder form.

A contrary situation is reported in another study, which shows that the use of glass (irrespective of its color) reduces the flow capacity, showing reductions inverse to the increase of RGA used, this being due to the irregular shape and lower density of its particles [16]. Another work coincides with this behavior, as the increase in substitution of RGA by UA, required higher doses of water to reach the same value by means of the flow table [17]. On the contrary, in another study, no significant differences were shown between the water requirements of the RGM and the reference UM; this behavior (compared with other studies) was attributed to the different types of glass used and the water absorption, which depends on the size of the aggregates [29].

Regarding the density property of the mortars in their fresh state, previous studies show reductions in the RGM, presenting an inverse behavior to the increased RGA substitution for UA; this is due to the lower specific gravity of the RGA compared with the UA [16, 30], as well as the more angular form of the glass compared with the sand, which results in a greater volume of spaces [29]. The opposite is shown in another study, in which 100% of RGA replaces the UA; the RGA is characterized by two different sizes (studied separately), and both of the final RGMs showed a greater density than the UM. The explanation for this behavior may be the same as in the previous study, in which the density in the fresh state is attributed to the density of the RGA itself compared with the UA, which in this case was due to the use of cathode ray tube glass [25].

Finally, there is very little information in the literature regarding the air content property; among the published works, the use of fine RGA is mentioned (of between 2.36 and 5 mm), which results in a greater air content in the mixture due to the irregular shape of the particles, leading to greater relative surface which in turn creates larger areas that retain air voids [19, 31]. Other studies have also reported greater air contents in RGM, showing increases when the RGA replacement is higher [30, 32], for example, reaching a maximum difference of 54% for mortar with 100% RGA (air content of 12.6%), compared with the reference mortar [29]. This behavior is attributed to the lower density of the particles and their shape, with sharper edges showing a greater relation of aspect which allows more air to be retained in the surfaces of the RGA particles; it has also been commented that microcracks have been noticed in the case of transparent glass particles, which could have affected the packing of the aggregates mainly in the case of high contents [16].

Based on the aforementioned, it can be asserted that documental evidence is scarce and therefore the information about the fresh properties of RGM is minimal, which in general makes it difficult to predict the behavior that these mortars can show before manufacture, as they do not present established behavioral tendencies. The aim of this study is to provide a triple interaction diagram with which to predict the w/c ratio, the density, or the air content that the RGM may show with different RGA percentages substituting the UA and with three different aggregate/cement (a/c) ratios. Moreover, it can be used for deciding what percentage of substitution is needed for use in the RGM to obtain a particular density; in general, the link between the fresh-state properties can easily be determined, as well as predicting the ratios shown by the different RGMs from the design phase of the mixtures. Both the diagram, to be presented later, and the regression equations are the result of numerical and statistical analysis performed on the data of an experimental campaign, which establishes the properties of the consistency, density, and air content of the different ratios of a/c, w/c, and the replacement factor (RF) in the RGM.

2. Materials

Transparent glass residues with a particle size of between 0 and 1.8 mm, obtained from a local waste glass processing company, were used as RGA; specifically, the glass used came from laminated glass (commonly used in windows, doors, and car windscreens). A UA composed of silica sand with the particle size of 0–4 mm was used to manufacture the reference mortars; the silica sand came from a local supplier of aggregates for construction.

The adjusted granulometric curves of both aggregates, as well as the limits established by the ASTM C144 standard [33], are shown in Figure 1. As the original granulometric profile of the UA was positioned outside the lower limit of the ASTM, and that of the RGA was located above the upper limit of the same, a granulometric adjustment was made to distribute the particle sizes in the two aggregates used; consequently, both granulometric curves conformed—as much as possible—to the limits, also achieving a similarity in profiles. The adjustment procedure involved separating both materials with a no. 30 sieve (0.59 mm), where the particles that passed through the sieve being established as fine fraction (FF) and those retained in the sieve as coarse fraction (CF); then mixtures of different FF and CF combinations were made until the proportions with the maximum compactness were obtained (RGA and UA separately). In the case of the RGA, the combination that fulfilled the requirement was that of 70% FF and 30% CF; and for UA, the combination was 50% of each fraction [34]. After making the adjustment, it can be seen in Figure 1 that the adjusted granulometric profile of the RGA is located very close to the upper limit, demonstrating that the majority of these aggregate's particles are composed of fine particles; in Table 1, the fineness modulus (FM) = 1.8, which places it outside the desired specified range for sands (2.3 to 3.1, according to the ASTM C33 standard). The opposite is seen in the case of the adjusted profile of UA, generally found within the established limits, with the exception of 4% of the material which comes from the lower limit (sieve no. 8 (2.38 mm)); the MF of the UA (2.4) locates within the desired range for sands. For both aggregates, the percentage of fine particles with sizes less than 75 μm is below 10% (requirement specified by the standards). To sum up, it can be said that the adjustment allows both aggregates to be compared with minimum profile modifications, meaning that the replacement of both materials does not lead to variations of their properties due to the effect of the distribution of their particle sizes; in addition, the adjustment permits their profiles to be closely aligned to that recommended by the standards, thereby guaranteeing better behavior of the mixtures.

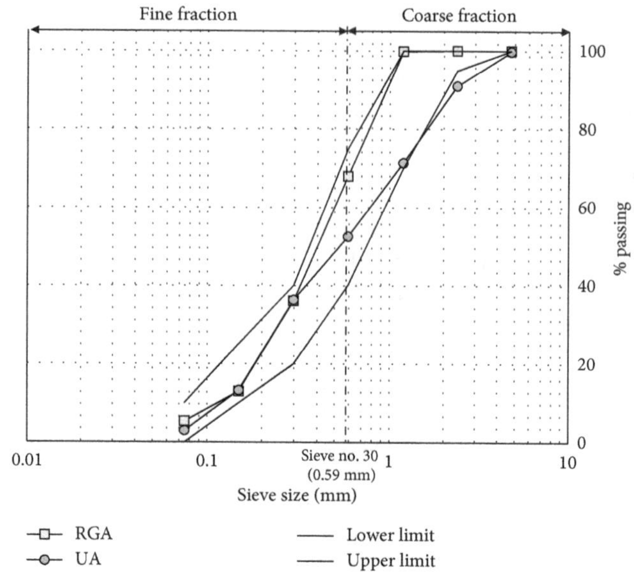

FIGURE 1: Granulometric distribution curves.

Consequently, the procedures specified by the ASTM C128 [35], C136 [36], and C117 standards [37] were carried out in order to determine the general physical properties of the aggregates to be used (Table 1). Significant differences between both aggregates were established for the water absorption coefficient, which for the RGA meant a difference of 1% with respect to that obtained by the UA. Regarding the dry-state density (OD) and the saturated surface dry (SSD) state, the UA shows slightly higher values than the RGA (107.76 and 134.8 kg/m^3, respectively).

Finally, Portland cement with the usual properties and components, classified as CEM I 42.5 N/SR (UNE EN 197-1: 2011 [38]), was used as a binding material in making the different UM and RGM, along with tap water.

3. Experimental Campaign

The experimental campaign was divided into two sections: the first was for obtaining the experimental data, in which

TABLE 1: Physical properties of the aggregates.

Property	RGA	UA
Density (OD) (kg/m^3)	2473.84	2581.6
Density (SSD) (kg/m^3)	2488.8	2623.6
Water absorption (%)	0.6	1.6
Fineness modulus	1.8	2.4
Materials finer than 75 μm (%)	5.3	2.9

the procedure and tests for obtaining data were explained and the second was the data analysis, which established the numerical data analysis procedure that allows the regression coefficients and the triple interaction diagram proposed for the unified behavior of the RGM to be obtained.

3.1. Obtaining the Experimental Data. Several mixtures of RGM were made, composed of different percentages of RGA as a UA substitute (15, 30, 60, and 100%); for each, three different a/c ratios were specified (3.25, 4, and 4.75). In order to simplify the identification of the RGM studied, the following agreement of nomenclature was established: "*XX*% RGM-*A*/*B*/*C*," where XX% is the percentage of RGA (or RF) and *A*/*B*/*C* represents the three different ratios of a/c studied ($A = 3.25$, $B = 4$, and $C = 4.75$). The reference mortars UM (0% of RGA) for each of the a/c ratios studied were identified as UM-*A*, UM-*B*, and UM-*C*. All the mixtures were designed for an initial w/c ratio of 0.5. Table 2 shows the amount of materials needed to make the RGM and UM.

The mixing process for each of the mortars implied introducing the RGA and/or the UA along with the total water (obtained from the initial w/c ratio) in the mixing recipient (one minute at rest) to obtain the saturation of the aggregates, introducing the cement later. Then, the recipient containing the materials was placed in the mixer (Mod. E93, Matest brand,) and the mixing procedure was carried out in accordance with point 8.1.3 of the ASTM C305-06 standard [39]. When the mixing was finished, the tests on the mortar behavior in the fresh state were carried out.

3.2. Consistency. Once the mixing process of the mortars had ended, the consistency of the RGM was determined by means of a mortar flow test according to the ASTM C1437 standard [40], using a flow table with specifications conforming to the C230/C230M standard [41]. As a mandatory step in making a feasible comparison of the different mixtures studied, it was decided that they should all reach a flow of $110 \pm 5\%$ according to the specifications of the ASTM C109/C109M standard [42] (standardized mortar); as a standard mold with a 10 mm base was used, the diameter sought at the end of the test was 210 ± 5 mm, which was determined by the average measurement of the diameters of the principal axes passing through the center of the extended sample. If the sample did not present the required flow, more water was added repeatedly, in small amounts, until the desired flow was obtained and the mortar could be considered standardized.

3.3. Density. The density of the previously obtained RGM was determined; for this test, the equipment used was a standard recipient of the one liter, which was also part of the equipment needed for the air content test. The standard mixture was put into the recipient in three successive layers, each individually compacted (20 distributed blows) with a metal tamper with a flat circular section; then in the final layer, a metal ruler was used to skim off any excess mixture. The resulting specimen was then weighed (Mod. ST-4000 Series ST, GRAM *Precision* brand, accurate to 0.01 grams). When the mixture weight was obtained (deducting the weight of the mold) and considering the volume of the standardized mold to be a known constant, the density of the RGM was determined by means of the following equation:

$$\rho = \frac{M}{V}, \quad (1)$$

where ρ = density (g/cm^3), M = mass (g), and V = volume (cm^3).

3.4. Air Content. To finish, the content of the air included in the RGM study samples was tested; the standard equipment for determination being a 1-liter TESTING brand Air Entrainment Meter which included the recipient was also used in the density tests. The test was based on the pressure method (ASTM C231 standard [43]), with the air content of each of the studied RGM being obtained with a calibrated manometer. As validation of the test, the air content was also determined by means of the method described in the ASTM C185-99 standard [44]; this uses the density values of the mortars and materials and the mixture proportions. In this second method, the air content is obtained by means of

$$\text{air content, volume \%} = 100\left[1 - \frac{W_a}{W_c}\right], \quad (2)$$

where W_a = actual mass per unit of volume as determined by this test method (g/cm^3) and W_c = theoretical mass per unit of volume, calculated on an air-free basis and using the values of the quantities of materials and specific gravities from each of them—see the complete equation in the standard ASTM C185-99.

The reason for the need to validate this property through its determination by two different procedures was in order to be able to differentiate, or corroborate, the values obtained by the different RGM studied; this was necessary when using aggregates with densities and absorption of different values to make RGM or to determine whether this property was sensitive to variation.

3.5. Regression Coefficients and Trend Equations. The results obtained from the tests were used in the statistical analysis software SPSS v24.0.0.0 of IBM for Windows, into which the data were introduced with the aim of finding the regression coefficients that would allow the trend equations to be determined; consequently, the ratios between the RF and the different properties of RGM in the fresh state could be established. It was noticed in the dispersion graphs for each

Table 2: Dosage of RGM (kg/m³).

Materials (kg/m³)		Classification														
		UM-*A*	15RGM-*A*	30RGM-*A*	60RGM-*A*	100RGM-*A*	UM-*B*	15RGM-*B*	30RGM-*B*	60RGM-*B*	100RGM-*B*	UM-*C*	15RGM-*C*	30RGM-*C*	60RGM-*C*	100RGM-*C*
RGA[a]	FF	0	154	287	501	684	0	146	284	513	758	0	159	283	526	780
	CF	0	66	123	215	293	0	63	122	220	325	0	68	121	225	334
UA[a]	FF	781	623	478	239	0	800	593	474	244	0	811	643	472	250	0
	CF	781	623	478	239	0	800	593	474	244	0	811	643	472	250	0
Cement		480	451	420	367	301	400	349	339	305	271	342	318	284	263	234
Water		327	315	300	298	278	334	294	303	302	304	334	316	286	303	328

[a]Dry condition; FF: fine fraction; CF: coarse fraction.

ratio studied that the variables were closer to regressions with curvilinear estimates; additionally, in all the ratios studied, they did not comply with the requisites of linear or simple distribution (distribution of the normal dependent variable, for each value of the independent variable; the variance of the dependent variable distribution is a constant for all values of the independent variable; there is a linear relationship between the dependent variable and each independent variable; and all observations are independent). On the contrary, all the necessary requirements for determining a curvilinear regression are satisfied; in particular, the second grade quadratic regressions adjust better, in accord with the acquirement of the highest R^2 parameter (measurement of the goodness of the fit or reliability of estimated data model, $R^2 \leq 1.0$). The requirements to be met for a quadratic regression were as follows:

(1) The dependent and independent variables are quantitative
(2) The model's residues are distributed normally and randomly
(3) The quadratic model was adapted to the studied series with a "detached" or damped distribution

In each pair of variables studied, it was considered that the RF represented the independent variable—because it has no precedent in this area of knowledge—and density, and the w/c ratio and air content were the dependent variables; this allowed the obtained regression equations to be established for the mathematical formulation which best predicted the group of data for the dependent variables. The trend curves were determined by calculating the study points (grouping of independent and dependent variables) by means of adjusting the squared minimums and using the following Equation as the structure of the mathematical formula:

$$y = ax^2 + bx + c, \tag{3}$$

where x = value x_i of the independent variable; y = value y_i of the dependent variable; and a, b, c = constant coefficients of any value, except for $a = 0$.

4. Results and Discussion

The experimental results of the fresh-state properties of the studied RGM corresponding to the consistency (quantity of water and final w/c ratio), density, and air content are presented in Table 3.

The regression coefficients that defined the different behavior equations of this research were obtained from the previous values by means of quadratic regressions; Figure 2 shows the experimental results of the density, the w/c ratio, and the air content with regard to RF—for the three different a/c ratios (A, B, and C)—as well as the equations with their regression coefficients.

It can be seen that, in the case of the RGM density with regard to RF, a damping ratio is established, which means that the higher the independent variable (RF), the lower the value of the dependent variable (RGM density). The previous behavior is similar (parallel trends) for the different a/c ratios studied, following an initial increasing order in the a/c ratio (A followed by B and then C), which causes decreases in the density. When RF is close to 40%, the order between B and C is inverted, which may be due to a balancing or better fit of the particles (RGA and UA) in the RGM with the variable C ($a/c = 4.75$), improving their compactness and, therefore, their density.

With regard to the RF ratio and the dependent variable w/c, this shows an ascendant distribution; this should be understood in terms of the need for more water in the RGM as the RF is increased. This is related to the irregular shape of the RGA particles, which generates a greater need for water in order to obtain a standard comparable consistency. In order to corroborate this, the shape of the particles forming the RGA was observed, using a JEOL JSM-6510 scanning electron microscope (SEM) to study the samples previously embedded in a resin and then with their surfaces polished. Figure 3 shows SEM images in which the wavy shape of UA can be seen in Figure 3(a); while Figure 3(b) shows the RGA particles (more angular and of longer proportions).

Regarding the a/c ratio, the established experimental order was ascendant (A, B, and C); this is a natural response, as any increase in the aggregate content of a mixture also requires an increase of water (angled and elongated particle shape of the RGA).

As for the representative equations of the air content, it can be seen that the increase of RF causes damping, with two important differential zones: one comprising the values lower than RF 30% to 60%, which results in a damping of the curves proportional and incremental, and following it, a decelerated damping that tends to a constant (similar behavior to the density property but in the inverse sense).

Using the regression equations presented in Figure 2, the theoretical values that make up the curves of each ratio can be obtained by numerical simulation, thereby establishing the triple action diagram (Figure 4) which provides with the unified global behavior of the RGM in the fresh state. The following are some of the observations obtained from the graph:

(i) The behavior of the density reduction due to the effect of the RF might be caused by the low density of the RGA, having an effect on the matrix of the RGM. However, for RF $\geq$ 53.82%, and with a high a/c ratio (C), the density values of the RGM are similar (in curve B), thereby losing the effect of the RF factor in reducing density.
(ii) The bigger the RF, the greater the w/c ratio; this situation is caused by the increase in the need for water in the RGM to obtain the same standardized consistency in the mixtures. The previous behavior is also valid for the a/c ratio, as a mortar mixture containing more RGA (with respect to an equal amount of cement) will need more water (and therefore a higher w/c ratio), which is necessary to give more "mobility" to the aggregates in the mixture (workability).

Table 3: Results of the fresh-state properties of RGM.

	Water needed for 1 dm^3 (g)	Final w/c ratio	Density (g/cm^3)	Air content (%) ASTM C231	Air content (%) ASTM C185
UM-*A*	327.35	0.68	2.18	2.0	0.2
15RGM-*A*	314.63	0.70	2.02	8.3	7.0
30RGM-*A*	300.14	0.71	1.87	13.7	13.1
60RGM-*A*	297.70	0.81	1.63	22.1	22.3
100RGM-*A*	278.39	0.93	1.57	22.3	22.6
UM-*B*	334.44	0.84	2.16	2.0	0.2
15RGM-*B*	293.83	0.84	1.96	10.1	8.9
30RGM-*B*	303.43	0.90	1.78	16.7	16.2
60RGM-*B*	301.54	0.99	1.57	23.6	24.1
100RGM-*B*	304.32	1.12	1.48	25.0	26.1
UM-*C*	334.28	0.98	2.15	1.8	0
15RGM-*C*	316.49	0.99	1.91	11.7	10.4
30RGM-*C*	286.15	1.01	1.77	17.4	16.8
60RGM-*C*	302.84	1.15	1.59	22.4	22.7
100RGM-*C*	327.67	1.40	1.50	22.9	23.8

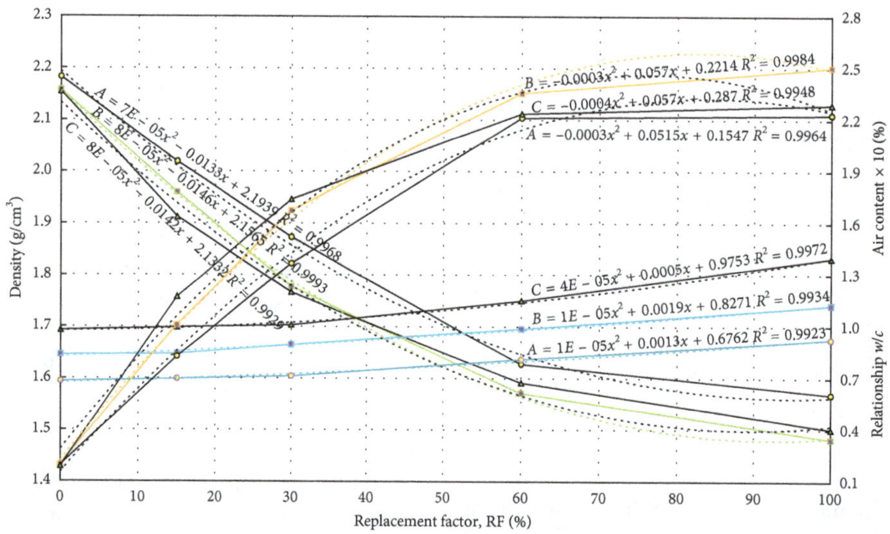

Figure 2: Experimental values of the UM and RGM, and regression equations with their resulting reliability.

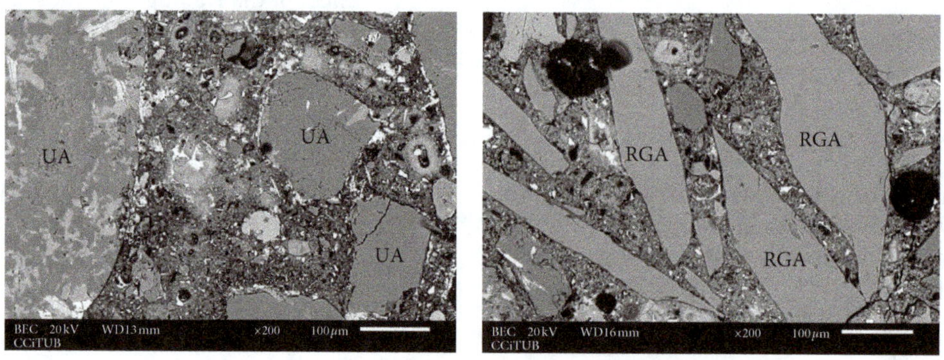

(a) (b)

Figure 3: SEM images. (a) UM-*B* (UM with 100% UA). (b) 30RGM-*B* (RGM with 30% RGA).

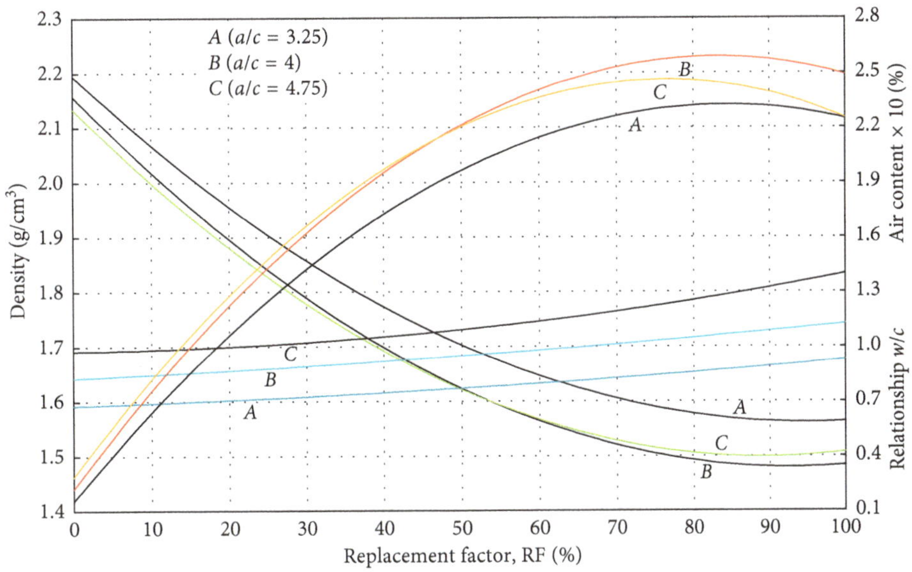

FIGURE 4: Triple interaction diagram of the fresh-state properties of the RGM.

(iii) The effect of the increase in REF leads to greater air content in the RGM; this is due to the irregular shape of the RGA particles, which cannot obtain the best fit to improve the mixture's compactness, while also causing more air to be trapped between them. This effect also holds true for a higher a/c ratio; however, mortars with RF ≥ 47.06% and ratio C show a similar behavior to that reported in the densities, with the difference that here it is in the opposite, or decreasing, sense; the air content in the mortar is maintained even when the quantity of RGA is increased.

The triple interaction diagram obtained has two different forms of use, which allow the reference values of design and fresh-state properties to be obtained simply and quickly from different RGM mixtures with any value of RF. Examples of the diagram's use might be as follows:

(iv) *For predictive use in the design of an RGM mixture.* One example might be defined as follows: what content range of RF should be used to obtain an RGM with a determined density or air content?

(v) *For estimation or revision of an RGM mixture.* For example, what density and air content does an RGM show when starting from the data of the RF and the a/c ratio?

With the aim of calibrating the degree of certainty in the obtained diagram, Table 4 presents the theoretical values of the different properties studied and approximation errors (which were determined by the previously prepared regression equations). Identical RF contents were established for a comparison of the estimated and experimental values. In the case of the properties of the w/c ratio and the RGM density, the maximum approximation error was of only 0.02%, with a general average error of only 0.0067%, thereby clearly validating a correct application. Regarding the air content, the maximum approximation error established was of 0.53% (UM-C), with a general average error of 0.0847%.

With the aim of revising the proposed diagram's adaptability strength, the results presented in previous works were included in order to observe their relative graphic positioning. The resulting graph is shown in Figure 5, with the addition of previous research (all broken lines); the green lines refer to the density (values on left axis), the blue lines show the w/c ratio, and the red line shows the air content.

Regarding the density properties, the values of Tan and Du's 2013 research [16] establish densities higher than those of this study. However, they are similar for RF = 0%, which indicates that these variations do not originate (at least directly) in the difference between the density of the UA in their study and that used in this research. It should be pointed out that, in this research, the RF effect has very little emphasis on the change in this property, which could be explained by the low a/c ratio used (2.75); as they contain fewer aggregates and therefore a lower amount of RGA, they cause no significant losses of density. Finally, the direction of their damping curves does coincide with that shown in this research, although in this case it is closer to a straight sloping line than to a curve typical of a quadratic equation.

In the case of the data of Penacho et al. [29], at up to RF = 20%, they show density values inferior to the curves of the proposed diagram; nevertheless at higher RF values, the density is greater and the trend decrease regarding the RF is also weakened, although at a lower rate of change and with a similar a/c to the diagram proposed for B. In this last case, the deviation may be explained by the influence of the RGA density (an average of 1030 kg/m^3), being 42% lower than that used in this research; the former also leads to decreases in density for high RF values, although with fewer accentuated changes than those shown in the diagram (effect of the RGA density).

Table 4: Theoretical values of the RGM properties and approximation errors.

Mortars of study	Property					
	Final w/c ratio		Density (g/cm^3)		Air content (%)	
	Estimated value	Approximation error[a] (%)	Estimated value	Approximation error[a] (%)	Estimated value	Approximation error[a,b] (%)
UM-*A*	0.68	0.01	2.19	0.01	1.55	0.21
15RGM-*A*	0.70	0.00	2.01	0.00	8.58	0.04
30RGM-*A*	0.73	0.02	1.86	0.01	14.25	0.04
60RGM-*A*	0.80	0.01	1.65	0.01	21.45	0.03
100RGM-*A*	0.93	0.00	1.56	0.00	22.49	0.01
Approximate mean error[c] (%)		0.01		0.01		0.07
UM-*B*	0.83	0.01	2.16	0.00	2.21	0.12
15RGM-*B*	0.86	0.02	1.96	0.00	9.99	0.01
30RGM-*B*	0.89	0.00	1.79	0.01	16.22	0.03
60RGM-*B*	0.98	0.01	1.56	0.01	24.05	0.02
100RGM-*B*	1.13	0.00	1.48	0.00	24.91	0.01
Approximate mean error[c] (%)		0.01		0.00		0.04
UM-*C*	0.98	0.00	2.13	0.01	2.87	0.53
15RGM-*C*	0.99	0.00	1.94	0.01	10.58	0.10
30RGM-*C*	1.02	0.01	1.78	0.01	16.61	0.04
60RGM-*C*	1.14	0.01	1.57	0.02	23.63	0.06
100RGM-*C*	1.40	0.00	1.51	0.00	22.52	0.02
Approximate mean error[c] (%)		0.01		0.01		0.15

[a]Approximation error (%) = |(experimental value − estimated value)/(experimental value)|. [b]Experimental value = (experimental value ASTM C231 + experimental value ASTM C185)/2. [c]Average error (%) = $(\sum_{i=1}^{n}$ Approximation error$)/n$.

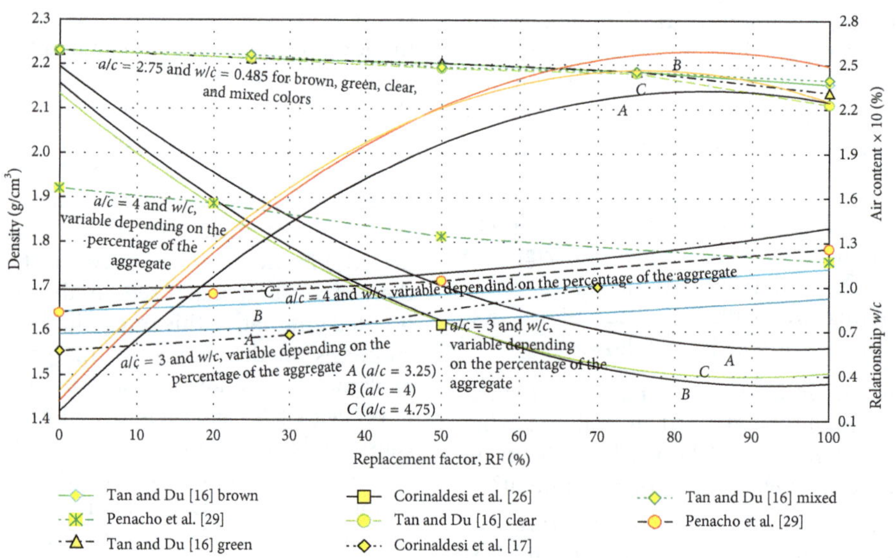

Figure 5: Triple interaction diagram and results of previous studies.

In regard to the particular case of the w/c ratio, two previous studies provide results applicable to the diagram of this research: Corinaldesi et al. [17] and Corinaldesi et al. [26], both allowing the curves proposed for this property in the diagram to be validated. Both studies used an $a/c = 3$ ratio, close to the curve *A* of the diagram proposed here. In the first study, when RF ≤ 40%, the order, position, and precision of the data can be considered acceptable and congruent with the proposed diagram; this is also the case of the second study, with the only value provided being RF = 50%. Beyond RF > 40%, the data are positioned above the curve *A* ($a/c = 3.25$) with a type of the ascendant curve very similar to those found in this research.

Finally, given the scarcity of previous information about the properties of the air content of RGM, only the data of one study were suitable for inclusion in the proposed diagram: the work of Penacho et al. [29], with the RFs of 0, 20, 50, and 100%, showing an ascendant curve with a distribution like those shown in this research; however, the air content figures were lower than those proposed. The nondirect application of these data to the proposed diagram may be explained by the characteristics of the aggregates used, described as significantly more angular than the UA [29]. In a paper present in the literature, with the substitution percentages of 25, 35, 50, 75, and 100% of replacement and with similar experimental campaign, they report that the incorporated air content increased with the replacement of UA by RGA in almost linear relation ($R^2 \approx 0.99$), similar to that presented in this paper [30].

Once the previous studies had been analyzed and compared with the target diagram of this study, it was possible to validate its use in predicting the fresh-state properties of RGM that contains RGA of similar characteristics to those used. As a necessary future objective, it justifies the need to prepare a more universal or extensive diagram in which additional experimental data with more variables can be included; the analysis should also include values of the continuity of this research as well as those of other studies. Obtaining this universal diagram will improve the prediction and application of RGM use.

5. Conclusions

The following conclusions were obtained from the work presented:

(i) Having the experimental data of the RGM permitted the triple interaction diagram to be made, which allowed the behavior of the properties in the fresh state to be unified; this focused both on the design phase of the mixtures and the prediction of their properties, with the RF being an important and decisive parameter

(ii) The data obtained from the diagram may be useful in the composition of other new diagrams in which the hardened-state properties of RGM, such as compressive strength, can be included

(iii) In order to obtain more reliable predictions, it is recommended to verify that the characteristics of the RGA (particle size, density, etc.), to be evaluated, are similar to those used in the creation of the diagram obtained in this investigation

(iv) More types of tests should be carried out on RGM, which will provide a broader knowledge of its properties; consequently, diagrams containing different variables for multiple alternative uses can be prepared

Conflicts of Interest

The authors declare that there are no conflicts of interest regarding the publication of this paper.

Acknowledgments

The authors thank the Barcelona School of Building Construction-UPC, the Department of Architecture Technology-EPSEB-UPC, the School of Engineering Mochis-UAS, and finally the Civil Engineering School-UAM.

References

[1] C. Shi, Y. Wu, C. Riefler, and H. Wang, "Characteristics and pozzolanic reactivity of glass powders," *Cement and Concrete Research*, vol. 35, no. 5, pp. 987–993, 2005.

[2] A. Behrens, S. Giljum, J. Kovanda, and S. Niza, "The material basis of the global economy. Worldwide patterns of natural resource extraction and their implications for sustainable resource use policies," *Ecological Economics*, vol. 64, no. 2, pp. 444–453, 2007.

[3] C. B. Farinha, J. de Brito, R. Veiga, J. M. Fernández, J. R. Jiménez, and A. R. Esquinas, "Wastes as aggregates, binders or additions in mortars: Selecting their role based on characterization," *Materials*, vol. 11, no. 3, p. 453, 2018.

[4] I. B. Topçu, A. R. Boğa, and T. Bilir, "Alkali-silica reactions of mortars produced by using waste glass as fine aggregate and admixtures such as fly ash and Li_2CO_3," *Waste Management*, vol. 28, no. 5, pp. 878–84, 2008.

[5] Y. Shao, T. Lefort, S. Moras, and D. Rodriguez, "Studies on concrete containing ground waste glass," *Cement and Concrete Research*, vol. 30, no. 1, pp. 91–100, 2000.

[6] S.-B. Park and B.-C. Lee, "Studies on expansion properties in mortar containing waste glass and fibers," *Cement and Concrete Research*, vol. 34, no. 7, pp. 1145–1152, 2004.

[7] T.-C. Ling, C.-S. Poon, and S.-C. Kou, "Feasibility of using recycled glass in architectural cement mortars," *Cement and Concrete Research*, vol. 33, no. 8, pp. 848–854, 2011.

[8] P. S. Bhandari and K. M. Tajne, "Use of waste glass in cement mortar," *International Journal of Civil and Structural Engineering*, vol. 3, no. 4, pp. 704–711, 2013.

[9] A. Khmiri, M. Chaabouni, and B. Samet, "Chemical behaviour of ground waste glass when used as partial cement replacement in mortars," *Construction and Building Materials*, vol. 44, pp. 74–80, 2013.

[10] N. Degirmenci, A. Yilmaz, and O. A. Cakir, "Utilization of waste glass as sand replacement in cement mortar," *Indian Journal of Engineering and Materials Sciences*, vol. 18, no. 4, pp. 303–308, 2011.

[11] R. Idir, M. Cyr, and A. Tagnit-Hamou, "Use of fine glass as ASR inhibitor in glass aggregate mortars," *Construction and Building Materials*, vol. 24, no. 7, pp. 1309–1312, 2010.

[12] C. S. Poon and T. C. B. Ling, "Use of Recycled Glass in Architectural Mortars," *Journal of Wuhan University of Technology*, vol. 32, no. 17, pp. 231–234, 2010.

[13] A. Karamberi and A. Moutsatsou, "Participation of coloured glass cullet in cementitious materials," *Cement and Concrete Composites*, vol. 27, no. 2, pp. 319–327, 2005.

[14] T.-C. Ling and C.-S. Poon, "Feasible use of large volumes of GGBS in 100% recycled glass architectural mortar," *Cement and Concrete Composites*, vol. 53, pp. 350–356, 2014.

[15] T.-C. Ling and C.-S. Poon, "Utilization of recycled glass derived from cathode ray tube glass as fine aggregate in cement mortar," *Journal of Hazardous Materials*, vol. 192, no. 2, pp. 451–456, 2011.

[16] K. H. Tan and H. Du, "Use of waste glass as sand in mortar: part I-fresh, mechanical and durability properties," *Cement and Concrete Composites*, vol. 35, no. 1, pp. 109–117, 2013.

[17] V. Corinaldesi, G. Gnappi, G. Moriconi, and A. Montenero, "Reuse of ground waste glass as aggregate for mortars," *Waste Management*, vol. 25, no. 2, pp. 197–201, 2005.

[18] G. Lee, T.-C. Ling, Y.-L. Wong, and C.-S. Poon, "Effects of crushed glass cullet sizes, casting methods and pozzolanic

materials on ASR of concrete blocks," *Construction and Building Materials*, vol. 25, no. 5, pp. 2611–2618, 2011.

[19] T.-C. Ling and C.-S. Poon, "Properties of architectural mortar prepared with recycled glass with different particle sizes," *Materials & Design*, vol. 32, no. 5, pp. 2675–2684, 2011.

[20] T.-C. Ling and C.-S. Poon, "Effects of particle size of treated CRT funnel glass on properties of cement mortar," *Materials and Structures*, vol. 46, no. 1, pp. 25–34, 2012.

[21] T.-C. Ling and C.-S. Poon, "A comparative study on the feasible use of recycled beverage and CRT funnel glass as fine aggregate in cement mortar," *Journal of Cleaner Production*, vol. 29-30, pp. 46–52, 2012.

[22] A. M. Matos and J. Sousa-Coutinho, "Durability of mortar using waste glass powder as cement replacement," *Construction and Building Materials*, vol. 36, pp. 205–215, 2012.

[23] R. B. C. Sales, F. A. Sales, E. P. Figueiredo, W. J. Dos Santos, N. D. S. Mohallem, and M. T. P. Aguilar, "Durability of Mortar Made with Fine Glass Powdered Particles," *Adv. Mater. Sci. Eng.*, vol. 2017, article 3143642, 9 pages, 2017.

[24] L. A. Pereira-de-Oliveira, J. P. Castro-Gomes, and P. M. S. Santos, "The potential pozzolanic activity of glass and red-clay ceramic waste as cement mortars components," *Construction and Building Materials*, vol. 31, pp. 197–203, 2012.

[25] Z. Hui and W. Sun, "Study of properties of mortar containing cathode ray tubes (CRT) glass as replacement for river sand fine aggregate," *Construction and Building Materials*, vol. 25, no. 10, pp. 4059–4064, 2011.

[26] V. Corinaldesi, A. Nardinocchi, and J. Donnini, "Reuse of recycled glass in mortar manufacturing," *European Journal of Environmental and Civil Engineering*, vol. 20, no. 1, pp. 140–151, 2016.

[27] Z. Hui, C. S. Poon, and T. C. Ling, "Properties of mortar prepared with recycled cathode ray tube funnel glass sand at different mineral admixture," *Construction and Building Materials*, vol. 40, pp. 951–960, 2013.

[28] H. Öznur Öz, H. Erhan Yücel, and M. Güneş, "Comparison of glass powder and fly ash effect on the fresh properties of self-compacting mortars," *IOP Conference Series: Materials Science and Engineering*, vol. 245, article 032036, 2017.

[29] P. Penacho, J. de Brito, and M. R. Veiga, "Physico-mechanical and performance characterization of mortars incorporating fine glass waste aggregate," *Cement and Concrete Composites*, vol. 50, pp. 47–59, 2014.

[30] M. Bentchikou, S. Bouriah, N. Hamdane, F. Debieb, and O. Boukendakdji, "Comparative study of the properties of mortars with recycled glass aggregates incorporated by addition and substitution," *Energy Procedia*, vol. 139, pp. 499–504, 2017.

[31] H. Du and K. H. Tan, "Concrete with recycled glass as fine aggregates," *ACI Materials Journal*, vol. 111, no. 1, pp. 47–58, 2014.

[32] A. R. G. de Azevedo, J. Alexandre, E. B. Zanelato, and M. T. Marvila, "Influence of incorporation of glass waste on the rheological properties of adhesive mortar," *Construction and Building Materials*, vol. 148, pp. 359–368, 2017.

[33] ASTM International, *C144-99, Standard Specification for Aggregate for Masonry Mortar*, ASTM International, West Conshohocken, PA, USA, 1999.

[34] F. G. Cabrera-Covarrubias, J. M. Gómez-Soberón, J. L. Almaral-Sánchez, R. Corral-Higuera, A. Navarro-Ezquerra, and M. Tous-Coll, "Characterization of three recycled materials for alternative use of mortars," *International Journal of Sustainable Materials, Processes & ECO-Efficient*, vol. 1, no. 3, pp. 13–17, 2014.

[35] ASTM International, *C128-04a, Standard Test Method for Density, Relative Density (Specific Gravity), and Absorption*, ASTM International, West Conshohocken, PA, USA, 2004.

[36] ASTM International, *C136-06, Standard Test Method for Sieve Analysis of Fine and Coarse Aggregates*, ASTM International, West Conshohocken, PA, USA, 2006.

[37] ASTM International, *C117-95, Standard Test Method for Materials Finer than 75-μm (No. 200) Sieve in Mineral*, ASTM International, West Conshohocken, PA, USA, 1995.

[38] AENOR, *UNE-EN 197-1, Cemento Parte 1: Composición, especificaciones y criterios de conformidad de los cementos comunes*, AENOR, Madrid, Spain, 2011.

[39] ASTM International, *C305-06, Mechanical Mixing of Hydraulic Cement Pastes and Mortars of Plastic Consistency*, ASTM International, West Conshohocken, PA, USA, 2006.

[40] ASTM International, *C1437-99, Standard Test Method for Flow of Hydraulic Cement Mortar*, ASTM International, West Conshohocken, PA, USA, 1999.

[41] ASTM International, *C230/C230M-03, Standard Specification for Flow Table for Use in Tests of Hydraulic Cement*, ASTM International, West Conshohocken, PA, USA, 2003.

[42] ASTM International, *C109/C109M-05, Standard Test Method for Compressive Strength of Hydraulic Cement Mortars*, ASTM International, West Conshohocken, PA, USA, 2005.

[43] ASTM International, *C231-08b, Standard Test Method for Air Content of Freshly Mixed Concrete by the Pressure Method*, ASTM International, West Conshohocken, PA, USA, 2008.

[44] ASTM International, *C185-99, Standard Test Method for Air Content of Hidraulic Cement Mortar*, ASTM International, West Conshohocken, PA, USA, 2001.

7

Influences of Process Parameters and Vibration Parameters on the Forming Force in the Ultrasonic-Assisted Incremental Forming Process

Lang Bai, Yan Li, Mingshun Yang, Zimeng Yao, and Zhiyuan Yao

School of Mechanical and Precision Instrument Engineering, Xi'an University of Technology, Xi'an, Shaanxi 710048, China

Correspondence should be addressed to Yan Li; jyxy-ly@xaut.edu.cn

Academic Editor: Hongchao Kou

In the field of plastic forming, ultrasonic vibration has the advantages of small forming force and high forming quality, and it has been introduced into a single-point incremental plastic forming technique with high flexibility and high precision. The ultrasonic vibration single-point incremental compound forming technology with all the above advantages has been achieved. To reveal the variation tendency of the forming force under ultrasonic vibration and single-point incremental coupling, the process parameters (layer spacing, tool head radius, and feed rate) and vibration parameters (frequency and amplitude) on the forming force of the composite technology were quantitatively analyzed by theory, simulation, and experiment. The simulation and experimental results showed that ultrasonic vibration can significantly reduce the forming force of the composite technology. Compared with the simulation results, the theoretical results are not only closer to the experimental results but also have a shorter computation time and better prediction effect on the forming force. The change in the process parameters has a linear effect on the forming force, the nonlinearity of the vibratory parameters influences the forming force, and the frequencies and amplitudes in a suitable range allow the forming force to reach a minimum value. These conclusions have some significance for further studies on the ultrasonic vibratory single-point incremental composite forming technology.

1. Introduction

Single-point incremental forming (SPIF), which is a new type of sheet metal-plastic forming technology, solves the problems of the long manufacturing cycle, high cost, and low flexibility associated with traditional sheet metal stamping and has become the subject to scholars' wide attention [1]. However, in the process of SPIF, the existence of sheet springback, sheet bending, tool head deflection, and other factors leads to some defects that affect the surface quality [2], dimensional accuracy [3], and forming limit [4]. The defects are always accompanied by changes in the state of stress. Therefore, the magnitude and direction of the forming force on the contact area between the tool head and the sheet have important influences on the above defects. Therefore, domestic and foreign scholars have put forward some new composite technologies, such as two-point incremental forming [5], electric auxiliary single-point incremental forming [6], and laser-assisted single-point incremental forming [7], in order to avoid defects by changing the magnitude and action mode of the forming force.

The ultrasonic vibratory plastic forming technology is superior to the ordinary plastic forming technology in regard to both difficult-to-form materials and precision forming and is specifically expressed in two aspects: first, it can significantly reduce the yield stress and flow stress; second, it can improve the lubrication effect and change the friction mechanism. The single-point incremental forming process assisted by ultrasonic vibration is taken as the research object. Through theory and experiments, the stress of the sheet in the ultrasonic vibration single-point incremental forming (UV-SPIF) process is analyzed.

The technical principle is shown in Figure 1. The target is a frustum cone with an upward opening. The sheet metal to

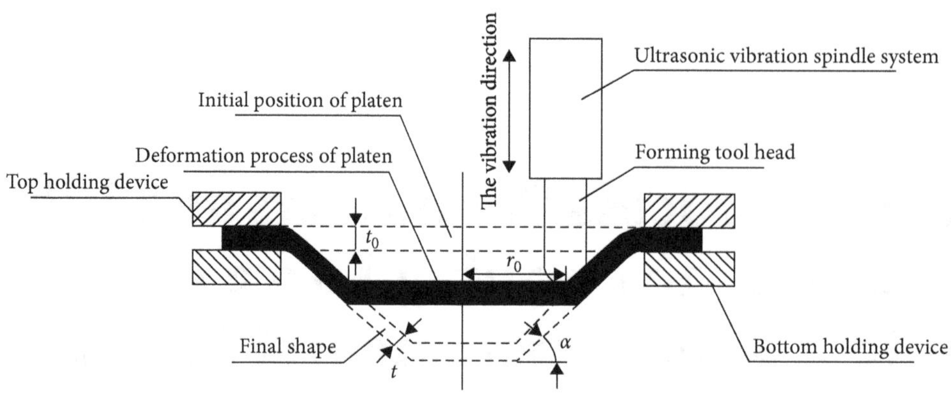

Figure 1: UV-SPIF technology schematic.

be formed is a suitable magnitude metal circular sheet, which is fixed on the workbench with the top and bottom holding devices. The initial position of the sheet is shown with the dashed line above, with an initial thickness of t_0. The tool head forms the sheet metal along the trajectory of the contour line layer by layer. The trajectory is generated discretely in the height direction based on the dimension of the frustum cone. The geometry of the frustum cone determines the magnitude of the forming angle α. The black area in the figure is the current deformation state of the sheet. At this time, the tool head with a radius of r_0 moves in a circular track. The final shape is shown with the lower dashed line in Figure 1.

When the tool head is driven by the ultrasonic vibratory spindle, the sheet can be formed along the trajectory by the tool head with axial ultrasonic vibration. The vibration direction is shown in Figure 1. In this case, the contact state of the tool head and the sheet is not continuous rolling contact under SPIF but high-frequency impact point pressure contact under UV-SPIF. In addition, compared to SPIF, the contact type, stress condition, lubrication state, and plastic flow law of the metal are changed.

Many advantages of ultrasonic vibration in the field of plastic processing are under consideration [8–11], and related scholars have also studied the UV-SPIF technology. Mundo et al. [12] applied high-frequency vibration to a metal sheet for SPIF. It was determined that the vibration improved the surface quality and reduced the distortion of the sheet material. Vahdati et al. [13] demonstrated the effects of the ultrasonic vibration on the reduction of the axial force, springback, and surface roughness of SPIF. Amini et al. [14] applied an axial ultrasonic vibration with a high frequency and low amplitude on the tool head to form a straight flute of 1050 aluminum sheet. It was verified that the ultrasonic vibration could reduce the forming force and improve the formability. Obikawa and Hayashi [15] conducted experiments of UV-SPIF with aluminum alloy, steel, and titanium sheets. It was determined that the ultrasonic vibration could significantly improve the geometric accuracy and forming limit of the parts. Li et al. [16] studied the deformation mechanism of sheets under UV-SPIF via a finite element simulation, obtaining the effect of the amplitude on the magnitude and distribution of the plastic strain.

For studying the variation law of the forming force in the UV-SPIF process, the forming force of SPIF was taken as the starting point of this research. Scholars have conducted significant research in this direction. Henrard [17] established the finite element model and constitutive equation of the forming force in SPIF on the basis of tensile and shear experiments. This approach successfully predicted the force under specific process parameters. Liu [18] explored the influences of the forming force on the forming limit and a sheet thickness of AA7075-O parts. Bagudanch [19] designed special equipment and online control strategies to study the effects of the forming force on the forming angle under different process parameters. Finally, it was confirmed that the forming force increased with an increase in the tool head dimension. Centeno et al. [20] analyzed the influences of the process parameters on the forming properties of the AISI 304 sheet metal via experiments. Additionally, the axial force in the forming process was analyzed. Ankush et al. [21] proposed a predictive model for the forming force in SPIF. The prediction results were compared with the experimental results. As a consequence, the correctness and validity of the predictive model were proven.

Combining the previous research on UV-SPIF by our team [22, 23] with the existing research on the theory of the forming force, an analytical model of the forming force of the UV-SPIF was constructed, and a corresponding simulation model and experimental force measurement platform were built. From the three perspectives of theory, simulation, and experiment, the mapping relationship between the process parameters and the vibration parameters and the forming force is revealed. The variation laws of the forming force under different process parameters and vibration parameters are obtained.

2. Methods

2.1. UV-SPIF Force Analysis. As a common method of analyzing physical problems, the microelement method adopts the concept of a "part to whole." This method can simplify complex physical deformation problems. To explore the force of the sheet in the UV-SPIF technology, the deformation area of the sheet metal was analyzed by means of the microelement method. On this basis, the mechanical equilibrium equations and the geometric equations were established. In addition, the mapping relationships between the forming force, the forming parameters, and the vibration parameters were calculated.

2.1.1. Mechanical Equilibrium Equations. As the contact area of the tool head on the sheet is hemispherical, the sheet contact area S is close to the spherical shallow pit when the tool head vibrates and impacts the plastic deformation of the sheet, as shown in Figure 2. F_r, F_θ, and F_φ are the vertical, circumferential, and radial forces on the S surface, respectively. v_1 and v_2 are the impact velocity of the tool head in the vibration direction and the feed speed under the control of the numerical control program, respectively. The two velocities through the principle analysis were obtained using the following equations:

$$v_1 = \dot{y} = \frac{d[A \cdot \sin(2\pi f t)]}{dt} = 2\pi f A \cdot \cos(2\pi f t), \qquad (1)$$
$$v_2 = v,$$

where y is the axial displacement of the tool head; A and f are the amplitude and frequency, respectively, which are constants; and v is the feed speed of the tool head.

In the sheet contact area S, the sheet microelement of the spherical coordinate system was taken as the analysis object, as shown in Figure 3. R is the tool head radius. $d\theta$ and $d\varphi$ are the angles of the opposite sides in the height direction of the microelement. σ_r, σ_θ, and σ_φ are the stresses of the microelements in the vertical, circumferential, and radial directions, respectively. In addition, t is the thickness of the microelement sheet. The spherical coordinate stress equilibrium equation was vertically established in the microelement, as shown in the following equation:

$$\begin{aligned}
&\left(\sigma_r + \frac{\partial \sigma_r}{\partial r} dr\right) \cdot (R+t)^2 \cdot d\theta \cdot d\varphi - \sigma_r \cdot (R \cdot d\theta) \cdot (R \cdot d\varphi) \\
&+ \sigma_\varphi \cdot \sin\frac{d\varphi}{2} \cdot \left(\pi \cdot t^2 \cdot \frac{d\theta}{2\pi} + 2\pi \cdot t \cdot R \cdot \frac{d\theta}{2\pi}\right) \\
&+ \left(\sigma_\varphi + \frac{\partial \sigma_\varphi}{\partial \varphi} d\varphi\right) \cdot \sin\frac{d\varphi}{2} \cdot \left(\pi \cdot t^2 \cdot \frac{d\theta}{2\pi} + 2\pi \cdot t \cdot R \cdot \frac{d\theta}{2\pi}\right) \\
&+ \sigma_\theta \cdot \sin\frac{d\theta}{2} \cdot \left(\pi \cdot t^2 \cdot \frac{d\varphi}{2\pi} + 2\pi \cdot t \cdot R \cdot \frac{d\varphi}{2\pi}\right) \\
&+ \left(\sigma_\theta + \frac{\partial \sigma_\theta}{\partial \theta} d\theta\right) \cdot \sin\frac{d\theta}{2} \cdot \left(\pi \cdot t^2 \cdot \frac{d\varphi}{2\pi} + 2\pi \cdot t \cdot R \cdot \frac{d\varphi}{2\pi}\right) = 0.
\end{aligned} \qquad (2)$$

The third-order minute term was eliminated, so the following equation could be obtained:

$$\sigma_r \cdot d\theta \cdot d\varphi + \sigma_\varphi \cdot \sin\frac{d\varphi}{2} \cdot d\theta + \sigma_\theta \cdot \sin\frac{d\theta}{2} \cdot d\varphi = 0. \qquad (3)$$

Based on the following two considerations, Equation (3) could be simplified. The first one was that the

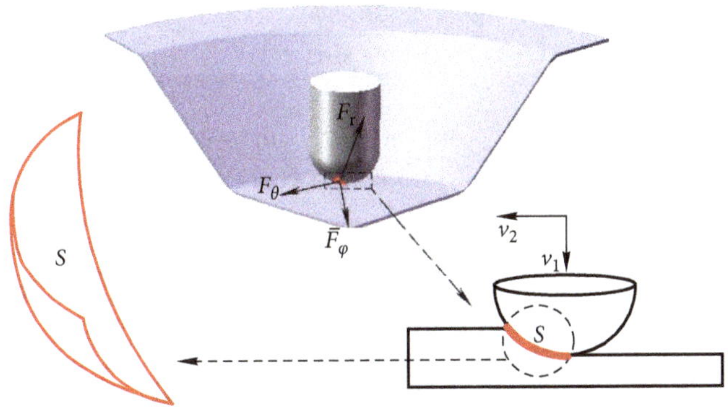

FIGURE 2: The contact area of the sheet and tool head.

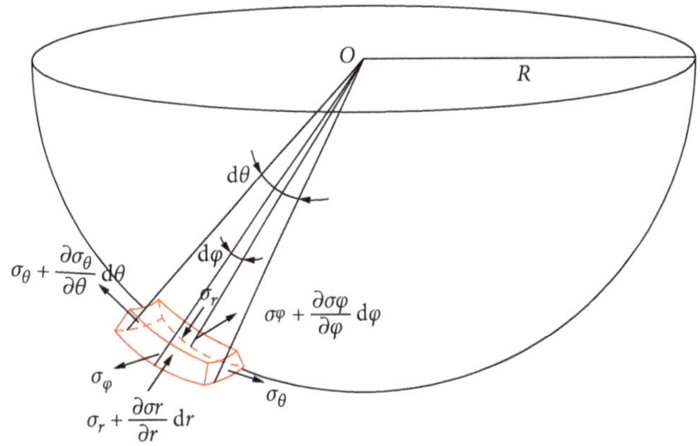

FIGURE 3: Stress components of the sheet microelement in spherical coordinates.

microelement was arbitrary on the contact surface. Therefore, the triaxial stresses in the contact surface satisfied the relationship of the above formula at every point. The second consideration was that the actual deformation of the sheet in the circumference was far less than the radial stretching and vertical thinning [4]. Therefore, the circumferential strain ε_θ is approximately zero, and the circumferential stress σ_θ is the average stress of the radial stress σ_φ and the vertical stress σ_r. σ_φ and σ_r are opposite to each other, and σ_θ is zero. Therefore, Equation (3) could be simplified to the following equation:

$$\sigma_r \cdot \varphi + \sigma_\varphi \cdot \sin\frac{\varphi}{2} = 0, \qquad (4)$$

where φ is the angle between the front and rear ends of the contact surface S.

According to the yield criterion of the Mises plane deformation, the following relation is known:

$$\sigma_r - \sigma_\varphi = \beta \cdot \sigma_s = \frac{2}{\sqrt{3}}\sigma_s, \qquad (5)$$

where σ_s is the yield limit of the sheet.

Equations (4) and (5) were solved simultaneously to obtain the values of σ_φ and σ_r, as shown in the following equations:

$$\begin{cases} \sigma_\varphi = \dfrac{2\sigma_s \cdot \varphi}{\sqrt{3}\,(\varphi + \sin(\varphi/2))}, \\ \sigma_r = \dfrac{4\sigma_s \cdot \varphi + 2\sigma_s \cdot \sin(\varphi/2)}{\sqrt{3}\,(\varphi + \sin(\varphi/2))}, \end{cases} \tag{6}$$

2.1.2. The Geometric Equation of the Contact Region. The geometry and contact area of the tool head are shown in Figure 4. It can be seen that the trajectory of the tool head is a sinusoid in one vibration cycle with a period of T and an amplitude of A. The forming force is P when the amplitude reaches its maximum. P is the combination of the vertical force, the circumferential force, and the radial force, as shown in Equation (7). According to Equation (8), it can be seen that the pressing time of the tool head is very short in one cycle when the vibratory frequency of 20 kHz is taken as an example. Therefore, P can be considered a constant force. Time 1 and time 2 are separated by one vibratory period, and their distance is ΔL. The relationship between the tool feed speed v and the vibratory frequency f is shown in Equation (9):

$$P = \sqrt{F_r^2 + F_\theta^2 + F_\varphi^2}, \tag{7}$$

$$t = \frac{T}{2} = \frac{1}{2f} = \frac{1}{2 \times 20000} = 2.5 \times 10^{-5}\,(\text{unit: s}), \tag{8}$$

$$\Delta L = CE = \frac{v}{f}. \tag{9}$$

According to the geometric relationship of the contact area in Figure 4, the length of OE could be calculated in Equation (10). Then, Equation (11) was derived by the relationship between OE and $\varphi_{\max}$:

$$OE = \sqrt{R^2 + \frac{v^2}{f^2} - \frac{2v}{f}\sqrt{(\Delta z + A) \cdot (2R - \Delta z - A)}}, \tag{10}$$

$$\varphi_{\max} = 2\arccos\frac{OE}{R} = 2\arccos\frac{\sqrt{R^2 + (v^2/f^2) - (2v/f)\sqrt{(\Delta z + A) \cdot (2R - \Delta z - A)}}}{R}. \tag{11}$$

Since the feed rate was in mm/min and the unit of frequency was hertz, the unit of nonuniformity made Equation (11) unable to be calculated. Therefore, it was processed to obtain

$$\varphi_{\max} = 2\arccos\frac{\sqrt{R^2 + (v^2/f^2) \cdot (1/3600) - (v/f) \cdot (1/30)\sqrt{(\Delta z + A) \cdot (2R - \Delta z - A)}}}{R}. \tag{12}$$

2.1.3. Forming Force Analytical Model. Since the constant interval of φ is $[0, \varphi_{\max}]$ and Equation (8) is a function of σ_φ and σ_r with respect to φ, after integration of the independent variables, two radial and vertical forming forces were obtained, as shown in the following equations:

$$F_\varphi = \int_0^{\varphi_{\max}} \frac{2\sigma_s \cdot \varphi}{\sqrt{3}(\varphi + \sin(\varphi/2))} d\varphi, \tag{13}$$

$$F_r = \int_0^{\varphi_{\max}} \frac{4\sigma_s \cdot \varphi + 2\sigma_s \cdot \sin(\varphi/2)}{\sqrt{3}(\varphi + \sin(\varphi/2))} d\varphi. \tag{14}$$

2.2. UV-SPIF Force Simulation. Based on the analytical model of the above deformation force, the finite element model of the UV-SPIF process was developed. The model is shown in Figure 5. The effects of the process parameters and vibration parameters on the forming force were explored by the simulation results.

According to the preconditions of the analytical model, the upper and lower platens and the tool head were set as analytically rigid, so they did not deform themselves. In this paper, AL1060 aluminum alloy, which is commonly used in automotive body construction, is selected as the research material, and its yield limit σ_s is 90.2 MPa. Since the thickness of the sheet element was consistent on all sides during the analysis process, it was set as a deformable shell element. The dimension is a square sheet with a side length of 140 mm and an initial thickness of 1 mm. The mesh was divided by an S4R four-node reduced integral shell element with a cell size of 1 mm × 1 mm. The outer radius of the clamp platen is 70 mm, and the inner ring has a radius of 55 mm. Material model selection, contact conditions, boundary conditions, meshing, and other specific settings are shown in Table 1. In addition, the Coulomb friction model was used to describe the friction relationship between the contact pairs. The friction coefficient between the tool head and the sheet was set to 0.1. The coefficients between the sheet and the upper and lower platens were set to 0.2. The translation and rotation of the upper and lower platens in all directions were constrained, and the sheet was clamped by the upper and lower platens.

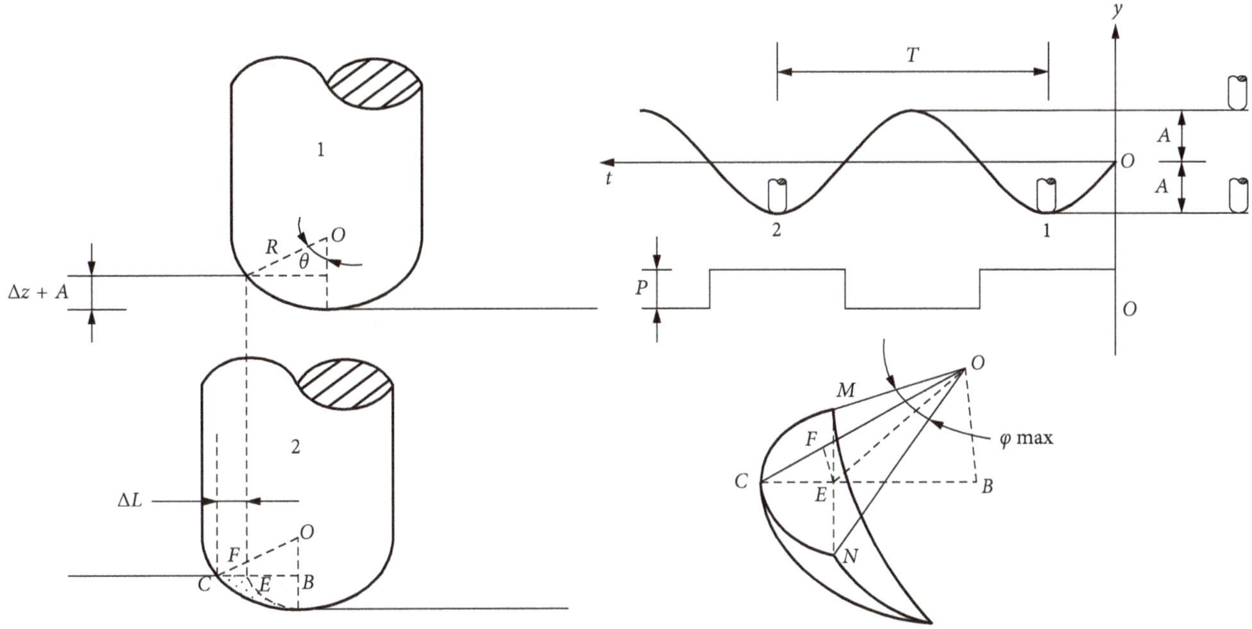

FIGURE 4: Geometric diagram of the tool head movement.

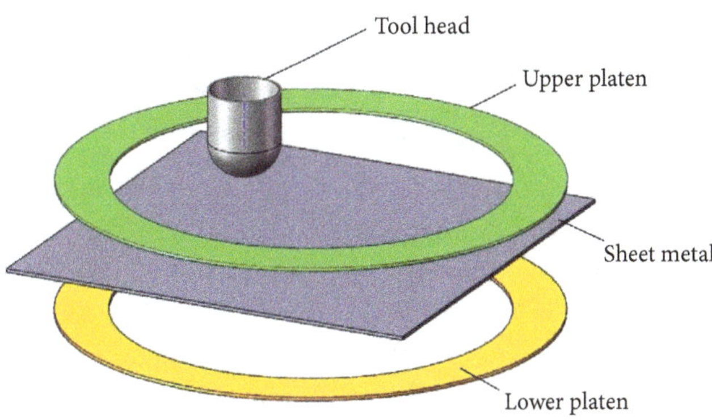

FIGURE 5: US-SPIF finite element simulation model.

TABLE 1: Simulation model settings.

Meshing	Workpiece	Grid type: S4R Number of units: 19600
	Tool head	Analytically rigid
Material model selection	Workpiece	Model: plastic model of the shell metal
	Tool head	Material: 1060 aluminum sheet (elastic modulus is 70 GPa; Poisson's ratio is 0.3; density is 2710 kg/m^3)
Contact conditions		Contact type of tool head to sheet: surface-to-surface
Boundary conditions	Workpiece	6 degrees of freedom on the surface were constrained
	Tool head	Rotation: rotate anticlockwise around the sheet

The amplitude was used as a parameter for the boundary conditions and load constraints. By defining it, it is possible not only to define the movement of the tool head in the horizontal direction but also to apply vertical ultrasonic vibration to the tool head during the UV-SPIF process. The movement of the tool head on each layer is a circular trajectory. This motion can be achieved by a periodic amplitude of the Fourier series. The Fourier series is shown as follows:

$$a = A_0 + \sum_{n=1}^{N} \left[A_n \cos n\omega (t - t_0) + B_n \sin n\omega (t - t_0)\right], \tag{15}$$

where A_0 is the initial amplitude, N is the number of Fourier series terms, ω is the circular frequency (rad/s), and t_0 is the initial time. Through the Fourier series, the displacement of the tool head in the x and y directions of the horizontal plane can be realized by the periodic amplitude, as shown in the following equation:

$$x = A\cos 2\pi t,$$
$$y = A\sin 2\pi t, \quad (16)$$

where A is the radius of each circular layer and t is the forming time of each layer. The trajectories in the x and y directions are combined, that is, the circular trajectory when the tool head forms each layer.

Due to the introduction of ultrasonic vibration, the tool head not only performs a circular motion in the plane but also performs short displacements and high-frequency impact motion in the vertical direction. This motion was constrained by the periodic amplitude so that the displacement of the tool head in the z direction is as follows:

$$z = a\sin[2\pi f(t)], \quad (17)$$

where a is the ultrasonic vibration amplitude and f is the ultrasonic vibration frequency.

2.3. UV-SPIF Force Experiment. UV-SPIF experiments were performed on an Hass high-speed NC milling machine. An ultrasonic vibration spindle and sheet clamping system were designed and developed. A Kistler forming force measurement system was established. Through a large number of experiments, the variation in radial and vertical forces under different processes and vibration conditions was obtained.

2.3.1. UV-SPIF Experiment. The Hass CNC milling machine was used as the UV-SPIF experimental platform, as shown in Figure 6. The upper platen and the lower platen are used to hold the sheet metal. The lower platen is connected with the dynamometer, as shown in Figure 7. Kistler 9257B was used as a force sensor with three directions. The force collected by the sensor was passed to the charge amplifier. Then, the signal was sent to the data acquisition instrument by the charge amplifier. Finally, the force signal was read by the VIBSYS software on the terminal computer.

AL1060 aluminum alloy, which is commonly used in lightweight automobiles, was selected as the experimental material. The material is synthesized by adding a trace amount of copper element to pure aluminum. Therefore, the material has excellent plastic forming properties, good electrical conductivity, better welding property, and strong corrosion resistance [24]. The main chemical composition of the AL1060 aluminum alloy is shown in Table 2. The tool head was grounded from a tungsten steel rod. It has high hardness, rigidity, and wear resistance to ensure that it is not easily deformed during the forming process. A sheet metal was formed by the tool head along a predetermined trajectory under the command of a numerical control program.

To quantitatively study the influence of the frequency and amplitude on the forming force, a stepless frequency modulation ultrasonic generator was selected as the vibration generator. The vibratory frequency of the generator is continuously adjustable from 13 kHz to 75 kHz, and the adjustment is accurate to 0.1 kHz. The generator was

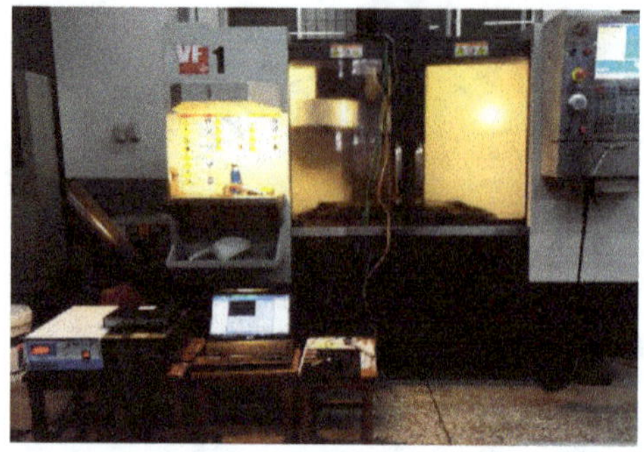

FIGURE 6: Hass CNC milling machine.

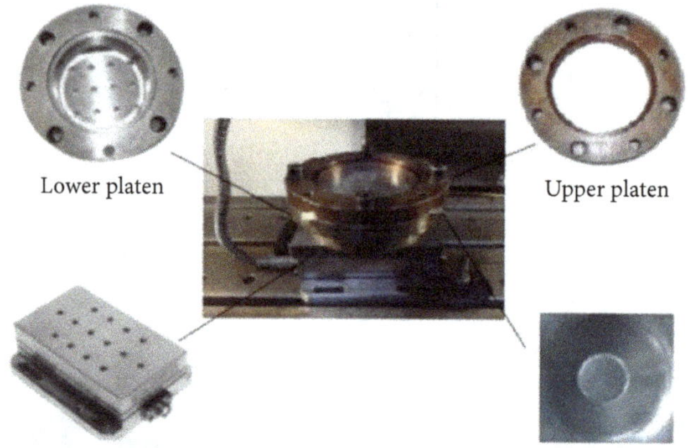

FIGURE 7: Sheet holding system.

TABLE 2: The main chemical composition of the AL1060 aluminum alloy (%, mass fraction).

Chemical element	Al	Cu	Ti	Zn	Si	Mg	Mn	Fe
Content (%)	99.63	0.03	0.02	0.03	0.12	0.02	0.01	0.14

connected to the vibration spindle system via a conductor. The spindle system was composed of a transducer, amplitude transformer, and spindle shell. The transducer and amplitude transformer were mounted inside the spindle shell as internal components. One end of the vibratory spindle was connected to the machine tool holder, and the other end was connected to the tool head. The connection between the vibratory spindle and the tool holder was realized through a coupling, which is a stepped sleeve. The upper small circle and the tool holder were connected coaxially through the connecting thread. The lower large circle was connected through fixing screws and locknuts to ensure the coaxiality of the connection between the tool holder and the vibratory spindle in four directions, as shown in Figure 8. The vibratory spindle and the tool head were connected by a fine thread with an outer diameter of 10 mm and a pitch of 1 mm. Since the forming process was accompanied by a certain vibration, a locknut was also installed between the tool head and the spindle. After assembling the vibratory system and the force-measuring system, a corresponding numerical control program was established according to the shape and magnitude of the target. The UV-SPIF experiments were

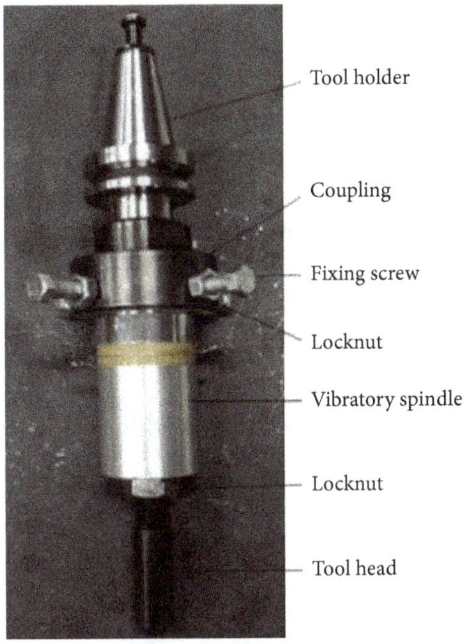

FIGURE 8: Vibratory spindle system.

performed on the AL1060 aluminum alloy sheet. The experimental process is shown in Figure 9.

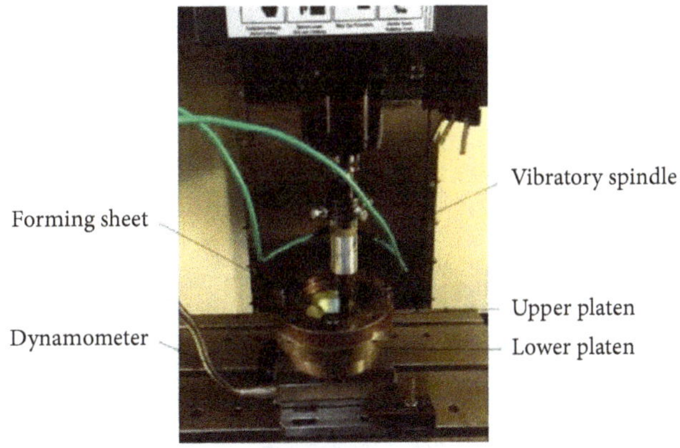

FIGURE 9: Forming process.

3. Results and Discussion

3.1. Volume and Surface Effects of Ultrasonic Vibration. Based on the SPIF simulation process, an ultrasonic vibration (f = 20 kHz and A = 0.02 mm) was applied to the tool head to obtain a UV-SPIF stress nephogram, as shown in Figure 10(b). The SPIF and UV-SPIF experiments were performed under the same process parameters. The SPIF and UV-SPIF parts were obtained, as shown in Figure 11. According to the simulated numerical values of the nephogram and experimental components, it can be seen that the surface quality of the workpiece was improved by the ultrasonic vibration. At the same time, the forming stress was effectively reduced. This result confirms the volume effect and surface effect of ultrasonic vibration on the sheet metal forming process.

The experimental results of UV-SPIF (f = 20 kHz and A = 0.02 mm) and SPIF were compared and analyzed (the forming force was the axial force F_z, which was taken as the analysis object), as shown in Figure 12. A frustum cone was used as the target component, with a total of 20 layers being formed. The average axial force of each layer was taken to generate its law curve about the number of forming layers. It can be seen that the axial force of UV-SPIF and SPIF increased with an increase in the number of layers and reached stability on the 10th lap. It can also be seen that the ultrasonic vibration did not affect the variation law of the axial force with respect to the number of layers. After the analysis and calculation, the axial force of UV-SPIF was reduced by 23.61% compared with that of SPIF.

3.2. Analysis of Theoretical, Simulation, and Experimental Results. The analytical modeling coordinate system was based on a deformation ramp S. The different direction from the direction of the coordinate system was established by the experiment and simulation, which was based on the initial sheet plane. Therefore, the three results could not be compared directly. It was necessary to uniformly process the directions of the forming force in advance.

3.2.1. Processing the Coordinate System Directions. The original sheet plane was used as the x-y plane and the perpendicular direction as the z-axis in the experimental and simulated coordinate systems. However, the analytical modeling coordinate system was based on the spatial geometry of the contact area between the tool head and the sheet, as shown in Figure 2. Although there were differences in the directions of the two coordinates, there was a certain relationship in terms of geometric relations, as shown in Figure 13.

When the tool head forms the sheet via UV-SPIF technology with the feed rate F, the forces in the analytical modeling coordinate system are as follows: radial force F_φ, circumferential force F_θ, and vertical force F_r. The forces in the experimental force measurement coordinate system are as follows: F_x, F_y, and F_z. The microelement on the contact area S of the sheet with the tool head (Figure 2) was used as the analysis target (the left side of Figure 13). In the analytical modeling process, the sheet material was considered to be only affected by the radial force F_φ in the x-y plane. In addition, F_φ is the resultant force of F_x and F_y, as shown in Equation (18). The tool head was used as the study object (the right side of Figure 13). The OH is perpendicular to the CD, and the foot point is H. Equation (19) can be obtained from the geometry of the tool head. Therefore, the relationship between the vertical force F_r and the axial force F_z could be calculated by Equation (20). In addition, the analytical model was an ideal model, which did not consider friction. Therefore, the friction F_t downward from the contact area S was not considered.

$$F_\varphi = \sqrt{F_x^2 + F_y^2}, \tag{18}$$

$$\frac{F_r}{F_z} = \frac{OH}{OD} = \frac{CB}{CD} = \sqrt{1 - \frac{\Delta z + A}{2R}}, \tag{19}$$

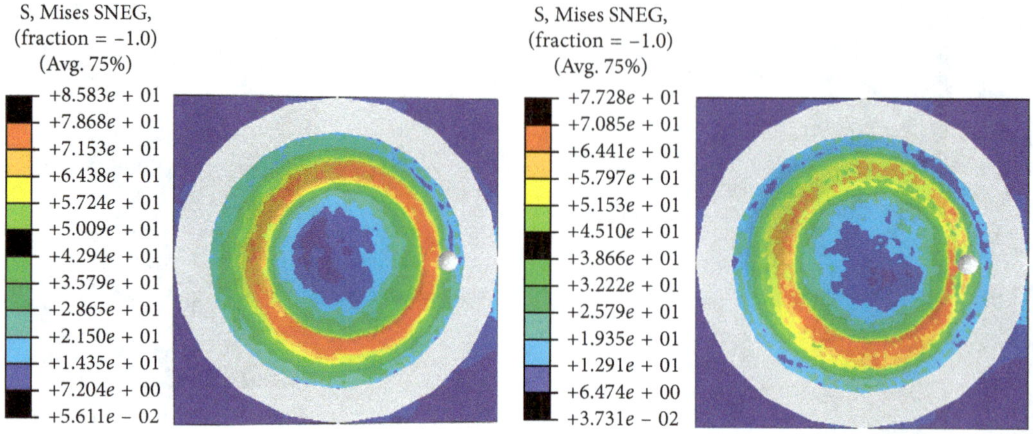

(a) (b)

FIGURE 10: Nephogram of SPIF (a) and UV-SPIF (b).

(a) (b)

FIGURE 11: Workpieces of SPIF (a) and UV-SPIF (b).

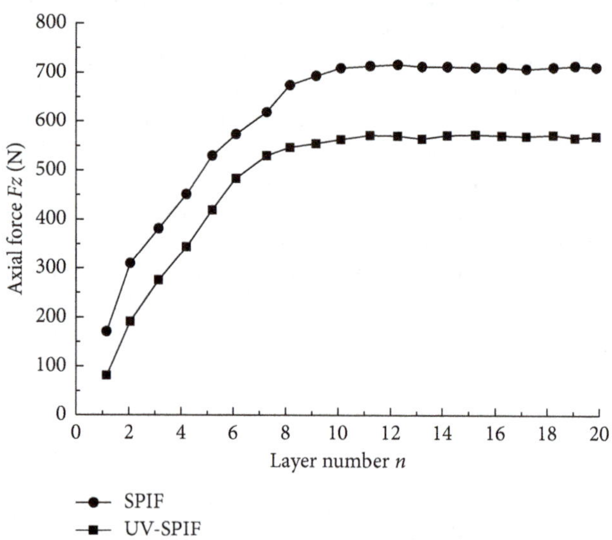

FIGURE 12: Comparison of axial force in SPIF and UV-SPIF

$$F_r = F_z \cdot \sqrt{1 - \frac{\Delta z + A}{2R}}, \tag{20}$$

3.2.2. Comparative Analysis of the Three Results. Adopting theory, simulation, and experiment as analysis methods, the relationships between the process parameters, vibration parameters, and forming force were studied using the control variable method. As shown in Figures 14–18, the influences of the amplitude, frequency, radius of the tool head, layer spacing, and feed speed on the forming force are shown in order. It can be seen that the theoretical and simulation results followed the experimental results, and their performance trends were consistent. Among the results, simulation F_r < theory F_r < experiment F_r, so it could be considered that the prediction ability of the analytical model was stronger than the simulation results. Due to the partial idealization of the two models, the resulting values of the theory and simulation were lower than the experimental

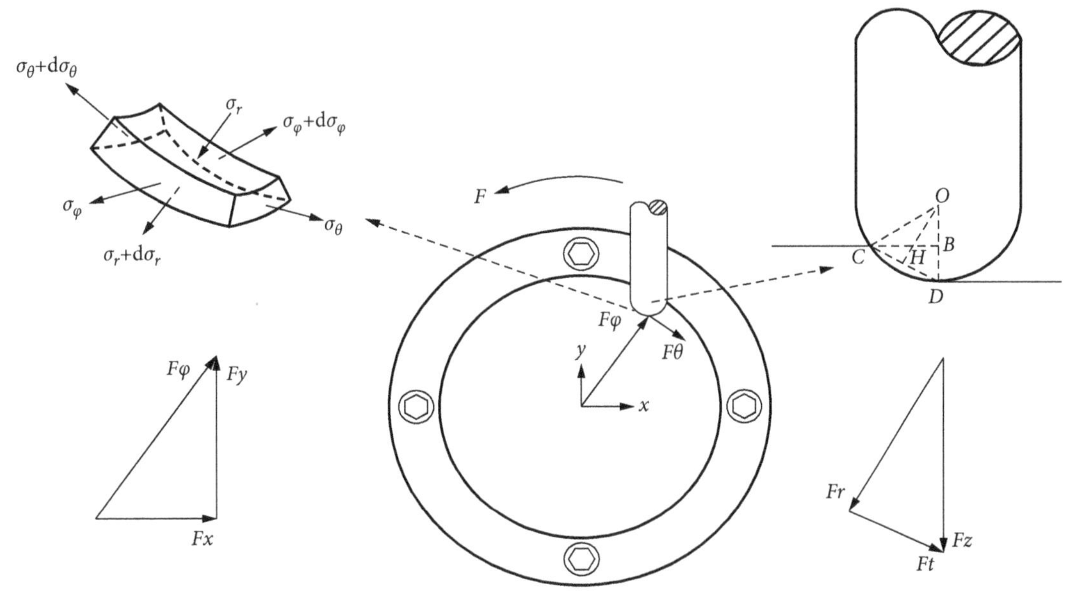

FIGURE 13: Relationship of the forming forces in different coordinate systems.

values. The theoretical F_φ and simulation F_φ values were also lower than the experimental F_φ values for the same reason as mentioned above. The circumferential force was ignored during the derivation of the theoretical analytical model; therefore, the theoretical F_φ value was smaller than the simulation F_φ value.

In Figure 14, $f = 20$ kHz, $\Delta z = 1.0$ mm, $R = 5$ mm, and $v = 100$ mm/min were constant, and the amplitude was variable. The variation laws of the forming forces with respect to amplitude were obtained in theory, simulation, and experiment. F_r is nonlinearly related to A, and F_φ is linearly related to A. When $A < 0.03$ mm, the sheet was not sufficient to be fully yielded by the tool head. When the tool head was separated from the sheet, the elastic energy was released. The instantaneous springback maintained the contact between the tool head and the sheet, making F_r increase. When A was 0.03 mm, the instantaneous impact force of the tool head exceeded the yield limit of 1060AL. Therefore, the springback was suppressed, and the moment of separation F_r was zero. When $A > 0.03$ mm, the press distance of the tool head was increased to within a vibration cycle. The deformation energy increased with the deformation of the material, which led to F_r increase. As the amplitude increased, the radial deformation increased and F_φ also increased.

In Figure 15, $A = 0.01$ mm, $\Delta z = 1.0$ mm, $R = 5$ mm, and $v = 100$ mm/min were constant, and the frequency varied. The variation laws of the forming forces with respect to frequency were obtained in the three aspects of theory, simulation, and experiment. F_r is nonlinearly related to f, and F_φ is linearly related to f. When $f < 30$ kHz, the hit rate of the tool head increased with an increase in f. The activity of the sheet material was increased due to the vibration of the particles in the interior sheet. The sheet temperature increased, which led to softening of the material. Plastic reinforcement made the sheet easily deform, and it made F_r decrease. When $f > 30$ kHz, excessive frequency increased the instantaneous deformation of the sheet, making the corresponding F_r increase. The increase in frequency led to a reduction in the radial deformation. Therefore, F_φ decreased.

In Figure 16, $A = 0.01$ mm, $f = 20$ kHz, $\Delta z = 1.0$ mm, and $v = 100$ mm/min were constant, and the tool head radius was variable. The variation laws of the forming forces about the radius of the tool head were obtained in terms of theory, simulation, and experiment. Both F_r and F_φ are linearly related to R because an increase in the radius of the tool head directly affects the deformation of the sheet material. However, when R was too small, a stress concentration tended to occur, so the parts were prone to cracks. When R was too large, it was not conducive to forming complex parts. Therefore, the different magnitudes of the tool heads should be selected for parts with different complexity. Within the scope of satisfying the forming ability, the tool head with a larger radius should be selected.

In Figure 17, $A = 0.01$ mm, $f = 20$ kHz, $R = 5$ mm, and $v = 100$ mm/min were constant, and the layer spacing in the forming process was variable. The variation laws of the forming force about the layer spacing were obtained in theory, simulation, and experiment. Both F_r and F_φ are linearly related to Δz. The degree of correlation is greater than the effect of R on the force, which is because a greater sheet deformation was caused by an increase in Δz. However, if Δz was too small, the forming efficiency was low. When Δz was too large, the sheet was prone to bending and springback. Therefore, Δz should be increased under the requirements of satisfying the accuracy. Δz should also be reduced under the formation of complex features.

In Figure 18, $A = 0.01$ mm, $f = 20$ kHz, $R = 5$ mm, and $\Delta z = 1.0$ mm were constant, and the feed speed of tool head forming was variable. The variation laws of the forming force with respect to the feed speed were obtained in the three aspects of theory, simulation, and experiment. Both F_r and F_φ are linearly related to v. The slope of the curve is the lowest.

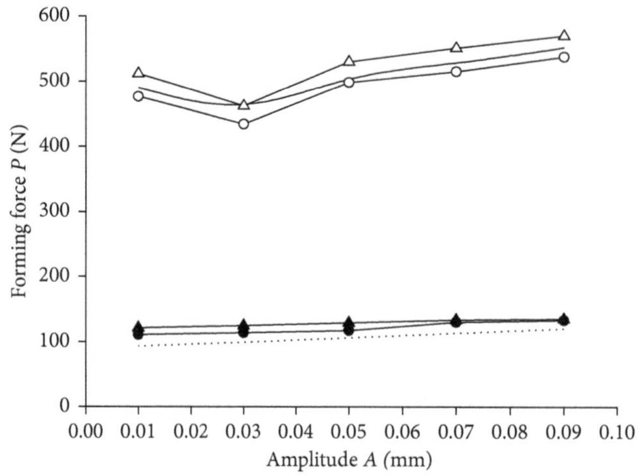

FIGURE 14: Effect of amplitude on the forming force.

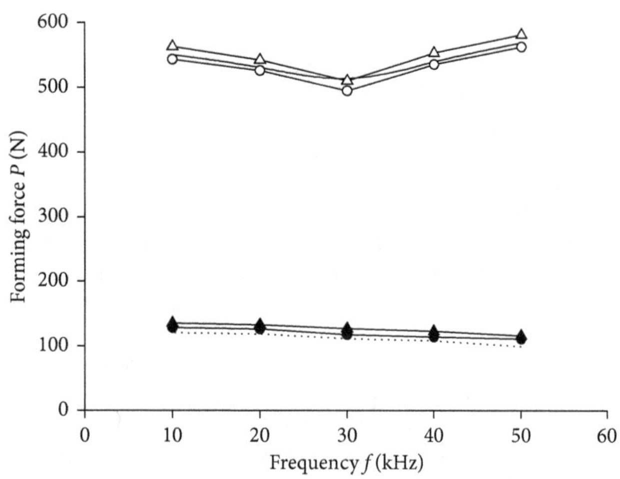

FIGURE 15: Effect of frequency on the forming force.

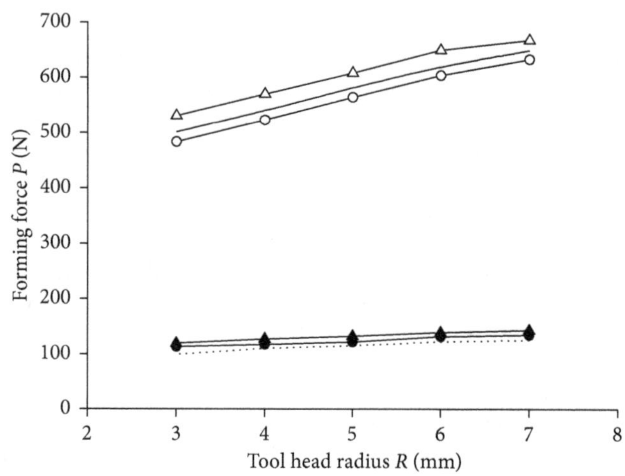

FIGURE 16: Effect of the tool head radius on the forming force.

Because the instantaneous deformation time of the sheet was only 2.5×10^{-5} seconds, the change in the feed speed had minimal effect on the instantaneous deformation of the sheet. For improving the forming efficiency, v should be increased as much as possible. However, when v was too large, the metal in the interior sheet could not be recovered or recrystallized in time. At this time, the plasticity would become lower, and stress concentrates and cracks could occur.

4. Conclusion

The forming principle and movement rule of the UV-SPIF technology were analyzed. The stress conditions at the sheet deformation area were clarified and interpreted analytically. Furthermore, a mathematical analytical model adapting to the UV-SPIF technology was deduced. A corresponding simulation model was established to simulate the force of the sheet in the UV-SPIF process. The experimental platform and force measurement system were set up to study the influences of various process parameters and vibration parameters on the forming force of UV-SPIF. By comparing and analyzing the results of the three aspects, the following conclusions are drawn:

(1) The axial force on the sheet material is reduced by 23.61% after the axial ultrasonic vibration ($f = 20$ kHz and $A = 0.02$ mm), confirming the volume effect of the ultrasonic vibration in metal-plastic forming. In addition, the changes in the forming force cause the macroscopic springback and the microscopic lattice of

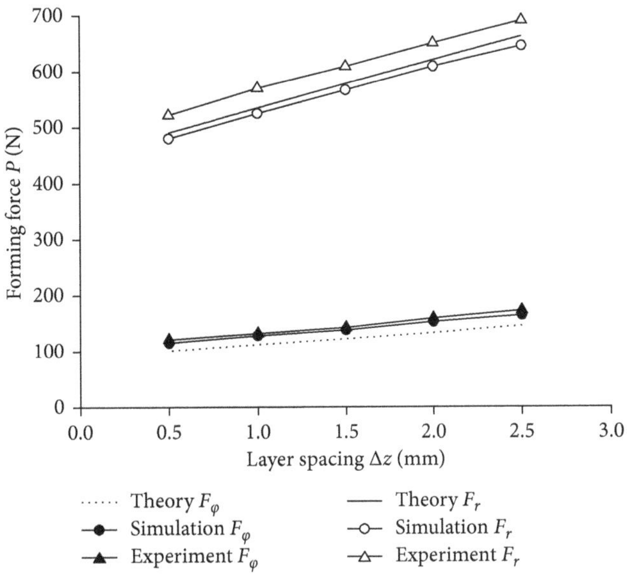

FIGURE 17: Effect of layer spacing on the forming force.

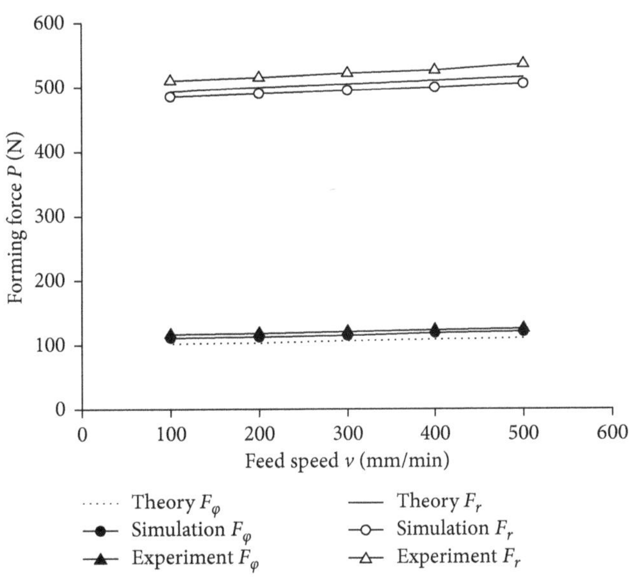

FIGURE 18: Effect of feed speed on the forming force.

the sheet to change accordingly, which has a certain influence on the surface quality, dimensional accuracy, and forming limit of the final piece.

(2) It is determined that the influence of A and f on F_φ is linear, and the influences on F_r are nonlinear. A and f can reduce the forming force by changing the deformation energy and microstructure of the sheet, and this changing relationship is nonlinear. Therefore, there is a range of frequencies and amplitudes that can minimize the forming force.

(3) When the vibration parameters are constant, the influences of R, Δz, and F on F_φ and F_r are linear. Compared with previous studies [25], the ultrasonic vibration has minimal interference with the mapping relationship between the process parameters and the forming forces. Therefore, previous research conclusions about the effects of the process parameters on the forming force can be applied to the study of UV-SPIF.

Conflicts of Interest

The authors declare that they have no conflicts of interest.

Acknowledgments

The authors would like to acknowledge the National Natural Science Foundation of China (51475366 and 51475146), the Science and Technology Planning Project of Shaanxi Province (Grant No. 2016JM5074), and the Key Laboratory of Scientific Research Projects of Shaan'xi Educational Committee (Grant No. 12JS072).

References

[1] A. K. Behera, R. A. de Sousa, G. Ingarao, and V. Oleksik, "Single point incremental forming: an assessment of the progress and technology trends from 2005 to 2015," *Journal of Manufacturing Processes*, vol. 27, pp. 37–62, 2017.

[2] S. Kurra, N. H. Rahman, S. P. Regalla, and A. K. Gupta, "Modeling and optimization of surface roughness in single point incremental forming process," *Journal of Materials Research and Technology*, vol. 4, no. 3, pp. 304–313, 2015.

[3] Y. Li, H. Lu, W. J. T. Daniel, and P. A. Meehan, "Investigation and optimization of deformation energy and geometric accuracy in the incremental sheet forming process using response surface methodology," *International Journal of Advanced Manufacturing Technology*, vol. 79, no. 9–12, pp. 2041–2055, 2015.

[4] C. Raju, N. Haloi, and C. Sathiya Narayanan, "Strain distribution and failure mode in single point incremental," *Journal of Manufacturing Processes*, vol. 30, pp. 328–335, 2017.

[5] M. A. Davarpanah, Z. Zhang, S. Bansal, J. Cao, and R. Malhotra, "Preliminary investigations on double sided incremental forming of thermoplastics," *Manufacturing Letters*, vol. 8, pp. 21–26, 2016.

[6] R. Liu, B. Lu, D. Xu et al., "Development of novel tools for electricity-assisted incremental sheet forming of titanium alloy," *The International Journal of Advanced Manufacturing Technology*, vol. 85, no. 5–8, pp. 1137–1144, 2016.

[7] A. Mohammadi, H. Vanhove, A. Van Bael, and J. R. Duflou, "Towards accuracy improvement in single point incremental forming of shallow parts formed under laser assisted conditions," *International Journal of Material Forming*, vol. 9, no. 3, pp. 339–351, 2016.

[8] Z. Xie, Y. Guan, J. Zhai, L. Zhu, and C. Zhong, "Study on ultrasonic vibration assisted upsetting of 6063 aluminum alloy," *Procedia Engineering*, vol. 207, pp. 490–495, 2017.

[9] S. A. Sajjady, H. Nouri Hossein Abadi, S. Amini, and R. Nosouhi, "Analytical and experimental study of topography of surface texture in ultrasonic vibration assisted turning," *Materials and Design*, vol. 93, pp. 311–323, 2016.

[10] M. M. Molaie, J. Akbari, and M. R. Movahhedy, "Ultrasonic assisted grinding process with minimum quantity lubrication

using oil-based nanofluids," *Journal of Cleaner Production*, vol. 129, pp. 212–222, 2016.
[11] B. Yang and M. Yang, "The influence of superimposed ultrasonic vibration on surface asperities deformation," *Journal of Materials Processing Technology*, vol. 229, pp. 367–374, 2016.
[12] D. Mundo, G. Gatti, G. Ambrogio et al., "Considerations on process performance in incremental forming by inducing high frequency vibration," in *Proceedings of EUCOMES 08*, pp. 523–530, Dordrecht Netherlands, January 2009.
[13] M. Vahdati, R. Mahdavinejad, and S. Amini, "Investigation of the ultrasonic vibration effect in incremental sheet metal forming process," *Proceedings of the Institution of Mechanical Engineers, Part B: Journal of Engineering Manufacture*, vol. 231, no. 6, pp. 971–982, 2015.
[14] S. Amini, A. H. Gollo, and H. Paktinat, "An investigation of conventional and ultrasonic-assisted incremental forming of annealed AA1050 sheet," *International Journal of Advanced Manufacturing Technology*, vol. 90, no. 5, pp. 1569–1578, 2017.
[15] T. Obikawa and M. Hayashi, "Ultrasonic-assisted incremental microforming of thin shell pyramids of metallic foil," *Micromachines*, vol. 8, no. 5, p. 142, 2017.
[16] Y. Li, X. Chen, J. Sun, J. Li, and G. Zhao, "Effects of ultrasonic vibration on deformation mechanism of incremental point-forming process," *Procedia Engineering*, vol. 207, pp. 777–782, 2017.
[17] C. Henrard, C. Bouffioux, P. Eyckens et al., "Forming forces in single point incremental forming: prediction by finite element simulations, validation and sensitivity," *Computational Mechanics*, vol. 47, no. 5, pp. 573–590, 2010.
[18] Z. Liu, Y. Li, and P. A. Meehan, "Experimental investigation of mechanical properties, formability and force measurement for AA7075-O aluminum alloy sheets formed by incremental forming," *International Journal of Precision Engineering and Manufacturing*, vol. 14, no. 11, pp. 1891–1899, 2013.
[19] I. Bagudanch, G. Centeno, C. Vallellano, and M. L. Garcia-Romeu, "Forming force in single point incremental forming under different bending conditions," *Procedia Engineering*, vol. 63, pp. 354–360, 2013.
[20] G. Centeno, I. Bagudanch, A. J. Martínez-Donaire, M. L. García-Romeu, and C. Vallellano, "Critical analysis of necking and fracture limit strains and forming forces in single-point incremental forming," *Materials and Design*, vol. 63, pp. 20–29, 2014.
[21] A. Bansal, R. Lingam, S. K. Yadav, and N. Venkata Reddy, "Prediction of forming forces in single point incremental forming," *Journal of Manufacturing Processes*, vol. 28, pp. 486–493, 2017.
[22] P. Li, J. He, Q. Liu et al., "Evaluation of forming forces in ultrasonic incremental sheet metal forming," *Aerospace Science and Technology*, vol. 63, pp. 132–139, 2017.
[23] Y. Li, L. Bai, M. Yang et al., "Research on forming force in single point incremental forming with ultrasonic vibration," *Mechanical Science and Technology for Aerospace Engineering*, vol. 37, no. 2, pp. 270–275, 2018.
[24] Z. J. Wang, M. Ma, Z. X. Qiu et al., "Microstructure, texture and mechanical properties of AA 1060 aluminum alloy processed by cryogenic accumulative roll bonding," *Materials Characterization*, vol. 139, pp. 269–278, 2018.
[25] H. Xu, P. Yi, W. An et al., "Simulation and experimental study of sheet metal single point incremental forming force," *Chinese Journal of Applied Mechanics*, vol. 32, no. 6, pp. 1062–1068, 2015.

8

Assessment of Real-Time Compaction Quality Test Indexes for Rockfill Material Based on Roller Vibratory Acceleration Analysis

Tianbo Hua[1,2] **Xingguo Yang**[1,2] **Qiang Yao**[1,2] **and Hongtao Li**[1,2]

[1]*State Key Laboratory of Hydraulics and Mountain River Engineering, Sichuan University, Chengdu, Sichuan 610065, China*
[2]*College of Water Resource and Hydropower, Sichuan University, Chengdu, Sichuan 610065, China*

Correspondence should be addressed to Hongtao Li; htl@scu.edu.cn

Academic Editor: Antonio Gilson Barbosa de Lima

Compaction quality is directly related to the structure and seepage stability of a rockfill dam. To timely and accurately test the compaction quality of the rockfill material, four real-time test indexes were chosen to characterize the soil compaction degree based on the analysis of roller vibratory acceleration, including acceleration peak value (a_p), acceleration root mean square value (a_{rms}), crest factor value (CF), and compaction meter value (CMV). To determine which of these indexes is the most appropriate, a two-part field compaction experiment was conducted using a vibratory roller in different filling zones of the dam body. Data on rolling parameters, real-time test indexes, and compaction quality indexes were collected to perform statistical regression analyses. Combined with the spectrum analysis of the acceleration signal, it was found that the CF index best characterizes the compaction degree of the rockfill material among the four indexes. Furthermore, the quantitative relations between the real-time index and compaction quality index were established to determine the control criterion of CF, which can instruct the site work of compaction quality control in the rockfill rolling process.

1. Introduction

The layered filling compaction of dam materials is an important process in the construction of a rockfill dam. An effective compaction quality control is the key to ensure the safe and stable operation of the dam. Nowadays, the compaction quality control in rockfill dam projects mainly depends on the manual control of rolling parameters and random spot tests after construction, which is normally called the "dual control method" [1]. With the expansion of the filling scale, this traditional compaction quality control method has been unable to meet the requirements of modern mechanized construction. Therefore, it has been quite necessary to develop a fast, real-time, and accurate compaction quality test method with continuous, automatic, and high precision characteristics.

To provide a quality-inspection process that is more reliable and timely, previous studies in the field of transportation engineering focused on correlating compaction quality to soil properties and construction-operating parameters that can be obtained easily and quickly from the field. Geodynamik and Dynapac of Sweden characterized compaction quality by compaction meter value (CMV), which is the ratio between the amplitude of the first harmonic and that of the fundamental frequency [2, 3]. Previous studies verified that CMV is closely related to the fundamental reaction and physical properties of the compacted soil [4–7]. According to Forssblad [8–11], CMV has been noted to range from 40 to 70 for gravel, from 25 to 40 for sand, and from 20 to 30 for silt.

The compaction control value (CCV) adopted by Sakai company is similar to CMV, which is also determined from the measured acceleration data but based on more harmonic frequency components, including 0.5, 1, 1.5, 2, 2.5, and 3 fundamental frequency and harmonic components [12, 13]. Caterpillar used machine drive power (MDP) to determine the compaction characteristics in road work [14, 15], which originated from the study of vehicle-terrain interaction [16]. Researchers found that MDP is highly correlated with compaction density and elastic modulus of the soil [17–19].

Bomag from Germany used a software algorithm which derived from a mechanical model to calculate the dynamic modulus E_{vib} of the material to describe the compacted condition of soil [20]. Mooney and Rinehart [21] and Rinehart and Mooney [22] proposed that the total harmonic distortion (THD) is a highly sensitivity index to evaluate the soil compaction state. The larger the THD, the stiffer the soil will be. Ammann calculated the soil stiffness, K_B, as a measure of compaction quality [23] and verified the strong correlation between K_B and soil rigidity [24, 25]. China Southwest Jiaotong University used vibratory compaction value (VCV), a dynamic subgrade structural reaction index, to evaluate the soil compaction quality and achieved good results in a roadbed continuous compaction quality test [26].

However, the abovementioned researches are mainly carried out on the subgrade filling material. In fact, there are great differences between the rockfill in earth-rock dams and the subgrade filler in road engineering in terms of the soil particle size and shape. At present, the maximum particle size of the rockfill in hydropower projects has reached 1 m, and many particles are solid block stones with sharp and jagged edges. In contrast, the maximum particle size of the subgrade filler is generally less than often 150 mm, and the main contents are spherical particles with few edges [13]. This makes the compaction characteristics of the two materials quite different. Therefore, further study is needed to know whether the index that applied well on subgrade filler can also achieve excellent test results on the rockfill material of earth-rock dams.

The objective of this study is to obtain an index that can characterize the compaction degree of the rockfill material timely and accurately. The rockfill material tested in the experiment belongs to cohesionless coarse-grained soil; therefore, the compaction degree in the study actually refers to the physical indicator of relative density. The following section first analyzes the relationship between the roller vibratory acceleration and the soil compaction degree. Then, combined with the abovementioned study, four real-time test indexes derived from acceleration signal were chosen to characterize the rockfill compaction degree. To analyze and judge the detection effect of each index on the rockfill material, comparison experiments were conducted in different filling zones of the dam body with different rolling parameters. Statistical regression analysis of test data and spectrum analysis of acceleration signal were then performed to get the final conclusions.

2. Methodology

This section first introduces the terrain-vehicle dynamical model in the existing literature and analyzes the relationship between roller vibratory acceleration and the compacted condition of soil. Next, the four real-time test indexes utilized in the experimental test are introduced, with their calculating methods and physical significances elaborated. Finally, the testing method and equipment are presented.

2.1. Vibration Model. According to the study of terrain-vehicle systems [16] and the research on the vibratory roller from Yoo and Selig [27–29], it can be known that the acceleration of the vibratory roller is closely related to the compaction condition of the pressed material. The terrain-vehicle vibration model [29] can be described as in Figure 1.

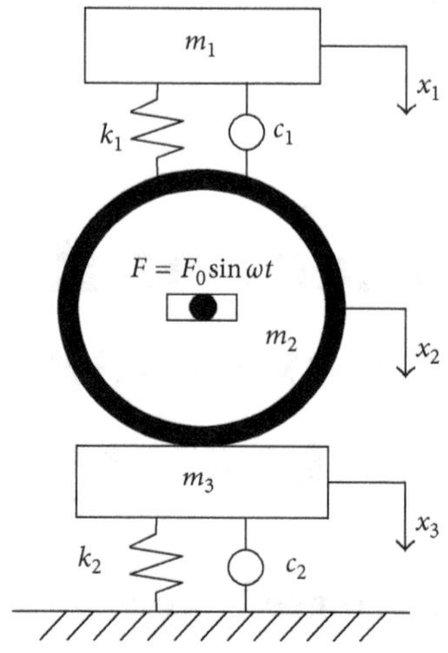

FIGURE 1: Terrain-vehicle dynamical model. m_1: frame mass, m_2: roller mass, m_3: soil mass, k_1: roller-frame stiffness, c_1: roller-frame damping, k_2: soil stiffness, c_2: soil damping, x_1: vertical displacement of frame, x_2: vertical displacement of roller, x_3: vertical displacement of soil, F: excitation force, $\ddot{x}$: acceleration.

To facilitate the calculation and analysis, the model is simplified by not taking the horizontal displacement and jump vibration of the vibratory roller into account. In such case, m_2 and m_3 always remain in contact, that is, $x_2 = x_3$. The force equations between the layers are expressed as

$$\begin{aligned} m_1\ddot{x}_1 + c_1(\dot{x}_1 - \dot{x}_2) + k_1(x_1 - x_2) &= 0, \\ (m_2 + m_3)\ddot{x}_2 + (c_1 + c_2)\dot{x}_2 + (k_1 + k_2)x_2 & \\ - c_1\dot{x}_1 - k_1 x_1 &= F_0 \sin \omega t. \end{aligned} \tag{1}$$

The equations can be expressed into matrix form as

$$\begin{aligned} &\begin{bmatrix} m_1 & 0 \\ 0 & m_2 + m_3 \end{bmatrix}\begin{bmatrix} \ddot{x}_1 \\ \ddot{x}_2 \end{bmatrix} + \begin{bmatrix} c_1 & -c_2 \\ -c_1 & c_1 + c_2 \end{bmatrix}\begin{bmatrix} \dot{x}_1 \\ \dot{x}_2 \end{bmatrix} \\ &\quad + \begin{bmatrix} k_1 & -k_2 \\ -k_1 & k_1 + k_2 \end{bmatrix}\begin{bmatrix} x_1 \\ x_2 \end{bmatrix} = \begin{bmatrix} 0 \\ F_0 \sin \omega t \end{bmatrix}. \end{aligned} \tag{2}$$

By analytic calculation, we can obtain

$$|\ddot{x}_2| = \omega^2 |x_2| = \omega^2 F_0 \left(\frac{A_1^2 + B_1^2}{C^2 + D^2}\right)^{1/2} = f(k_2, c_2), \tag{3}$$

where $A_1 = k_1 - m_1\omega^2$, $B_1 = c_1\omega$, $C = (m_2 + m_3)m_1\omega^4 - (m_2 + m_3)k_1\omega^2 - m_1k_2\omega^2 - c_1c_2\omega^2 + k_1k_2 - m_1k_1\omega^2$, and $D = k_2c_1\omega + k_1c_2\omega - (m_2 + m_3)c_1\omega^3 - m_1c_2\omega^3 - m_1c_1\omega^3$.

It can be seen from (3) that the vertical vibratory acceleration of the roller ($\ddot{x}_2$) is related only to the stiffness k_2 and damping c_2 of the pressed soil. With the stiffness and damping of the soil directly reflecting the compactness of the soil, the roller vibratory acceleration (hereinafter referred to as a) and the soil compaction degree are closely related. However, due to the complexity of (3), the relationship between k_2, c_2, and a cannot be expressed with a definite function formula. The purpose of the experiment is to gain a dataset of roller vibratory acceleration and soil compactness, then finding out the definite relationship between them.

2.2. Real-Time Test Indexes. The study of Zhong et al. [30, 31] and Liu et al. [32, 33] indicates that describing the soil compactness by directly using vibratory acceleration index a causes much uncertainty and volatility. Therefore, the following four derived indexes are selected by analyzing the roller acceleration signal.

2.2.1. Acceleration Peak Value (a_p). This index describes the amplitude variation of roller acceleration signal and is calculated as

$$a_p = \max\{|a_i|\} \quad (i = 1, 2, \ldots, n), \tag{4}$$

where a_i is a random acceleration measurement and n is the number of collected samples within a certain period.

2.2.2. Acceleration Root Mean Square Value (a_{rms}). This index reflects the effective vibratory acceleration in the rolling process, which is calculated as

$$a_{\mathrm{rms}} = \sqrt{\frac{1}{n}\sum_{i=1}^{n} a_i^2} = \sqrt{\frac{a_1^2 + a_2^2 + \cdots + a_n^2}{n}}. \tag{5}$$

2.2.3. Crest Factor Value (CF). This index is a common evaluation index in alternating current, which is used to describe the ability of the AC power to output peak load current [34]. The CF value of a standard sinusoidal wave is 1.414. When the wave is distorted, the CF changes significantly. It is used by the authors to describe the waveform variation law of the roller vibratory acceleration signal and is defined as

$$\mathrm{CF} = \frac{a_p}{a_{\mathrm{rms}}}. \tag{6}$$

The formula shows that CF is a dimensionless acceleration index that integrates the variation trend of a_p and a_{rms}. The physical meaning of the index in this experiment is to describe the ability of the roller to output peak acceleration, namely, the ability of the soil to produce the greatest reaction force to the roller.

2.2.4. Compaction Meter Value (CMV). The index shows the ratio between the amplitude of the first harmonic and that of the fundamental frequency, which is obtained by the tuning analysis of the acceleration signal. The definition of CMV is expressed as

$$\mathrm{CMV} = C \cdot \frac{A_1}{A_0}, \tag{7}$$

where C is the amplification coefficient and usually sets as 300, A_1 refers to the amplitude of the first harmonic of vibratory acceleration, and A_0 refers to the amplitude of the fundamental frequency of vibratory acceleration.

2.3. Testing Method. The experiment adopts the DHDAS dynamic signal acquisition system of Donghua (Figure 2), which mainly includes the acceleration sensor, the vibration data acquisition instrument, and the monitoring and analysis software for clients. The main physical quantity tested in the experiment is the vertical vibratory acceleration of the roller and its main vibration frequency. The roller used is the 32 t single-drum vibratory roller of Jointark mechanical YZ series (Figure 3). The sensors were installed vertically on the inner frame of the roller, and the data were collected simultaneously with the dual channel.

The a_p index is directly read by DHDAS software, and a_{rms} index is calculated by (5) with the tested acceleration data in a certain period. The CF is worked out by (6) with the measured a_p and a_{rms}. CMV is calculated by (7) with the amplitude of the first harmonic and that of the fundamental frequency in the spectrum of acceleration signal, which is obtained through fast Fourier transformation (FFT) analysis.

3. Experimental Testing

3.1. Testing Site and Materials. The experiment was carried out at the construction site of Chang-he Dam hydropower station, which located in Dadu River, China. The Chang-he Dam is a 240-meter-high core rockfill dam, and the structural design of it is shown in Figure 4. The experimental material includes the main rockfill, secondary rockfill, and filter material. Figure 5 shows the field experiment on the main rockfill, and the grain size distribution of the material is shown in Figure 6(a). The main compaction parameters of the test material are shown in Table 1.

In Table 1, the parameter "rolling times" means the compaction pass of the vibratory roller on each strip. The index relative density is defined as

$$D_r = \frac{e_{\max} - e}{e_{\max} - e_{\min}} = \frac{(\rho_d - \rho_{d\min})\rho_{d\max}}{(\rho_{d\max} - \rho_{d\min})\rho_d}, \tag{8}$$

where $e_{\max}$ is the maximum void ratio, $e_{\min}$ is the minimum void ratio, e is the compacted void ratio, $\rho_{d\max}$ refers to the maximum dry density, $\rho_{d\min}$ refers to the minimum dry density, and ρ_d refers to the compacted dry density.

The index ρ_d is calculated by measuring the volume and dry weight of the soil sample collected from the field. The index $\rho_{d\max}$ and $\rho_{d\min}$ need to be measured in the laboratory individually. As the maximum particle size that can be tested by the device in the lab is 60 mm, the maximum grain size of the rockfill in the project has reached 800 mm. The oversize particles were replaced by the soil whose particle size is smaller than 60 mm and larger than 5 mm with equal mass, and at the ratio calculated by (9), which is called the equivalent weight replacement method [35]. The grading curve of the main rockfill after replacement is shown in Figure 6(b).

$$P_i = \frac{P_{\mathrm{oi}}}{P_5 - P_{d\max}} P_5, \tag{9}$$

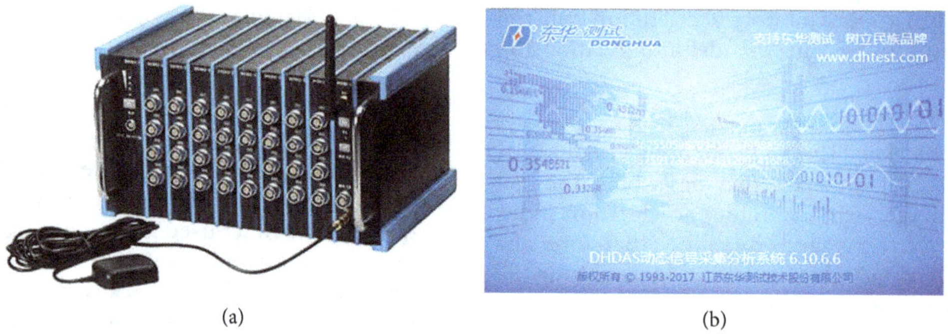

FIGURE 2: DHDAS dynamic signal acquisition and analysis system. (a) Data acquisition instrument. (b) Data analysis system.

FIGURE 3: The vibratory roller and setting of sensors. (a) Vibratory roller. (b) Setting of sensors.

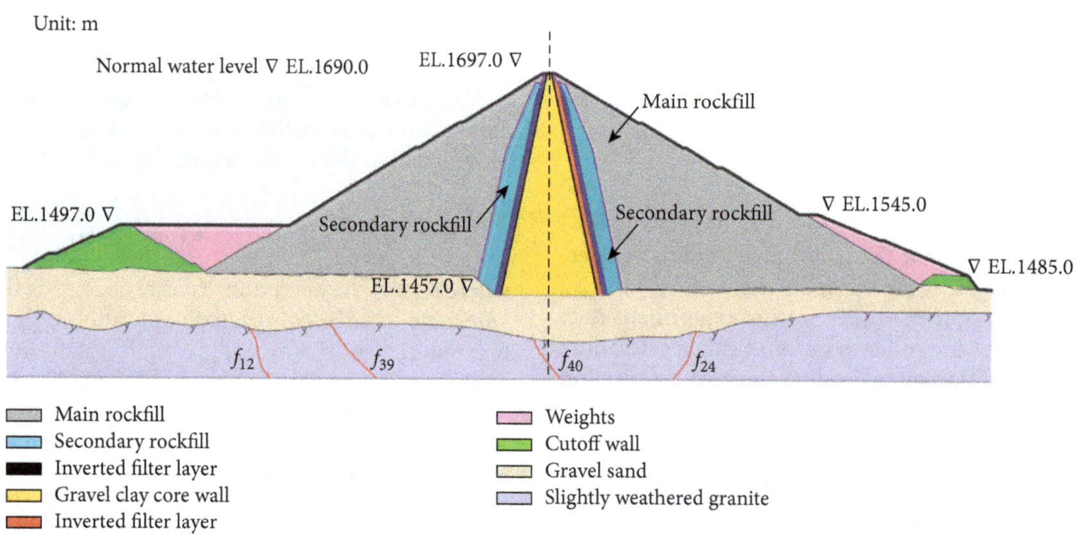

FIGURE 4: Structural diagram of Chang-he Dam.

where P_i refers to the weight percentage of some particle size group in the soil after replacement, P_{oi} refers to the weight percentage of some particle size group in the original soil, P_5 refers to the weight percentage of the soil whose particle size is larger than 5 mm, and $P_{d\max}$ refers to the weight percentage of the oversize particles in the original soil.

After the replacement work to the soil sample, $\rho_{d\min}$ and $P_{d\max}$ can be calculated out with the volume and weight of the dried soil in the sample cylinder under the state of fluffy and the state after 8 min vibration on the platform vibrator respectively, which is called the loose filling test method and the platform vibrator experiment method [35], respectively.

(a) (b)

Figure 5: Field experiment on the main rockfill. (a) Main rockfill. (b) Field experiment.

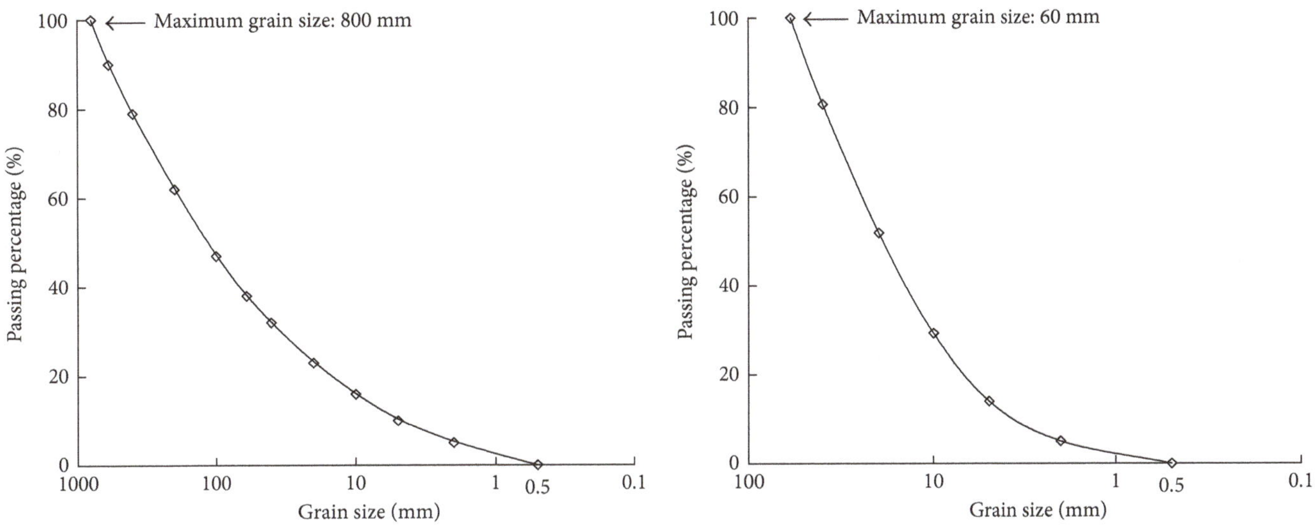

(a) (b)

Figure 6: Grain size distribution of the main rockfill. (a) The original. (b) After replacement.

Table 1: Compaction parameters of the test material.

Testing material	Standard rolling parameters			Requirements of compaction quality indexes		
	Rolling times	Layer thickness (cm)	Driving speed (km/h)	Relative density	Dry density (g/cm^3)	Porosity
Main rockfill	Static roll (2), vibratory roll (8)	100	2.7 ± 0.2	—	≥2.22	≤21%
Secondary rockfill	Static roll (2), vibratory roll (8)	50	2.7 ± 0.2	≥0.90	≥2.33	≤20%
Filter material	Static roll (2), vibratory roll (8)	30	2.7 ± 0.2	≥0.85	≥2.08	—

3.2. Testing Program. To evaluate the test effects of indexes a_p, a_{rms}, CF, and CMV on rockfill and other dam materials comprehensively, a two-part field experiment was performed. Part 1 was used to evaluate the relationship between the four real-time test indexes and the rolling parameters. Part 2 was targeted at studying the correlation between real-time test indexes and compaction quality indexes. The results of experimental testing will provide a dataset for the subsequent modeling analysis in Section 4.

3.2.1. Experiment Part 1. Experiment part 1 was carried out with the rolling parameters of rolling times, running speed, and compaction thickness changed. Since the roller keeps in a vibration state of low frequency and high amplitude in the rolling process, the parameter of roller excitation force remains unchanged in the experiment.

For the cohesionless soil of the rockfill in the study, a high compacted dry density can be achieved under the condition of completely dry or fully saturated in the Proctor

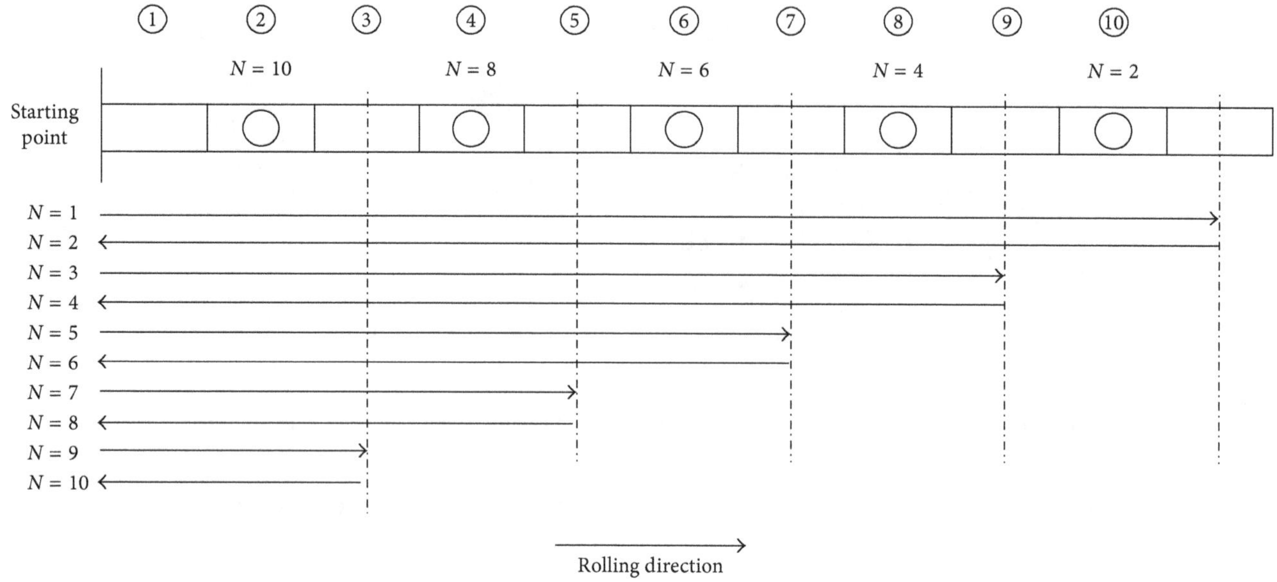

FIGURE 7: Rolling route of the vibratory roller (*N*: rolling times).

compaction test. However, due to the high hydraulic conductivity of the rockfill material, it is very hard for the material to keep high water content in the field compaction. And it is easy to raise dust if the dam material is completely dry, which is adverse to the construction environmental protection. Based on the comprehensive consideration of field compaction effect and construction cost and conditions, the water content of the rockfill material was kept at 1%–2% (weight ratio) in the field. The control method is to calculate and add water to the dam material automatically by using the intelligent water adding system developed by Sinohydro Bureau 5 Co., Ltd. before paving and supplement water to the paved soil with the sprinkling truck to maintain the water content before compaction.

Combined with the analysis above, the specific test plan can be made as follows by not taking the parameters of excitation force and water content into account.

(1) Change in rolling times: one rolling strip was selected in the filling zone of each test material with a length of 60 m and width of 2.2 m (width of the roller), respectively. As the dam material in the rolling strip cannot be completely even in the paving process, the data measured in different regions of the same strip change greatly. Therefore, the test strips are divided into small test areas with the grid of 5 m long and 2.2 m wide (Figure 5(b)), and the data on each small test area are analyzed separately. The standard vibratory rolling times of the test material is eight times (Table 1). In order to increase the sample volume and observe the variation trend of acceleration signal when the compaction pass exceeds standard, the number of vibratory rolling times is set at ten times.

(2) Change in running speed and compaction thickness: three adjacent strips were selected in the main and secondary rockfill zones, respectively, to analyze the change of each index brought by different running speeds, where the vibratory roller proceeded with low speed ($v = 1.8$ km/h), medium speed ($v = 2.2$ km/h), and high speed ($v = 2.6$ km/h). The same goes to the number of test strips in the experiment of change in compaction thickness. The thickness of the test strip on main rockfill was set to be 80 cm, 100 cm, and 120 cm, respectively, and that of the secondary rockfill to be 30 cm, 60 cm, and 90 cm, respectively.

3.2.2. Experiment Part 2. To conduct the correlation test of real-time test index and compaction quality index, experiment part 2 selected four adjacent strips with a length of 50 m and width of 2.2 m in the main and secondary rockfill zones, respectively. Each test strip was divided into ten small test areas with the grid of 5 m long and 2.2 m wide and numbered ①–⑩, as shown in Figure 7. The area with odd number is the buffer area in the rolling process, and we take the test data on the even-numbered area for analysis. When the required rolling process (Figure 7) for each strip is finished, the digging test method is used to obtain the conventional compaction quality indexes in the even-numbered areas.

4. Results and Discussion

This section focuses on analyzing the correlations among rolling parameters, real-time test indexes, and compaction quality indexes according to the dataset from the two-part experiment described above. In the analysis of data correlation, the linear model and hyperbolic model are compared in a bid to determine the correlation degree of the data and establish a more appropriate correlation formula. In addition, the spectrum analysis and discussion is conducted to make an in-depth analysis on the frequency distribution regularity of acceleration signal and compare the test effects of the four indexes on different test materials.

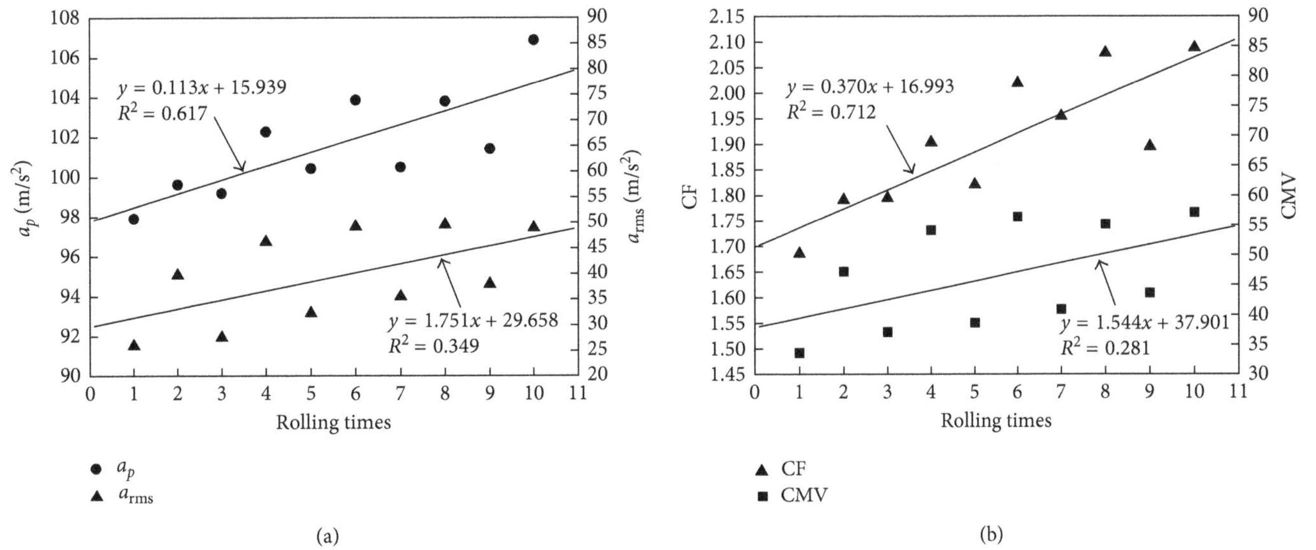

FIGURE 8: Linear relationships between real-time test indexes and rolling times. (a) a_p, a_{rms}. (b) CF, CMV.

4.1. Result Analysis of Experiment Part 1

4.1.1. Correlations between Rolling Times and the Four Indexes. A total of 360 groups of valid data were obtained in the experiment of change in rolling times. The scatter diagram of the indexes measured on each small test area was plotted with the Origin software so as to analyze the correlation between the real-time test indexes and rolling times.

The linear model is commonly used for the correlation analysis in the study field [30–33], and the function expression of which is

$$y = a + bx, \tag{10}$$

where a and b are the regression coefficients.

Firstly, the linear model was used in the regression analysis, and the result of a certain area on the main rockfill is shown in Figure 8.

From the scatter diagram, the following two characteristics are identified:

(1) The variable y increases with the rise of x, but its growth rate gradually slows down.

(2) When variable x increases further, y gradually comes to a constant, which indicates that the curve may have a horizontal asymptote.

Thus, it can be assumed that the trend line of the scatter plot is a hyperbolic curve and the function formula of which can be expressed as

$$y = \frac{x}{ax + b}, \tag{11}$$

where a and b are the regression coefficients.

Then, the hyperbolic model was used to analyze the scatter plot described above, and the result is shown in Figure 9.

The determination coefficients (R^2) of the regression models on the small test areas of each strip are evaluated using the trimmed mean to reduce the effect of an accidental error. The results are shown in Table 2. From this table, it can be seen that there are some correlations between all the four test indexes and rolling times on each tested dam material. Specifically, the linear correlation between CF and rolling times is the strongest, with all the determination coefficients (R^2) being over 0.7.

When considering the practical application of a test index, both its correlation degree and stability of the data in different test areas need to be taken into account. Therefore, the standard deviation coefficients (V_σ) of the 120 groups of data measured on each testing material were calculated to analyze the discrete degree of the dataset. It can be seen from the results (Table 3) that with the increase of the particle size in the filter material, the secondary rockfill and main rockfill, the V_σ of each test index also gradually increase, while the V_σ of a_p, a_{rms}, and CF stay at a low level, indicating that the data stability of a_p, a_{rms}, and CF is better.

4.1.2. Correlations between Running Speed, Compaction Thickness, and the Four Indexes. In the experiment of change in running speed and compaction thickness, 24 sets of valid data were obtained on the main and secondary rockfill, respectively. The changes of real-time indexes brought by the two parameters on secondary rockfill are shown in Figures 10 and 11, respectively.

It can be inferred from Figures 11 and 12 that the a_p and CF show a significant downward trend with the increase of running speed and compaction thickness, respectively, and the two indexes increase regularly with the growth of rolling times. Though a_{rms} and CMV rise along with the number of rolling times in Figure 11(b), they do not show the tendency in Figure 12(b). And there is no obvious regularity in the tendencies of a_{rms} and CMV when the roller running speed and soil compaction thickness change. The results of the test on the main rockfill are the same as

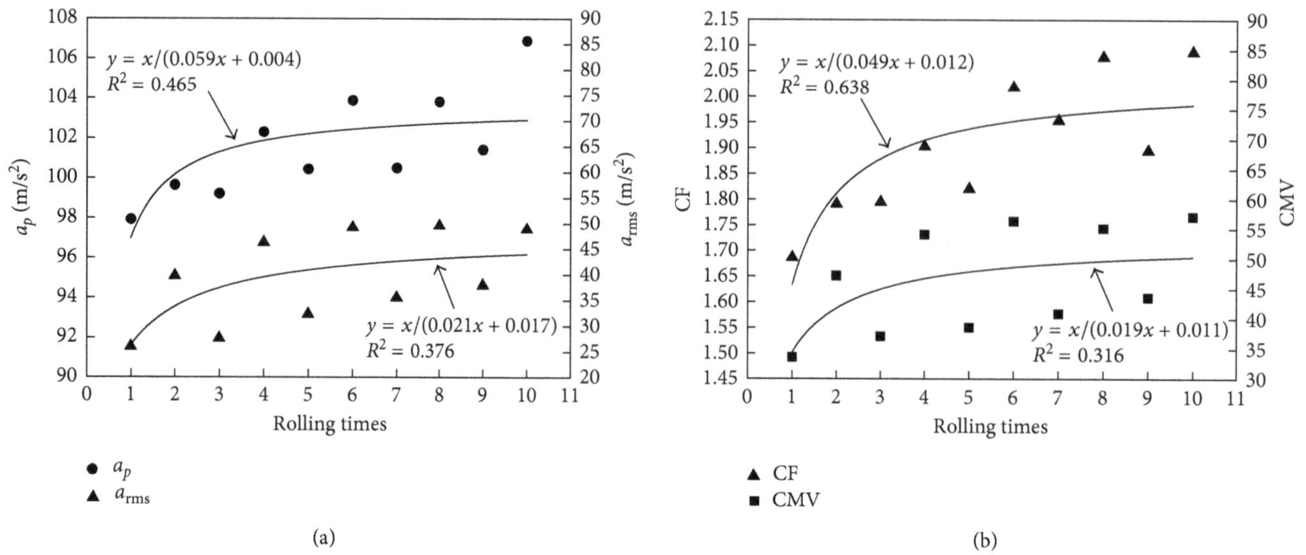

FIGURE 9: Hyperbolic relationships between real-time test indexes and rolling times. (a) a_p, a_{rms}. (b) CF, CMV.

TABLE 2: Determination coefficients (R^2) for regression analysis between real-time test indexes and rolling times.

Testing material	a_p		a_{rms}		CF		CMV	
	Linear model	Hyperbolic model	Linear model	Hyperbolic model	Linear model	Hyperbolic model	Linear model	Hyperbolic model
Main rockfill	0.601	0.628	0.584	0.486	0.731	0.679	0.496	0.498
Secondary rockfill	0.663	0.687	0.595	0.523	0.784	0.721	0.535	0.579
Filter material	0.672	0.704	0.630	0.572	0.796	0.752	0.670	0.604

TABLE 3: Standard deviation coefficients (V_σ) of real-time test indexes (%).

Testing material	a_p	a_{rms}	CF	CMV
Main rockfill	5.61	4.35	2.44	14.27
Secondary rockfill	2.90	2.88	1.27	13.28
Filter material	2.15	1.59	1.03	10.03

those on the secondary rockfill, which will not be repeated hereby.

For energy-related analysis, when the roller running speed and soil compaction thickness increase, the energy absorbed by the soil in unit and volume will decrease, respectively. With the same rolling times, the soil compaction degree is reduced. The same goes to the reaction force of the soil to the vibratory roller, which results in a decrease of the roller vibratory acceleration. Therefore, all the four derived indexes should show a declining trend, and only the tendency of a_p and CF among them are consistent with the theoretical situation.

4.2. Result Analysis of Experiment Part 2. Experiment part 2 obtained twenty groups of real-time test index and compaction quality index data on the main and secondary rockfill, respectively. For the main and secondary rockfill in the project, the main compaction quality control indexes are porosity (P) and relative density (D_r), respectively [36]. The correlation analysis of real-time test indexes and the two compaction quality indexes is conducted in this section.

The linear regression model and hyperbolic regression model were utilized to analyze the scatter diagram plotted by the Origin software. The results of linear regression analysis on the main rockfill are shown in Figure 12, and the results on the secondary rockfill are shown in Figure 13. The determination coefficients (R^2) of the regression models are shown in Table 4.

As can be seen from Table 4, in the correlation analysis of the four test indexes and porosity, the hyperbolic model of CF and the index shows the highest R^2 of 0.823. As for relative density, the linear model of CF and the index registers the highest R^2 of 0.820. This means that the correlation between CF and the compaction quality indexes is the strongest among the four test indexes, and the regression functions can be expressed as

$$\text{CF} = \frac{P}{0.802P - 6.430}, \tag{12}$$

$$\text{CF} = 0.284D_r + 1.725. \tag{13}$$

The compaction quality control criteria on the main and secondary rockfill are $P \le 21\%$ and $D_r \ge 0.90$, respectively. The criterion of $\text{CF} \ge 2.017$ can be got by taking $P \le 21\%$ into (12), and $\text{CF} \ge 1.981$ can be obtained by plugging $D_r \ge 0.90$

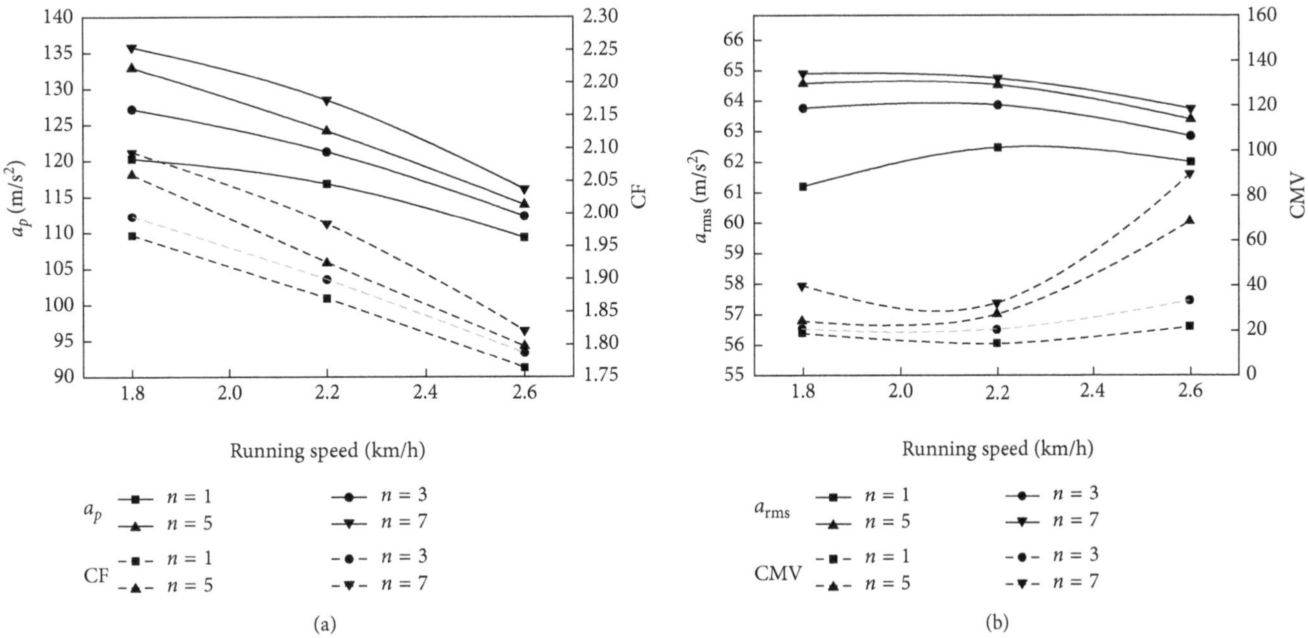

FIGURE 10: Tendency of real-time indexes with the change of running speed. (a) a_p, CF. (b) a_{rms}, CMV.

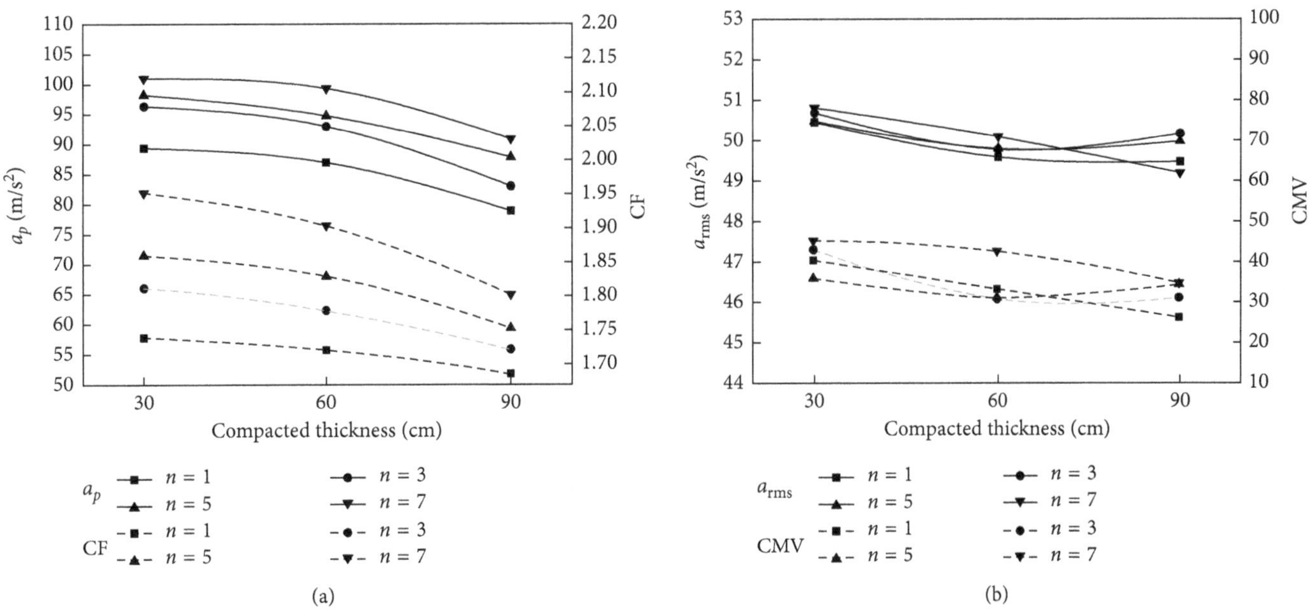

FIGURE 11: Tendency of real-time indexes with the change of compaction thickness. (a) a_p, CF. (b) a_{rms}, CMV.

into (13). It is thus tentatively believed that the compacted soil achieves the required compaction quality when the CF values reach 2.017 and 1.981 on the main and secondary rockfills, respectively. The more precise control criteria of CF values need to be determined based on the quantitative relations among the soil compaction degree, the reaction force of the soil (characterized by the acceleration indexes discussed in the paper), and the strength and stiffness of the rockfill material, which will be studied in the future work.

4.3. Spectrum Analysis and Discussion. In order to analyze the application scope of the four indexes, this section conducted a spectrum analysis of the acceleration signal and had a further discussion about the test effects of the four indexes on different test materials. Figures 14–16 show the comparison of the acceleration signal and indexes in the eighth compaction pass of the main rockfill and filter material. The 3D spectra in Figure 15 are obtained through the fast Fourier transformation (FFT) analysis of the

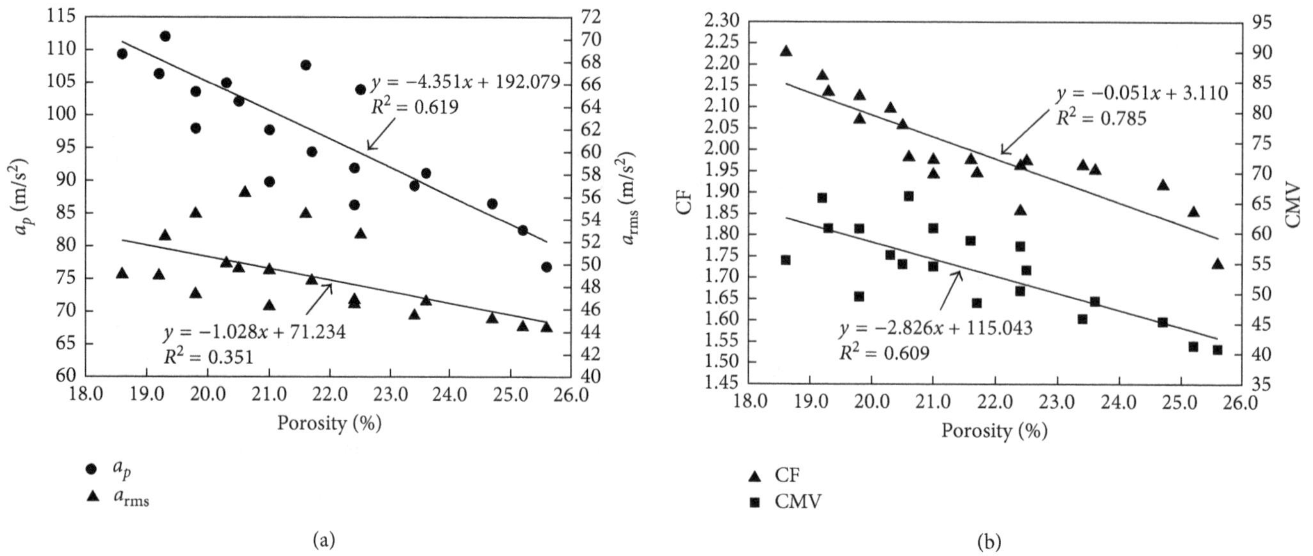

Figure 12: Linear relationships between porosity and real-time indexes on the main rockfill. (a) a_p, a_{rms}. (b) CF, CMV.

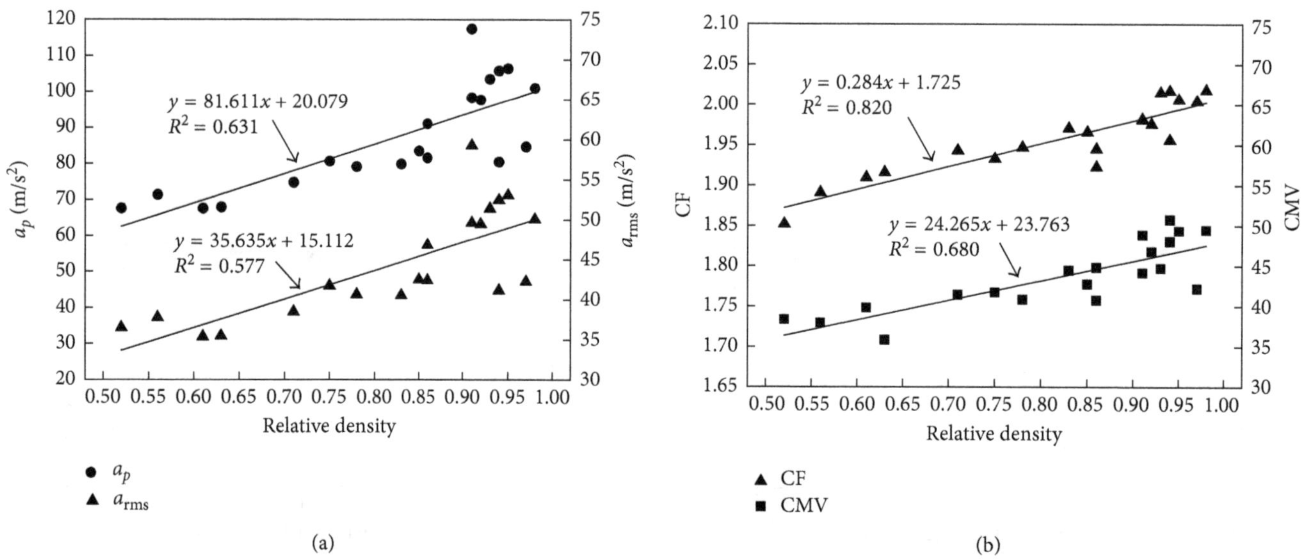

Figure 13: Linear relationships between relative density and real-time indexes on the secondary rockfill. (a) a_p, a_{rms}. (b) CF, CMV.

Table 4: Determination coefficients (R^2) for regression analysis between real-time test indexes and compaction quality indexes.

Compaction quality index	a_p		a_{rms}		CF		CMV	
	Linear model	Hyperbolic model	Linear model	Hyperbolic model	Linear model	Hyperbolic model	Linear model	Hyperbolic model
Porosity	0.619	0.580	0.351	0.308	0.785	0.823	0.609	0.528
Relative density	0.631	0.622	0.577	0.564	0.820	0.804	0.680	0.650

acceleration signal. The CF value in Figure 16 is magnified tenfold to display its tendency clearly.

As can be seen from Figure 14, the acceleration waveform is regularly distorted when the filter material get compacted, while the acceleration wave changes chaotically on the main rockfill. The spectra of Figure 15 show that the frequencies of the acceleration signal on the filter material are regularly distributed with stable fundamental frequency, first and second harmonic components, and few higher frequency components in the rolling process. While the spectrum on the main rockfill shows that the frequency distribution of the acceleration signal is irregular and disorderly with unsystematic frequency components.

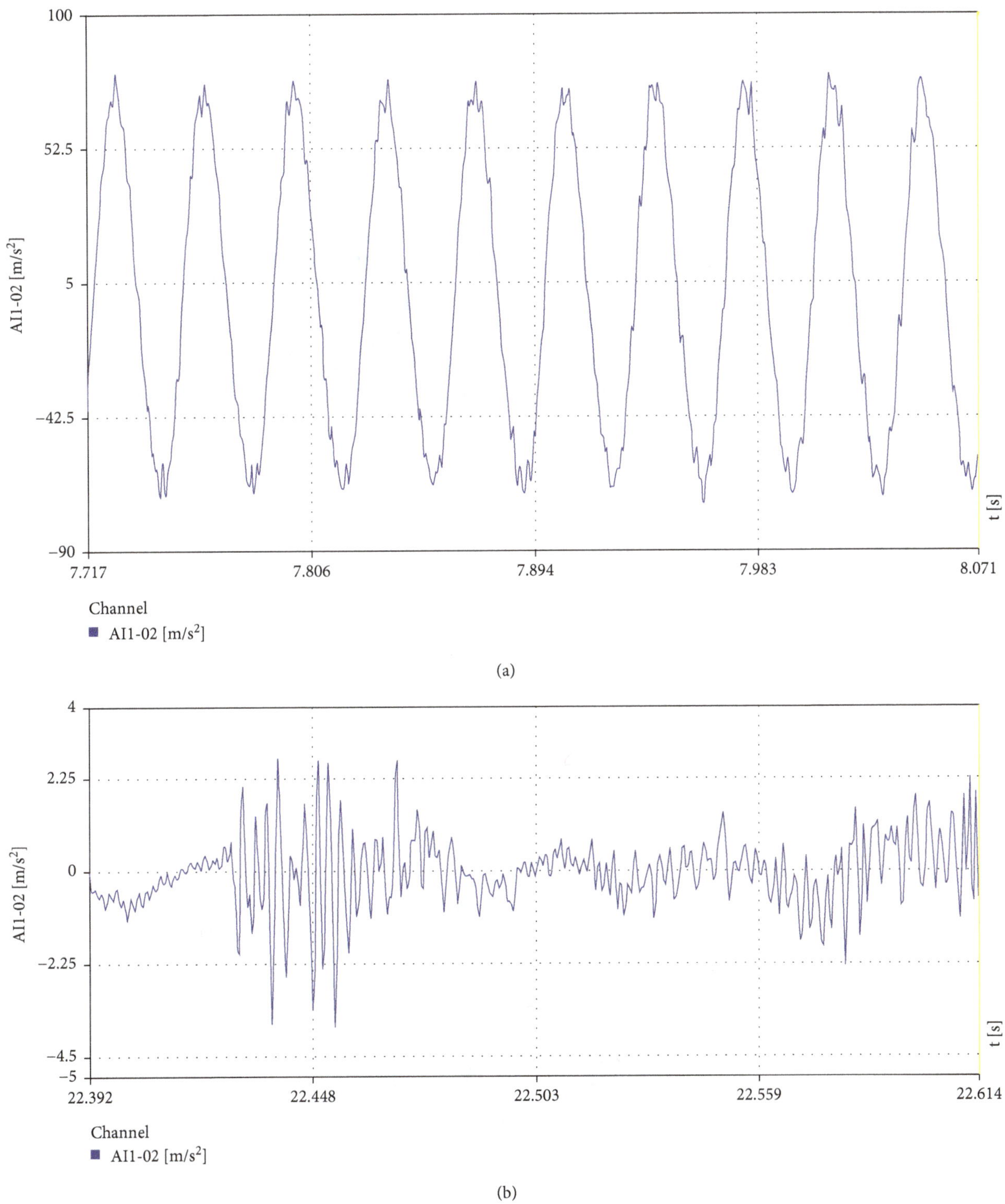

FIGURE 14: Time-history curve of acceleration signal. (a) Filter material. (b) Main rockfill.

Analysis of the reasons shows that since the main contents of the filter material are spherical particles with few edges, the contact between the soil particles and roller is relatively even, and there is not much difference among the reaction forces of the soil to the roller in unit areas. With the rolling moving on, the soil becomes more compacted and the reaction force to the roller increases, which causes a regular increase in the amplitude of the harmonic components and a regular wave distortion of the acceleration signal. Therefore, CMV and other acceleration indexes have a regular change in the rolling process, and all the four indexes can well reflect the compaction degree of the fine-grained soil.

However, in the case of the rockfill material which contains many sharp-edged stones, the contact between the

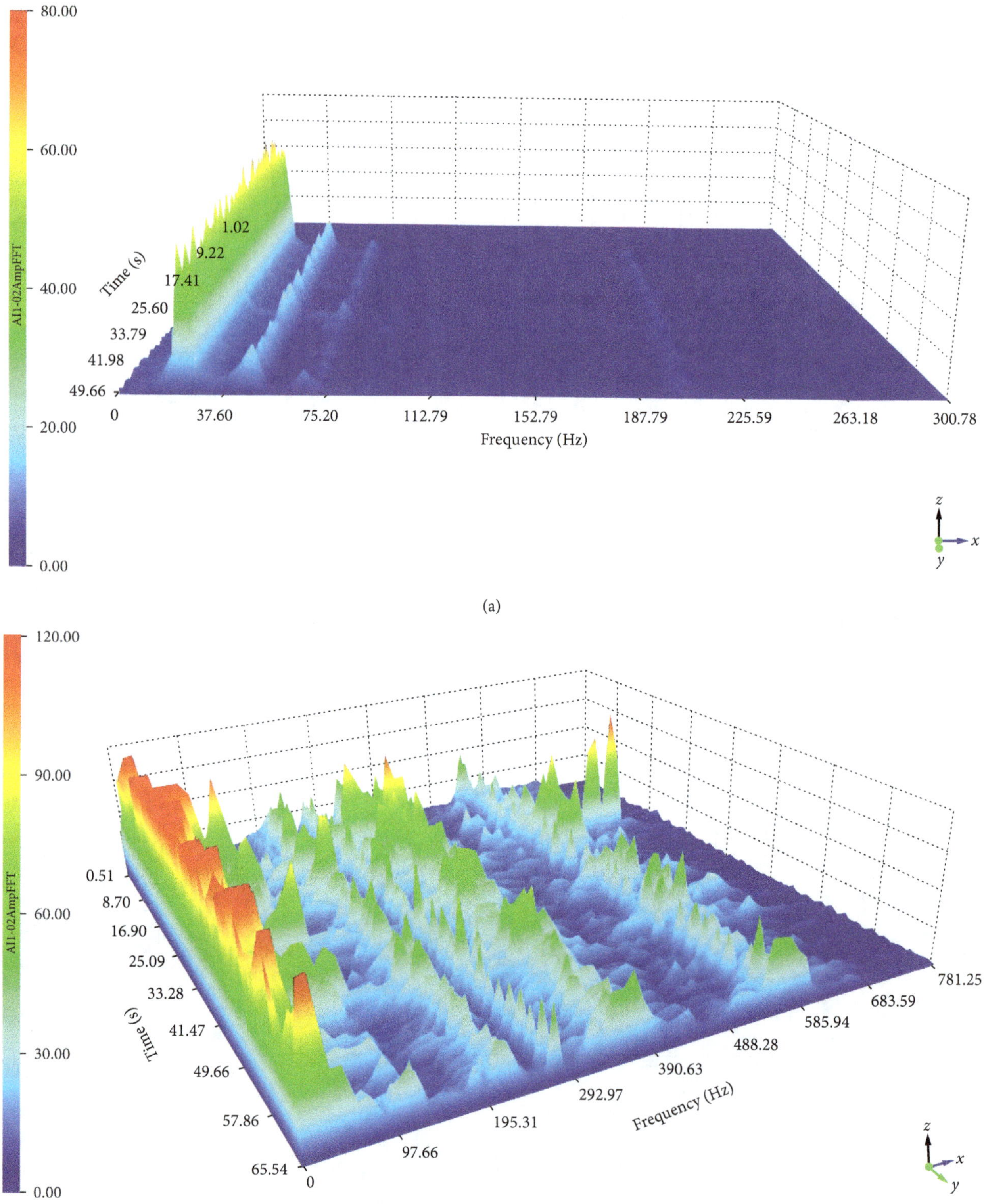

Figure 15: Three-dimensional spectra of acceleration signal. (a) Filter material. (b) Main rockfill.

soil particles and the roller is uneven, resulting in significant differences among the reaction forces of the soil to the roller in unit areas. The energy in the reaction force of the soil is unevenly transmitted to the roller, which leads to an irregular frequency distribution and a clutter waveform of the acceleration signal on the rockfill material. Due to the unstable change of the amplitude of fundamental frequency and first harmonic component in the rolling process, CMV

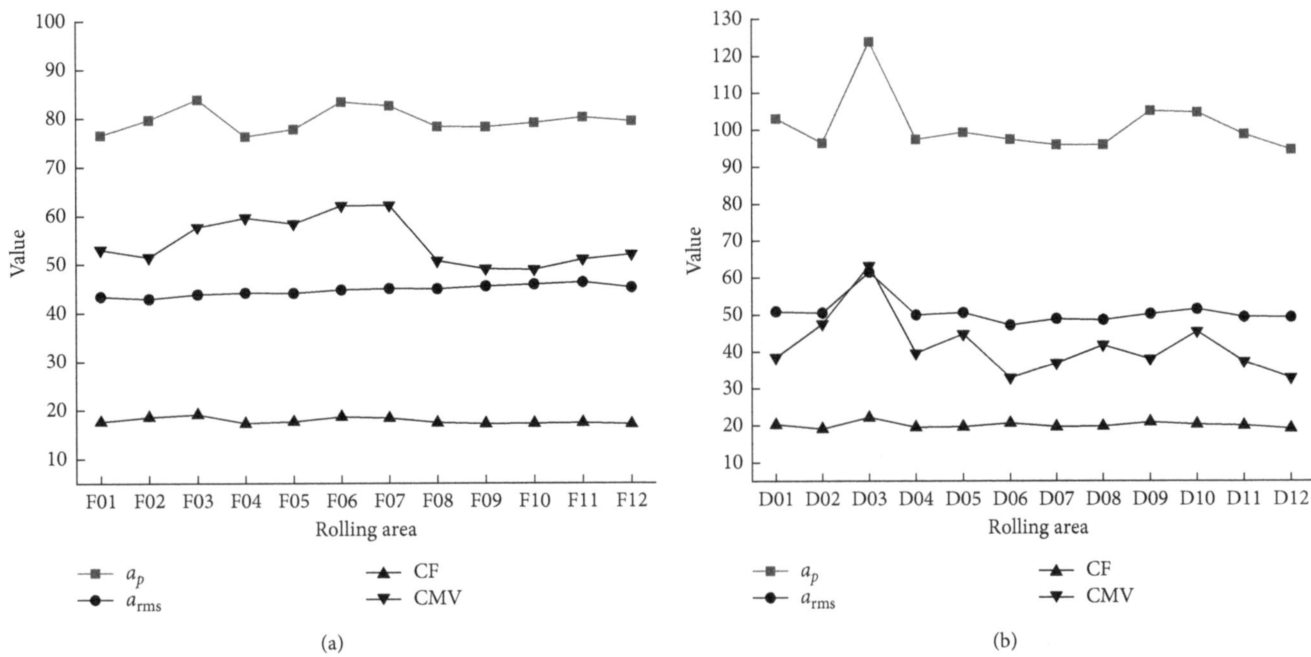

FIGURE 16: Tendency of acceleration indexes in the eighth compaction pass. (a) Filter material. (b) Main rockfill.

is disadvantaged in characterizing the compaction degree of the rockfill.

Judging from the physical meaning, the a_p and a_{rms} can only reflect part of the characteristics of the acceleration signal, while the CF combines the variation trend of them and describes the ability of the soil to produce the greatest reaction force to the roller. Therefore, the effect of the unevenness of the dam material on CF is relatively small compared to a_p and a_{rms}, which is consistent with the test result that the V_σ of CF is smaller than that of a_p and a_{rms} on all test materials (Table 3). The result can be proved by Figure 16 that the CF value changes stably while other indexes have greater volatility in the rolling process of the same dam material.

Combined with the analyses above, it can be known that the CF index best characterizes the compaction degree of the rockfill material among the four indexes. In the future, the authors will further establish a more detailed vibration model from the theoretical perspective, derive the formula, and clearly distinguish the applicability of CF index and other three acceleration indexes on the rockfill material.

5. Conclusions

In this study, the authors utilized four derived acceleration indexes to characterize the soil compaction degree, including a_p, a_{rms}, CF, and CMV. A two-part field compaction test was performed to analyze and judge the test effects of the four indexes on the rockfill and other dam materials. After the data correlation analysis and signal spectrum analysis, the following conclusions can be drawn:

(1) The data of a_p, a_{rms}, and CF are stable with low V_σ of them, and only the tendencies of a_p and CF with the change of roller running speed and compacted soil thickness are consistent with the theoretical situation. Besides, the linear correlation between CF and rolling times is the strongest, with the trimmed mean of R^2 being over 0.7 on all test materials. Thus, the correlation between CF and the rolling parameters is the strongest among the four indexes.

(2) The correlation between CF and the two compaction quality indexes is the strongest among the four indexes, with the highest R^2 of 0.823 and 0.820 in the regression analysis, respectively. The tentative control criteria of CF $\geq$ 2.017 on the main rockfill and CF $\geq$ 1.981 on the secondary rockfill are given by establishing the quantitative relations between CF and compaction quality indexes, which can instruct the field compaction quality control in the rolling process.

(3) The signal spectrum analysis shows that the frequencies of the acceleration signal on filter material are regularly distributed, and all the four indexes can well reflect the compaction degree of the fine-grained soil. The frequency distribution of the acceleration signal on main rockfill is irregular, and CMV is disadvantaged in characterizing the compaction degree of the rockfill. The effect of the unevenness of the dam material on CF is relatively small compared to a_p and a_{rms}. Combined with the results of statistical regression analyses, it can be concluded that the CF index best characterizes the compaction

degree of the rockfill material among the four indexes.

Conflicts of Interest

The authors declare that they have no conflicts of interest.

Acknowledgments

The authors wish to thank the great help of Sinohydro Bureau 5 Co., Ltd., in the field test. This work was supported by the National Science and Technology Support Program of China (no. 2014BAB03B04), the Research Foundation for Young Teachers of Sichuan University (no. 2016SCU11039), and the Science and Technology Program of the Power Construction Corporation of China (no. ZDZX-05).

References

[1] Trade Standard of P. R. China, "Specifications for rolled earth-rockfill dam construction," Tech. Rep. DL/T 5129–2013, National Energy Administration of P. R. China, Beijing, China, 2013.

[2] H. Thurner and A. Sandstrom, "A new device for instant compaction control," in *Proceedings of the International Conference on Compaction*, vol. 2, pp. 611–614, Paris, France, 1980.

[3] A. Sandstrom and J. C. Pettersson, "Intelligent systems for QA/QC in soil compaction," in *Proceedings of the 83rd Annual Transportation Research Board Meeting, Transportation Research Board*, pp. 1–17, Washington, DC, USA, 2004.

[4] D. Adam, "Continuous compaction control (CCC) with vibratory rollers," in *Proceedings of the 1st Australia-Newzealand Conference on Environmental Geotechnics–GeoEnvironment*, pp. 245–250, Melbourne, VIC, Australia, 1997.

[5] H. Brandl and D. Adam, "Sophisticated continuous compaction control of soils and granular materials," in *Proceedings of the 14th International Conference on Soil Mechanics and Foundation Engineering*, pp. 1–6, Hamburg, Germany, 1997.

[6] D. J. White, "Report of the workshop on intelligent compaction for soils and HMA," in *Proceedings of the Workshop on Intelligent Compaction for Soils and HMA*, pp. 1–129, Des Moines, IA, USA, April 2008.

[7] G. Chang, Q. W. Xu, J. Rutledge et al., *Accelerated Implementation of Intelligent Compaction Technology for Embankment Subgrade Soils, Aggregate Base, and Asphalt Pavement Materials*, Federal Highway Administration, Washington, DC, USA, 2011.

[8] L. Forssblad, "Compaction meter on vibrating rollers for improved compaction control," in *Proceedings of the International Conference on Compaction*, vol. 2, pp. 541–546, Paris, France, 1980.

[9] L. Forssblad, *Vibratory Soil and Rock FillCompaction*, Swedish National Road and Transport Research Institute (VTI), Dynapac Maskin, Karlskrona, Sweden, 1981.

[10] L. Forssblad, "Roller-Mounted compaction meters," in *Proceedings of the 10th International Road Federation*, pp. 1–8, RJ, Brazil, 1984.

[11] L. Forssblad, "Soil compaction and compaction control," in *Proceedings of the 3rd International Geotechnical Seminar*, pp. 1–8, Singapore, 1985.

[12] H. Fujii, "Comparison of various control methods in situ," in *Proceedings of the 1st Asian-Pacific Conference of International Society Terrain-Vehicle Systems*, pp. 10–18, Beijing, China, 1986.

[13] X. L. Guo and D. Wen, *Field Compaction of Coarse Aggregates*, China Water & Power Press, Beijing, China, (Chinese), 1999.

[14] G. Komandi, "An evaluation of the concept of rolling resistance," *Journal of Terramechanics*, vol. 36, no. 3, pp. 159–166, 1999.

[15] D. J. White, T. D. Rupnow, and H. Ceylan, "Influence of subgrade/subbase non-uniformity on PCC pavement performance," in *Proceedings of the Geotechnical Engineering for Transportation Projects (GeoTrans2004)*, pp. 1058–1065, ASCE, Los Angeles, CA, USA, July 2004.

[16] M. G. Bekker, *Introduction to Terrain-Vehicle Systems*, University of Michigan Press, Ann Arbor, MI, USA, 1969.

[17] M. J. Thompson and D. J. White, "Field calibration and spatial analysis of compaction-monitoring technology measurements," *Transportation Research Record: Journal of the Transportation Research Board*, vol. 2004, pp. 69–79, 2007.

[18] M. J. Thompson and D. J. White, "Estimating compaction of cohesive soils from machine drive power," *Journal of Geotechnical and Geoenvironmental Engineering*, vol. 134, no. 12, pp. 1771–1777, 2008.

[19] D. J. White and M. J. Thompson, "Relationships between in situ and roller-integrated compaction measurements for granular soils," *Journal of Geotechnical and Geoenvironmental Engineering*, vol. 134, no. 12, pp. 1763–1770, 2008.

[20] W. Krober, R. Floss, and W. Wallrath, "Dynamic soil stiffness as quality criterion for soil compaction," *Geotechnics for Roads, Rail Tracks and Earth Structures*, pp. 189–199, Balkema, Rotterdam, Netherlands, 2001.

[21] M. A. Mooney and R. V. Rinehart, "Field monitoring of roller vibration during compaction of subgrade soil," *Journal of Geotechnical and Geoenvironmental Engineering*, vol. 133, no. 3, pp. 257–265, 2007.

[22] R. V. Rinehart and M. A. Mooney, "Instrumentation of a roller compactor to monitor vibration behavior during earthwork compaction," *Automation in Construction*, vol. 17, no. 2, pp. 144–150, 2008.

[23] R. Anderegg and K. Kaufmann, "Intelligent compaction with vibratory rollers: feedback control systems in automatic compaction and compaction control," *Transportation Research Record: Journal of the Transportation Research Board*, vol. 1868, pp. 124–134, 2004.

[24] R. Anderegg, D. A. von Felten, and K. Kaufmann, "Compaction monitoring using intelligent soil compactors," in *Proceedings of the Geotechnical Engineering in the Information Technology Age (GeoCongress2006)*, pp. 41–46, ASCE, Atlanta, GA, USA, April 2006.

[25] K. Kaufmann and R. Anderegg, "3D-construction applications III: GPS-based compaction technology," in *Proceedings of the 1st International Conference on Machine Control & Guidance*, pp. 1–10, Zurich, Switzerland, 2008.

[26] G. H. Xu and Y. Cai, "Dynamics monitor in the process of road structure forming," in *Proceedings of the 15th International Road Federation (IRF) World Meeting*, Thailand, 2005.

[27] T. S. Yoo, *A Theory for Vibratory Compaction of Soil*, pp. 44–52, State University of New York at Buffalo, New York, NY, USA, 1975.

[28] E. T. Selig and T. S. Yoo, "Fundamentals of vibratory roller behavior," in *Proceedings of the 9th International Conference*

on Soil Mechanics and Foundation Engineering, pp. 375–380, Tokyo, Japan, 1977.

[29] T. S. Yoo and E. T. Selig, "Dynamics of vibratory-roller compaction," *Journal of the Geotechnical Engineering Division*, vol. 105, no. 10, pp. 1211–1231, 1979.

[30] D. H. Zhong, B. Cui, D. H. Liu, and D. Tong, "Theoretical research on construction quality real-time monitoring and system integration of core rock-fill dam," *Science in China Series E: Technological Sciences*, vol. 52, no. 11, pp. 3406–3412, 2009.

[31] D. H. Zhong, D. H. Liu, and B. Cui, "Real-time compaction quality monitoring of high core rockfill dam," *Science China Technological Sciences*, vol. 54, no. 7, pp. 1906–1913, 2011.

[32] D. Liu, J. Sun, D. Zhong, and L. Song, "Compaction quality control of earth–rock dam construction using real-time field operation data," *Journal of Construction Engineering and Management*, vol. 138, no. 9, pp. 1085–1094, 2012.

[33] D. Liu, Z. Li, and Z. Lian, "Compaction quality assessment of earth–rock dam materials using roller-integrated compaction monitoring technology," *Automation in Construction*, vol. 44, pp. 234–246, 2014.

[34] R. B. W. Heng and M. J. M. Nor, "Statistical analysis of sound and vibration signals for monitoring rolling element bearing condition," *Applied Acoustics*, vol. 53, no. 1-3, pp. 211–226, 1998.

[35] Trade Standard of P. R. China, "Specification of soil test," Tech. Rep. SL237–1999, Ministry of Water Resources of P. R. China, Beijing, China, 1999.

[36] Trade Standard of P. R. China, "Specification on rockfill dam built with gravel and soil core wall in hydroelectric and hydraulic engineering," Tech. Rep. DL/T 5269–2012, National Energy Administration of P. R. China, Beijing, China, 2012.

9

Combustion and Mechanical Properties of Polymer-Modified Cement Mortar at High Temperature

Hyung-Jun Kim[1] and Won-Jun Park[2]

[1]*Hazard Mitigation Evaluation Technology Center, Korea Conformity Laboratories, Cheongju 28115, Republic of Korea*
[2]*Department of Building System Engineering, Kangwon National University, Samcheok 25913, Republic of Korea*

Correspondence should be addressed to Won-Jun Park; wjpark@kangwon.ac.kr

Academic Editor: Michele Zappalorto

Although polymer-modified cement mortar (PCM) has been extensively used as finishing and repairing material, it is necessary to understand its combustion properties and mechanical properties at high temperature. This study evaluated the combustion characteristics with varying experimental parameters such as polymer type, unit polymer content, polymer-cement ratio, and thickness of the specimen. This study also evaluated the compressive strength and elastic modulus of PCMs with focus on the effects of the type of polymer, unit polymer content, and polymer-cement ratio. As a result, the incombustibility of the PCM was in the following order: SBR < VVA < EVA. After heating end, the mass loss rate of the PCM was less than 30%, regardless of the polymer type, unit polymer content, and W/C. In heat release rate test, the higher the unit polymer content, the greater the total heat release, regardless of the difference in W/C. The PAE series showed excellent fire resistance in the temperature range of 200~400°C. PCMs with a unit polymer content of 5%, modified with EVA and PAE, showed outstanding compressive strength at 200~400°C. At elevated temperature, the modulus of elasticity of PCM declined with an increase in the unit polymer content, regardless of the polymer type.

1. Introduction

Repair and strengthening are necessary to extend the service life of existing buildings. Polymer-modified cement mortar (PCM) has been extensively used for repairing works because PCM has superior performances as a high-performance material, particularly for finishing and repairing works in concrete buildings [1–4]. PCMs are also commonly used in civil infrastructures, bridges, external wall insulation mortar, self-leveling mortar, and concrete repaired fields due to their excellent strength, environmental protection, adhesion, waterproofing, resistance to chemical attack, and workability [5–8]. Additionally, PCM has outstanding performance compared to ordinary mortar as concrete repair materials in case of applying them to deteriorated reinforced concrete structure [9–12]. However, it is very difficult to choose the appropriate patch repair materials due to insufficient experimental data on the fire resistance and safety of the PCM [12, 13]. With the increasingly widespread use of PCM in buildings, concern has developed regarding the behavior of this material in the case of fire because PCM is generally combustible and susceptible to deterioration of the mechanical and bonding properties at elevated temperature [12–14]. Therefore, it is necessary to clarify its behavior at high temperature upon exposure to fire; however, insufficient data are available in this regard [14–16]. To meet fire endurance requirements, the performance of these materials at elevated temperature must be understood based on assessment to the mechanical properties [14]. However, when repair is required, it is very difficult to select the appropriate patch repair materials because of insufficient experimental data on the fire resistance and safety of PCM [7–9].

As PCM contains an organic polymer, it is necessary to understand its combustion properties and mechanical properties at high temperature such as those encountered during fires, for which there is not sufficient available data.

TABLE 1: Physical properties of polymers.

Type of polymer	Volatile portion (%)	Particle diameter* (%)	Acid value (mgKOH/mg)	Apparent density (g/mL)
EVA	Less than 2.0	Less than 2.0	Less than 2.0	0.50 ± 0.10
VVA	Less than 2.0	Less than 2.0	Less than 2.0	0.53 ± 0.10
PAE	Less than 2.0	Less than 2.0	Less than 2.0	0.50 ± 0.10
Polymer dispersion	Solid ratio (%)	pH (20°C)	Viscosity (mPa·s)	Density (g/mL)
SBR	44.6	8.0–9.0	500–1500	1.0

*Residual mass ratio after passing through a 300 μm sieve.

Conventional technologies and existing studies are mainly focusing on the mechanical properties such as adhesion and compressive strength, mix proportions of PCM, and durability for use before repairing and after repairing using PCMs in concrete structure.

This study evaluated the combustion characteristics of PCM by means of cone calorimeter, noncombustibility, and flammability tests with varying experimental parameters such as polymer type, unit polymer content, polymer-cement ratio, and thickness of the specimen. The effects of polymer type, unit polymer content, and water-cement ratio (W/C) were evaluated from the results of the above-mentioned tests. In addition, ignition and heat release rate tests were performed for various PCM thicknesses. Additionally, the mechanical properties of PCM are evaluated through the hot pressing test. The effect of the types of polymer and the unit polymer content are evaluated by analyzing the compressive strength and modulus of elasticity.

2. Experimental Procedure

This study evaluated the combustion characteristics of PCMs by means of cone calorimeter, noncombustibility, and flammability tests with varying experimental parameters such as polymer type, unit polymer content, polymer-cement ratio, and thickness of the specimen [15, 16]. This study also evaluated the compressive strength and elastic modulus of PCMs containing four types of polymers (ethylene-vinyl acetate copolymer (EVA), vinyl acetate-vinyl versatate (VVA), styrene butadiene rubber (SBR), and polyacrylic ester (PAE)) with focus on the effects of the type of polymer, unit polymer content, and polymer-cement ratio.

2.1. Combustion Properties of PCM

2.1.1. Materials and Composition. Portland cement and fine aggregate (fineness modulus 2.97, density 2.63 g/cm^3, and water absorption 1.81%) were used to produce the mortar. Polymers specified in JIS A 6203 (polymer dispersions and redispersible polymer powders for cement modifiers), EVA, VVA, PAE, and SBR were used for cement modifiers. Meanwhile, 1% of a blowing agent was added to the redispersible polymer powders. 1% of the blowing agent was added to the polymer dispersion at different total solid ratios. Sixty compositions were used for the tests, with a fixed cement-fine aggregate (mass ratio of 1 : 3) and various polymer types, polymer amounts, and W/C values. In this study, the amount of the polymer is expressed as weight per unit volume. For example, 10 kg/m^3 of the unit polymer weight in the PCM composition corresponds to 2% polymer-cement ratio. Flat specimens (100 × 100 × 10 (mm)) were used for the heat release rate tests and cylindrical specimens (diameter 44 × 50 (mm)) were used for the noncombustibility tests. The 100 × 100 × 400 (mm) specimen used for the ignition test was cut into three types of test pieces, 100 × 100 × 10 (mm), 100 × 100 × 30 (mm), and 100 × 100 × 50 (mm), after water curing (20°C) for four weeks. The physical properties of the polymers used are presented in Table 1. Table 2 shows the PCM composition and experimental parameters. All the specimens were cured for four weeks in water at 20°C and then subjected to air curing at 20°C and 60% RH for nine weeks. Finally, the samples were dried at 60°C and cooled to room temperature in a desiccator for three days before testing. Two specimens with 0.5%–2.0% water absorption were subjected to the heat release rate and ignition tests. Three specimens with 2.0%–4.0% water absorption were subjected to the noncombustibility test.

2.1.2. Test Methods. The heat release rate test was conducted according to ISO 5660-1 (Cone Calorimeter Standard Test Method) [17]. The source of ignition was an electric spark generated by a cone-type electric. The heat release amount and heat release rate were measured by oxygen-consumption calorimetry [15–17]. An incombustible material shows a total calorific value of less than 8 MJ/m^2 for 20 min after heating and a maximum heat release rate of less than 200 kW/m^2 for 10 s continuously. In addition, such a material is devoid of cracks or holes on the surface, which render it hazardous when used as a fire retardant. Noncombustibility tests were conducted according to ISO 1182 (reaction to fire tests for building products: noncombustibility test) [18]. The temperature of the heating furnace and weight change of the specimen were measured. As part of the test procedure, a cylindrical specimen was placed in a heating furnace at 750°C, and the temperature increase inside the furnace was measured.

As an important requirement in the noncombustibility test, the difference between the temperature of the incombustible material and the maximum temperature inside the

Table 2: PCM composition.

Type of polymer	Polymer content (kg/m^3)	Mortar flow (mm) W/C 45%	W/C 50% 10 (mm)	W/C 50% 30 (mm)	W/C 50% 50 (mm)	W/C 60%
None*	0	113	150	159	159	195
EVA	10	118	166			215
	20	128	180	158	158	213
	30	140	183			225
	40	138	190			215
	50	153	160	182	182	193
	75		169			
	100	148	168	172	172	198
VVA	10		185			
	20	145	185			213
	30		190			
	40		185			
	50		195			
	100		180			
SBR	10		183			
	20	155	180	180	180	228
	30		195			
	40		197			
	50		223	223	223	
	75		—			
	100		—	—	—	
PAE	10		190			
	20		177			
	30		191			
	40		193			
	50		190			
	75		203			
	100		208			

*Without polymer addition; blank implies no parameter; em dash implies that measurement was impossible.

furnace must be 20°C within 20 minutes after heating is commenced, and the average temperature 1 min after heating is ceased must be less than 20°C, with less than 30% weight loss of the specimen.

Moreover, as a mortar contains water and is hence not a uniform material, more than 20 min would be necessary to stabilize the temperature inside the furnace. In this study, the heating time was set as 30 min from the preliminary experiment. In addition, some specimens (unit polymer content: more than 50 kg/m^3) that could undergo ignition or spalling were tested by JIS A 9523 (loose fill thermal insulation materials) in the preliminary experiment [19]. The temperatures of the sample and interior of the furnace were measured using a K-type thermoelectric couple installed at both inside and outside of specimen. Figure 1 shows the setup for the flammability and temperature measurement tests.

2.2. Mechanical Properties of PCM

2.2.1. Materials and Composition. The mortar specimen for mechanical property test contained the same materials as shown in Table 1. The experimental parameters and PCM composition were adjusted based on the polymer type, polymer-cement ratio, and hot pressing test, as summarized in Table 3. Four proportions were used for the tests with a fixed water-cement ratio of 50. In this study, the amount of polymer is presented as the weight per unit volume, where 10 $kg{\cdot}m^{-3}$ unit polymer weight in the PCM composition corresponds to a polymer-cement ratio of 2%.

The specimens for the compressive strength test and the elastic modulus test were cylindrical (100 mm in diameter and 200 mm in height) and more than three specimens were prepared for each test. The results presented hereinafter are

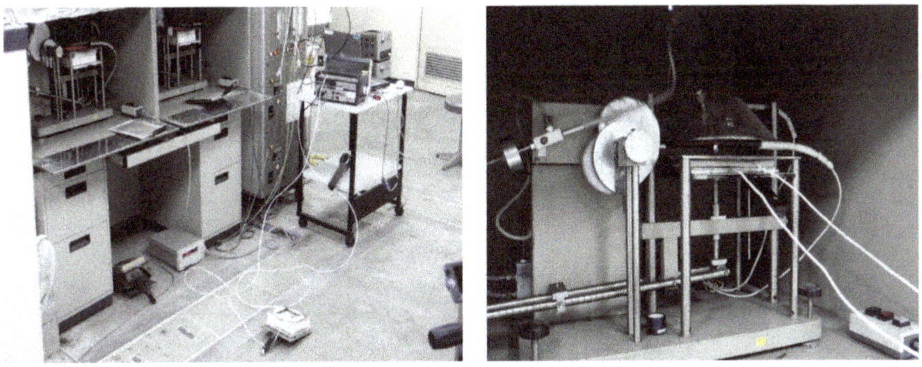

FIGURE 1: Flammability test and temperature measurement test.

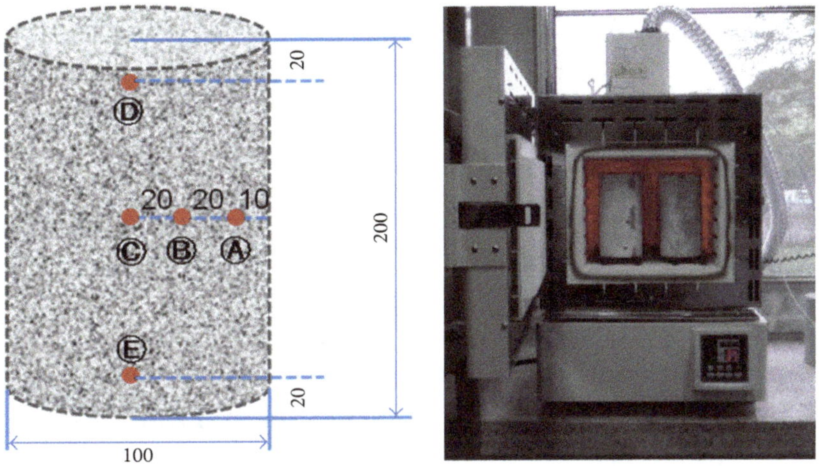

(a) Thermocouple

(b) Heating furnace

FIGURE 2: Setup for temperature measurement of PCM.

TABLE 3: Composition of PCM mixtures.

Type of polymer	Polymer-cement ratio (%)	Water-cement ratio (%)	Air content (%)	Flow (mm)
None*	0	50	4.9	171
	5		6.2	207
EVA	10	50	7.6	217
	20		8.7	222
	5		7.7	193
VVA	10	50	7.6	201
	20		7.7	205
	5		6.8	194
PAE	10	50	8.2	203
	20		9.2	224

*Without polymer addition.

averages. Each specimen was cured under conditions of 90% relative humidity at 20°C for fourteen days. All specimens were dried under conditions of 60% relative humidity at 20°C for sixty-three days. Finally, all specimens were dried in an oven at 60°C for seven days and were kept in a desiccator. Using these procedures, the water absorption rate for each specimen was controlled in range of 1.5~2.0%.

2.2.2. Test Methods. A hot pressing test method was used to evaluate the mechanical properties of PCM at elevated temperature. As illustrated in Figure 2, five thermocouples were placed into the specimen to measure the temperature changes. Each specimen was placed in the heating furnace for the hot pressing test, as shown in Figure 3. Based on previous data, the rate and amount of heat released from PCM depended on the polymer content regardless of the W/C. In this study, the amount of heat released followed the order VVA < EVA < PAE < SBR. This indicated that PCM modified with VVA and EVA at polymer contents of less than 75 $kg{\cdot}m^{-3}$ and PCM with PAE at a polymer content of less than 30 $kg{\cdot}m^{-3}$ meet the requirements as incombustible materials.

In the conventional test method at a heating rate of 150$^{\circ}C{\cdot}h^{-1}$, the specimen generated spalling after 3 h of heating with a water absorption rate of 3%. In this study, the water absorption rate was controlled in the range of 1.5~2.0%, and all specimens were heated at a rate of 100$^{\circ}C{\cdot}h^{-1}$.

FIGURE 3: Setup for heating compression test.

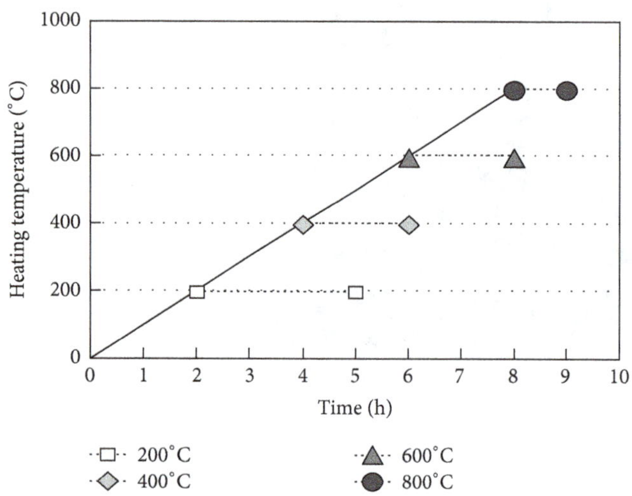

FIGURE 4: Heating program.

The diameter of the test specimen used in the experiment is 100 mm. As shown in Figure 2(a), the thermocouples were embedded in the specimen to check the temperature of the surface, inside, and center of the specimen. When the surface of each specimen reached a predetermined temperature, the heating temperature was maintained until the center of the specimen reached the target temperature. The time for the inside and outside of the PCM to reach the same temperature depends on the heating temperature and time. The heating program used for the furnace is shown in Figure 4. Figure 4 shows the time required to reach the total internal temperature of the specimen in each heating temperature range. In this study, heating temperatures were set to 200°C, 400°C, 600°C, and 800°C. The surface temperature of each specimen reaches 2, 4, 6, and 8 hours, respectively. Thereafter, the time at which the center temperature of each specimen reaches the corresponding temperature is 3, 2, 2, and 1 hour, respectively. In the case of heating at 200°C, the time for reaching the inner and outer temperatures to 200°C was 5 hours, and for 800°C,

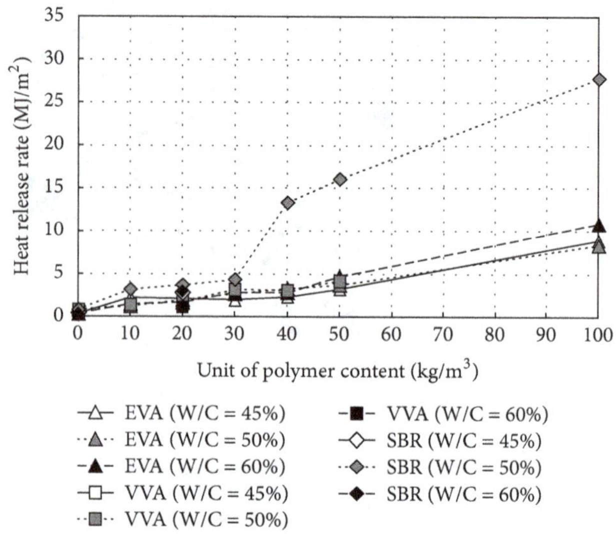

FIGURE 5: Total heat release amount of PCM for 20 min after start of heating.

it took 9 hours for the inner and outer temperatures to reach 800°C.

As it illustrated in Figure 3, the compressive strength was measured after the temperature at the center of the specimen reached the target temperature.

3. Results and Analysis

3.1. Combustion Properties

3.1.1. Heat Release Rate Test Results. Specimen fuming was generated in every SBR series and EVA series with 40 kg/m^3 unit polymer content. PCM in the SBR series with more than 40 kg/m^3 polymer content and EVA series containing 100 kg/m^3 polymer content underwent ignition. Spalling was particularly evident in the case of the VVA series specimens with 100 kg/m^3 polymer content. These observations indicate that the fuming, ignition state, and spalling characteristics of the PCMs differ with the polymer type, unit polymer content, and W/C. It was also observed that the specimens without spalling had no cracks or holes, regardless of the presence of the polymer.

Figure 5 illustrates the total heat release amount of the PCM for 20 min after the start of the heating. In case of the EVA and VVA series, the higher the the unit polymer content, the greater the total heat release, regardless of the difference in W/C. In the case of the samples in the SBR series with more than 40 kg/m^3 unit polymer content, the total heat release amount increased with an increase in the polymer content upon ignition. The polymer content and release rate exceeded 40 kg/m^3 and 8 MJ/m^2, respectively.

SBR is decomposed into styrene and butadiene-latex at around 200°C and is vaporized by pyrolysis at around 360°C. The temperature at the surface layer of the specimen exceeded the maximum temperature of decomposition and vaporization 5 min after heating was started. It is thought that

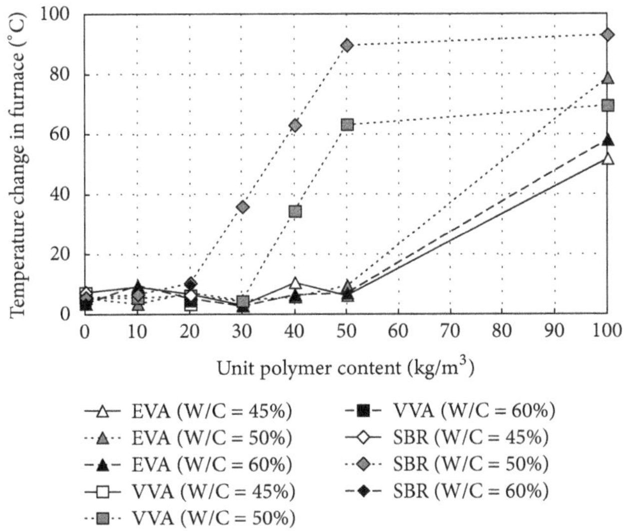

Figure 6: Relationship between unit polymer content in PCM and temperature change inside the furnace.

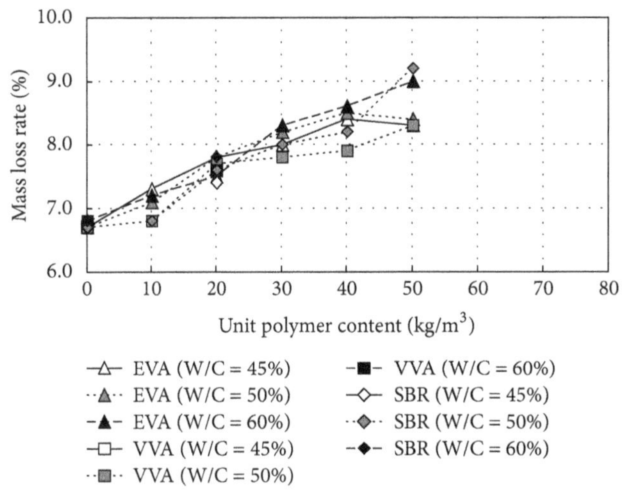

Figure 7: Relationship between unit polymer content in PCM and mass loss rate after heating.

the rapid increase in the heat release amount and heat release rate resulted from the oxygen consumption upon ignition.

3.1.2. Noncombustibility Test Results. Table 4 presents the observation results of the specimens during the test. Spalling occurred between 80 and 180 s; the higher the unit polymer content in the PCM, the faster the spalling. Ignition and smoke generation occurred between 90 and 320 s; again, the higher the unit polymer content in the PCM, the faster the ignition of the sample. Figure 6 also shows the relationship between the polymer content in the PCM and the temperature change inside the furnace.

Temperature change indicates the temperature difference between the maximum temperature inside the furnace 30 min after the start of heating and the average temperature for 1 min at the end of heating. The EVA series specimens with a unit polymer content of less than 50 kg/m^3 showed a temperature change of less than 20°C, while the EVA series with 100 kg/m^3 polymer content showed a temperature change of more than 50°C. The VVA series with a unit polymer content more than 30 kg/m^3 and the SBR series with a unit polymer content exceeding 20 kg/m^3 showed a temperature change of more than 20°C. The higher the unit polymer content in the PCM, the greater the temperature change. However, the VVA and SBR series with 100 kg/m^3 unit polymer content showed similar temperature change as the series with a unit polymer content of 50 kg/m^3.

The mass of each PCM series specimen without spalling was also measured after heating end. Figure 7 indicates the relationship between the unit polymer content in the PCM and the mass loss rate after heating. The mass loss rate shows a linear increase with an increase in the unit polymer content.

The mass loss rate of the PCM was less than 30%, regardless of the polymer type, unit polymer content, and W/C. It is thought that the mass reduction of the PCM was caused by free water in the specimen, some combined water in the cement matrix, and the decomposed polymer. Moreover, it is thought that the increase in mass loss with increasing polymer content was caused by the decomposition of the polymer.

When the polymer content was 10 kg/m^3 and the specimen mass was varied in the range 0.4%–0.5%, the mass loss rate of the SBR and VVA series increased with the unit polymer content. Additionally, the results of the noncombustibility test showed that decomposition pattern of the polymer differed between polymer types. In particular, it was thought that the SBR series underwent combustion more readily than the other PCM series, because of the combustion of the polymer.

3.1.3. Evaluation of Combustion Properties Based on PCM Composition and Specimen Thickness. Figure 8 presents the relationship between the unit polymer content in the PCM and the total heat release. Ignition was generated in the case of the PAE series with more than 40 kg/m^3 unit polymer content and specimens with 20, 50, and 100 kg/m^3 unit polymer content in the SBR series. Spalling was generated in the case of the EVA sample with 100 kg/m^3 unit polymer content, while no cracks or holes were found in the other samples of the same series.

As shown in Figure 8, both the PAE series and SBR series show a proportionate relationship to the unit polymer content. It is necessary to consider PCM ignition due to the ignition possibility, especially in the SBR series.

Figure 9 presents the relationship between the thickness of the specimen and the total heat release amount. Ignition occurred in the case of the specimens with 10 mm thickness, in the SBR and EVA series with 50 kg/m^3 unit polymer content. Spalling was also observed for the samples in the EVA series having 30 and 50 mm thickness, with 50 kg/m^3 unit polymer content. The thicker the specimen, the lower the total heat release amount because the absolute polymer content increases with the specimen thickness.

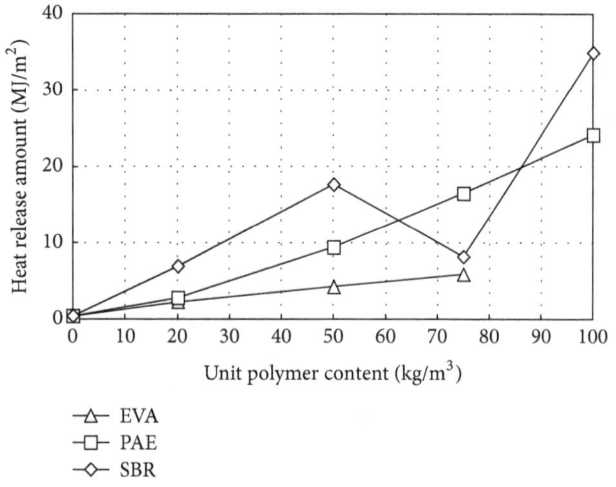

FIGURE 8: Relation between unit polymer content in PCM and total heat release amount.

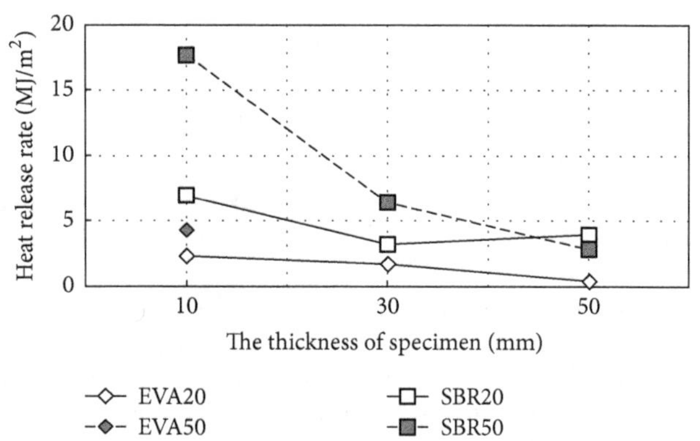

FIGURE 9: Relationship between specimen thickness and total heat release amount.

Figure 10 shows the temperature change for the specimens in the EVA series with 20 kg/m^3 unit polymer content and different thicknesses. For a thickness of 10 mm, the surface temperature increased after 3 min as compared to that for the other samples; the internal temperature of the specimen increased upon heating, possibly because of the accumulation of heat inside the specimen due to continuous heating. In cases of the specimens with 30 mm and 50 mm thickness, the maximum temperature was less than 200°C.

3.2. Mechanical Properties

3.2.1. Mechanical Properties at Room Temperature. The mechanical properties of PCM at room temperature were evaluated based on the polymer type and content. The compressive strength of PCM is presented in Figure 11 and the modulus of elasticity of PCM is shown in Figure 12. The series of EVA- and PAE-modified PCM showed an increase in the compressive strength with increasing polymer content, whereas the VVA-modified PCM series showed little increase. On the other hand, the VVA- and PAE-modified samples showed a decline in the modulus of elasticity with

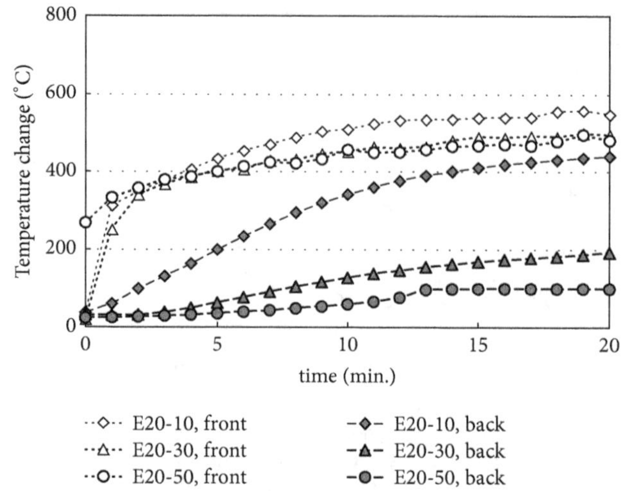

FIGURE 10: Temperature change inside and outside the specimen with specimen thickness.

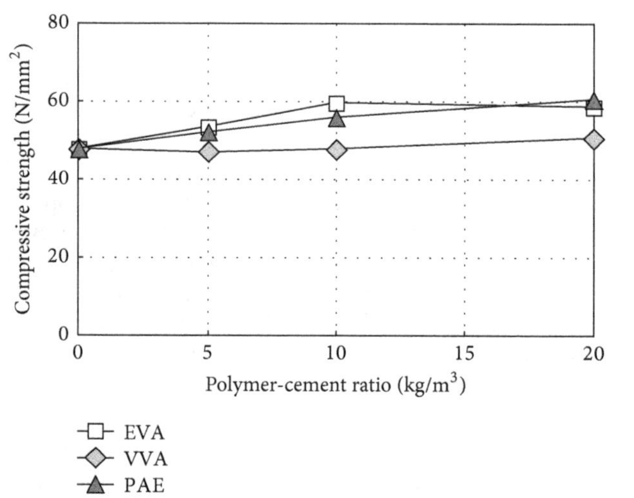

FIGURE 11: Compressive strength of PCM with various polymers at room temperature.

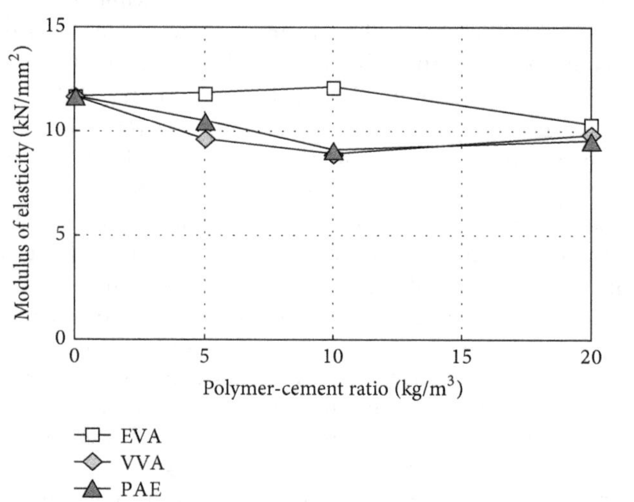

FIGURE 12: Modulus of elasticity of PCM with various polymers at room temperature.

Table 4: Observation of specimens.

State	Type of polymer	Unit polymer content in PCM (kg/m^3)						
		0	10	20	30	40	50	100
Spalling	EVA					△	△	○
	VVA				△	△		○
	SBR						△	○
Ignition and smoke	EVA							○
	VVA						○	○
	SBR			△	○	○	○	○

○: generation at all specimens; △: generation at part of specimens.

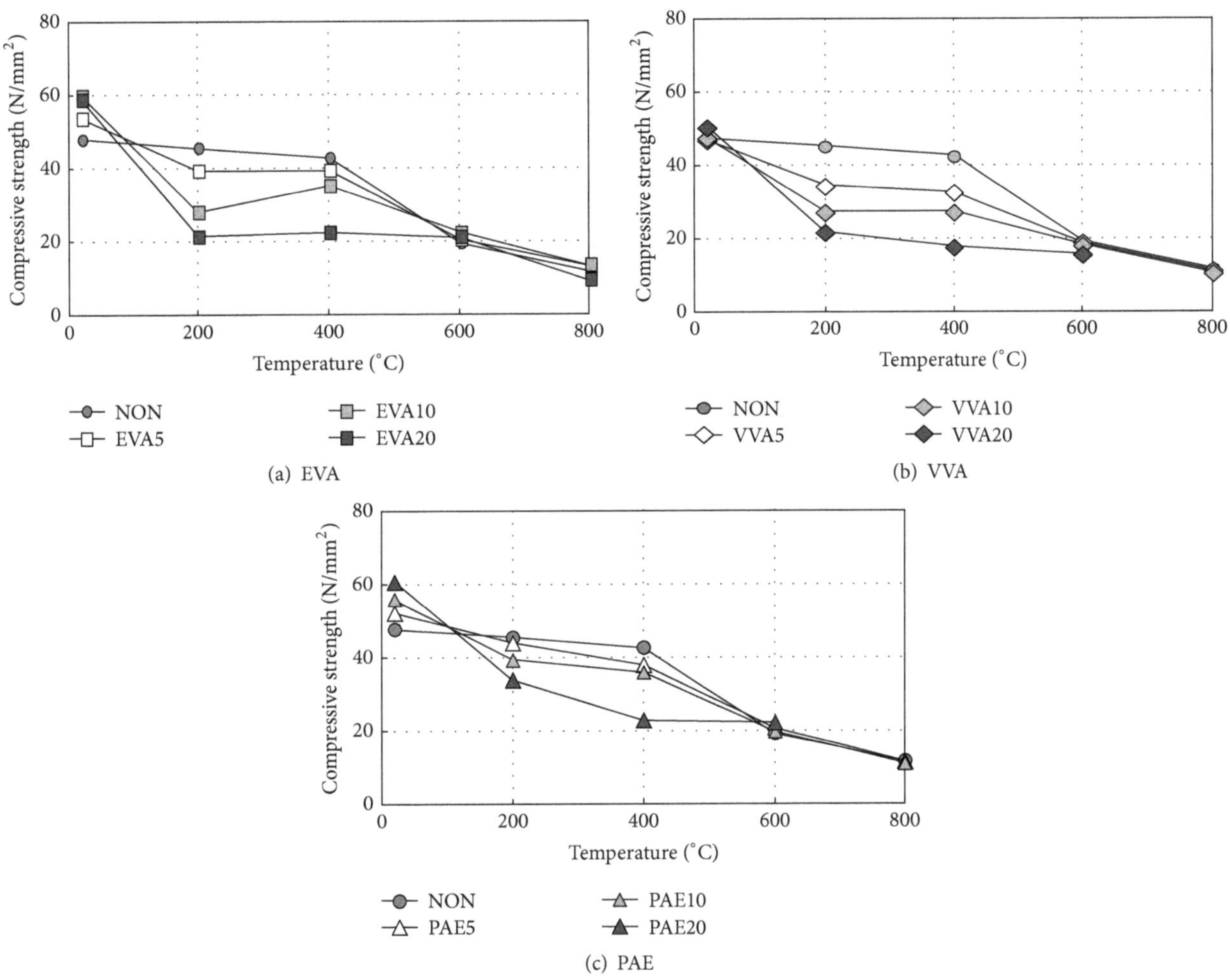

Figure 13: Compressive strength of PCM with various polymers at high temperature.

increasing polymer content. PCM with EVA 10% and the PAE series showed good mechanical properties.

3.2.2. Mechanical Properties at High Temperature. The mechanical properties of PCM were also evaluated at elevated temperature with variation of the polymer type and content. The compressive strength of the PCM series is shown in Figure 13 and the elastic modulus of PCM is shown in Figure 14. As shown in Table 4, VVA had a higher spalling risk than the other polymers. The compressive strength test showed that the VVA series had lower strength. According to a previous study, the incombustibility of PCM varies depending on the polymer type and unit polymer content. The compressive strength decreased with increasing polymer content. The PAE series showed excellent fire resistance in the temperature range of 200~400°C. On the other hand, the EVA

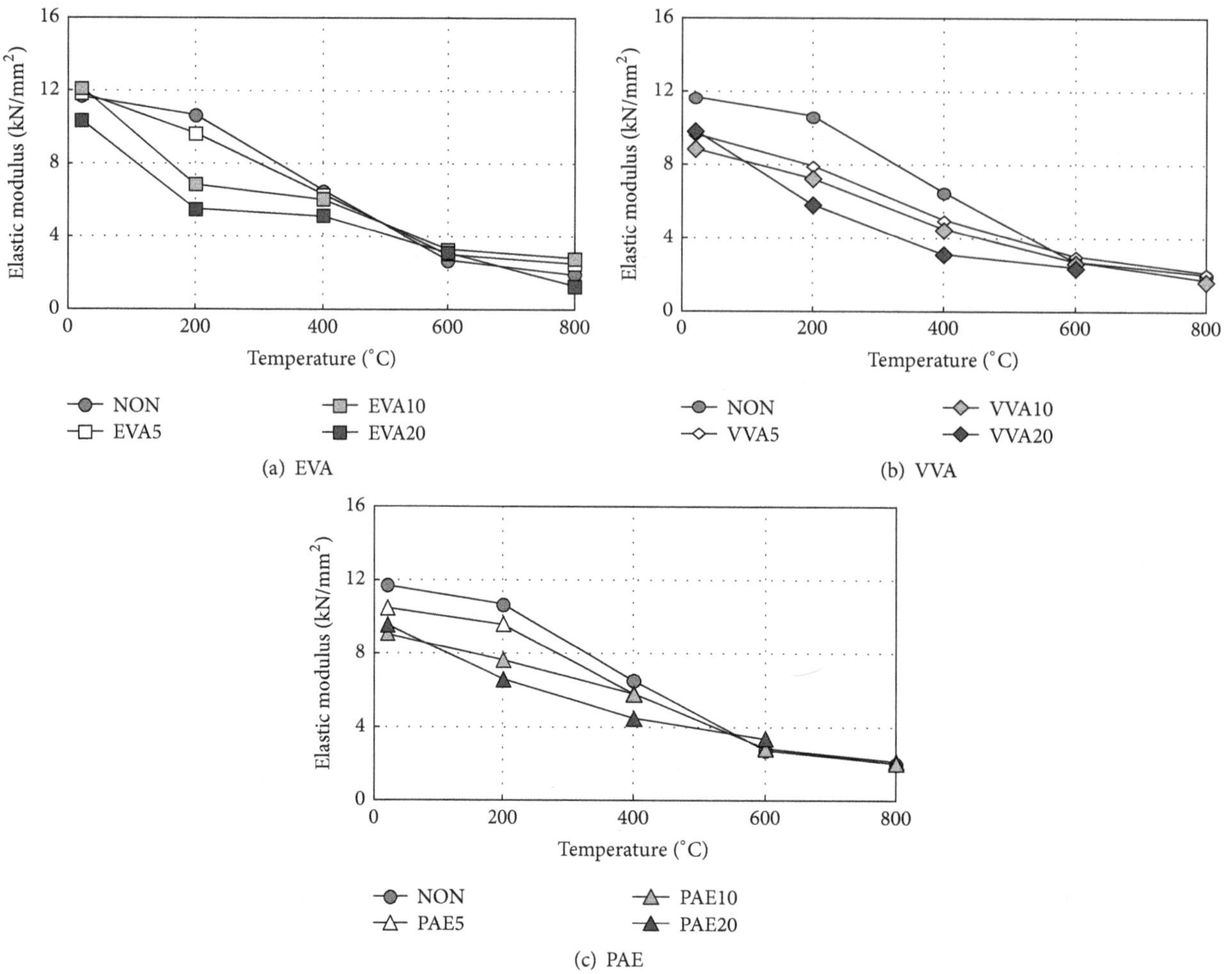

FIGURE 14: Modulus of elasticity of PCM with various polymers at high temperature.

and VVA series showed a rapid decrease in the compressive strength when the polymer content exceeded 5%. Most of the specimens showed a decline in the elastic modulus with increasing polymer content, with the exception of the 5% EVA and 5% PAE species. PCM with EVA 5% and the PAE series showed good mechanical properties in temperature range of 200~400°C.

4. Conclusions

The rate and amount of heat release from a PCM depended on the polymer content, regardless of the W/C. In this study, the heat release amount was found to follow the order VVA < EVA < PAE < SBR.

Specimens in the VVA and EVA series with polymer contents less than 75 kg/m^3, as well as the specimens in the PAE and SBR series with polymer contents less than 30 kg/m^3, met the requirements for an incombustible material. The noncombustibility of the PCM differed depending on the polymer type and unit polymer content. In this study, the incombustibility of the PCM was in the following order: SBR < VVA < EVA.

Given the insufficient experimental data on the fire resistance and safety of PCM, the mechanical properties of PCM at elevated temperature were investigated using three types of polymers as modifiers, namely, EVA, VVA, and PAE.

The test method employed involves spalling risk due to the heating speed and water absorption rate of PCM. This study focused on a water absorption rate of 1.5~2.0% and the polymer-cement ratio required to prevent spalling. The hot pressing test demonstrated that PCM with EVA 5% and the PAE series showed good mechanical properties in the temperature range of 200~400°C.

Conflicts of Interest

The authors declare that there are no conflicts of interest regarding the publication of this paper.

Acknowledgments

This research was supported by the Fire Fighting Safety & 119 Rescue Technology Research and Development Program funded by the Ministry of Public Safety and Security (NEMA-Next Generation-2014-58).

References

[1] Y. Ohama, "Polymer-based admixtures," *Cement and Concrete Composites*, vol. 20, no. 2-3, pp. 189–212, 1998.

[2] ACI Committee 548, *State-Of-The-Art Report on Polymer Modified Concrete (548.3R-95)*, Farmington Hills, MI, USA, American Concrete Institute.

[3] Y. Ohama, *Handbook of Polymer-Modified Concrete and Mortars, Properties and Process Technology*, Noyes Publications, Park Ridge, NJ, USA, 1995.

[4] K. P. Grosskurth and A. Konietzko, "Structure and mechanical behaviour of polymer modified cement concrete," in *Proceeding of 5th International Congress on Polymers in Concrete*, pp. 171–174, 1987.

[5] M. Wang, R. Wang, H. Yao et al., "Research on the mechanism of polymer latex modified cement," *Construction and Building Materials*, vol. 111, pp. 710–718, 2016.

[6] L. K. Aggarwal, P. C. Thapliyal, and S. R. Karade, "Properties of polymer-modified mortars using epoxy and acrylic emulsions," *Construction and Building Materials*, vol. 21, no. 2, pp. 379–383, 2007.

[7] D. Park, J. Ahn, S. Oh, H. Song, and T. Noguchi, "Drying effect of polymer-modified cement for patch-repaired mortar on constraint stress," *Construction and Building Materials*, vol. 23, no. 1, pp. 434–447, 2009.

[8] D. Park, S. Park, Y. Seo, and T. Noguchi, "Water absorption and constraint stress analysis of polymer-modified cement mortar used as a patch repair material," *Construction and Building Materials*, vol. 28, no. 1, pp. 819–830, 2012.

[9] J. Mirza, M. S. Mirza, and R. Lapointe, "Laboratory and field performance of polymer-modified cement-based repair mortars in cold climates," *Construction and Building Materials*, vol. 16, no. 6, pp. 365–374, 2002.

[10] A. Yahia, "Shear-thickening behavior of high-performance cement grouts - Influencing mix-design parameters," *Cement and Concrete Research*, vol. 41, no. 3, pp. 230–235, 2011.

[11] C. A. Anagnostopoulos, G. Sapidis, and E. Papastergiadis, "Fundamental properties of epoxy resin-modified cement grouts," *Construction and Building Materials*, vol. 125, pp. 184–195, 2016.

[12] J.-P. Won, S.-W. Choi, C.-G. Park, and C.-I. Jang, "High strength polymer-modified repair cementitious composite for fire protection," *Polymers and Polymer Composites*, vol. 15, no. 5, pp. 379–388, 2007.

[13] J.-P. Won, H.-B. Kang, S.-J. Lee, and J.-W. Kang, "Eco-friendly fireproof high-strength polymer cementitious composites," *Construction and Building Materials*, vol. 30, pp. 406–412, 2012.

[14] Y. Qiao, J. K. Deliwala, S. S. Chakravarthula, and X. Kong, "High-temperature tensile properties of a polymer intercalated/exfoliated cement," *Materials Letters*, vol. 59, no. 28, pp. 3616–3619, 2005.

[15] H. Hamasaki, H. J. Kim, and T. Noguchi, "Study on heat release properties of polymer-modified cement mortar," in *Proceedings of the Part 1 Effect of the Unit Polymer Weight and Mixed Design*, pp. 435-436, 2009.

[16] H. J Kim, H. hamasaki, and T. Noguchi, "Study on heat release properties of polymer-modified cement mortar," in *Proceedings of the Part 2 Effect on the Thickness of Specimen on Annual Conference of Architectural Institute of Japan*, pp. 437-438, 2009.

[17] ISO, ISO 5660-1:2015 Reaction-to-fire tests – Heat release, smoke production and mass loss rate - Part 1: Heat release rate (cone calorimeter method) and smoke production rate (dynamic measurement), Geneva. Switzerland: International Organization for Standardization; 2015.

[18] ISO, ISO 1182: Reaction to fire tests for building products - Non-combustibility test, Geneva. Switzerland: International Organization for Standardization; 2010.

[19] JIS, JIS A 9523, Loose fill thermal insulation materials, Japanese Standards Association; 2016.

10

Formation and Properties of the Ta-Y_2O_3, Ta-ZrO_2, and Ta-TaC Nanocomposites

J. Jakubowicz,[1] M. Sopata,[1] G. Adamek,[1] P. Siwak,[2] and T. Kachlicki[1]

[1]*Institute of Materials Science and Engineering, Poznan University of Technology, Jana Pawla II 24, 61-138 Poznan, Poland*
[2]*Institute of Mechanical Technology, Poznan University of Technology, Piotrowo 3, 60-965 Poznan, Poland*

Correspondence should be addressed to J. Jakubowicz; jaroslaw.jakubowicz@put.poznan.pl

Academic Editor: Akihiko Kimura

The nanocrystalline tantalum-ceramic composites were made using mechanical alloying followed by pulse plasma sintering (PPS). The tantalum acts as a matrix, to which the ceramic reinforced phase in the concentration of 5, 10, 20, and 40 wt.% was introduced. Oxides (Y_2O_3 and ZrO_2) and carbides (TaC) were used as the ceramic phase. The mechanical alloying results in the formation of nanocrystalline grains. The subsequent hot pressing in the mode of PPS results in the consolidation of powders and formation of bulk nanocomposites. All the bulk composites have the average grain size from 40 nm to 100 nm, whereas, for comparison, the bulk nanocrystalline pure tantalum has the average grain size of approximately 170 nm. The ceramic phase refines the grain size in the Ta nanocomposites. The mechanical properties were studied using the nanoindentation tests. The nanocomposites exhibit uniform load-displacement curves indicating good integrity and homogeneity of the samples. Out of the investigated components, the Ta-10 wt.% TaC one has the highest hardness and a very high Young's modulus (1398 HV and 336 GPa, resp.). For the Ta-oxide composites, Ta-20 wt.% Y_2O_3 has the highest mechanical properties (1165 HV hardness and 231 GPa Young's modulus).

1. Introduction

Refractory materials of the melting point higher than 3000°C are the most desired in design and manufacturing of heavy load-bearing components where resistance to high temperature and wear plays a crucial role. Additionally, these materials usually have high corrosion resistance in very aggressive environments as well as high mechanical properties [1]. The examples of refractory materials are pure metals such as Ta, W, or Mo and theirs alloys [2]. Other most commonly used refractory materials are ceramics such as oxides (ZrO_2 and Y_2O_3), carbides (TaC, ZrC, and WC), or nitrides (TiN and Si_3N_4) [3–5]. Both types of refractory materials, that is, metals and ceramics have found applications in the design of bulk parts or coatings. Due to their high hardness, refractory materials (particularly ceramics) are brittle. Both materials can be joined together in the form of composites, which usually constitutes a combination of the best properties of both the metals and the ceramics [6–8]. Particularly, the high brittleness of ceramics can be limited by the addition of a metallic phase and vice versa, and the addition of ceramic phase into the metallic matrix leads to the improvement of the hardness and wear resistance of refractory metals. Refractory materials require high temperature processes for the formation of materials and products. For example, powder metallurgy requires the sintering temperature of at least 1500°C (usually above 2000°C) for proper microcrystalline powder consolidation [9]. Conventional high temperature and longtime sintering processes can be applicable for coarse-grained materials of micrometer size grains. Nanomaterials, compared to microcrystalline ones, can be consolidated for a shorter time and at significantly lower temperatures to achieve optimum properties. The consolidation processes used for nanocrystalline powders are usually different than conventional powder metallurgy used for microcrystalline powders. For example, for the consolidation of nanomaterials, the hot pressing working in the heating mode of the spark plasma sintering (SPS) or pulse plasma sintering (PPS) gives the best results [10, 11]. In these processes of consolidation, both the pressure and the temperature increase simultaneously, which results in a shortening of the time for which the material is kept at a given high sintering temperature, and this process can be done at a significantly lower consolidation temperature compared to conventional pressureless sintering [12]. Both factors (temperature and time) are crucial for the reduction of the grain growth and the maintenance of the nanostructure or ultrafine structure [13]. Differences in the absence of wetting and the densities of the melted metal and ceramic components result in their segregation, which requires special casting techniques [14]. Therefore, powder metallurgy is very useful for the formation of homogeneous composites [15]. In the process of preparation of the refractory composites, the powders of metallic and ceramic components of the designed chemical composition are mixed together and then consolidated using hot pressing, SPS, PPS, or other relevant techniques [16–18]. For the formation of nanocomposite powders, the mechanical alloying process can be applied, in which the reduction of microcrystalline into nanocrystalline grains is provided by high-energy impacts of the balls in the milling vial [19]. In the mechanical alloying process, the final powders' mixture comes in the form of agglomerates of the micrometer or submicrometer size composed of nanometer size grains of metallic as well as ceramic phases uniformly distributed in the entire volume of the material [20].

New prospects for refractory nanomaterials are related to their outstanding mechanical properties [21], whereas high-temperature applications are limited due to excess grain growth at elevated temperatures [22]. At high temperatures, the nanostructure is unstable and grows up, which leads to deterioration of the mechanical properties.

In this work, the authors focus on the preparation and properties of tantalum-based nanocomposites, reinforced by ceramic Y_2O_3, ZrO_2, and TaC. Ta has the melting point of 3017°C and the density of 16.4 g/cm^3. The ceramics have the melting point of 2690, 2715, and 3985°C for Y_2O_3, ZrO_2, and TaC, respectively. The density of ceramics is 5.03, 5.68, and 14.5 g/cm^3 for Y_2O_3, ZrO_2, and TaC, respectively. The nanocomposites having 5, 10, 20, and 40 wt.% of the ceramic phase were formed using mechanical alloying and PPS. The paper studies the formation of nanocomposites and their structure, microstructure, and mechanical properties.

2. Materials and Methods

In this work, nanocrystalline Ta-xY_2O_3, Ta-xZrO_2, and Ta-xTaC composites ($x = 5$, 10, 20, and 40 wt.% of the ceramic phase) were synthesized using mechanical alloying (MA) followed by hot pressing in the mode of pulse plasma sintering (PPS). In the MA process, the tantalum powder (<44 μm, purity > 99.97%; Alfa Aesar) was intensely mixed and milled with the Y_2O_3 powder (<50 nm, purity >99.9%; Sigma-Aldrich) as well as the ZrO_2 powder (0.1–2 μm, stabilized with 5.4% of Y_2O_3; Goodfellow) and TaC (<45 μm, purity >99.5%; Goodfellow). The mixture of a total of 5.5 g of the metallic and ceramic powders was loaded and reloaded into the milling vial in the Unilab glove box (MBraun) providing a high purity Ar 5.0 atmosphere. For each composite composition, several syntheses were performed to provide material for 5 consolidated samples with 8 mm in diameter and 4 mm in height. In the mechanical alloying process (SPEX 8000M Mixer/Mill; SpexSamplePrep), the two types of powders (Ta + selected ceramic one) were high-energy mixed and milled for 48 h at the room temperature in the Ar 5.0 atmosphere. The steel-hardened vial with ball bearings (>62 HRC) was used for proper mixing. Due to the milling of the ceramic phase, the Fe impurity was introduced to the nanocomposites, but, its content did not exceed 2 wt.%. The as-milled powders were axially hot-pressed (Elbit) at 4 Pa vacuum. The graphite die and the graphite movable punches were coated by boron-nitride lubricant spray (HeBoCoat) during the process. The pressure of the punches directed at the powder was 50 MPa. The pulse plasma sintering mode (PPS) was used for the heating. The heating rate, sintering temperature, and time were set at 650°C/min, 1300°C, and 5 s. The other setup parameters for the PPS process were automatically selected to maintain proper heating rate and sintering temperature.

For comparison, the nanocrystalline pure tantalum was made using high-energy ball-milling (HEBM) and PPS. In the MA process, at least two different powders (Ta and ceramic powders) are mixed and milled together, whereas in HEBM, only the one component (Ta powder) is mixed and milled. Both processes were conducted in SPEX mill at the same conditions.

The structure and microstructure were investigated using Empyrean XRD (Panalytical) with CuKα radiation, SEM Vega 5135 (Tescan) with EDS PGT Prism 200 Avalon (Princeton Gamma-Tech), AFM Q-Scope 250 (Quesant), and TEM CM 20 Super Twin (Philips). For the AFM and SEM observations, all the bulk samples were grinded up to 1000 grit, polished in the Al_2O_3 suspension, and etched in an $H_2SO_4 + HNO_3 + HF$ mixture to reveal the grain morphology. More details regarding the above equipment used by the authors have been provided in [23].

The mechanical properties were measured using a Picodentor HM500 (Fischer) nanoindentation tester. The following parameters were measured: HV, Vickers hardness; E_{IT}, indentation modulus; C_{IT1}, indentation creep; W_t, total mechanical work of indentation; and N_{plast}, plastic deformation portion. The indentation force load was 300 mN for 20 s. The load-displacement curves were recorded. For the measurement of the mechanical properties, the authors used unetched samples to avoid incorrect measurement results.

3. Results

The tantalum metal of a cubic structure (Figures 1(a), 2(a), and 3(a)) was mixed and milled with ceramic Y_2O_3 (cubic), ZrO_2(monoclinic), and TaC(cubic) powders (Figures 1(b), 2(b), and 3(b), resp.) of different crystallographic parameters. During the intense milling, the ceramic phase is

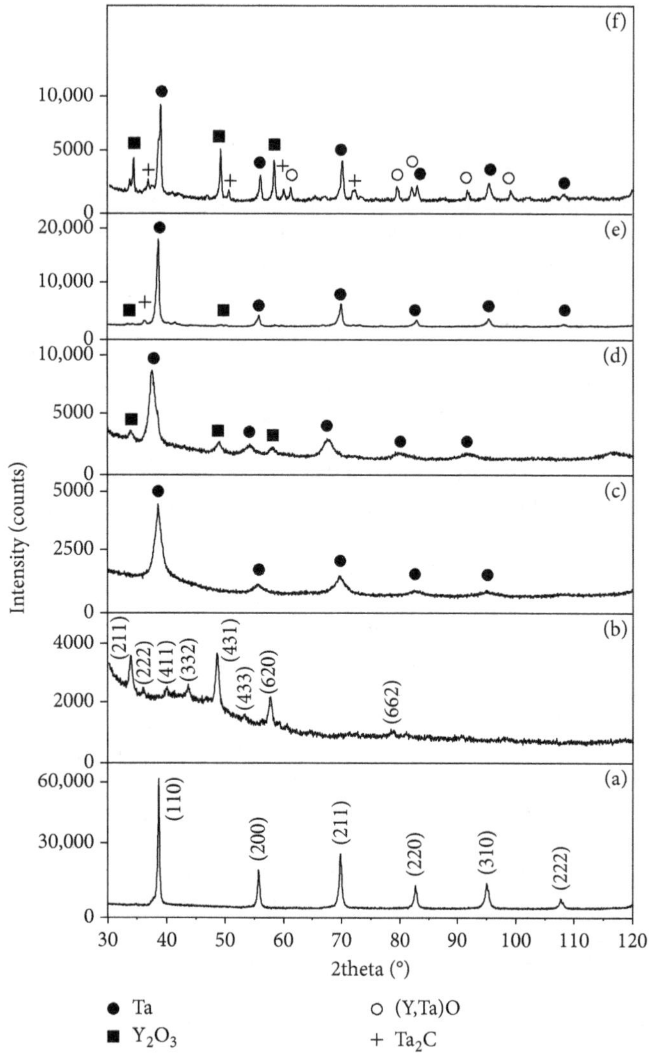

FIGURE 1: XRD of the pure Ta (a) and Y_2O_3 (b) powders; Ta-5Y_2O_3 (c) and Ta-40Y_2O_3 (d) nanocomposites' powder mixture after mechanical alloying; and Ta-5Y_2O_3 (e) and Ta-40Y_2O_3 (f) bulk nanocomposites after consolidation.

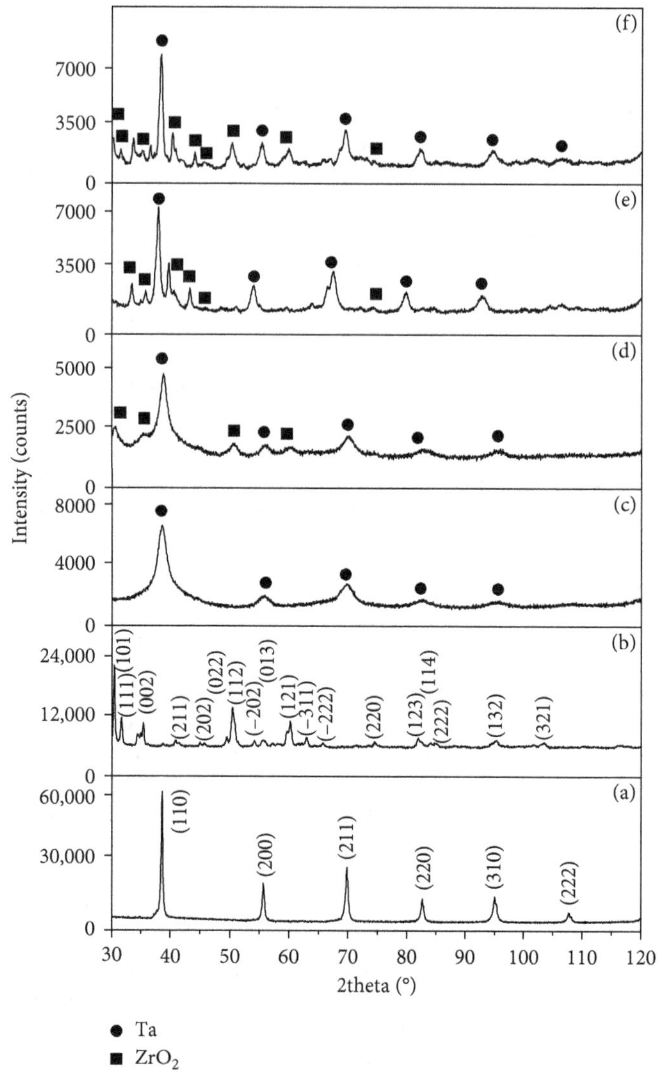

FIGURE 2: XRD of the pure Ta (a) and ZrO_2 (b) powders; Ta-5ZrO_2 (c) and Ta-40ZrO_2 (d) nanocomposites' powder mixture after mechanical alloying; and Ta-5ZrO_2 (e) and Ta-40ZrO_2 (f) bulk nanocomposites after consolidation.

homogenously distributed in the mixture of powders. For the Ta composites of low ceramic content (Figures 1(c), 2(c), and 3(c)), the MA process results mainly in the formation of a solid solution or a highly dispersed ceramic phase in the tantalum matrix. No visible peaks of the oxide ceramic phase are present (Figures 1(c) and 2(c)); however, carbides are visible (Figure 3(c)). An increase in the content of the ceramic phase leads to a typical composite structure, showing peaks (tantalum and the ceramic phase) (Figures 1(d), 2(d), and 3(d)). The consolidation process conducted at an elevated temperature in a graphite die can lead to additional carburization and formation of additional tantalum carbides (Figures 1(e) and 1(f)). High temperature can lead to diffusion of oxygen from the oxides and formation of (Y, Ta) O at a higher ceramic phase content (Figure 1(f)). The made materials are classified as composites, in which the metallic phase coexists with the ceramic phase.

During the milling process, the ceramic phase affects the crystallographic parameters of the tantalum (Figure 4(a)) and vice versa—the tantalum affects the crystallographic parameters of the ceramic phase (Figure 4(b)). The ceramic phase has significantly different lattice parameters compared to tantalum. All the used materials have a cubic-type structure, but the lattice constant for the Ta, Y_2O_3, ZrO_2, and TaC is 3.306, 10.604, 5,065, and 4.456 Å, respectively. The volume of the unit cell for these materials is 35.127, 1192.365, 129.939, and 88.478 Å^3, respectively. Only Y_2O_3 of the highest lattice constant leads to an increase in the lattice and volume of the unit cell of the tantalum matrix, whereas ZrO_2 and TaC participate in a slight reduction of these parameters for tantalum (Figure 4(a)). Alternatively, the tantalum can affect the lattice constant of the ceramic phase (Figure 4(b)) leading to a decrease in the lattice constant and volume of the unit cell, especially for a high Y_2O_3 content.

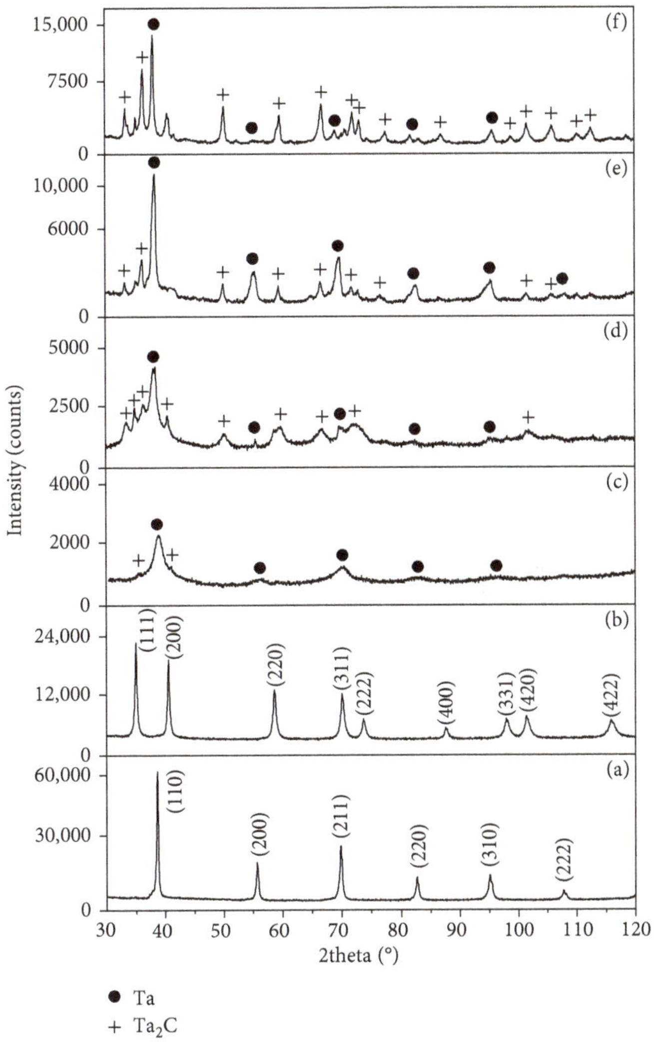

FIGURE 3: XRD of the pure Ta (a) and TaC (b) powders; Ta-5TaC (c) and Ta-40TaC (d) nanocomposites' powder mixture after mechanical alloying; and Ta-5TaC (e) and Ta-40TaC (f) bulk nanocomposites after consolidation.

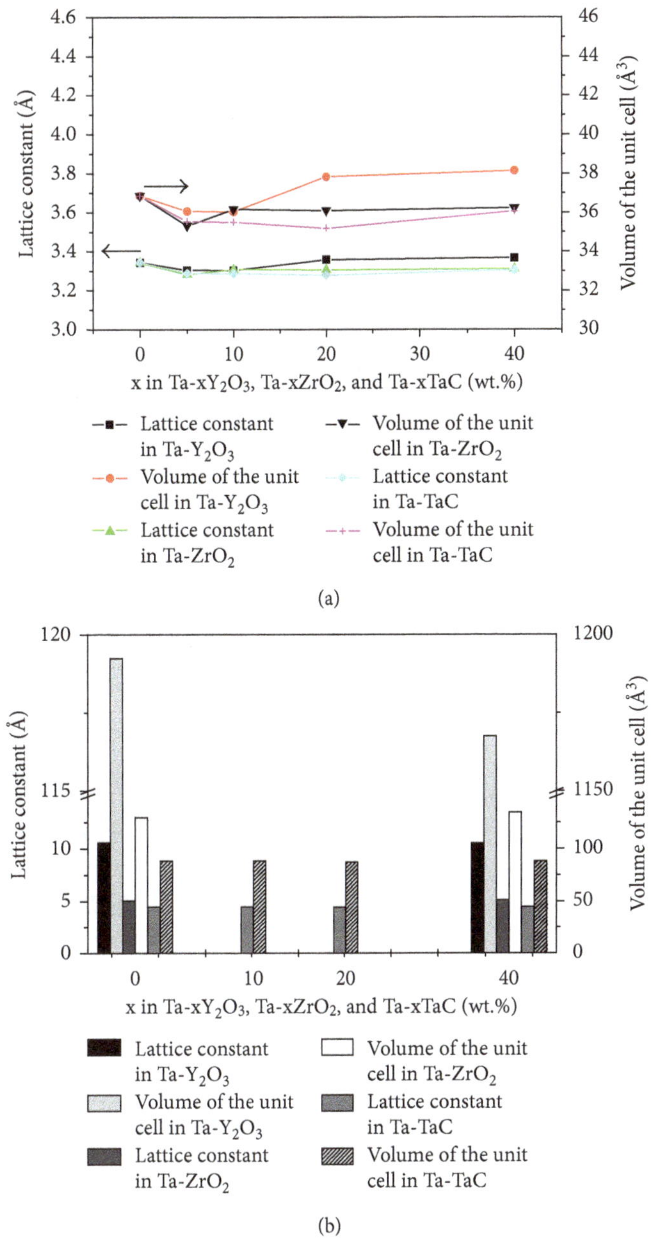

FIGURE 4: The effect of the Y_2O_3, ZrO_2, and TaC content in the Ta-xY_2O_3, Ta-xZrO_2, and Ta-xTaC composites on the Ta lattice constant and the Ta volume of the unit cell (after MA process) (a) and the effect of the Ta content in the Ta-xY_2O_3, Ta-xZrO_2, and Ta-xTaC composites on the Y_2O_3, ZrO_2, and TaC lattice constant and the volume of the unit cell (b); $x = 0$, 5, 10, 20, and 40 wt.%.

The example TEM images of Ta and Ta-Y_2O_3 have been shown in Figure 5. The results confirm the two-phase nanostructure of the composite powders. Ta and the presented Ta-Y_2O_3 have a grain size of approximately 40–100 nm. The dark spots, clearly visible in Figures 5(b)–5(d), belong to the Y_2O_3 grains that are uniformly distributed in the Ta matrix. Generally, an increase in the Y_2O_3 content leads to a tantalum matrix grain size reduction.

The grain size reduction with the increased Y_2O_3 content was confirmed in the AFM measurements (Figure 6). The grains of nanocrystalline Ta (170 nm) were estimated earlier [24]. The introduction of nanocrystalline Y_2O_3 significantly shifts the grain size towards lower values. For 5% Y_2O_3, the average grain size was estimated at 76 nm, whereas for 10, 20, and 40 wt.% of Y_2O_3 it changed to 55, 70, and 42 nm, respectively. The increases in the Y_2O_3 concentration lead to a narrower grain size distribution as well as a smaller size of the largest grains.

The comparison of the Ta-based composite microstructure with different ceramic phases (all 40 wt.% for Y_2O_3, ZrO_2, and TaC) has been shown in Figure 7. The smallest grains have their composites reinforced with nanocrystalline Y_2O_3 (a, b), whereas the largest grains have the composites reinforced with TaC (e, f), but are the most homogenous. As for the tantalum-based composites reinforced with ZrO_2, they are inhomogeneous and concentrated ZrO_2 precipitations are present (c).

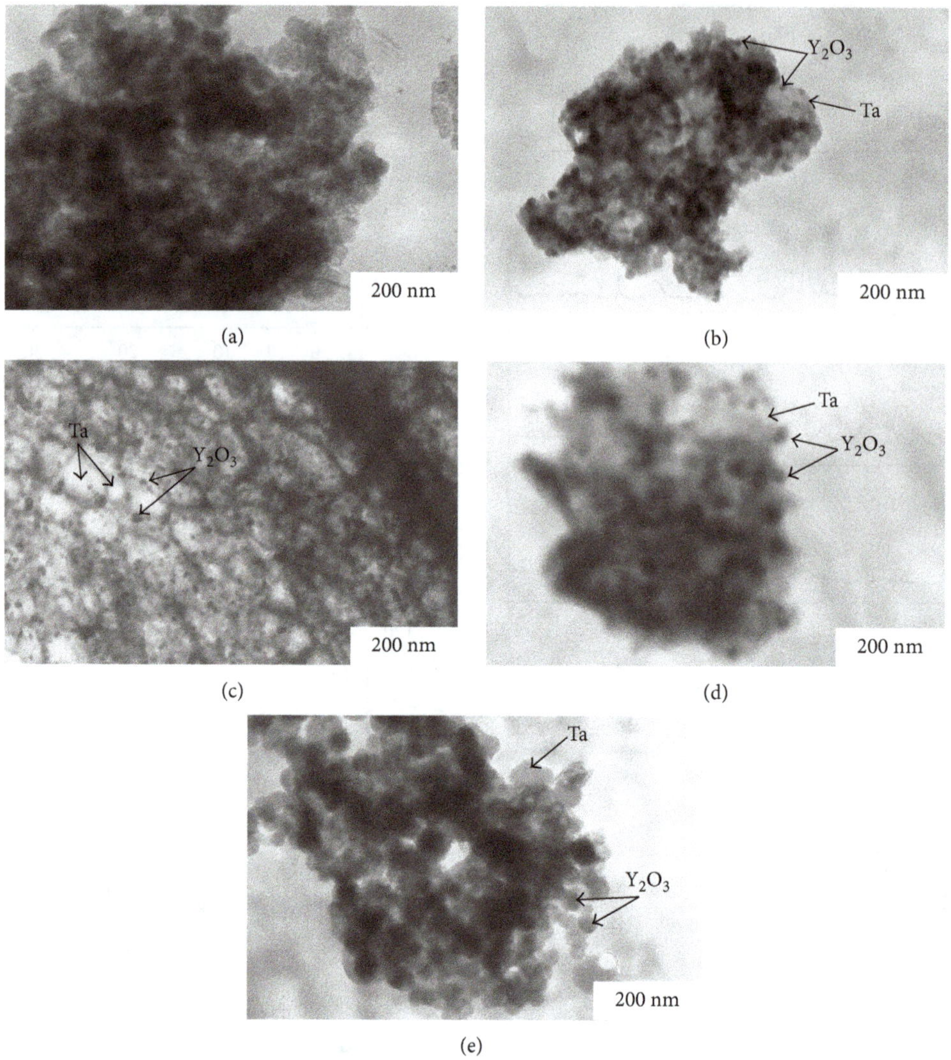

FIGURE 5: TEM images of the mechanically milled nanocrystalline Ta (a) and mechanically alloyed Ta-5Y_2O_3 (b), Ta-10Y_2O_3 (c), Ta-20Y_2O_3 (d), and Ta-40Y_2O_3 (e) nanocomposite powders; the smaller and darker particles correspond to ceramic phase, whereas bigger and lighter to tantalum matrix; the example grains are marked.

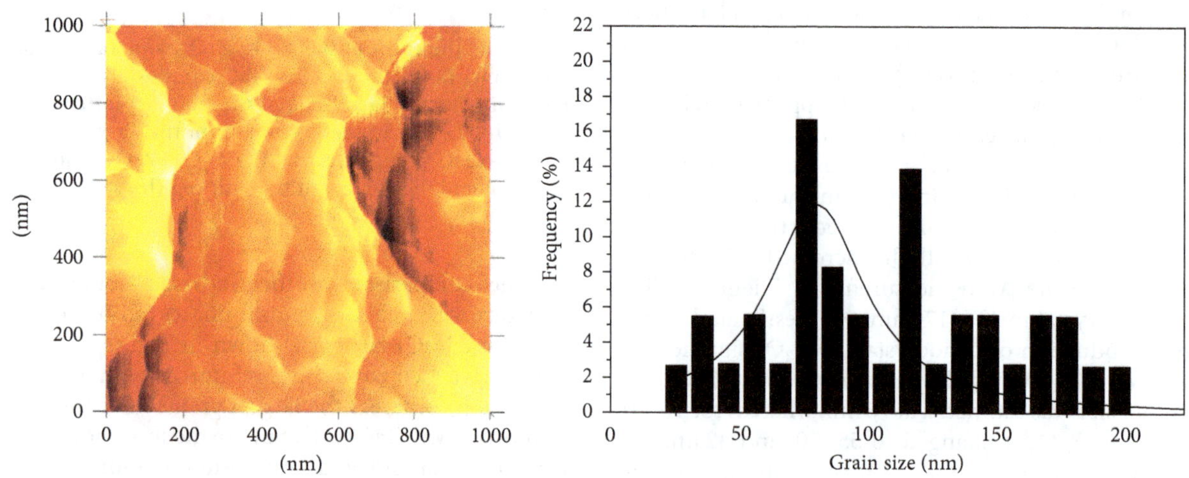

FIGURE 6: Continued.

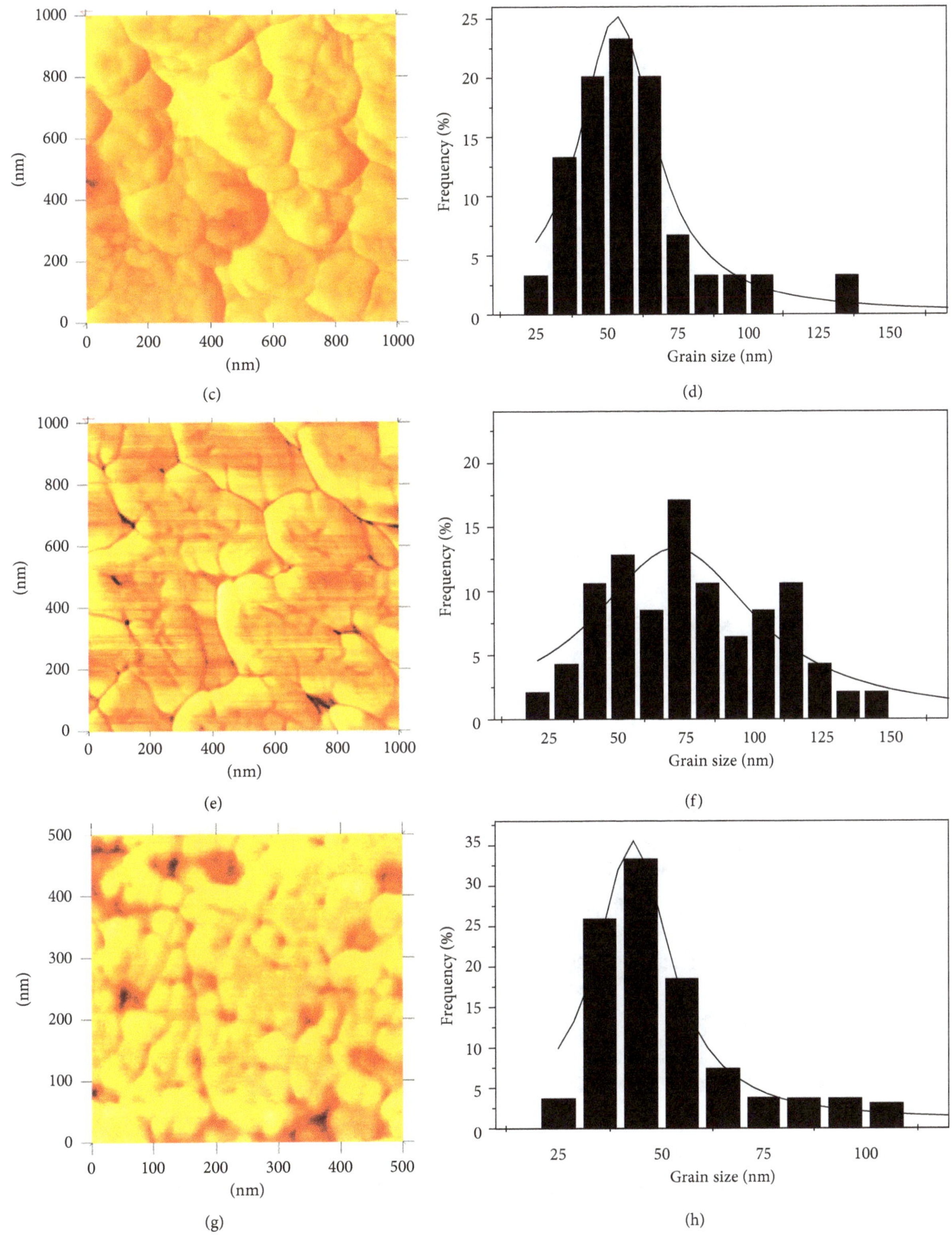

FIGURE 6: AFM pictures (a, c, e, g) and grain size distribution (b, d, f, h) of the mechanically alloyed and consolidated bulk nanocomposites: Ta-5Y_2O_3 (a, b), Ta-10Y_2O_3 (c, d), Ta-20Y_2O_3 (e, f), and Ta-40Y_2O_3 (g, h).

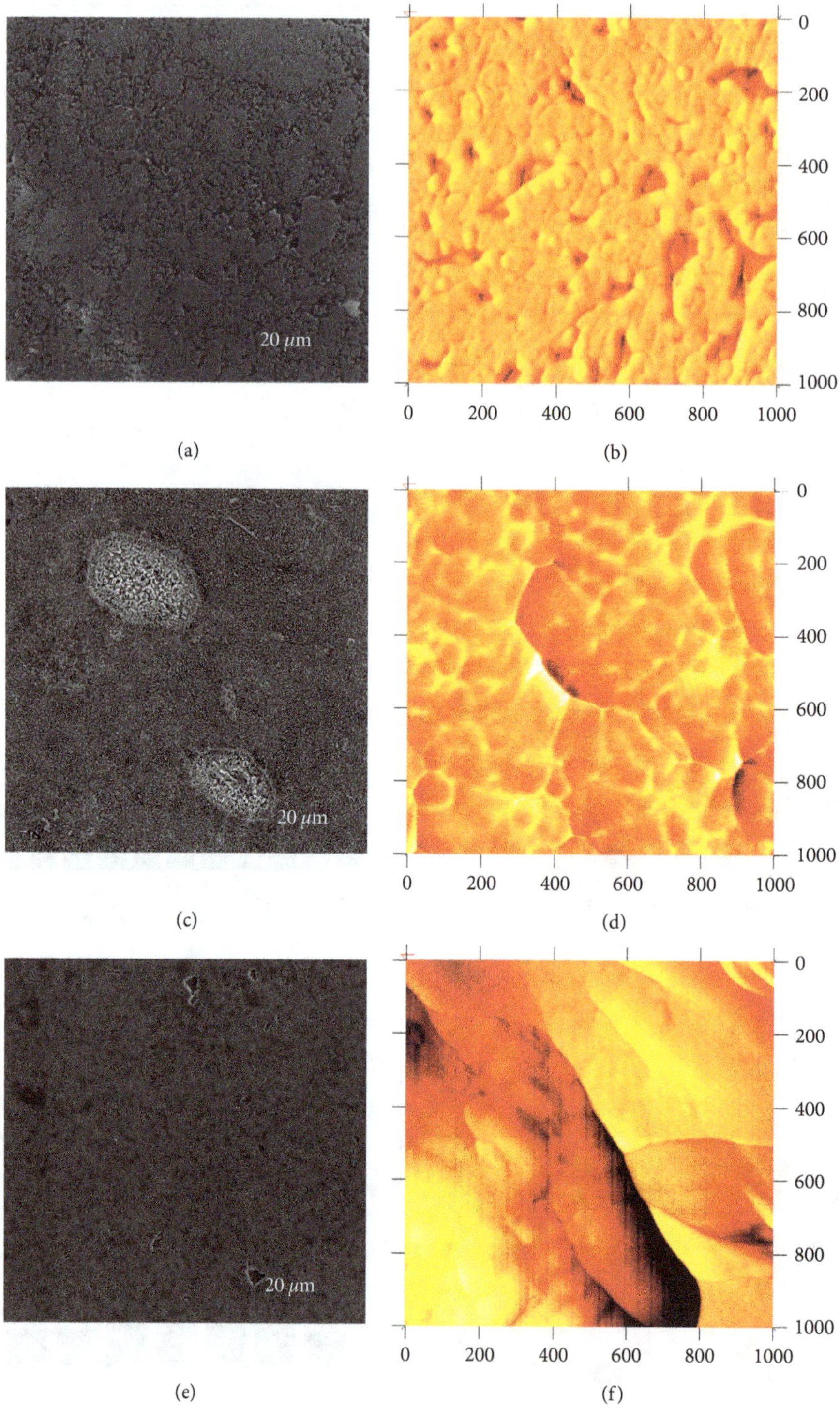

Figure 7: Example of SEM (a, c, e) and AFM (b, d, f) images of the consolidated bulk Ta-40Y_2O_3 (a, b), Ta-40ZrO_2 (c, d), and Ta-40TaC (e, f) nanocomposites.

The average grain size (estimated using AFM) of all the investigated bulk composites has been shown in Figure 8. In all the composites, the grains are smaller in comparison to pure nanocrystalline tantalum. The smallest grains have the composites reinforced with Y_2O_3 and ZrO_2. Generally, the majority of the composites have the grain size significantly below 100 nm.

The ceramic phase grains are well visible on the TEM images of the powders (Figure 5) and are well identified by AFM in the bulk samples (Figure 9). The example

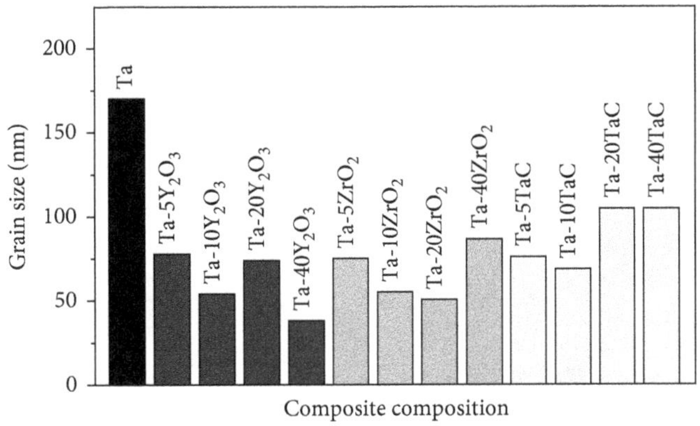

FIGURE 8: Average grain size for the Ta-xY_2O_3, Ta-xZrO_2, and Ta-xTaC bulk nanocomposites ($x = 0$, 5, 10, 20, and 40 wt.%); data measured using AFM (not all AFM images and grain size distributions have been shown in this paper).

microstructure of the Ta-20ZrO_2 bulk nanocomposite shows two-phase morphology (Figure 9(a)), with slightly larger Ta grains (Figure 9(b)) compared to the ZrO_2 grains (Figure 9(c)). The corresponding EDS spectra show the grain composition of both the metallic and the ceramic phases. On the lower magnified image (Figure 9(a)), the ceramic phase grains are well visible, because after chemical etching, they are flatter compared to the Ta grains (the samples for the microscopic observations were grinded, polished, and chemically etched to reveal the grains, hence, the etched craters among the sintered agglomerate particles).

The mechanical properties (Figure 10, Table 1) show that the nanocrystalline materials have high strength. The hardness of pure microcrystalline Ta (447 HV) is lower compared to pure nanocrystalline Ta (584 HV). An introduction of the reinforced ceramic phase results in an increase in the hardness up to 1398 HV for Ta-10TaC. Up to the 20 wt.% content of the ceramic phase, the hardness remains very high for all the investigated materials: the highest for the Ta-TaC and the lowest for the Ta-ZrO_2 composites. As for the composites of the 40 wt.% content of the ceramic phase, their content is too high to achieve full material integration at the sintering temperature of 1300°C (this needs further investigation at a higher consolidation temperature); therefore, the hardness is significantly lower compared to other composites. Young's modulus increases from 164 GPa for nanocrystalline Ta to the highest value of 346 GPa for Ta-40TaC nanocomposites. The other parameters for composites (W_t, total mechanical work of indentation; N_{plast}, plastic deformation portion; and C_{IT}, indentation creep) are the highest for the 40 wt.% content of the ceramic phase, pointing to the too low sintering temperature (or time) to achieve full integrity of the samples (especially in the case of oxide-reinforced composites). The load-displacement curves (5 indents made on each sample at different spots) in most cases overlap one another indicating very good homogeneity of the microstructure and uniform material deformation during force loading-unloading. Consequently, the mechanical properties have a low value of standard deviation. At the highest oxide ceramic phase content, the curves have a relatively broad spectrum, which is reflected in the low mechanical properties of these composites.

4. Discussion

The mechanical alloying applied to the tantalum and the ceramic phase particles leads to the formation of nanocrystalline mixture of both tantalum and ceramic grains. The grains have the size of several nanometers. The hot pressing in the PPS mode (performed at an elevated but relatively low temperature) leads to diffusion processes, which is necessary for strong particle bonding. The elevated temperature, however, is also the driving force for the grain growth. The relatively fast heating rate, short sintering time, and low sintering temperature result in a limited grain growth. The large volume of the grain boundaries of the nanocrystalline material should improve the densification process through the sliding mechanism. The agglomerates (formed in the mechanical alloying), which have significant voids between them work against high densification [25]. The pressure acting on the powders when the temperature increases (to the constant sintering temperature) improves the densification and reduces the voids between the consolidated powders. In general, for the densification of the nanocrystalline material, the consolidation temperature can be significantly lower compared to the microcrystalline material [13, 25]. The ceramic phase in the tantalum-based composites can suppress grain growth during consolidation. Hence, all the nanocomposites have a significantly lower grain size compared to pure nanocrystalline tantalum. The higher content (40 wt.%) of the oxide phase acts as a diffusion barrier; thus, in the Ta-40ZrO_2 and Ta-40Y_2O_3 composites, full microstructure integration during hot pressing at given conditions was not obtained. This is reflected in the mechanical properties, which are the worst for the Ta-40ZrO_2 and Ta-40Y_2O_3 composites. The high affinity of tantalum to oxygen leads to the formation of grain boundary oxides that cause embrittlement and grain size growth and the high temperature intensifies this effect [26]. In the case of the Ta-40TaC composite, the aforementioned process is limited. The high-temperature contact of the Ta-based powders with the graphite die and the punches leads to the diffusion of carbon and the

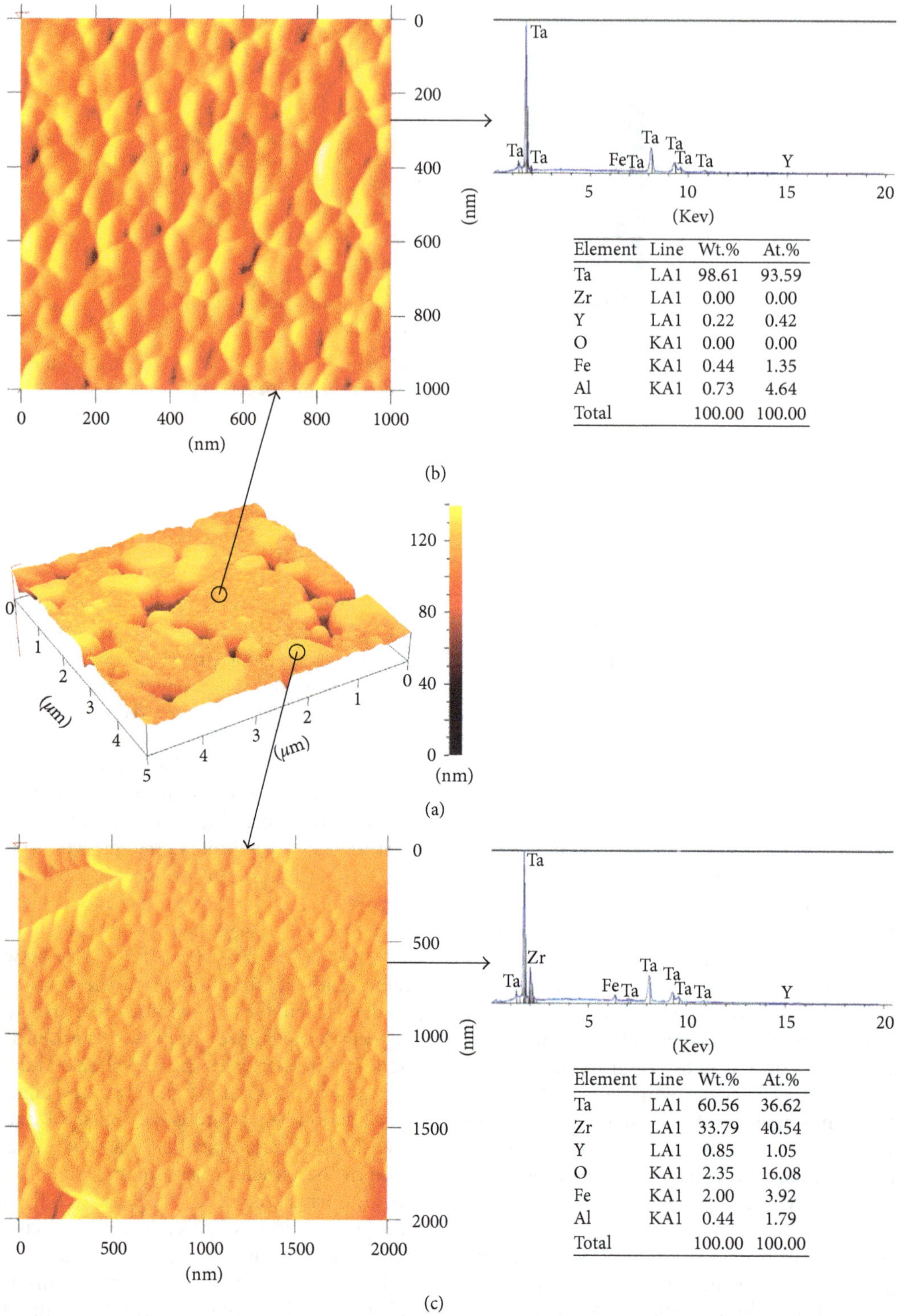

Element	Line	Wt.%	At.%
Ta	LA1	98.61	93.59
Zr	LA1	0.00	0.00
Y	LA1	0.22	0.42
O	KA1	0.00	0.00
Fe	KA1	0.44	1.35
Al	KA1	0.73	4.64
Total		100.00	100.00

Element	Line	Wt.%	At.%
Ta	LA1	60.56	36.62
Zr	LA1	33.79	40.54
Y	LA1	0.85	1.05
O	KA1	2.35	16.08
Fe	KA1	2.00	3.92
Al	KA1	0.44	1.79
Total		100.00	100.00

FIGURE 9: Example AFM images and EDS analysis showing the phase distribution in the consolidated bulk Ta-20ZrO_2 nanocomposite: larger area view (a), magnified area of the Ta grains (b), and ZrO_2 grains (c) with the corresponding EDS analysis.

formation of tantalum carbides, which was also observed by other authors [27].

The nanocomposites (composed of refractory tantalum as the metal matrix with the embedded refractory ceramic phase such as Y_2O_3, ZrO_2, and TaC) have very high mechanical properties, particularly the nanocrystalline Ta-TaC materials, which are promising in terms of heavy load conditions.

5. Conclusions

In this work, nanocomposites based on the Ta matrix reinforced by the ceramic phase of the Y_2O_3, ZrO_2, and TaC particles were developed. The mechanical alloying followed by hot pressing working in the pulse plasma sintering mode was applied for the bulk nanocomposite formation. We

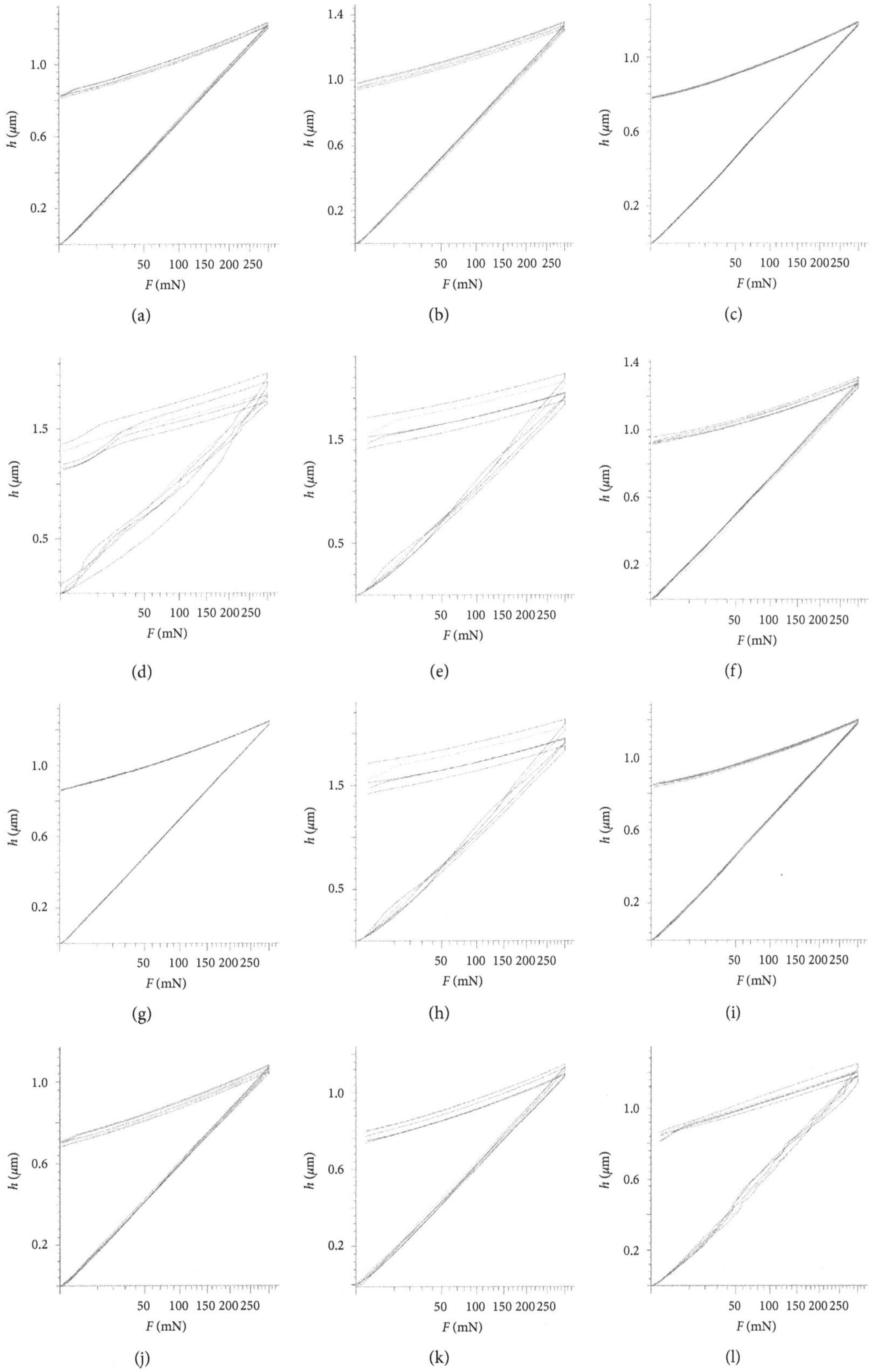

FIGURE 10: Continued.

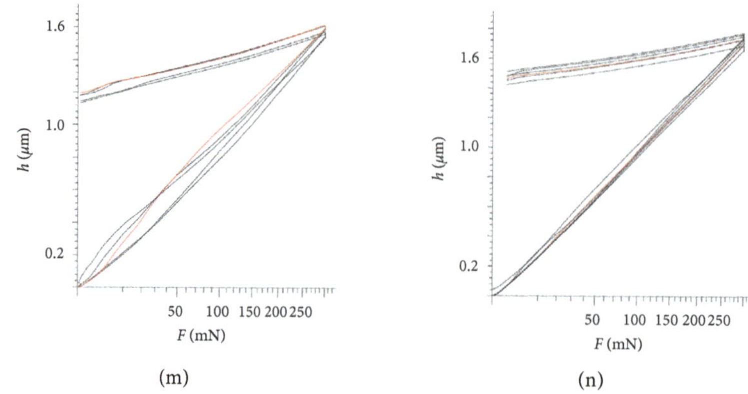

FIGURE 10: Load-displacement curves for consolidated bulk nanocrystalline composites: Ta-5Y_2O_3 (a), Ta-10Y_2O_3 (b), Ta-20Y_2O_3 (c), Ta-40Y_2O_3 (d), Ta-5ZrO_2 (e), Ta-10ZrO_2 (f), Ta-20ZrO_2 (g), Ta-40ZrO_2 (h), Ta-5TaC (i), Ta-10TaC (j), Ta-20TaC (k), and Ta-40TaC (l); for comparison bulk consolidated nanocrystalline Ta (m) and microcrystalline Ta (n).

TABLE 1: Mechanical properties of the investigated hot-pressed Ta-ceramic nanocomposites (data for micro- and nanocrystalline Ta for comparison).

	Mechanical properties				
	HV	E_{IT} (GPa)	W_t (μJ)	N_{plast} (%)	C_{IT} (%)
Ta micro	447 ± 17	211 ± 5	0.17 ± 0.01	81.6 ± 0.5	1.92 ± 0.08
Ta nano	584 ± 19	164 ± 8	0.16 ± 0.01	71.1 ± 2.5	1.54 ± 0.19
Ta nanocomposites:					
Ta-5Y_2O_3	1055 ± 23	248 ± 2	0.12 ± 0.01	63.6 ± 0.7	1.14 ± 0.08
Ta-10Y_2O_3	835 ± 23	234 ± 2	0.14 ± 0.01	68.7 ± 0.5	1.56 ± 0.07
Ta-20Y_2O_3	1165 ± 8	242 ± 1	0.12 ± 0.01	60.0 ± 0.2	1.26 ± 0.03
Ta-40Y_2O_3	436 ± 52	114 ± 10	0.20 ± 0.04	70.3 ± 3.5	1.70 ± 0.48
Ta-5ZrO_2	821 ± 24	231 ± 4	0.14 ± 0.01	69.6 ± 1.0	1.66 ± 0.08
Ta-10ZrO_2	920 ± 24	240 ± 8	0.13 ± 0.01	67.3 ± 0.5	1.72 ± 0.09
Ta-20ZrO_2	1003 ± 3	246 ± 1	0.13 ± 0.01	64.7 ± 0.1	1.30 ± 0.03
Ta-40ZrO_2	366 ± 44	115 ± 3	0.21 ± 0.02	75.5 ± 2.6	2.15 ± 0.17
Ta-5TaC	984 ± 21	273 ± 3	0.13 ± 0.01	67.8 ± 0.4	1.29 ± 0.05
Ta-10TaC	1398 ± 46	336 ± 5	0.11 ± 0.01	61.6 ± 0.9	1.01 ± 0.05
Ta-20TaC	1252 ± 69	303 ± 11	0.12 ± 0.01	63.5 ± 1.2	1.17 ± 0.07
Ta-40TaC	994 ± 48	346 ± 12	0.12 ± 0.01	70.1 ± 1.6	2.17 ± 0.69

HV, Vickers hardness; E_{IT}, indentation modulus; W_t, total mechanical work of indentation; N_{plast}, plastic deformation portion; C_{IT}, indentation creep.

made Ta composites with addition of up to 40% of the ceramic phase. The nanocomposite materials have the average grain size of approximately 40–100 nm, significantly lower than nanocrystalline Ta (170 nm). The lowest grain size has the composites reinforced by oxides. Both reinforced factors, the fine-grained nanostructure and the ceramic phase, lead to significant increase in the hardness as well as Young's modulus. For Ta-10TaC, the hardness and the Young's modulus reach 1398 HV and 336 GPa, respectively. For comparison, pure nanocrystalline Ta has the value of 584 HV and 164 GPa, respectively, whereas microcrystalline Ta has the value of 447 HV and 211 GPa, respectively. The combination of the MA and PPS processes has a potential in developing nanocomposites of high mechanical properties.

Conflicts of Interest

The authors declare that there are no conflicts of interest regarding the publication of this paper.

Acknowledgments

The work has been financed by the National Science Centre, Poland, under project identification DEC-2015/19/B/ST5/02595.

References

[1] G. Samsono, *Handbook of Refractory Compounds*, Springer-Verlag, New York, NY, USA, 2012.

[2] C. L. Briant, "The properties and uses of refractory metals and their alloys," *MRS Proceedings*, vol. 322, pp. 305–314, 1993.

[3] H. O. Pierson, *Handbook of Refractory Carbides and Nitrides. Properties, Characteristics, Processing and Applications*, Noyes Publ., New York, NY, USA, 1996.

[4] S. A. Ghaffari, M. A. Faghihi-Sani, F. Golestani-Fard, and M. Nojabayy, "Diffusion and solid solution formation between the binary carbides of TaC, HfC, ZrC," *International Journal of Refractory Metals and Hard Materials*, vol. 41, pp. 180–184, 2013.

[5] E. P. Simonenko, N. P. Simonenko, Y. S. Ezhov, V. G. Sevastyanov, and N. T. Kuznetsov, "Study of the synthesis of nanocrystalline mixed tantalum–zirconium carbide," *Physics of Atomic Nuclei*, vol. 78, no. 12, pp. 1357–1365, 2015.

[6] L. Xu, S. Wei, J. Li, G. Zhang, and B. Dai, "Preparation, microstructure and properties of molybdenum alloys reinforced by in-situ Al_2O_3 particles," *International Journal of Refractory Metals and Hard Materials*, vol. 30, no. 1, pp. 208–212, 2012.

[7] F. Xiao, L. Xu, Y. Zhou et al., "Preparation, microstructure, and properties of tungsten alloys reinforced by ZrO_2 particles," *International Journal of Refractory Metals and Hard Materials*, vol. 64, pp. 40–46, 2017.

[8] M. Battabyal, R. Schäublin, P. Spätig, and N. Baluc, "W-2 wt.% Y_2O_3 composite: microstructure and mechanical properties," *Materials Science and Engineering A*, vol. 538, pp. 53–55, 2012.

[9] S. A. Ghaffari, M. A. Faghihi-Sani, F. Golestani-Fard, and S. Ebrahimi, "Pressureless sintering of Ta0.8Hf0.2C UHTC in the presence of $MoSi_2$," *Ceramics International*, vol. 39, no. 2, pp. 1985–1989, 2013.

[10] M. A. Hussein, C. Suryanarayana, and N. Al-Aqeeli, "Fabrication of nano-grained Ti–Nb–Zr biomaterials using spark plasma sintering," *Materials and Design*, vol. 87, pp. 693–700, 2015.

[11] A. Michalski and D. Siemiaszko, "Nanocrystalline cemented carbides sintered by the pulse plasma method," *International Journal of Refractory Metals and Hard Materials*, vol. 25, no. 2, pp. 153–158, 2007.

[12] R. Chaim, "Superfast densification of nanocrystalline oxide powders by spark plasma sintering," *Journal of Materials Science*, vol. 41, no. 23, pp. 7862–7871, 2006.

[13] Z. Fang, P. Maheshwari, X. Wang, H. Y. Sohn, A. Griffo, and R. Riley, "An experimental study of the sintering nanocrystalline WC-co powders," *International Journal of Refractory Metals and Hard Materials*, vol. 25, no. 4–6, pp. 249–257, 2005.

[14] M. K. Surappa and P. K. Rohatgi, "Preparation and properties of cast aluminium-ceramic particle composites," *Journal of Materials Science*, vol. 16, no. 4, pp. 983–993, 1981.

[15] L. Olmos, C. L. Martin, and D. Bouvard, "Sintering of mixtures of powders: experiments and modelling," *Powder Technology*, vol. 190, no. 1-2, pp. 134–140, 2009.

[16] R. Orrù, R. Licheri, A. M. Locci, A. Cincotti, and G. Cao, "Consolidation/synthesis of materials by electric current activated/assisted sintering," *Materials Science and Engineering R*, vol. 63, no. 4–6, pp. 127–287, 2009.

[17] M. S. Yurlova, V. D. Demenyuk, Y. L. Lebedev, D. V. Dudina, E. G. Grigoryev, and E. A. Olevsky, "Electric pulse consolidation: an alternative to spark plasma sintering," *Journal of Materials Science*, vol. 49, no. 3, pp. 952–985, 2014.

[18] J. Q. Li, W. A. Sun, W. Q. Ao, K. M. Gu, and P. Xiao, "Al_2O_3–FeCrAl composites and functionally graded materials fabricated by reactive hot pressing," *Composites Part A*, vol. 38, no. 2, pp. 615–620, 2007.

[19] S. S. Nayak, M. Wollgarten, J. Banhart, S. K. Pabi, and B. S. Murty, "Nanocomposites and an extremely hard nanocrystalline intermetallic of Al–Fe alloys prepared by mechanical alloying," *Materials Science and Engineering A*, vol. 527, no. 9, pp. 2370–2378, 2010.

[20] G. Zhang and D. Gu, "Synthesis of nanocrystalline TiC reinforced W nanocomposites by high-energy mechanical alloying: microstructural evolution and its mechanism," *Applied Surface Science*, vol. 273, pp. 364–371, 2013.

[21] O. B. Zgalat-Lozinskii, "Nanocomposites based on refractory compounds, consolidated by rate-controlled and spark-plasma sintering (review)," *Powder Metallurgy and Metal Ceramics*, vol. 53, no. 1-2, pp. 19–30, 2014.

[22] T. Niu, W.-W. Chen, H.-W. Cheng, and L. Wang, "Grain growth and thermal stability of nanocrystalline Ni–TiO_2 composites. Transactions of Nonferrous Metals," *Society of China*, vol. 27, no. 10, pp. 2300–2309, 2017.

[23] J. Jakubowicz, G. Adamek, M. Sopata, J. K. Koper, T. Kachlicki, and M. Jarzebski, "Microstructure and electrochemical properties of refractory nanocrystalline Tantalum-based alloys," *International Journal of Electrochemical Science*, vol. 13, pp. 1956–1975, 2018.

[24] J. Jakubowicz, G. Adamek, and M. Sopata, "Characterization of high-energy ball-milled and hot-pressed nanocrystalline tantalum," *IOP Conference Series: Materials Science and Engineering*, vol. 216, p. 012006, 2017.

[25] S. H. Yoo, T. S. Sudarshan, K. Sethuram, G. Subhash, and R. J. Sowding, "Consolidation and high strain rate mechanical behavior of nanocrystalline tantalum powder," *Nanostructured Materials*, vol. 12, no. 1–4, pp. 23–28, 1999.

[26] M. Bischof, S. Mayer, H. Leitner et al., "On the development of grain growth resistant tantalum alloys," *International Journal of Refractory Metals and Hard Materials*, vol. 24, no. 6, pp. 437–444, 2006.

[27] Y. Kim, E. P. Kim, J. W. Noh, S. H. Lee, Y. O. Kwon, and I. S. Oh, "Fabrication and mechanical properties of powder metallurgy tantalum prepared by hot isostatic pressing," *International Journal of Refractory Metals and Hard Materials*, vol. 48, pp. 211–216, 2015.

11

Numerical Simulations for Large Deformation of Geomaterials Using Molecular Dynamics

Ziyang Zhao[1,2] **and Jun Zhang**[1,2]

[1]*Key Laboratory of Transportation Tunnel Engineering, Ministry of Education, Chengdu 610031, China*
[2]*Key Laboratory of Highway Construction and Maintenance Technology in Loess Region, Shanxi Transportation Research Institute, Taiyuan 030006, China*

Correspondence should be addressed to Jun Zhang; zhangjun_hust@qq.com

Academic Editor: Francesco Ruffino

From the microperspective, this paper presents a model based on a new type of noncontinuous theoretical mechanical method, molecular dynamics (MD), to simulate the typical soil granular flow. The Hertzian friction formula and viscous damping force are introduced in the MD governing equations to model the granular flow. To show the validity of the proposed approach, a benchmark problem of 2D viscous material flow is simulated. The calculated final flow runout distance of the viscous material agrees well with the result of constrained interpolated profile (CIP) method as reported in the literature. Numerical modeling of the propagation of the collapse of three-dimensional axisymmetric sand columns is performed by the application of MD models. Comparison of the MD computational runout distance and the obtained distance by experiment shows a high degree of similarity. This indicates that the proposed MD model can accurately represent the evolution of the granular flow. The model developed may thus find applications in various problems involving dense granular flow and large deformations, such as landslides and debris flow. It provides a means for predicting fluidization characteristics of soil large deformation flow disasters and for identification and design of appropriate protective measures.

1. Introduction

As a kind of typical granular material, soil causes the common flow forms mainly for earthwork excavation, flow of soil slope by filling, and sudden slip on a weak foundation soil. The large deformation disasters caused by the flow of soil particles have received considerable attention and researches both at home and abroad.

Molecular dynamics originated in the 50s and began to be widely concerned in the mid-70s. In 1957, the state equations of gas and liquid were studied, firstly using molecular dynamics in the hard sphere model, thus setting a precedent for studying the macroscopic properties in the molecular dynamics method [1]. Molecular dynamics started to develop after 70s, gradually perfected its system in the late 80s and got rapid development along with the flourish of computer hardware in the 90s. Molecular dynamics is a connecting link between macroscopic characteristics and microstructure, which can make certain microscopic interpretation for theoretical analysis and experimental researchdifficult to understand. Especially in chemistry, biology, physics, and material science or related fields, it gets better application [2–4]. In recent years, molecular dynamics is also applied to the research on granular flow of geomaterials. The inducing factors of shallow landslide were studied using the molecular dynamics method [5]. Factors affecting the occurrence of microsliding in the sand pile were discussed with the molecular dynamics method [6]. Molecular dynamics simulations are particularly well suited to investigate the effects of materials properties of liquid-solid interfaces on flow boundary conditions [7].

Molecular dynamics method, which is a combination of physics, mathematics, and chemistry, mainly relies on Newtonian mechanics to simulate the movement of molecular system and can simulate the movement of a single particle in a large number of particle collection system (solid, gas, or liquid). Its key is the concept of movement, that is, to compute the evolution of the position, velocity, and orientation of particles over time. The particle in molecular dynamics can be atom, molecule, or larger set of particles [8]. The basic principle of molecular dynamics is the application of Newton's classical mechanics to firstly calculate the trajectories of all particles in the physical system and then by using relevant statistical methods to calculate mechanics, thermodynamics, and dynamics characteristics of the system. That is to say, firstly, the system of particles is abstracted to many interacting particles, and the interaction force between particles is obtained by using the potential energy function derivation method. And then, the acceleration and velocity of each particle are calculated according to the Newtonian mechanics, and the particles have the corresponding trajectories after a certain integral steps, which must be preserved by setting a certain time interval. Finally, calculation results of the interested quantities are extracted based on statistical principle [9–11].

At present, the molecular dynamics simulation researches mostly focus on physics, biology, and chemistry, but the application in the field of geotechnical engineering is rare. The molecular dynamics approach is suitable to model granular material and to observe the trajectory of a single particle, so as to possibly identify its dynamical properties. Thus, the molecular dynamics method is tried to be applied to the granular flow characteristics research, and a 3D MD model that simulates granular flow is presented in this work, which is dedicated to provide a strong scientific basis for solving geotechnical engineering problem.

2. Numerical Approach

The accuracy of the model depends on the accuracy of description of external forces acting on the particle and the treatment of collisions. The particles can be either rigid and/or soft. A soft particle model [12] is used in analyzing particle-particle and particle-wall interactions in this paper. The particles are considered to be spherical and identical. The collisions are assumed to be central and involve both linear elastic and damping forces. In the study, 3D ordered granular packings made up of spherical particles are considered. The motion equation for the sphere i is written in a general form as

$$\ddot{r}_i = \frac{F_i}{m_i}, \quad i = 1, 2, \ldots, N, \tag{1}$$

where $\ddot{r}_i$ is the displacement, F_i is the force, m_i is the mass, and N is the total number of particles. F_i is the composition of contact force between spheres j and i, where the interaction force between any contacting spheres under compression is given by the Hertzian law [13]:

$$F_{\text{hz}}(t) = \sqrt{\delta}\sqrt{\frac{R_i R_j}{R_i + R_j}} F_{\text{hk}}(t), \tag{2}$$

where δ is the overlap distance of two particles and the R_i and R_j stand for the radius of the particles i and j, respectively. The expression of $F_{\text{hk}}(t)$ is defined as [13]

$$F_{\text{hk}}(t) = \left(k_n \delta \mathbf{n}_{ij}(t) - m_{\text{eff}} \gamma_n v_n(t)\right) - \left(k_t \Delta \mathbf{s}_t(t) + m_{\text{eff}} \gamma_t v_t(t)\right), \tag{3}$$

where k_n and k_t are the elastic constants for normal contact and tangential contact, respectively; γ_n and γ_t are the viscoelastic damping constants for normal contact and tangential contact, respectively; m_{eff} is the effective mass of two particles of mass m_i and m_j; $\Delta \mathbf{s}_t$ is the tangential displacement vector between two spherical particles which is truncated to satisfy a frictional yield criterion; $\mathbf{n}_{ij}$ is the unit vector along the line connecting the centers of the two particles; and v_n and v_t are the normal and tangential components of the relative velocity of the two particles, respectively.

Viscous damping force calculation formula of the particle is based on the Stokes drag law of fluid mechanics [14], which can be expressed as

$$f = -3\pi\eta d v, \tag{4}$$

where η is the dynamic viscosity of the frictional fluid, d is the diameter of particle, and v is the velocity.

At present, in the field of molecular dynamics, the integral of the motion equation mainly adopts Verlet algorithm [15]. The integral algorithm putted forward by Verlet is the most widely used in the molecular dynamics. The numerical simulations are performed using the molecular dynamics package LAMMPS [10] in the work, by considering spheres as point masses connected by nonlinear springs.

The calculation formulas of Verlet algorithm are as follows:

$$r_i(t+\Delta t) = r_i(t) + \Delta t \cdot \dot{r}_i(t) + (\Delta t)^2 \cdot \frac{\ddot{r}_i(t)}{2} + O\left((\Delta t)^3\right), \tag{5}$$

$$r_i(t-\Delta t) = r_i(t) - \Delta t \cdot \dot{r}_i(t) + (\Delta t)^2 \cdot \frac{\ddot{r}_i(t)}{2} + O\left((\Delta t)^3\right), \tag{6}$$

$$r_i(t+\Delta t) = -r_i(t-\Delta t) + 2r_i(t) + \frac{F_i(\Delta t)}{m_i}(\Delta t)^2 + O\left((\Delta t)^3\right), \tag{7}$$

$$v_i(t) = \frac{dr_i}{dt} = \frac{1}{2(\Delta t)}\left[r_i(t+\Delta t) - r_i(t-\Delta t)\right] + O\left((\Delta t)^2\right). \tag{8}$$

3. Validation of the MD Model

The performance of the new model will be verified in this section. It is essential to create an input script containing the desired commands before running LAMMPS. The Verlet algorithm is applied to the integral of motion equation in the input scripts. When numerical calculation starts, LAMMPS reads the input script, and the displacement, velocity, and acceleration of every granular particle are calculated and updated, thus obtaining the evolution

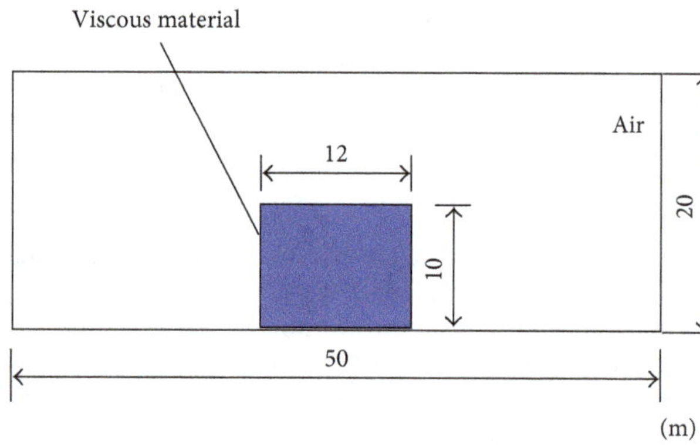

FIGURE 1: Calculation model [16].

TABLE 1: Calculation parameters.

Parameters	Value
Grain diameter (m)	0.05 [16]
Density (kg/m^3)	2000.0 [16]
Elastic constant for normal contact (MPa)	1 [17]
Elastic constant for tangential contact (MPa)	0.29
Damping constant for normal contact	0.001 [18]
Damping constant for tangential contact	0.0005
Cohesion (Pa)	800 [16]
Time step (s)	1.0×10^{-4}
Particle number	57,251

trajectory of the whole calculation system over time. A two-dimensional simulation of viscous material flow was conducted to verify the reliability of the MD method.

Viscous materials flow on the plane under its own gravity, and a literature [16] has detailed calculation and analysis on this flow process using constrained interpolated profile (CIP). CIP method was developed for prediction of large deformations associated with a geomaterial flow, so it can be used as a contrast with the MD method in this work. The model [17] is shown in Figure 1.

Based on the molecular dynamics model mentioned above, the viscous material flow process was simulated. In MD simulation, the calculation area of 50 m long and 20 m wide and the viscous material size of 12 m long and 10 m wide were set up. The particles were idealized as circle with ignoring particle size distribution. The boundary conditions were fixed, and the interaction between particles in the system was calculated according to the Hertzian friction formula. The initial velocity was set to zero.

Calculation parameters and simulation results are shown in Table 1 and Figure 2, respectively. The molecular dynamics simulation results are compared with the CIP results of the existing literature [16], to verify the validity and correctness of molecular dynamics algorithm in particle flow simulation. Contrast can be seen in Figure 3, and the red parts are the MD numerical simulations of flow profile, while black lines show CIP method calculation configurations in the literature [16].

Figure 3 shows that the MD results are consistent with the CIP results. The literature [16] discussed in detail the applicability and accuracy of the CIP method in describing

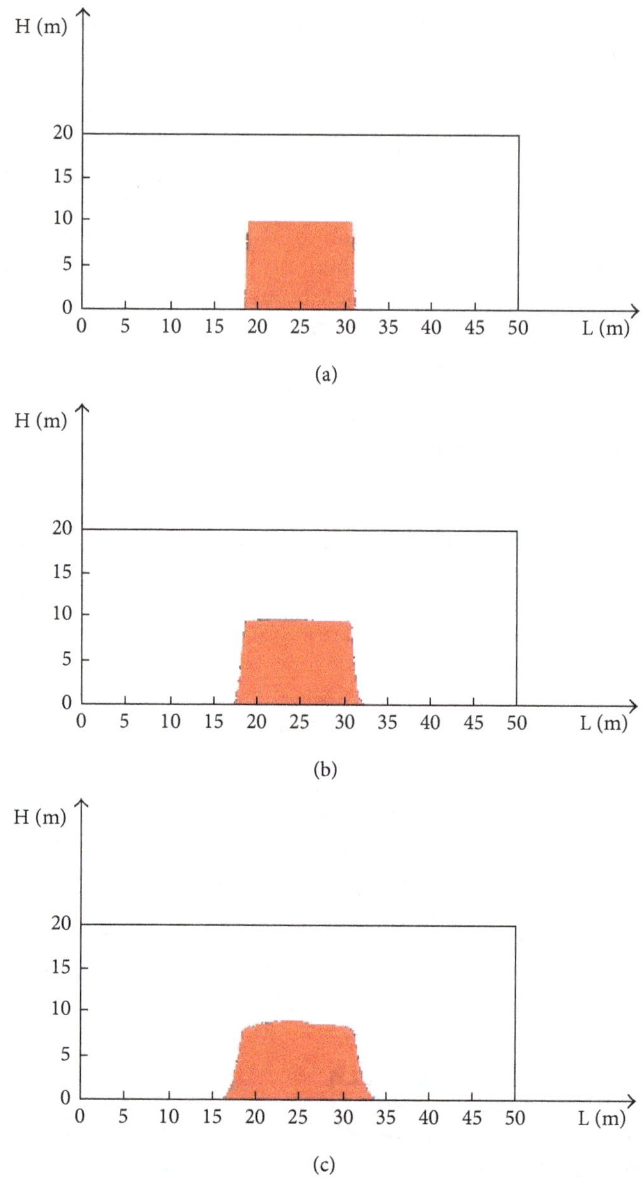

FIGURE 2: Simulated propagation of the viscous material flow. (a) $t = 0.1$ s. (b) $t = 0.5$ s. (c) $t = 1.0$ s.

the large deformation and fluidization motion characteristics of the viscous material. Therefore, the applicability and accuracy of the MD program in the description of fluidization motion characteristics of the viscous material have also been verified and validated.

4. Application of MD Modeling to Collapse Test and Numerical Simulation for Sand Column Collapse

This section discusses the three-dimensional simulation of the propagation of sand flow as the application of MD modeling to real large deformation of geomaterials. The motions of sands after collapse were represented in the MD simulation. This study can produce preliminary results for the prediction fluidization characteristics of large deformation flow disasters of geomaterials.

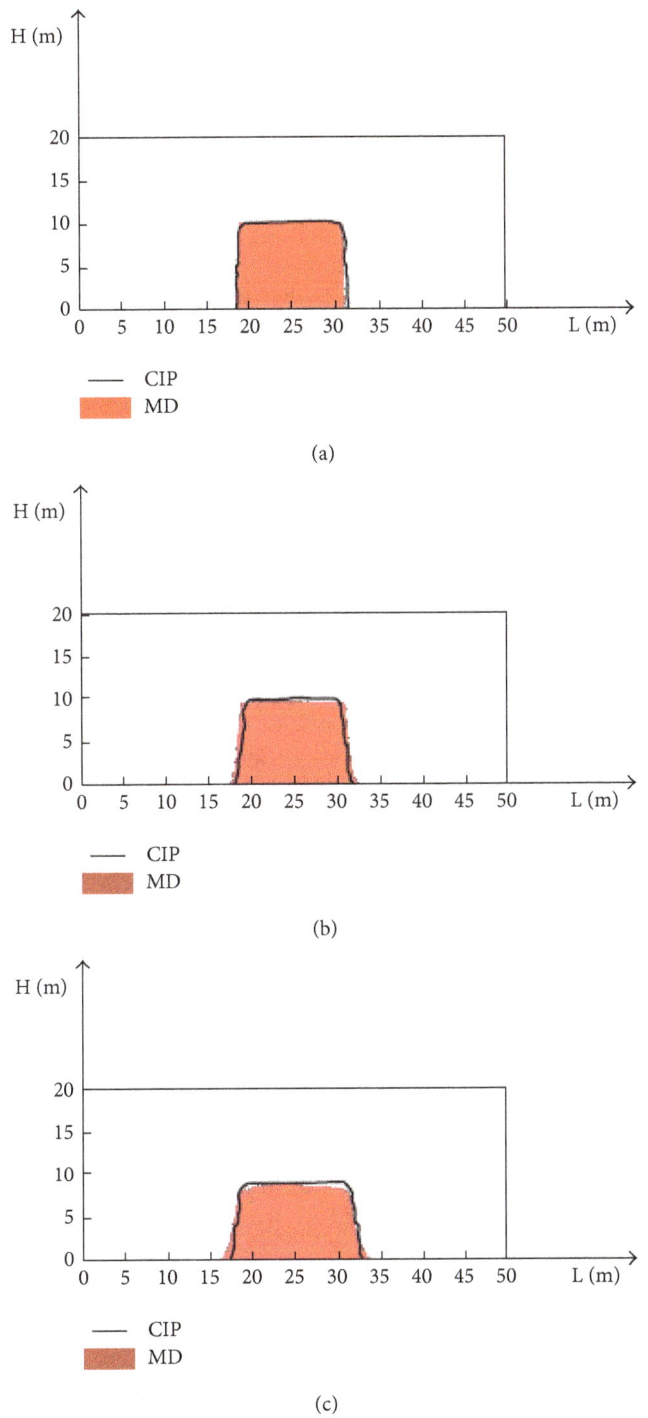

FIGURE 3: Comparisons of MD and CIP results. (a) $t = 0.1$ s. (b) $t = 0.5$ s. (c) $t = 1.0$ s.

FIGURE 4: The final configuration of sand flow.

4.1. Sand Column Collapse Experiment. The collapse test of three-dimensional axisymmetric sand columns was carried out on the basis of the apparatus developed independently, which was made of steel with the inner diameter of 10 cm, the height of 10 cm, and the thickness of 1 cm.

Firstly, the collapse device was placed on a smooth horizontal plane to ensure that there was no gap between the experimental device and the contact surface. Secondly, the dry sand was slowly sprinkled into apparatus to keep uniform distribution. When the sand sample reached the height of the design of 10 cm, it was stopped and made to stand for a period of time, until the sand was compacted under its own gravity. Then, the camera was opened, and the collapse device was vertically and quickly removed. Finally, configuration images of sand were collected and analyzed. In the experiment, the frictional effects between individual grains on granular flow play a role only in the last instant of the flow, as it comes to an abrupt halt.

The final configuration of sand after collapse is conical in the experiment, as shown in Figure 4. The average value of the cone diameter is 28.87 cm. The time-history curve of the cone diameter in the slump test was obtained and shown in Figure 5. It shows that the cone bottom diameter has a trend of increase with time. The slope of the polyline indicates the speed of growth and the diameter growth rates are obviously different in the whole collapse process. As the evacuation of the test device has a certain friction effect on sands in the experiment, the forces of sands are not stable in the whole process, which may bring certain influence on the flow rate and the final configuration of the sands.

4.2. Simulation of Granular Flows during the Collapse of Sand Columns. The developed and validated MD model was used to simulate the collapse propagation of an initially vertical 3D axisymmetric sand column in the work. The MD input file for 3D simulation of sand columns collapse was compiled, and then the diameter of the bottom circle formed by the final flowing configuration was calculated using the LAMMPS package. In the proposed model, the sand materials were discretized into a series of MD particles with a certain diameter. In the numerical simulation, sands were regarded as the uniformly distributed spherical particles, the calculation area was a cylinder with the diameter of 10 cm and the height of 10 cm, and the initial velocity of every sand particle was set to zero. The parameters used in the MD simulation were listed in Table 2. Calculating diameter was set 3 times of the expectation value of the particle diameter about 2 mm. The value of elastic constant for normal contact of the sand was 10 MPa,

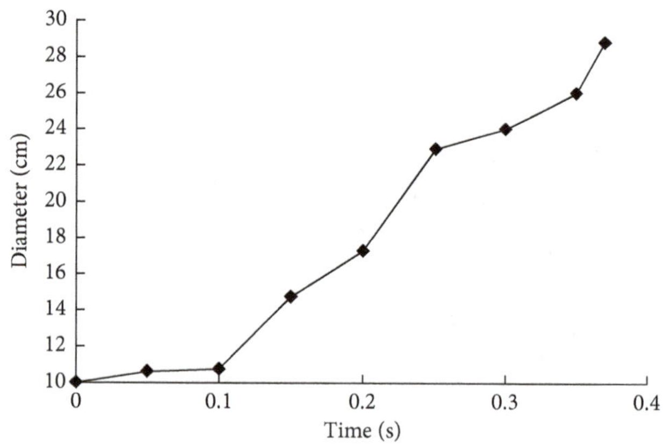

Figure 5: The time-history curve of the cone diameter.

Table 2: The parameters used in the MD simulation of a sand slump.

Parameters	Value
Grain diameter (mm)	2
Density (kg/m^3)	2650
Elastic constant for normal contact (MPa)	10
Elastic constant for tangential contact (MPa)	2.86
Damping constant for normal contact	0.1
Damping constant for tangential contact	0.05
Time step (s)	1.0×10^{-5}
Particle number	118,700

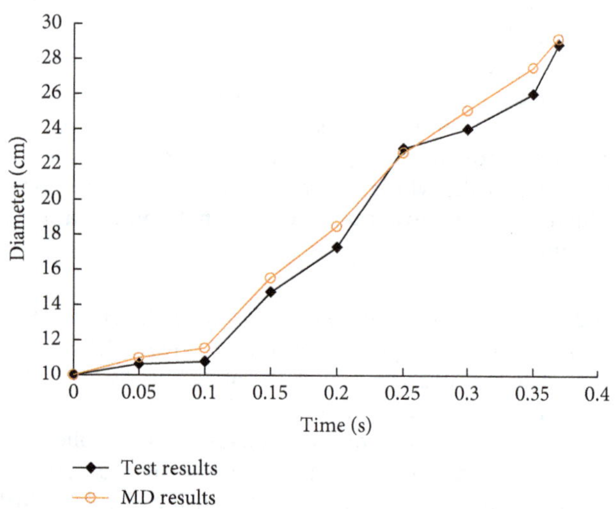

Figure 6: The change of the diameter of the cone bottom with time in MD simulation.

and the value of elastic constant for tangential contact was set two-sevenths of the former. The damping constant for normal contact of the sand was set as 0.1, and the damping constant for tangential contact was half of the former. The time step was 1.0×10^{-5} s, and there were about 118,700 MD particles used to represent the sand column in the simulations conducted. The final runout distance was investigated and compared with the experimental observation, and the result was presented subsequently.

Comparisons of MD numerical simulation results and the experimental results are shown in Figure 6. Excellent agreement is observed between the experiments and 3D MD simulations conducted with respect to the flow distance. Obviously, Figure 6 illustrates that there is a small error between the MD simulation results and the experimental results. The possible reason is that the MD numerical calculation process has ignored the friction effect on sand of experimental device.

Thus, the MD method can describe well the granular flow of sand columns after collapse, and the correctness and applicability of the model and algorithm of molecular dynamics are verified again.

5. Conclusions

Application of the MD method to the simulation of soil granular materials under large deformation is the subject of this paper. The Hertzian friction formula and viscous damping force are implemented in the MD framework to model the granular flow.

The model developed is validated by the collapse of viscous material flow as reported in the available literature. Excellent agreement is observed between the MD model simulations and CIP simulations with respect to the flow distance, thus verifying the performance of the MD modeling scheme.

Simulation of the collapse of 3D axisymmetric sand columns is also conducted. Numerical results for the granular flow pattern and final runout distance are in good agreement with the experimental observations. The flow velocity of the simulated deformed region within a sand column during the collapse is bigger than the experimental observation. The likely cause is that MD numerical simulations ignore the influence of the evacuation of collapse device on the sand, whereas the collapse device has created the friction on the sand particles when it is removed in the experiment process.

This study indicates that the MD model developed can be used to effectively simulate large deformation of granular materials, and geomaterials in general, if proper calculation models are implemented. This is due to the fact that the MD method is a particle-based mesh-free numerical method. In the MD method, the variables of particles are calculated through the integration procedure. If an appropriate integration step is chosen, the MD model developed is capable of presenting many discrete characteristics of dense granular flow for granular materials under large deformation. Thus, the MD model may find extensive applications in various problems involving granular flow such as landslides and debris flow.

Conflicts of Interest

The authors declare that they do not have any commercial or associative interest that represents a conflicts of interest in connection with the work submitted.

Acknowledgments

This work was supported by the National Natural Science Foundation of China (NSFC) (Grant no. 51608316), the Research Fund of the Key Laboratory of Transportation Tunnel Engineering, Ministry of Education (Grant no. TTE2014-05), and the Traffic Science and Technology Program in Shanxi Province (Grant no. 2017-1-20). The authors would like to express their gratitude for the financial support.

References

[1] B. J. Alder and T. E. Wainwright, "Phase transition for a hard sphere system," *The Journal of Chemical Physics*, vol. 27, no. 5, pp. 1208-1209, 1957.

[2] K. J. Oh and M. L. Klein, "A general purpose parallel molecular dynamics simulation program," *Computer Physics Communications*, vol. 174, no. 7, pp. 560–568, 2006.

[3] W. F. van Gunsteren and H. J. C. Berendsen, "Computer simulation of molecular dynamics: methodology, applications, and perspectives in chemistry," *Angewandte Chemie International Edition in English*, vol. 29, no. 9, pp. 992–1023, 1990.

[4] Y. Li, J. Xu, and D. Li, "Molecular dynamics simulation of nanoscale liquid flows," *Microfluidics and Nanofluidics*, vol. 9, no. 6, pp. 1011–1031, 2010.

[5] G. Martelloni, F. Bagnoli, and E. Massaro, "A computational toy model for shallow landslides: molecular dynamics approach," *Communications in Nonlinear Science and Numerical Simulation*, vol. 18, no. 9, pp. 2479–2492, 2013.

[6] F. Umar, V. Sparisoma, and Nurhasan, "Molecular dynamics simulation on particular grain weighting in a granular pile: an attempt to induce an artificial micro-landslide," in *Proceedings of the Symposium of the International Conference on Physics and its Applications*, vol. 1454, 1, pp. 95–98, Bandung, Indonesia, June 2012.

[7] L. Bocquet and J. L. Barrat, "Flow boundary conditions from nano- to micro-scales," *Soft Matter*, vol. 3, pp. 685–693, 2007.

[8] J. A. Barker, R. A. Fisher, and R. O. Watts, "Liquid argon: Monte Carlo and molecular dynamics calculations," *Molecular Physics*, vol. 21, no. 4, pp. 657–673, 1971.

[9] S. M. H. Karimian and S. Izadi, "Bin size determination for the measurement of mean flow velocity in molecular dynamics simulations," *International Journal for Numerical Methods in Fluids*, vol. 71, pp. 930–938, 2013.

[10] J. M. Haile, *Molecular Dynamics Simulation*, Wiley, New York, NY, USA, 1992.

[11] G. Karniadakis, A. Beskok, and N. Aluru, *Microflows and Nanoflows: Fundamentals and Simulation*, Springer Science & Business Media, New York, NY, USA, 2006.

[12] P. K. Haff, "Discrete mechanics," in *Granular Matter: An Interdisciplinary Approach*, A. Mehta, Ed., pp. 141–160, Springer-Verlag, New York, NY, USA, 1994.

[13] K. L. Johnson, *Contact Mechanics*, Cambridge University Press, Cambridge, UK, 1987.

[14] P. Y. Cheng and H. K. Schachman, "Studies on the validity of the Einstein viscosity law and Stokes' law of sedimentation," *Journal of Polymer Science*, vol. 16, no. 81, pp. 19–30, 1955.

[15] L. Verlet, "Computer "experiments" on classical fluids I: thermodynamical properties of Lennard-Jones molecules," *Physical Review*, vol. 159, no. 1, pp. 98–103, 1967.

[16] S. Moriguchi, *CIP-Based Numerical Analysis for Large Deformation of Geomaterials*, Gifu University, Gifu, Japan, 2005.

[17] B. Wu, *Research on Preparation and Mechanical Properties of Biomimetic Adhesive Materials*, Nanjing University of Aeronautics and Astronautics, Nanjing, China, 2010.

[18] W. Jin-peng, Y. Zhi-chun, L. Bin et al., "Research on material damping test method," *Journal of Vibration, Measurement & Diagnosis*, vol. 28, no. 3, pp. 220–224, 2008.

12

Spectral Variation and Corresponding Changing Mechanism of Suspended Particulate Material Absorption in Poyang Lake during Flood Periods

Yuandong Wang,[1,2] **Xibin You**,[1] **Lianfang Yu**,[1] **Lihong Meng**,[1] **Xiangming Xu**,[1] **and Guangxu Liu**[1]

[1]*College of Geography and Planning, Gannan Normal University, Ganzhou 341000, China*
[2]*Key Laboratory of Digital Earth Science, Institute of Remote Sensing and Digital Earth, Chinese Academy of Sciences, Beijing 100094, China*

Correspondence should be addressed to Yuandong Wang; ydwang@yic.ac.cn

Academic Editor: Jose R. B. Cantalice

Remote sensing accuracy of the dynamic water environment under a changing environment due to climate and anthropogenic impacts should be improved by a deep understanding of the absorption properties pertinent to the water body. Spectral variation of suspended particulate material (SPM) absorption and its changing mechanism during flood periods of Poyang Lake were analyzed in this study. Absorption coefficients were measured and determined by spectrophotometry using the quantitative filter techniques (QFTs) based on field samples from July 12 to September 06, 2017. Two main optical active constituents (OACs) of SPM including nonalgal and phytoplankton algal particulates were quantified and analyzed in detail. The results suggested that, during study periods, Poyang Lake exhibits an overall high level of suspended nonalgal detritus and algal concentration and lower water clarity, showing a strong absorption by total suspended sediments that are closely related with the land-derived nonalgal detritus material which varied significantly with diverse proportions of minerals and organic materials from multiple sources. Due to variations in phytoplankton community pigment composition, concentration, and package effect, there exists a difference between optical absorption capacity on blue (440) and red (675) portions of bands, the former of which had an obvious enhancement due to the increased effect of accessory pigments in suspended phytoplankton particles from July to September. Evidence also presented that the nonlinear dependency of specific phytoplankton particulate absorption on pigment concentration for various trophic statuses in different periods could be unstable due to relative contributions of the package effect and accessory pigments; this could bring uncertainties to the parameterization of optical models and remote sensing algorithms proposed for accurate applications in lake water environment monitoring.

1. Introduction

Absorption coefficients of optical active constituents (OACs, i.e., particulate and dissolved materials) play an important role in the inherent optical properties (IOPs, i.e., absorption and specific coefficients), largely determining the apparent optical properties (AOPs, i.e., water leaving radiance and remote sensing reflectance) in natural waters [1, 2]. Therefore, remote sensing inversion accuracy of the dynamic water environment under a changing environment due to climate and anthropogenic disturbances should be improved by a deep understanding of the absorption properties pertinent to the water body under consideration [3–5].

Over past decades, variability in absorption coefficients has been thoroughly documented in oceanic waters (the so-called "Case 1"), where phytoplankton and their derivative products (organic detritus and dissolved yellow substance, arising by zooplankton grazing, or natural decay of the algal cells) are optically dominant, and other components are often assumed to covary with phytoplankton pigments [6–10]. Due to the simple configuration of Case 1 waters, single OAC models and algorithms based solely upon phytoplankton have performed satisfactorily in oceanic waters [9, 11–13].

However, absorption coefficients can show significant variability in space and time in inland waters, where non-algal particles consist of minerals and nonliving organics detritus such as feccal material or degrading phytoplankton cells are not proportional to phytoplankton. Therefore, models and algorithms based solely upon chlorophyll-bearing biota are often invalid when applied to inland waters that spread across different geographic regions [14–16].

In recent years, there has been an increasing focus on the environment responses of inland waters due to climate and anthropogenic impacts, wherein spectral properties are very often dominated by land-derived inorganic or organic suspended particulate materials (SPMs) [17–22]. As the largest fresh water lake in China, Poyang Lake is in a very dynamic hydrological environment with the complex benthic topography, and its water level is further fluctuated due to frequent occurrence of extreme climate events such as heavy precipitation and storms during flood periods [25]; besides, anthropogenic activities also impact the lake environment [17]. According to previous research, water clarity of most of the lake region is between 0 and 2 m and sediment material concentration could vary between 2.6 and 220.19 mg/L throughout the year [18]. The water quality, quantity, and aquatic conditions of Poyang Lake have great impact on downstream, middle, and lower Yangtze River and the East China seashore [17].

Many researchers have performed studies in this region, and they provided valuable evidence of the optical properties of suspended substances [23–25]. Chen et al. revealed that the partition of organic and inorganic materials is helpful to accurately determine the optical property of SPM [22]. Wu et al. confirmed the temporal and regional variability of specific inherent optical properties due to different water constituents [24], and the development of the regional SPM model is fundamental for managing and protecting aquatic ecosystems [25].

Therefore, the complex spectral behavior of suspended sediments due to various OAC composition dynamics under evolving hydrological conditions in a changing climate still needs further investigation. Understanding more of variability in IOPs and the processes acting to change the optical behavior of suspended materials can be helpful to improve the accuracy of water quality parameters retrieval using remote sensing techniques for this complex water environment. To establish regional satellite algorithms aimed at evaluation of the content of suspended materials across different time spans for this dynamic lake, it is necessary to quantify the dependence of absorption coefficients on their OAC concentration at feature wavelengths, exploring corresponding contribution effects on the spectral change.

The main objectives of this study are (1) to separate absorption coefficients of suspended particulate materials in Poyang Lake through controlled laboratory experiments during the flood period, which may have its own characteristics at this time, contributing to a better understanding of the optical behavior of SPM absorption, and (2) to clarify the changing mechanism for corresponding components of absorption, which may facilitate improvements in the parameterization of remote sensing algorithms for the interpretation of remote sensing information.

2. Materials and Methods

2.1. Study Area. Poyang Lake (115°47′–116°45′E, 28°22′–29°45′N) is situated at the lower Yangtze River basin (Figure 1), with a subtropical monsoon climate and a mean annual precipitation of 1680 mm [26]. It is fed by tributaries of five rivers of Ganjiang, Fuhe, Xinjiang, Raohe, and Xiushui, connecting to and exchanging water with Yangtze River through the lake mouth in the north. The water level is highly dynamic with significant fluctuations throughout the year, with an extending water surface area (maximum to 4000 km^2) during the flood period between April and September and then shrinking during the dry period between October and March [27]. Meanwhile, the impact of anthropogenic activities on the lake water environment such as major hydrologic engineering and sand mining have caught attentions and been reported [28, 29]. Therefore, the lake water environment is under both natural and anthropogenic influences, providing a wide range of IOPs and OACs.

2.2. Sampling Details. The present work was carried out on boats, and a total of 55 water samples were collected in Lake Poyang from July 12 to 19 and September 01 to 06 in 2017, which is the representative time of flood periods when water level is at a status of increasingly high. Considering the lake geometry shape and hydrological flow directions, sample points were at about 5 km intervals and sampling locations were kept as same as possible between two survey periods; 15 samples were excluded in statistical analysis due to large distance discrepancies and outliers beyond 2 standard deviations of the mean. At each location, water samples prepared for the particulate absorption measurement were collected from the surface to about 20 cm below in the vertical direction, which were then stored in ice bags and transported to the laboratory for determinations in time. Water transparency was also recorded with the Secchi disk (SD), and geographical locations of sample sites were positioned and navigated using handheld GPS (Figure 1).

2.3. Laboratory Measurements. In the laboratory, the extraction and measurement of the chlorophyll a concentration (referred to as Chl.a) were made with acetone following the SM 10200 H spectrophotometric method [15]. Total suspended particulate material (SPM) concentrations were measured gravimetrically on preweighed Whatman GF/F filters after rinsing with distilled water.

Light absorption measurements were performed using the standard procedure of the quantitative filter technique (QFT) after the suspended particles concentrated onto Whatman GF/F filters [7]. Optical density of particulate matter was measured between 350 and 800 nm at 1 nm

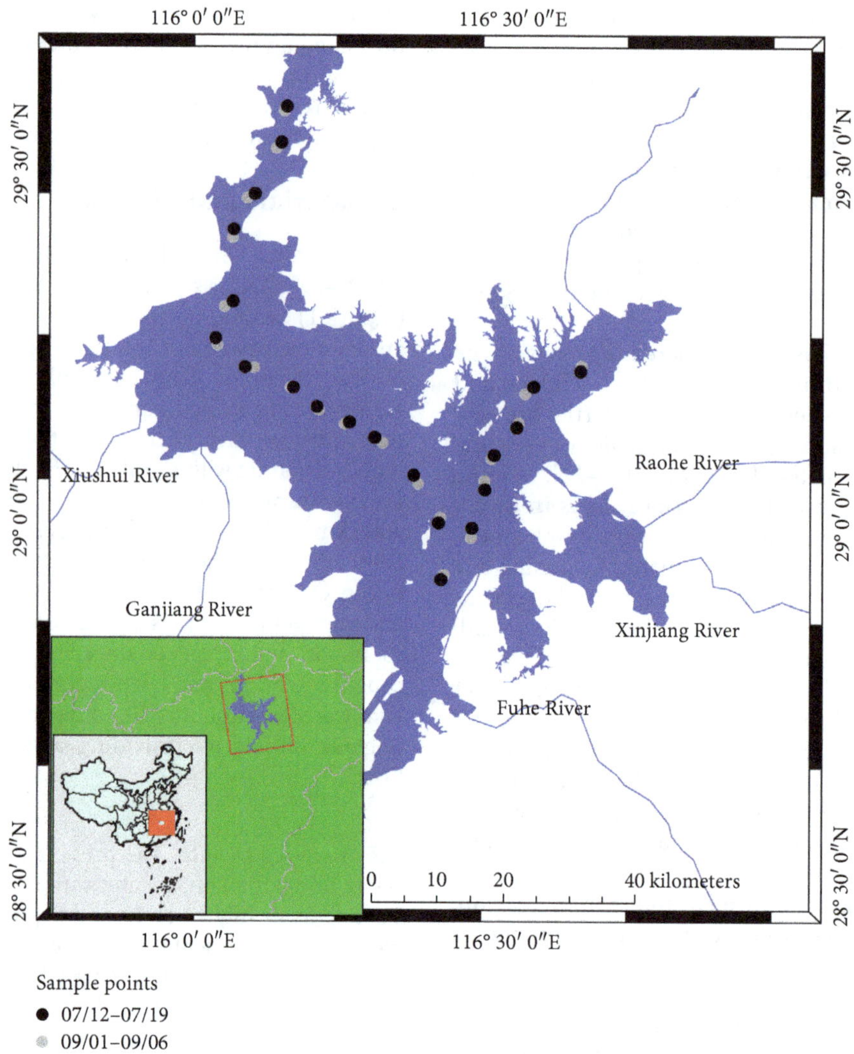

FIGURE 1: Location of Poyang Lake in China and sample sites over the lake during study periods.

intervals with a dual-beam scanning spectrophotometer with a blank filter saturated with distilled water as a reference. Nonalgal particle values were then determined after pigment extraction with pure methanol [7]. Statistical analysis was performed after measurements, descriptive statistics, one-way ANOVA, and correlation analysis were performed for Chl.a, SPM, and SD (Section 3), with null hypothesis stating the means of all levels are equal and alternative hypothesis stating the means of one or more levels are different at two periods. The correlation coefficient (R^2) is used to describe the strength of correlation using the guide that Evans (1996) suggested: 0.00–0.19 (very weak), 0.20–0.39 (weak), 0.40–0.59 (moderate), 0.60–0.79 (strong), and 0.80–1.0 (very strong) [30].

2.4. Analysis Methods. The absorption spectra of SPM (a_p) are decomposed into nonalgal (a_{NAP}) and phytoplankton (a_{ph}) algal components. a_p and a_{NAP} were calculated by converting total and extracted optical densities, respectively (Equation (1)). The absorption spectra for phytoplankton (a_{ph}) were computed as the difference between a_p and a_{NAP} (Equation (2)):

$$a_p(\lambda) = \frac{2.303\left[\mathrm{OD}_s(\lambda) - \mathrm{OD}_{(\mathrm{null})}\right]}{\beta(V_f)/A_f}, \tag{1}$$

$$a_{ph}(\lambda) = a_p(\lambda) - a_{NAP}(\lambda), \tag{2}$$

where $\mathrm{OD}_s(\lambda)$ is the sample absorbance, and the geometric absorption path length of the suspended material was calculated as V_f/A_f, where V_f is the volume filtered (200 ml) and A_f is the clearance area of the filter. The path-length amplification factor beta was assumed to be a constant value of 2 after the theoretical calculations of Roesler (1998), and null point correction ($\mathrm{OD}_{(\mathrm{null})}$) was performed by normalizing spectra to the average absorption measured between 740 and 750 nm assuming no wavelength dependence [31, 32].

The absorption spectrum of nonalgal particles for each sample was modeled from 380 to 600 nm with an exponential function using the least-squares method, and the absorption slope (S) of 40 samples was derived [33, 34]:

Table 1: Range of concentrations of the main OACs during two periods in 2017.

Time	Points	Range	SPM (mg/l)	Chl.a (μg/l)	SD (cm)
July	20	Mean	45.94	21.48	26
		Min	24.48	9.17	18
		Max	56.70	34.77	32
September	20	Mean	19.67	13.80	41
		Min	7.53	3.73	32
		Max	40.3	21.73	56
One-way ANOVA	F value		89.22	15.48	78.63
	Probability > F		4.93E − 10	3.91E − 4	3.16E − 10

Null hypothesis: the means of all levels are equal. Alternative hypothesis: the means of one or more levels are different.

$$a_{\mathrm{NAP}}(\lambda) = a_{\mathrm{NAP}}(440)e^{(-S(\lambda-440))}, \tag{3}$$

where S is the spectral slope, which can reveal relative proportions of minerals and organic materials for detritus in lake waters, because S values generally increase when mineral content decreases [34].

The absorption capacity of the phytoplankton algal component $a_{\mathrm{ph}}(\lambda)$ per mass unit of algae can be expressed as follows [33, 34]:

$$a^*_{\mathrm{ph}}(\lambda) = \frac{a_{\mathrm{ph}}(\lambda)}{\mathrm{Chl.a}\cdot \mathrm{m}^2/\mathrm{mg}}. \tag{4}$$

As $a^*_{\mathrm{ph}}(\lambda)$ usually changes due to package effects and pigment composition variations, the relationship between $a^*_{\mathrm{ph}}(\lambda)$ and Chl.a can be expressed as a power function [33, 34]:

$$a^*_{\mathrm{ph}}(\lambda) = A(\lambda) = [\mathrm{Chl.a}]^{-B(\lambda)}, \tag{5}$$

where A and B are the wavelength-dependent, positive parameters, with the exponent B representing the strength of variations of $a^*_{\mathrm{ph}}(\lambda)$.

3. Results and Discussion

The range, average level of the SPM and Chl.a, and results of one-way ANOVA are shown in Table 1. It could be seen that, at the 0.05 level, the population means of SPM and Chl. a were both significantly different between two periods.

SPM and Chl.a were higher in July than September, representing a decreasing level of suspended sediments in lake columns as water flooded during the study time. Correspondingly, lower water transparency (SD) was observed in July, indicating higher water absorption, which is in accordance with previous research [17, 23, 24].

Although the SPM is with a low composition of phytoplankton biomass (indicated by low Chl.a compared to SPM), SD exhibited some positive correlation with Chl. a ($R^2 = 0.56$, $p < 0.01$), compared to suspended sediments ($R^2 = 0.73$, $p < 0.01$) in July, suggesting that the increasing abundance of various phytoplankton species such as spirogyra, diatoms, and cyanobacteria may also be responsible for lower transparency.

Meanwhile, more evident positive correlation of SD with SPM ($R^2 = 0.69$, $p < 0.01$) than Chl.a ($R^2 = 0.39$, $p < 0.01$) was also found in September, indicating that the vast shallow lake is usually dominated by suspended sediments (with a maximum value of 40.3 mg/l in this work) during this period [17, 24]. And water clarity was still not very high (average SD of 41 cm), which might be induced by resuspended sediments due to vertical convection disturbance of the water layer because of seasonal water-level fluctuations and relevant anthropogenic disturbances [17].

3.1. Variation of Total Suspended Particulate Material Absorption. Absorption spectra of SPM (a_{p}) of Poyang Lake can be decomposed into nonalgal (a_{NAP}) and phytoplankton (a_{ph}) algal components, which had been illustrated in previous research [22, 24]. Spectra curves of total suspended particulate material absorption during the flood period are present in Figure 2 (solid bold line represents mean and dashed line represents standard deviation), suggesting that absorption coefficients of SPM ($a_{\mathrm{p}}(\lambda)$) were highly variable among sample points in both July and September, due to various SPM concentrations and compositions at different sample sites.

In July, the spectral shape of SPM resembled that of nonalgal particulates, with high absorption between 400 and 450 nm, ranging from 9.73 to 17.05 m^{-1} at 440 nm, indicating strong detrital particles dominance. Apparent phytoplankton absorption peaks could only be found at 675 nm, with much lower values between 1.93 and 5.83 m^{-1} and an average of 4.60 m^{-1} (Figure 2).

In September, magnitude of the whole spectrum (380–700 nm) decreased significantly, and the absorption feature of phytoplankton with accessory pigments appeared slightly at the short 400–450 nm wavelength region around 440 nm, with values between 2.47 and 4.69 m^{-1}. Meanwhile more evident peaks of phytoplankton absorption could be observed between 670 and 690 nm, but still with lower values 0.66–1.64 m^{-1} at 675 nm compared to shortwave bands (Figure 2). The growing peaks of phytoplankton absorption were also identified by Wu (2013) from August to October in their study of 2010 and 2011 [23].

To explore possible reasons accounted for SPM absorption variation, we select $a_{\mathrm{p}}(440)$ and $a_{\mathrm{p}}(675)$ as two feature bands to represent the blue portion (400–450 nm) and the red portion of the visible spectrum (670–690 nm), respectively. Regression analysis was performed between

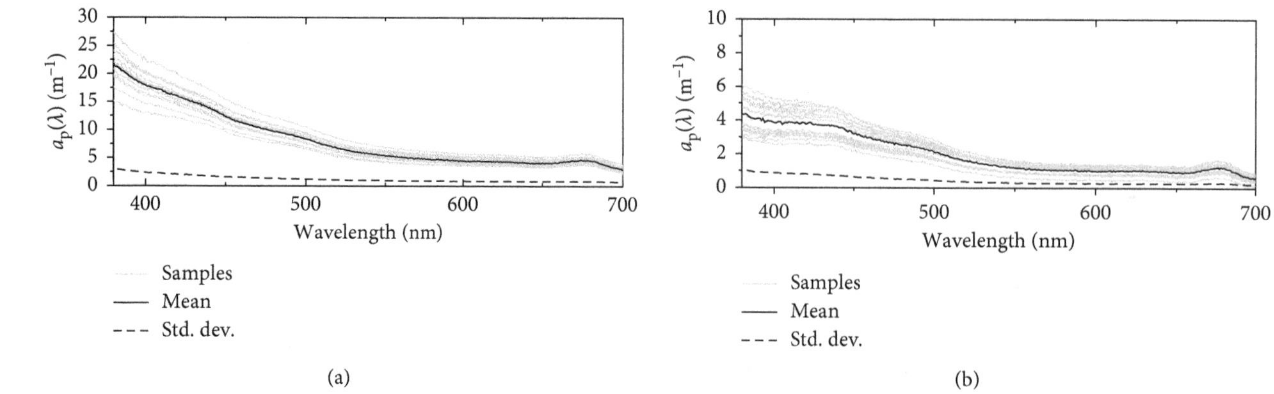

FIGURE 2: Total suspended particulate absorption spectra in Poyang Lake in July (a) and September (b).

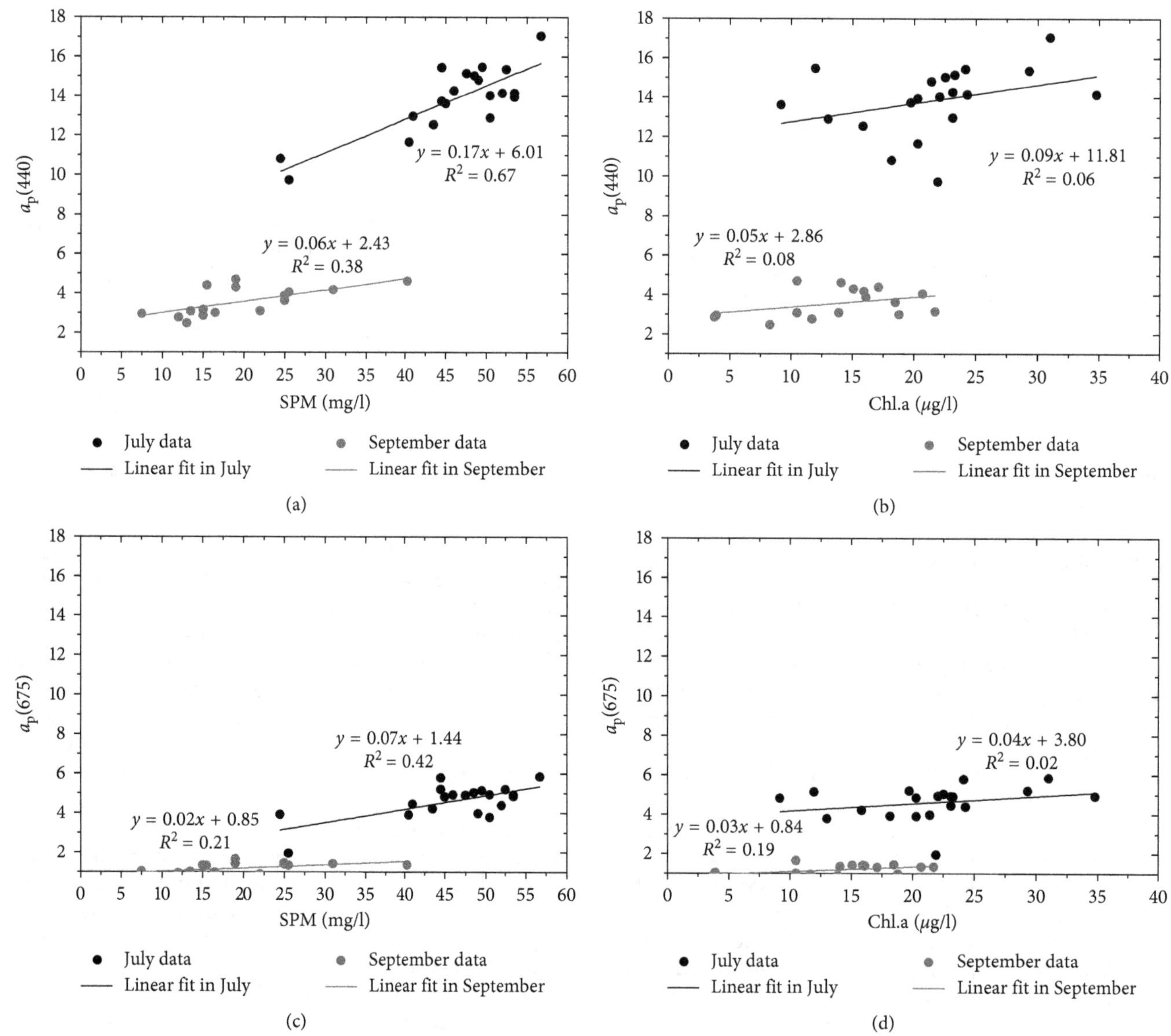

FIGURE 3: The linear relationships of $a_p(440)$ and $a_p(675)$ with SPM and Chl.a concentrations.

$a_p(440)$ and SPM and Chl.a, and $a_p(675)$ and SPM and Chl.a for each period (Figure 3).

It could be seen that the variation of $a_p(440)$ was more sensitive to SPM than to Chl.a for both July and September (Figures 3(a) vs 3(b)). The contribution of SPM to $a_p(440)$ tended to decrease from July to September (Figure 3(a)) as the effect of Chl.a remained at a slightly low level (Figure 3(b)). Thus, SPM absorption in blue portion of the visible spectrum was mainly dominated by nonalgal materials in lake water columns.

And the variation of $a_p(675)$ was also more sensitive to SPM than to Chl.a in July and September (Figures 3(c) vs 3(d)). The tendency of SPM contribution to $a_p(675)$ also decreased from July to September (Figure 3(c)). Although there was a slight increase of Chl.a effect in September, it was still small due to the low proportion of Chl.a in SPM composition.

Although the sediments' masking effect decreased in September, nonalgal particles still dominated the absorption at wavelengths around 400–450 nm, and contributions of phytoplankton at 600–690 nm were still weaker than those of nonalgal sediments. Thus, total suspended particulate material absorption during the flood period was mainly associated with detritus materials in both the blue and red spectra, which is in accordance with the results of previous research in Poyang Lake, Taihu Lake, and Chagan Lake, showing that the total SPM absorption is mainly dominated by nonalgal materials [23, 24, 35, 36].

Therefore, it is reasonable to be deduced that SPM absorption of Poyang Lake mainly has strong connection with nonalgal detritus, which will be elaborated in the next section.

3.2. Variation of Nonalgal Particulate Material Absorption. In this study, the absorption spectrum of nonalgal particles for each sample was modeled from 380 to 600 nm according to Equation (3) using the least-squares method, and the absorption slope (S) of 40 samples was derived, from which the relative proportions of minerals and organic materials in detritus could be revealed [32].

The results showed that, in July, S varied within 0.0060–0.0107 nm^{-1} around 0.0069 nm^{-1}, and in September, its magnitude dropped to 0.0049–0.0102 nm^{-1} with an average level of 0.0067 nm^{-1}. The variation of the detritus absorption slope could be clearly seen from absorption spectra normalized to 440 nm in Figure 4. It suggests that S values generally decreased from July to September, indicating an elevated proportion of inorganic materials in SPM composition.

Regression analysis was also conducted between $a_{NAP}(440)$ and SPM and Chl.a, and $a_{NAP}(675)$ and SPM and Chl.a for each period to reveal factors responsible for variation of nonalgal material absorption in the blue region (400–450) and red region of spectra, respectively (Figure 5).

It is shown that $a_{NAP}(440)$ had a much stronger relationship with SPM than with Chl.a in both July and September (Figures 5(a) vs 5(b)). From July to September, although there was slightly a decrease in SPM contribution (Figure 5(a)), the effect of Chl.a was still very low (Figure 5(b)), indicating that nonalgal sediments in the lake did not rely on phytoplankton degradation activities, which is often the case for ocean, but mainly derived from land-originated materials due to river runoff injections, such as soil surface inorganic mineral salts, reactants, nonpoint source pollutions, or decompositions of biodegradable organic compounds.

The variation of $a_{NAP}(675)$ could also be attributed to SPM in two periods (Figures 5(c) vs 5(d)). The decrease of the relationship between $a_{NAP}(675)$ and SPM can be observed due to the reduction of SPM concentration from July to September (Figure 5(c)). And because of the rather low proportion of phytoplankton degradation products in detritus, the significance of Chl.a effect on $a_{NAP}(675)$ remained at a low level (Figure 5(d)).

It is also verified by previous research of Wu (2013), who showed that there were higher correlations between a_{NAP} and SPM in 2010 and 2011. Our results again confirm previous analysis in Section 3.1: lake SPM absorption was mainly associated with land-derived nonalgal detritus. Despite there were some decreases in September, nonalgal particles still dominated the absorption in both the blue and red spectra.

3.3. Variation of Algal Particulate Material Absorption. Specific absorption represents absorption capacity or efficiency of algal particulates in SPM. Usually, it can be used to parameterize analytical algorithms to inverse the remote sensing signal into phytoplankton pigment centration, which is closely related with the carbon fixation rate at global or regional scales [27, 33, 34].

Spectral curves of phytoplankton particulate specific absorption derived from Equation (4) are present in Figure 6. It can be seen that $a^*_{ph}(\lambda)$ curves in July resembled the shape pattern of absorption curves in September, and there was more evident variability at 400–500 nm around 440 nm and at 650–700 nm around 675 nm.

In July, $a^*_{ph}(440)$ varied between 0.101 and 0.205 m^2/mg around an average of 0.145 m^2/mg, and $a^*_{ph}(675)$ was 0.056–0.111 m^2/mg with a mean value of 0.084 m^2/mg. In September, $a^*_{ph}(440)$ and $a^*_{ph}(675)$ increased to 0.115–0.424 m^2/mg and 0.067–0.169 m^2/mg, with an average of 0.229 m^2/mg and 0.107 m^2/mg, respectively.

Previous studies have demonstrated that fluctuations of $a^*_{ph}(\lambda)$ were mainly associated with relative contributions of accessory pigments (Chl.b, Chl.c, and carotenoids) and chlorophyll present in phytoplankton cells under different light and nutrient conditions [15, 17, 23, 32–35]. The increased abundance of accessory pigments tended to enhance the absorption of blue portion of the visible spectrum between 400 and 500 nm, while Chl.a constantly leads to a contribution focusing on 650–700 nm.

Therefore, the apparent variations of $a^*_{ph}(\lambda)$ here suggested that there might be evident changes in phytoplankton community or pigment compositions during study periods. To validate this speculation, $a^*_{ph}(\lambda)$ spectra were normalized to 440 nm in Figure 7. It could be clearly

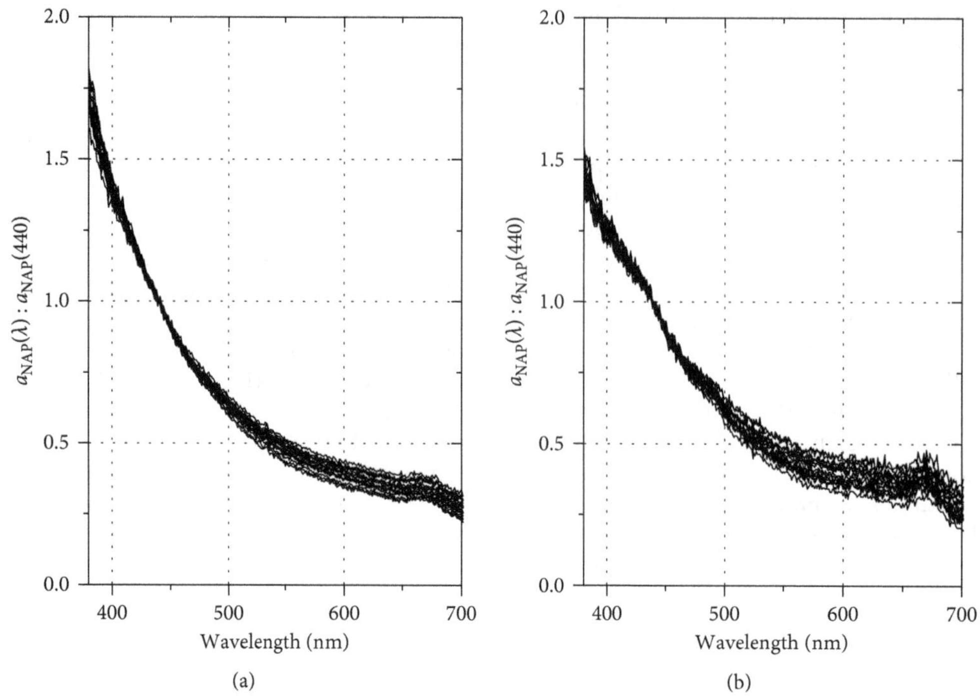

FIGURE 4: Nonalgal particulates of absorption spectra normalized to $a_{NAP}(440)$ in July (a) and September (b).

seen that compared to July, the ratio values at 675 nm ($a^*_{ph}(675) : a^*_{ph}(440)$) in September had decreased, exhibiting an increased effect of accessory pigments (Chl. b, Chl.c, and carotenoids).

While changes in optical absorption capacity in blue (at 440 nm) and red (at 675 nm) portions could be the consequence of pigment composition changes, the possible differences in the pigment concentration level and package effect in the specific trophic status during study periods may also be responsible for that.

Therefore, the variation of $a^*_{ph}(\lambda)$ due to pigment concentration (indicated by Chl.a) change was further analyzed using Equation (5), from which the strength of variations in $a^*_{ph}(\lambda)$ could be inferred by the exponent B (Figure 8).

Despite the fact that there was some dispersion, $a^*_{ph}(\lambda)$ exhibited a decreasing trend with increasing pigment concentration at 440 nm and 675 nm (Figure 8). To some extent, both the blue and red bands showed a decreasing trend because of an increasing package effect when pigment concentration in algal particles was elevated, but the decaying rate of optical absorption capacity was much faster on the blue band where optical absorption mainly associated with accessory pigments.

The variations of $a^*_{ph}(\lambda)$ were also found to be significant due to different phytoplankton constituents at different seasons by Wu (2013), which are similar to those found in Taihu Lake [35] and Chagan Lake [36]. It was also shown that the package effect might explain the higher a^*_{ph} values at 440 or 676 nm in Poyang Lake at a certain level when Chl. a was relatively lower compared with that of the most other water bodies [23].

With those observations and analysis illustrated in this section, it would be reasonable to confirm that the absorption capacity of algal particulates in SPM was dependent on the relative contributions of accessory pigments and chlorophyll in phytoplankton composition as well as the package effect and could fluctuate on the feature bands of the spectrum under various aquatic environments during flood time.

There is, however, some ambiguity in the partition of the relative contribution of the package effect and accessory pigments, which could be possibly intermingled with each other in the domain of 400–500 nm, and further research including more effective separation among algal compositions such as sodium hypochlorite bleaching or numerical separation method warranted to resolve these problems [23, 32, 34].

4. Conclusions

This study presents a case study of optical variations of suspended particulate material absorption during the flood time for Poyang Lake and provides more insight into the natural changes of the spectral parameters required for the parameterization of bio-optical models and remote sensing algorithms in this dynamic water environment under the background of climate change and anthropogenic impacts.

As one of the most representative inland waters in China, Poyang Lake exhibits an overall high level of suspended nonalgal detritus and algal concentration and lower water clarity during study periods, showing a strong absorption by total suspended sediments that are closely related with the land-derived nonalgal detritus material which varied

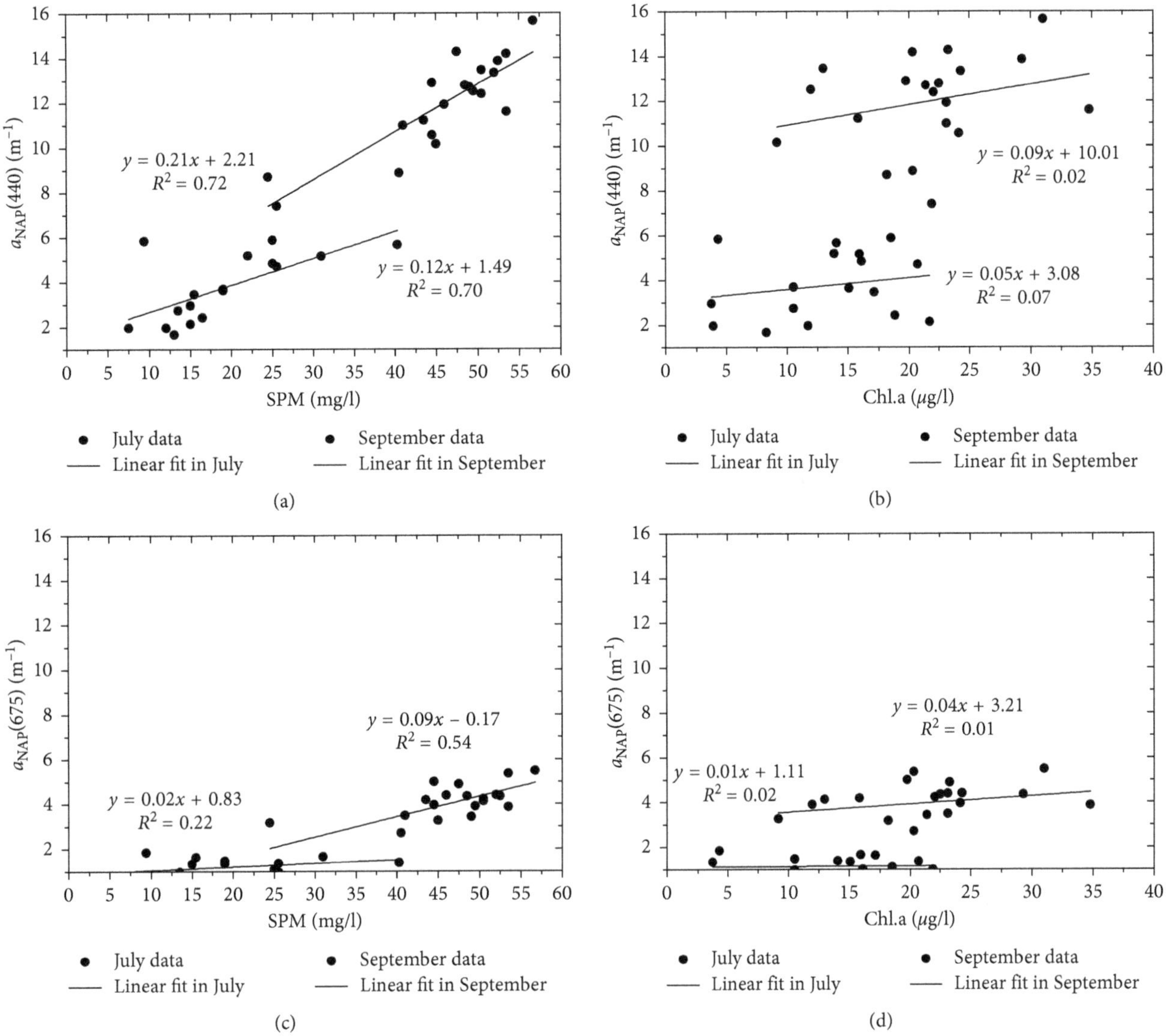

FIGURE 5: The linear relationships of $a_{NAP}(440)$ and $a_{NAP}(675)$ with SPM and Chl.a concentrations.

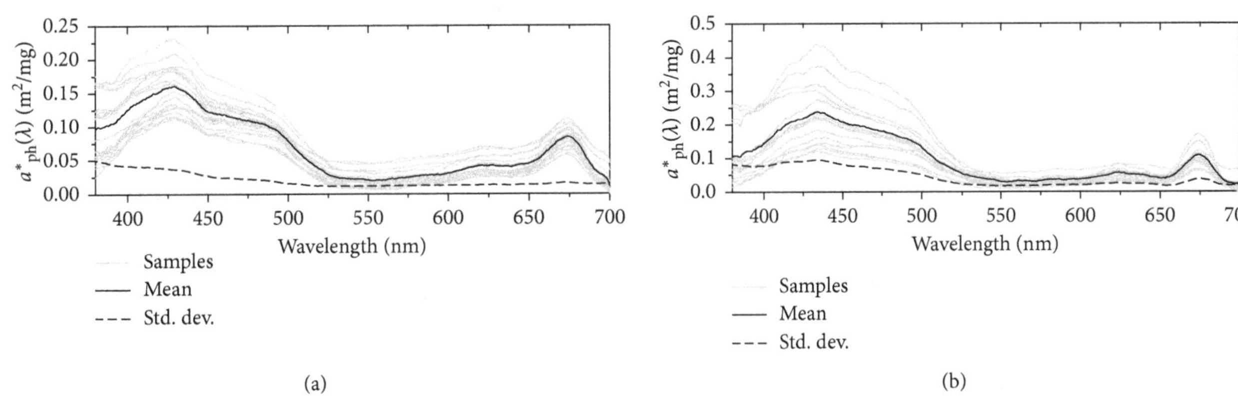

FIGURE 6: Phytoplankton specific absorption spectral curves in July (a) and September (b).

significantly with diverse proportions of minerals and organic materials from multiple sources loaded by river runoff injections, such as soil surface inorganic mineral salts [23, 24], reactants, nonpoint source pollutions, or decompositions of biodegradable organic compounds [4, 30, 32, 34].

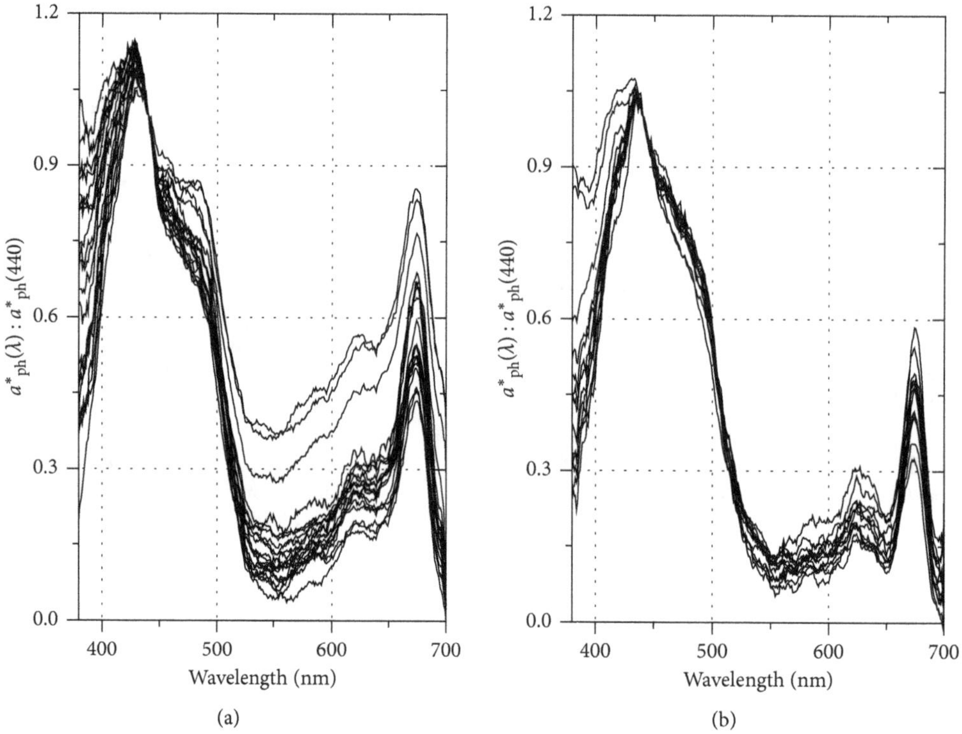

FIGURE 7: Specific absorption spectra normalized to $a^*_{ph}(440)$ in July (a) and September (b).

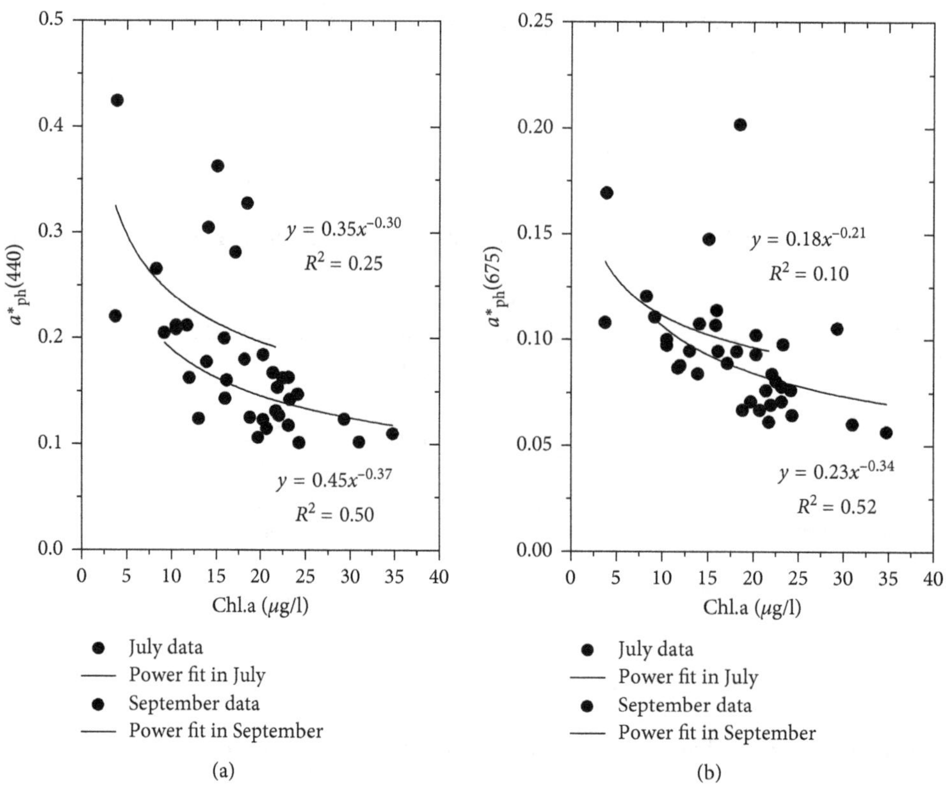

FIGURE 8: The power functions of $a^*_{ph}(440)$, $a^*_{ph}(675)$, and Chl.a.

Due to variations in phytoplankton community pigment composition, concentration, and package effect, there exists a difference between optical absorption capacity on blue (440) and red (675) portions of bands, the former of which had an obvious enhancement due to the increased effect of accessory pigments (Chl.b, Chl.c, and carotenoids) in suspended phytoplankton particles from July to September.

Evidence also presented that the nonlinear dependency of specific phytoplankton particulate absorption on pigment concentration for various trophic statuses in different periods could be unstable due to relative contributions of the package effect and accessory pigments, and this could bring uncertainties to the application accuracy of bio-optical models and remote sensing algorithms during parameterization.

There is still some ambiguity in the partition of the relative contribution of the package effect and accessory pigments, which might be intermingled in the domain of 400–500 nm, and further research including more effective separation among algal composition such as sodium hypochlorite bleaching or numerical separation method warranted to resolve these problems [23, 32, 34].

More works and studies are being carried out by researchers to learn about the spectral behavior of covaried optical active constituents present in this aquatic region, such as the identification of selective absorption of the dissolved organic material (DOM) onto sediments due to seasonal and dynamic biological activity (the presence of phytoplankton-derived DOM further processed by bacteria and/or by photo-oxidation) [27] and periodical covariation links between optical active materials [36], from which we hope to learn more about the impact of IOP variation on regional bio-optical model constructions and the interpretation of remote sensing observations in a changing environment.

Conflicts of Interest

The authors declare that they have no conflicts of interest.

Authors' Contributions

Yuandong Wang and Xibin You conceived and designed the experiments. Lianfang Yu performed the experiments. Yuandong Wang and Lihong Meng analyzed the data. Xiangming Xu contributed to the reagents and materials. Yuandong Wang and Guangxu Liu wrote and revised the paper.

Acknowledgments

This research was supported by the Science and Technology Planning Project of Jiangxi Provincial Education Department (No. GJJ151016) and the httpNational Natural Science Foundation of China (Grant Nos. 41601600, 41561020, and 41301226).

References

[1] H. R. Gordon, O. B. Brown, and R. H. Evans, "A semianalytic radiance model of ocean color," *Journal of Geophysical Research*, vol. 93, no. D9, p. 10909, 1988.

[2] T. T John, B. M Colleen, and S. M. Timthy, "Remote sensing of physical cycles in lake superior using a spatio temporal analysis of optical water typologies," *Remote Sensing of Environment*, vol. 171, pp. 149–161, 2015.

[3] P Ammenberg, P Flink, and T. Lindell, "Bio-optical modelling combined with remote sensing to assess water quality," *International Journal of Remote Sensing*, vol. 23, no. 8, pp. 1621–1638, 2002.

[4] L. Hubert, M. Antoine, and V. Vincent, "Variability of suspended particulate material concentration in coastal waters under the Mekong's influence from ocean color (MERIS) remote sensing over the last decade," *Remote Sensing of Environment*, vol. 150, pp. 218–230, 2014.

[5] S. C. Palmer, P. D. Hunter, and T. Lankester, "Validation of Envisat MERIS algorithms for chlorophyll retrieval in a large, turbid and optically-complex shallow lake," *Remote Sensing of Environment*, vol. 157, pp. 158–169, 2015.

[6] A. Morel, "Optical modeling of the upper ocean in relation to its biogenous material content (Case I waters)," *Journal of Geophysical Research*, vol. 93, no. C9, pp. 10749–10768, 1988.

[7] B. G. Mitchell, A. Bricaud, K. Carder et al., "Determination of spectral absorption coefficients of particles, dissolved material and phytoplankton for discrete water samples," in *Ocean Optics Protocols For Satellite Ocean Color Sensor Validation, Revision 2, NASA Technical Memorandum 2000-209966*, G. S. Fargion, J. L. Mueller, and C. R. McClain, Eds., pp. 125–153, NASA Goddard Space Flight Center, Greenbelt, MD, USA, 2000.

[8] R. A. Reynolds, D. Stramski, and E. B. G. Mitchell, "A chlorophyll-dependent semianalytical reflectance model derived from field measurement of absorption and backscattering coefficients within the Southern Ocean," *Journal of Geophysical Research*, vol. 106, no. C4, pp. 7125–7138, 2001.

[9] A. Morel and S. Maritorena, "Bio-optical properties of oceanic waters: a reappraisal," *Journal of Geophysical Research*, vol. 106, no. C4, pp. 7163–7180, 2001.

[10] C. J. Stephanie, T. K. Palmer, and D. H. Peter, "Remote sensing of inland waters: challenges, progress and future directions," *Remote Sensing of Environment*, vol. 157, pp. 1–8, 2015.

[11] S. Bagheri, C. Zetlin, and R. Dios, "Estimation of optical properties of nearshore water," *International Journal of Remote Sensing*, vol. 20, no. 17, pp. 3393–3397, 2010.

[12] B. P. David, F. R. James, and A. G. Gower, "A review of ocean color remote sensing methods and statistical techniques for the detection, mapping and analysis of phytoplankton blooms in coastal and open oceans," *Progress in Oceanography*, vol. 123, no. 4, pp. 123–144, 2014.

[13] D. Katja and O. Natascha, "Remote sensing for lake research and monitoring-Recent advances," *Ecological Indicators*, vol. 64, pp. 105–122, 2016.

[14] D. G. Bowers, G. E. L. Harker, and B. Stephan, "Absorption spectra of inorganic particles in the Irish Sea and its relation to tidal stirring," *International Journal of Remote Sensing*, vol. 17, no. 12, pp. 2789–2805, 1996.

[15] M. Babin, D. Stramski, G. M. Ferrari, and H. Claustre, "Variations in the light absorption coefficients of sssphyto-plankton, non-algal particles, and dissolved organic material in coastal waters around Europe," *Journal of Geophysical Research*, vol. 108, no. C7, p. 3211, 2003.

[16] B. M. Colleen, G. Steven, and A. Dirk, "Aquatic color radiometry remote sensing of coastal and inland waters: Challenges and recommendations for future satellite missions," *Remote Sensing of Environment*, vol. 160, pp. 15–30, 2015.

[17] W. Zhang, X. L. Chen, L. Q. Tian et al., "Temporal and spatial variations of TSM concentration in Poyang Lake: observations from HJ-1A/1B satellite CCD imagery," *Journal of Central China Normal University (Natural Sciences)*, vol. 47, no. 3, pp. 416–420, 2013.

[18] W. X. Cao, Y. Z. Yang, and X. Q. Xu, "Regional patterns of phytoplankton spectral absorption in the Pearl river estuary," *Chinese Science Bulletin*, vol. 48, no. 17, pp. 1876–1882, 2003.

[19] G. P. Wu and Y. B. Liu, "Capturing variations in inundation with satellite remote sensing in amorphologically complex, large lake," *Journal of Hydrology*, vol. 523, no. 6, pp. 14–23, 2015.

[20] D. Stramski, S. B. Wozniak, and P. J. Flatau, "Optical properties of Asian mineral dust suspended in seawater," *Limnology and Oceanography*, vol. 49, no. 3, pp. 749–755, 2004.

[21] L. Linhai, L. Lin, S. Kaishan, and L. Yunmei, "An inversion model for deriving inherent optical properties of inland waters: Establishment, validation and application," *Remote Sensing of Environment*, vol. 135, pp. 150–166, 2015.

[22] L. Q. Chen, X. L. Chen, and L. Q. Tian, "Partitioning of the suspended particulate spectral scattering coefficient in Poyang lake," *Spectroscopy and Spectral Analysis*, vol. 32, no. 3, pp. 729–733, 2012.

[23] G. F. Wu, L. J. Cui, H. T. Duan, T. Fei, and Y. L. Liu, "Specifc absorption and backscattering coefficients of the main water constituents in Poyang lake, China," *Environmental Monitoring and Assessment*, vol. 185, no. C13, pp. 4191–4206, 2013.

[24] G. F. Wu, L. J. Cui, J. J. He, H. T. Duan et al., "Comparison of MODIS-based models for retrieving suspended particulate matter concentrations in Poyang lake, China," *International Journal of Applied Earth Observation and Geoinformation*, vol. 24, no. 1, pp. 63–72, 2013.

[25] G. Claudia, B. Mariano, and V. Emiliana, "Airborne hyperspectral data to assess suspended particulate material and aquatic vegetation in a shallow and turbid lake," *Remote Sensing of Environment*, vol. 157, pp. 48–57, 2015.

[26] Z. Wu, Y. Cai, X. Liu, C. P. Xu, Y. Chen, and L. Zhang, "Temporal and spatial variability of phytoplankton in Lake Poyang: the largest freshwater lake in China," *Journal of Great Lakes Research*, vol. 39, no. 3, pp. 476–483, 2013.

[27] J. Xu, Y. Q. Wang, D. Gao, Z. R. Yan, C. Gao, and L. L. Wang, "Optical properties and spatial distribution of chromophoric dissolved organic matter (CDOM) in Poyang lake, China," *Journal of Great Lakes Research*, vol. 43, no. 4, pp. 700–709, 2017.

[28] X. Mei, Z. Dai, J. Du, and J. Chen, "Linkage between Three Gorges Dam impacts and the dramatic recessions in China's largest freshwater lake, Poyang Lake," *Scientific Reports*, vol. 5, 2015.

[29] S. Lei, X. P. Zhang, and R. F. Li, "Analysis the changes of annual for Poyang lake wetland-vegetation based on MODIS monitoring," *Procedia Environmental Sciences*, vol. 10, no. 4, pp. 1840–1846, 2011.

[30] A. Beldjazia and D. Alatou, "Precipitation variability on the massif Forest of Mahouna (North Eastern-Algeria) from 1986 to 2010," *International Journal of Management Sciences and Business Research*, vol. 5, no. 3, pp. 21–28, 2016.

[31] C. S. Roesler, "Theoretical and experimental approaches to improve the accuracy of particulate absorption coefficients derived from the quantitative filter technique," *Limnology & Oceanography*, vol. 43, pp. 1649–1660, 1998.

[32] B. G. Mitchel, M. Kahru, and J. Wieland, Determination of spectral absorption coefficients of particles, dissolved material and phytoplankton for discrete water samples. Ocean optics protocols for satellite ocean colour sensor validation, Revision 4, Volume IV: Inherent optical properties: Instruments, characterizations, field measurements and data analysis protocols, Goddard Space Flight Center Technical Memorandum, 01674, 2003.

[33] W. M. Mark, "Eutrophication and cyanobacterial blooms in South African inland waters: 10 years of MERIS observations," *Remote Sensing of Environment*, vol. 155, pp. 161–171, 2014.

[34] A. Bricaud, A. Morel, and M. Babin, "Variations of light absorption by suspended particles with chlorophyll a concentration in oceanic (case 1) waters: analysis and implications for bio-optical models," *Journal of Geophysical Research*, vol. 103, no. C13, pp. 31033–31044, 1998.

[35] R. H. Ma, D. L. Pan, H. T. Duan et al., "Absorption and scattering properties of water body in Taihu lake, China: backscattering," *International Journal of Remote Sensing*, vol. 30, no. 9, pp. 2321–2335, 2009.

[36] Y. D. Wang, D. W. Liu, K. S. Song, J. Du et al., "Characterization of water constituents spectra absorption in Chagan lake of Jilin province, Northeast China," *Chinese Geographical Science*, vol. 21, no. 3, pp. 334–345, 2011.

13

Simulation and Experimental Study on Properties of Ag/SnO_2 Contact Materials Doped with Different Ratios of Ce

Jingqin Wang, Ying Zhang, and Huiling Kang

State Key Laboratory of Reliability and Intelligence of Electrical Equipment, Hebei University of Technology, Tianjin 300130, China

Correspondence should be addressed to Ying Zhang; 13512474229@163.com

Academic Editor: Alfredo Juan

SnO_2 in the Ag/SnO_2 contact material is a kind of high hardness and almost insulated wide bandgap semiconductor material. In the process of use, the contact resistance is larger and the temperature rise is higher, which reduces the reliability of contacts and shortens the electrical life. In order to improve the properties of Ag/SnO_2 contact materials, based on the first principle of density functional theory, this paper presents a method of doping SnO_2 with the calculation of different proportions of rare earth Ce electrical and mechanical properties. The results of energy band, density of state, and elastic constant show that when the doping ratio of Ce is 0.125, the electron mobility is the highest, the conductivity is the best; the hardness decreases, and the universal elastic anisotropy index is the smallest. Finally, in the experiment, SnO_2 powders with different doping ratios are prepared by the sol-gel method, and Ag/SnO_2 contacts with different doping proportions are prepared by powder metallurgy. Arc energy, contact resistance, and hardness are measured; scanning electron microscopy was used to observe and analyze the surface morphology. The final simulation and experimental results are well matched.

1. Introduction

The contact resistance and the temperature rise in the process of using automotive relay contacts are increased, making life expectancy to decrease [1–3]. This is because SnO_2 in Ag/SnO_2 contact material is a kind of wide bandgap semiconductor material, which is almost insulator [4]. At the same time, SnO_2 is a high hardness brittle reinforcing phase, making contact material processing and composite molding extremely difficult. The study shows that the doping modification of Ag/SnO_2 with rare earth materials is a good method to solve the above problems [5].

For now, improving the performance of Ag/SnO_2 contact materials by doping is still in the experimental stage. How to get the best performance by doping with the proper amount of rare earth elements depends on the original experiments and experience, and there are some difficulties in the preparation of samples and the accuracy of experimental equipment, which restrict the development of Ag/SnO_2 materials.

Zhu Yancai and others in the study on new type of $Ag/SnO_2/CeO_2$ electrical contact material preparation and electrical properties found that CeO_2-doped Ag/SnO_2 improved electrical performance [6]. From the experimental point of view, they had only selected a single ratio of doping, not knowing the best proportion of doping to get better performance. In the study of the structure and optical properties of Ce-doped SnO_2, Shan Linting et al found that the Ce-doped SnO_2 makes conduction band bottom to move toward the low-energy side, the bandgap becomes smaller, and the conductivity increases [7]. The conductivity was only studied by simulation calculations.

Wang Qingwei et al found that CeO_2-doped SnO_2 when doping too little is not conducive to the sintering and growth of tin oxide crystals, while excessive doping will prevent the lattice growth and affects the stability of the lattice, easily leading to serious cracking of the product [8, 9]. So, the doping amount of CeO_2 should be controlled.

It is of great significance to understand the basic properties of Ag/SnO_2 contact material. In this paper, based on the first-principles calculation of density functional theory [10], different proportions of rare earth Ce are doped into the contact material of SnO_2, conducting geometric optimization, energy band, density of states and elastic constant, and then analyzing the conductivity and hardness to theoretical study the electronic structure and mechanical properties.

2. Cell Model and Calculation Method

2.1. Cell Model. In this paper, we had constructed a SnO_2 primary cell model, as shown in Figure 1(a). SnO_2 (cassiterite) has the rutile structure space group P42/mnm [11]. Each SnO_2 contains 2 Sn atoms and 4 O atoms. The atomic substitution method was used to construct the supercell model $Sn_{1-x}Ce_xO_2$ ($x = 0$, 0.083, 0.125, and 0.167). Table 1 shows the relationship between the doping ratio and the supercell. Taking the doping ratio of 0.125 as an example, the $1 \times 2 \times 2$ supercell model of SnO_2 is established and one of the Sn atoms is replaced by a rare earth atom Ce, as shown in Figure 1(b).

2.2. Calculation Method. In this article, we used Materials Studio software and completed the calculation by using the CASTEP module. The calculation process uses periodic boundary conditions based on DFT [12]. Since generalized gradient approximation (GGA) considers the effect of charge density near a certain location on the exchange-correlation energy, for example, taking into account the effect of the first-order gradient of density to correct local variations, it is possible to properly correct the effects of charge density regions exponential form; in the geometric optimization, energy band and density of state calculations have achieved good results [13].

The local density approximation (LDA) assumes that the nucleus spacing in the system is far apart, and the motion of electrons in its lattice background can be approximated as a behavior in a uniform field. It is suitable for the calculation of the ground state properties of various systems, and very good results are achieved when calculating the elastic properties of the system [14]. Therefore, the geometric optimization process uses the PBE method under GGA; the CA-PZ method under LDA is used to calculate the elastic constants in the mechanical properties on the basis of optimization. For geometric optimization, energy band, density of states, and elastic constants are calculated adopting the plane wave ultrasoft pseudopotential (Ultrasoft) method.

In order to compare the calculated results of SnO_2 doped with different proportions of Ce, the parameters of software simulation are set to be consistent. The calculated parameters are as follows: the cutoff energy of the plane wave is 370 eV, the SCF tolerance is 1.0×10^{-6} eV/atom, and the K point in the Brillouin region is $5 \times 5 \times 8$. The calculation process is carried out in the reciprocal space.

3. Results and Discussion

3.1. Crystal Structure. The crystal parameters of geometry optimization and experimentation of SnO_2 doped with different proportions of Ce are optimized, and the parameters after optimization are shown in Table 2.

According to the optimization results, the experimental values and calculated values of the lattice parameters of Ce-doped SnO_2 with different proportions are not much different. The lattice constant and volume of SnO_2 after doping have different growth degrees compared with SnO_2, which is consistent with Vegard's law [15], indicating that the calculation results are reliable. This shows that doping Ce can cause slight distortion of the crystal structure and thus can adjust the structural parameters of SnO_2, thereby improving the performance of SnO_2.

3.2. Conductivity

3.2.1. Density of States and Electron Concentration. The density of states of SnO_2 and SnO_2 doped with different proportions of rare earth Ce is shown in Figure 2. Figure 2(a) shows the density of states of SnO_2, and Figures 2(b)–2(d) show the density of states of SnO_2 with the Ce doping ratios of 16.7%, 12.5%, and 8.34%, respectively. The Fermi level is chosen to be zero of the energy scale. The energy level above the Fermi surface (0 eV) is the conduction band, and the energy level below the Fermi surface is the valence band.

It can be seen from Figure 2 that the density of states undoped and doped with different ratios of Ce shows that the Fermi surface of the SnO_2 supercell after doping moves to the valence band in different degrees, while the conduction band of undoped SnO_2 is about 0 eV–20 eV, and the conduction band of SnO_2 is about 0 eV–7 eV after doping. The conduction band is narrowed, which indicates that the SnO_2 model after different ratios of doping enhances the electronic state near the Fermi level, and more partial waves cross the Fermi level, the interaction between electrons is enhanced, and the conductivity is enhanced.

The Fermi surface of the Ce-doped SnO2 supercell with different ratios of rare earth elements enters the conduction band to a different extent, which indicates that the relative electron concentration entering the conduction band of SnO2 supercell is different. The relative number of electrons entering the conduction band can be obtained by integrating the density of states of Figures 2(a)–2(d) by Origin software. The obtained $Sn_{1-x}Ce_xO_2$ ($x = 0.083$, 0.125, 0.167) is shown in Table 3.

As can be seen from Table 3, the higher the doping concentration, the higher the relative electron number into the conduction band and the higher the concentration of relative electron.

The conductive performance of the material can be expressed by the conductivity δ_i. The conductivity of the material by semiconductor physics theory is [16]

$$\delta_i = n_i q \mu_i, \quad (1)$$

where q is the electron charge constant, and the conductivity is proportional to the electron concentration n_i and the electron mobility μ_i, which can be obtained from the above formula.

In addition to being affected by the relative number of electrons, the conductivity is also affected by the electron

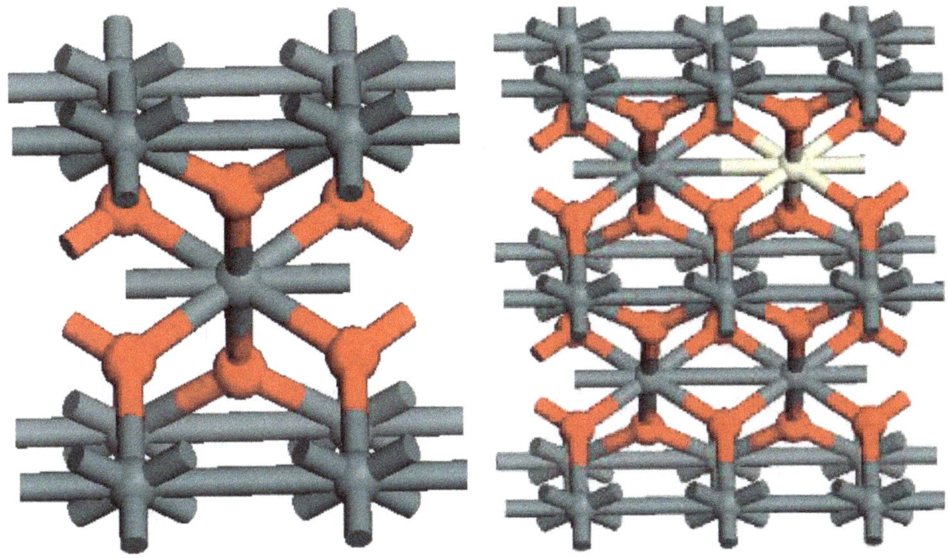

(a) (b)

FIGURE 1: Cell model (a) SnO_2 and (b) Ce-doped SnO_2.

TABLE 1: Doping ratio and supercell correspondence.

Supercell	Doping ratio
$1\times1\times3$	16.7%
$1\times2\times2$	12.5%
$1\times2\times3$	8.34%

TABLE 2: Doping ratio and crystal structures.

	Model	$1\times1\times3$ 16.7%	$1\times2\times2$ 12.5%	$1\times2\times3$ 8.3%
a	Experiment value	4.737	4.737	4.737
	Calculated value	4.958	4.951	5.007
b	Experiment value	4.737	4.737	4.737
	Calculated value	4.958	4.948	4.986
c	Experiment value	3.186	3.186	3.186
	Calculated value	3.346	3.327	3.317
Volume	Experiment value	214.202	285.965	428.947
	Calculated value	246.751	326.013	496.851

mobility. According to the characteristics of the CASTEP software. When applying the first-principles study, the software sets the temperature to a low temperature of 0 K. According to semiconductor theory, the crystal scattering is dominated by ionized impurities. Therefore, it is necessary to fully consider the influence of the relative electron number and electron mobility into the conduction band on the conductivity of SnO_2 to obtain a correct conclusion.

3.2.2. Band and Electronic Effective Mass. The band structure of SnO_2 and different ratios of Ce-doped SnO_2 shown in Figure 3(a) is the band of SnO_2, and Figures 3(b)–3(d) are the band of SnO_2 with the Ce doping ratios of 16.7%, 12.5%, and 8.34%, respectively.

From Figure 3, the bandgap decreases after doping. The band of SnO_2 after doping is introduced in the region of −35 eV~−30 eV new energy level, but this energy level is in the deep-rail region and has little effect on the conductivity. Therefore, we do not consider it. At the same time, new energy levels appear in the doping system near −15 eV and the distribution of electrons increases, providing more electronic states, and the valence band to the conduction band energy level jump probability is greatly increased. Band fluctuations become gentle, indicating that the doping of the electronic locality enhancement. Hence, when SnO_2 is doped with Ce, it makes metal properties to enhance to increase the conductivity of the material.

The electron mobility is inversely proportional to the effective electron mass, and the data derived from two derivation of the band by software are shown in Table 4.

The formula for the effective electron mass is as follows:

$$m_e^* = \frac{h^2}{4\pi^2 d^2 E/dk^2}, \tag{2}$$

where h is the Planck constant, k is the wave vector, and E is the electron energy at the wave vector.

3.2.3. Conductivity Analysis. The following equation is the relationship between the electron mobility μ_i, the doping concentration N_i, and the electron effective mass m_e^*:

$$\mu_i \propto \frac{q}{m_e^* \cdot N_i}. \tag{3}$$

The following equation is derived from the formulas (1)–(3), and the conductivity is proportional to the relative electron concentration and inversely proportional to the doping concentration and the electron effective mass [17]:

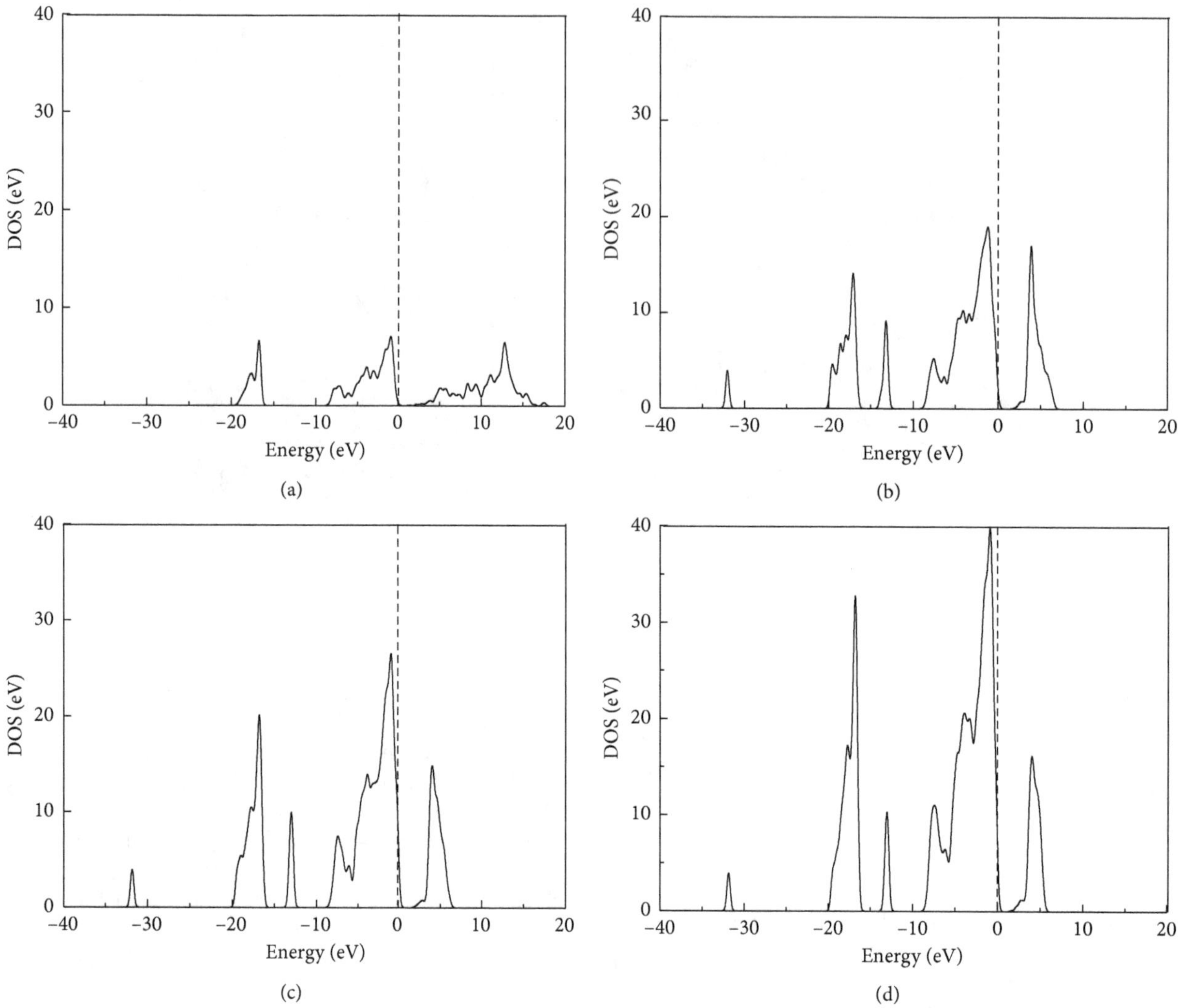

FIGURE 2: Density of states. (a) SnO_2. (b) 16.7% of Ce-doped SnO_2. (c) 12.5% of Ce-doped SnO_2. (d) 8.34% of Ce-doped SnO_2.

TABLE 3: Electron concentration.

Ratio	Relative electron number	Relative electron concentration (10^{21} cm^{-3})
16.7%	0.8	9.73
12.5%	1.2	14.7
8.34%	1.6	19.3

$$\delta_i \propto \frac{n_i}{m_e^*.N_i}, \qquad \frac{\delta_i}{\delta_j} = \frac{n_i \cdot m_{ej}^*.N_j}{n_j \cdot m_{ei}^*.N_i}. \tag{4}$$

The doping ratio of 12.5% is used as a reference to get the relative conductivity, as shown in Table 5.

It can be seen from Table 5 that the conductivity of SnO_2 with the Ce doping ratio of 12.5% is the highest.

3.3. Elastic Constant

3.3.1. The Formula for the Elastic Constant. For polycrystalline systems, Hill proved that VRH (Voigt–Reuss–Hill) is closer to the experimental results [18]. The formula for bulk modulus (B) and shear modulus (G) is as follows:

$$B = \frac{(B_V + B_R)}{2}, \qquad G = \frac{(G_V + G_R)}{2}, \tag{5}$$

where B_V and G_V are the elastic modulus and shear modulus of the Voigt model, respectively, and B_R and G_R are the elastic modulus and shear modulus of the Reuss model, respectively. The formula between bulk modulus, shear modulus, and elastic constant C_{ij} ($i, j = 1\sim6$) under the Voigt and Reuss model is as follows [19, 20]:

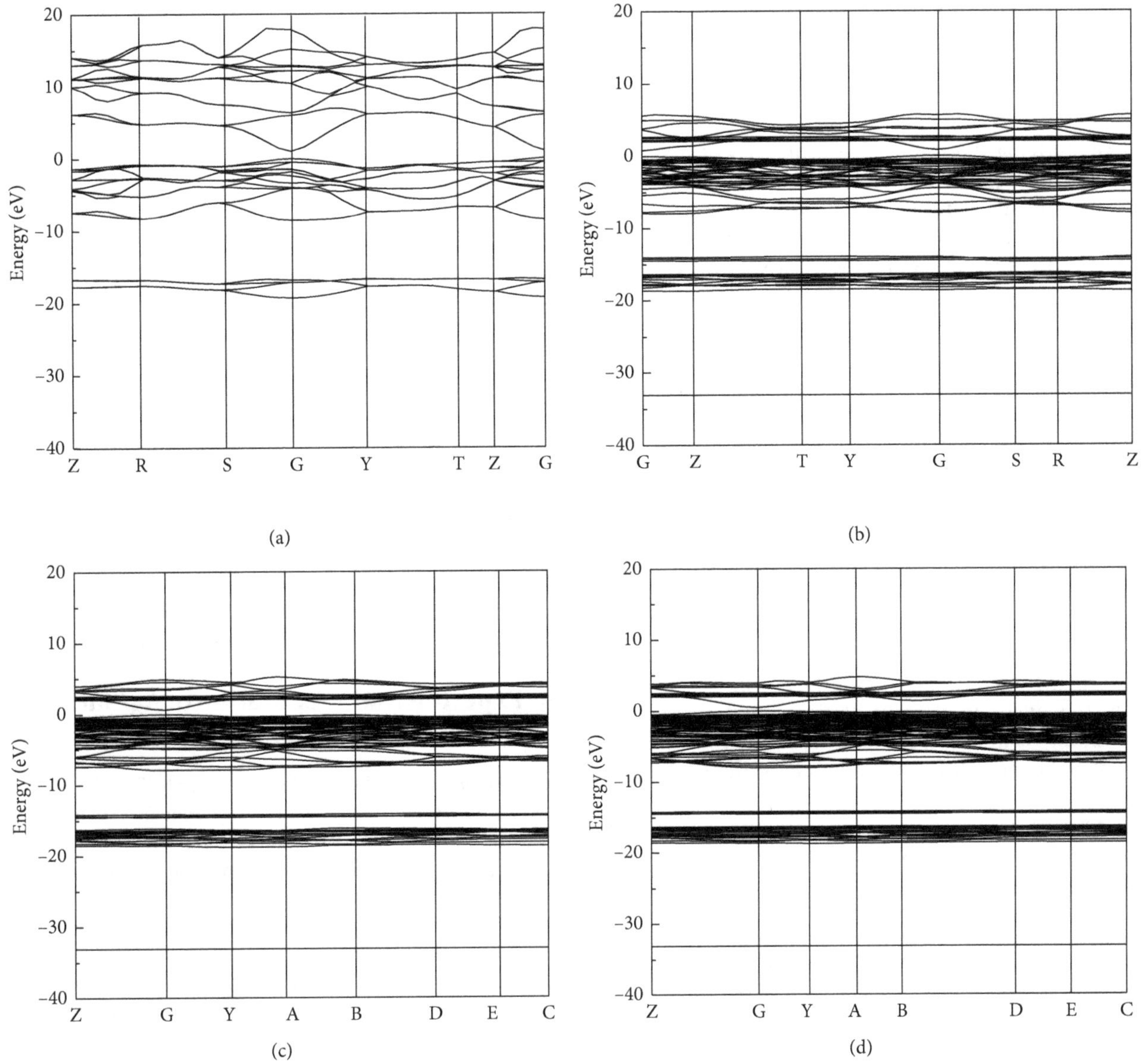

Figure 3: Band structure. (a) SnO_2. (b) 16.7% of Ce-doped SnO_2. (c) 12.5% of Ce-doped SnO_2. (d) 8.34% of Ce-doped SnO_2.

Table 4: Effective electron mass.

Ratio	Two derivation	Effective electron mass (10^{-32} kg)
16.7%	377.9	2.95
12.5%	310.5	3.60
8.34%	148.0	7.44

Table 5: Relative conductivity.

Ratio	Relative Conductivity
16.7%	0.60
12.5%	1
8.34%	0.95

$$B_V = \frac{1}{9}(2C_{11} + C_{33}) + \frac{2}{9}(2C_{13} + C_{12}),$$

$$G_V = \frac{1}{15}(2C_{11} + C_{33} - C_{12} - 2C_{13}) + \frac{1}{5}(2C_{44} + C_{66}),$$

$$B_R = \frac{(C_{11} + C_{12})C_{33} - 2C_{13}^2}{C_{11} + C_{12} + 2C_{33} - 4C_{13}},$$

$$G_R = 15\left(\frac{18B_V}{(C_{11} + C_{12})C_{33} - 2^2_{C_{13}}} + \frac{6}{C_{13} - C_{12}} + \frac{6}{C_{44}} + \frac{3}{C_{66}}\right)^{-1}. \quad (6)$$

Poisson's ratio (σ) is related to the bulk modulus and the shear modulus as calculated by Hill as follows:

$$\sigma = \frac{3B - 2G}{2(3B + G)}. \quad (7)$$

When studying the mechanical properties of materials, we usually consider the effects of microcracks and lattice distortion on the mechanical properties of materials, and the universal elastic anisotropy index A^U is usually the determinant of microcracks and lattice distortion [21]:

$$A^U = 5\frac{G_V + B_V}{G_R + B_R} - 6. \quad (8)$$

The Vickers hardness in GPa is evaluated based on the empirical formula of Tian (2012). When the hardness is greater than 5 GPa, the correlation between the empirical formula and the experimentally measured hardness value is quite good. The formula is as follows [22]:

$$H_V + 0.92K^{1.137}G^{0.708}. \quad (9)$$

where $k = G/B$.

3.3.2. Calculation and Analysis of Elastic Constants. In order to study the dynamic stability of the doped system, the corresponding relationship between the elastic constants (C_{ij} (GPa)) of SnO_2 and different ratios of Ce-doped SnO_2 is given in Table 6.

For tetragonal systems, if the dynamics are stable, the elastic constants satisfy the following requirements [23]: $C_{11} > 0\ C_{33} > 0\ C_{44} > 0\ C_{66} > 0,\ C_{11} - C_{12} > 0\ C_{11} + C_{33} - 2C_{13} > 0\ 2C_{11} + C_{33} + 2C_{12} + 4C_{13} > 0$.

The elastic constants of SnO_2 and different ratios of Ce-doped SnO_2 in Table 6 are calculated, and all satisfy the stability criterion. Therefore, SnO_2 and different ratios of Ce-doped SnO_2 are stable in dynamics.

In order to further predict theoretically the mechanical properties such as hardness (H_V) and crack, the bulk modulus (B), shear modulus (G), hardness (H_V), and universal elastic anisotropy index (A^U) values are shown in Table 7.

According to the Pugh criterion, it is generally considered that solid materials with $G/B > 0.57$ are brittle; if they are less than 0.57, it exhibits toughness [24]. From Table 7, the order of G/B values is $SnO_2 > Sn_{0.9166}Ce_{0.0834}O_2 > Sn_{0.875}Ce_{0.125}O_2 > Sn_{0.833}Ce_{0.167}O_2$, which indicates that doping can change the brittleness of the material, and the more the rare earth element Ce, the more obvious the improvement.

Covalent bond orientation directly affects the ability of the material to resist shear strain. Poisson's ratio can reflect the covalent bond orientation problem of internal structure of the material, the lower the value of the former, the stronger the ability of material to resist shear strain. Table 7 shows that Poisson's ratio becomes smaller after doping, indicating that their ability to resist shear strain is enhanced. SnO_2 with the Ce doping ratio of 12.5% has the smaller Poisson's ratio and the best ability to resist shear strain.

Table 6: The relationship between doping ratio and elastic constants

Ratio	C_{11}	C_{22}	C_{33}	C_{44}	C_{55}	C_{66}	C_{12}	C_{13}
0	108.0	108.0	286.1	88.2	88.2	146.8	26.8	41.8
16.7%	243.0	223.2	65.1	68.5	79.5	42.3	−28.3	92.2
12.5%	142.5	277.3	145.5	75.1	134.7	76.0	73.2	58.6
8.34%	147.1	290.6	146.1	79.1	135.9	80.5	74.1	58.1

Table 7: Bulk modulus (GPa), shear modulus (GPa), Poisson's ratio, and universal elastic anisotropy index.

Ratio	B	G	σ	A^U	G/B	H_V
0	72.6	82.2	0.26	1.39	1.13	24.0
16.7%	105.2	73.3	0.22	0.93	0.70	13.2
12.5%	103.0	75.5	0.21	0.93	0.73	14.1
8.34%	104.4	78.7	0.20	0.91	0.75	15.5

The index of pervasive elastic anisotropy, A^U, is often the decisive factor in the production of cracks. As can be seen from Table 7, the A^U values of SnO_2 after doping are smaller and closer than those of SnO_2, indicating that doping has an important effect on improving microcracks.

4. Experimental Preparation of Ag/SnO_2 and Ce-Doped Ag/SnO_2 Contact Materials

4.1. Preparation of Doped SnO_2 Powder by Sol-Gel Method and Ag/SnO_2-Ce Contact Material by Powder Metallurgy. The traditional powder metallurgy method only mixes the doping element powder, SnO_2 powder, and silver powder, which makes it difficult for the doping element to enter the SnO_2 crystal. In order to be consistent with the theoretical calculation model of this paper, the sol-gelation method was used to first prepare different proportions of Ce-doped SnO_2 powder. Then, Ag/SnO_2 and Ce-doped Ag/SnO_2 contact materials were prepared by the powder metallurgy method. The proportion of silver base in each sample was 88%.

4.2. Test of Contact Resistance and Arcing Energy. The electrical contact properties of test materials were tested by using the JF04C electrical contact material tester. 25000 on-off tests were conducted using each pair of contacts. The contact resistance and arcing energy measurement were done every 100 cycles under load 24 V, 13 A, and contact pressure was 86 cN for the on-off period of 0.4 seconds. Finally, the conductivity of the test material is evaluated by the contact resistance and the arcing energy. The arc energy and contact resistance of Ag/SnO_2 doped with different proportions of rare earth elements Ce are shown in Table 8.

Because the rare earth elements easily lose the outermost electrons to form a stable structure with a high melting point, they are not easily decomposed under the action of the arc but accumulate on the surface of the contact, thereby reducing the ablation of the contacts by the arc and reducing the arcing energy.

From Table 8, when the doping ratio is 12.5%, the arcing energy has the smallest variation range, average arcing

TABLE 8: Arc energy and contact resistance.

Ratio	Arc energy variation range (mJ)	Average arc energy (mJ)	Contact resistance variation range (mΩ)	Average contact resistance (mΩ)
16.7%	173.28~193.91	184.48	0.82~3.87	1.82
12.5%	173.90~189.33	179.31	0.56~2.58	1.24
8.3%	176.49~194.66	184.97	0.60~3.15	1.26
0	159.04~235.28	192.92	0.35~8.93	1.83

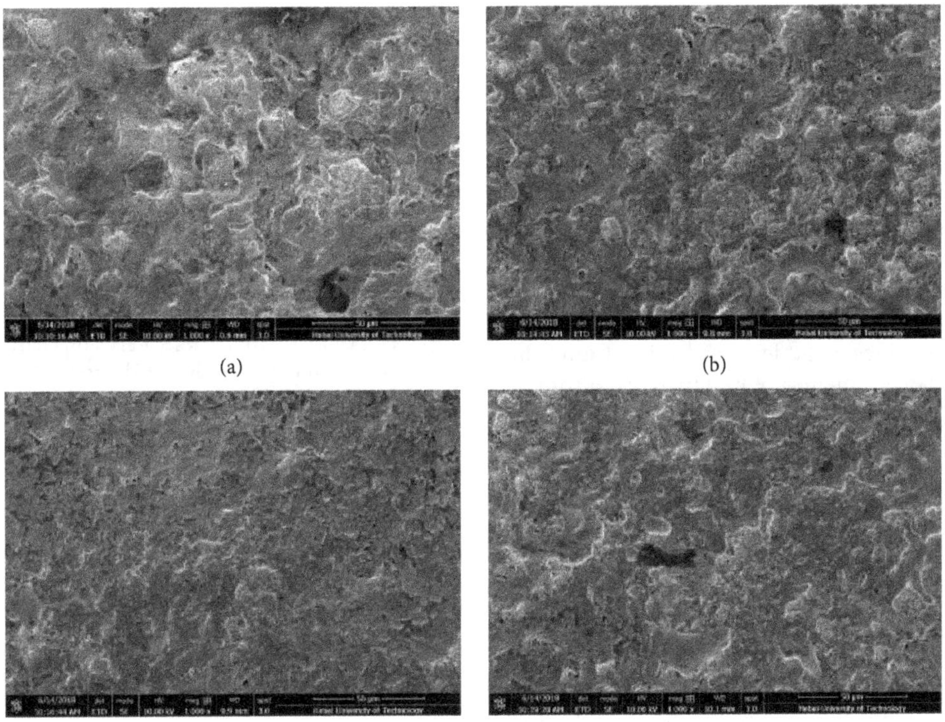

(a) (b) (c) (d)

FIGURE 4: SEM (a) SnO_2. (b) 16.7% of Ce-doped SnO_2. (c) 12.5% of Ce-doped SnO_2. (d) 8.34% of Ce-doped SnO_2.

energy, and the actual measured conductivity are also the smallest, indicating that Ce-doped SnO_2 has the smallest fluctuation and the most stable.

It is proved that when the rare earth element Ce doping ratio is 12.5%, the electrical properties of the Ag/SnO_2 contact material are the best under the applied test conditions here.

4.3. Microstructure Analysis of Contact Material Surface. Scanning electron microscopy (SEM) was used to observe the arc ablation morphology of the four types of Ag/SnO_2 contact materials, and the arc erosion resistance of the four types of contact materials was analyzed.

After 25,000 electrical contact performance tests of the contact material, the photomicrographs of the contact surface magnified 1000 times are shown in Figure 4, among them Figure 4(a) shows Ag/SnO_2 and Figures 4(b)–4(d) show the Ag/SnO_2 with the Ce doping ratios of 16.7%, 12.5%, and 8.34%, respectively.

The wettability between Ag and SnO_2 particles is poor, which causes the SnO_2 particles to be suspended in the liquid Ag molten pool when the Ag/SnO_2 contact material is in the arc. The surface is easy to form Ag-rich region and SnO_2-rich region. The formation of the Ag-rich region is likely to exacerbate the splash erosion of the contact Ag droplets, and the SnO_2-rich region tends to increase the contact resistance of the contacts. Doping can effectively improve the wettability between Ag liquid and SnO_2 particles so that SnO_2 particles are suspended in the molten system of silver, increasing the viscosity of Ag molten pool, reducing the probability of formation of the SnO_2 polymerization zone, and thus improving the resistance to arc erosion of the Ag/SnO_2 contact material.

Comparing the surface topography of the ablation edge regions of three different proportions of doping, when the doping is 12.5%, the contact surface is flat and without pits, which was because the contact surface shows signs of fluid metal activity after slight melting, and the gap and other phenomena were improved because of the movement of the flow metal. At this time, the surface is relatively flat and smooth, and the electrical properties are the best. The other two ratios are doped, the surface of the contact has holes, the surface of the molten layer exhibits a flowing state of liquid metal and a mountain pack forms after solidification, and some of the black nonconductive oxides are precipitated.

4.4. Hardness Measurement. The hardness of the sample was measured using the HXD-1000TM digital microhardness tester [17]. The test strength was selected, and the diamond was pressed into the test sample for 5 s. The two diagonal lengths of the diamond indentation on the surface of the sample were observed by a microscope. The hardness value was read, the sample position was changed, and the above process was repeated. Three measurements were made for each sample, and the average value of the hardness was calculated. The hardness values of Ag/SnO_2 and Ag/SnO_2 doped with different proportions of Ce are given in Table 9.

It can be seen from the above table that the more the doping, the lower the hardness. By comparing the theoretical and actual hardness values in Tables 7 and 9, it is found that the simulations and experiments are well matched. From the perspective of the application environment of the contact, high hardness plays an important role in improving the service life of the contact; from the perspective of the forming of the contact, high hardness affects the plasticity and reduces the yield. Therefore, the hardness value must be considered comprehensively.

5. Conclusions

Based on the first principle of density functional theory, the calculated and analyzed results of $Sn_{1-x}Ce_xO_2$ ($x = 0$, 0.083, 0.125, 0.167) show the following:

(1) When the doping proportion of Ce is 0.125, the electron mobility is the highest, and the conductivity and the ability to resist shear strain are the best

(2) When the doping is 0.834, the universal elastic anisotropy index is the smallest

Compared with all performances, 0.125 is the best doping ratio.

Finally, Ag/SnO_2 contacts with different doping ratios were prepared, and arcing energy, contact resistance, and hardness were measured. The final simulation and experimental results can be well matched.

Some people study on Ce-doped Ag/SnO_2 contact material to improve the performance is still in the experimental stage, and the control of the doping amount is also based on the original experience. The innovation of this paper is to obtain the most comprehensive performance by simulating different proportions of Ce doping. A good doping ratio is then verified by experimentation. In the experiment, we not only tested the contact resistance and arc energy but also performed a scanning electron microscope experiment to observe the arc ablation morphology of the contact material. The final theoretical model is well matched to the experiment, providing a method of saving manpower, material resources, and financial resources to improve the electrical properties of Ag/SnO_2 contact materials.

Table 9: Sample hardness measured.

Ratio	16.7%	12.5%	8.3%	0
Hardness (HV)	93.6	100.1	110.9	117.1

Conflicts of Interest

The authors have no conflicts of interest.

Acknowledgments

This study was funded by the National Natural Science Foundation (51777057).

References

[1] H. Li and Y. Hou, "Electro-erosion characteristics of Ag/SnO_2 contact material under DC lamp load," *Electrical Engineering Materials*, vol. 2, pp. 12–15, 2005.

[2] M. Zhang, "Development status of Ag/SnO_2 contact material and its application in relays," *Electrical Engineering Materials*, vol. 4, pp. 20–26, 2004.

[3] Z. Liu, *The Study on Nano-Doped Ag/SnO2 Contact material*, Tianjin University, Tianjin, China, 2007.

[4] S. Tingting, Z. Fuchun, and Z. Weihu, "Density functional theory study on the electronic structure and optical properties of SnO_2," *Rare Metal Materials and Engineering*, vol. 44, no. 10, pp. 2409–2414, 2015.

[5] Y. Zhu, *Study on the Ag/SnO_2 Contact Materials with the Addition of Rare Earth Oxides*, Hebei University of Technology, Tianjin, China, 2007.

[6] Y. Zhu, J. Wang, and L. An, "Preparation and electrical properties of new $Ag/SnO_2/CeO_2$ electrical contact materials," *Rare Metal Materials and Engineering*, vol. 34908, pp. 2011–2014, 2015.

[7] L. Shan, B. Dechun, Y. Lin et al., "The structure and optical properties of Ce-doped SnO_2 materials," *Vacuum*, vol. 1, pp. 25–28, 2014.

[8] Q. Wang, *Research on Tin Dioxide Electrode and its Strengthening and Toughening*, Donghua University, Shanghai, China, 2005.

[9] Q. Wang, *Preparation and Properties of Rare Earth Doped Tin Dioxide Ceramic Electrodes*, Donghua University, Shanghai, China, 2012.

[10] D. Guo and C. Hu, "First-principles study on the electronic structure and optical properties for SnO_2 with oxygen vacancy," *Applied Surface Science*, vol. 258, no. 18, pp. 6987–6992, 2012.

[11] S. F. Matar, D. Jung, and M. A. Subramanian, "The predominance of the rutile phase of SnO_2: first principles study," *Solid State Communications*, vol. 152, no. 5, pp. 349–353, 2012.

[12] M. D. Segall, P. J. D. Lindan, M. J. Probert et al., "First-principles simulation: ideas, illustrations and the CASTEP code," *Journal of Physics: Condensed Matter*, vol. 14, no. 11, pp. 2717–2744, 2002.

[13] K. G. Godinho, A. Walsh, and G. W. Watson, "Energetic and electronic structure analysis of intrinsic defects in SnO_2,"

Journal of Physical Chemistry C, vol. 113, no. 1, pp. 439–448, 2008.

[14] D. M. Ceperley and B. J. Alder, "Ground state of the electron gas by a stochastic method," *Physical Review Letters*, vol. 45, no. 7, pp. 566–569, 1980.

[15] J. A. Berger, L. Reining, and F. Sottile, "Efficient GW calculations for SnO_2, ZnO and rubrene: the effective-energy technique," *Physical Review B Condensed Matter*, vol. 85, no. 8, article 085126, 2012.

[16] E. Liu, B. Zhu, and J. Luo, *Semiconductor Physics*, Electronic Industry Press, Beijing, China, 6th edition, 2003.

[17] W. Jingqin, Z. Ying, and K. Huiling, "Study on properties of $AgSnO_2$ contact materials doped with rare earth Y," *Materials Research Express*, vol. 5, no. 8, article 085902, 2018.

[18] R. Hill, "The elastic behaviour of a crystalline aggregate," *Proceedings of the Physical Society*, vol. 65, no. 5, pp. 349–354, 2002.

[19] W. Voigt, *Lehrbuch Der Kristallphysik: Teubner*, Macmillan, Leipzig, Germany, 1928.

[20] A. Reuss, "Calculation of the flow limits of mixed crystals on the basis of the plasticity of monocrystals," *ZAMM—Zeitschrift für Angewandte Mathematik und Mechanik*, vol. 9, pp. 49–58, 1929.

[21] S. I. Ranganathan and M. Ostoja-Starzewski, "Universal elastic anisotropy index," *Physical Review Letters*, vol. 101, article 055504, 2008.

[22] Y. Tian, B. Xu, and Z. Zhao, "Microscopic theory of hardness and design of novel superhard crystals," *International Journal of Refractory Metals and Hard Materials*, vol. 33, pp. 93–106, 2012.

[23] F. Zhang, Z. Zhang, W. Zhang et al., "First-principles calculation of the electronic structure and optical properties of In_2O_3," *Journal of Chemistry*, vol. 66, no. 16, pp. 1863–1868, 2008.

[24] Z. Wu, E. Zhao, H. Xiang et al., "Crystal structures and elastic properties of superhard IrN2 and IrN3 from first principles," *Physical Review B*, vol. 76, no. 5, article 054115, 2007.

14

Evaluating the Dynamic Elastic Modulus of Concrete Using Shear-Wave Velocity Measurements

Byung Jae Lee,[1] Seong-Hoon Kee,[2] Taekeun Oh,[3] and Yun-Yong Kim[4]

[1]*R&D Center, JNTINC Co. Ltd., 9 Hyundaikia-ro, 830 Beon-gil, Bibong-Myeon, Hwaseong, Gyeonggi-do 18284, Republic of Korea*
[2]*Department of Architectural Engineering, Dong-A University, 37 Nakdong-Daero, 550 Beon-gil, Saha-gu, Busan 49315, Republic of Korea*
[3]*Department of Safety Engineering, Incheon National University, 119 Academy-ro, Yeonsu-gu, Incheon 22012, Republic of Korea*
[4]*Department of Civil Engineering, Chungnam National University, 99 Daehak-ro, Yuseong-gu, Daejeon 34134, Republic of Korea*

Correspondence should be addressed to Seong-Hoon Kee; shkee@dau.ac.kr

Academic Editor: Giorgio Pia

The objectives of this study are to investigate the relationship between static and dynamic elastic moduli determined using shear-wave velocity measurements and to demonstrate the practical potential of the shear-wave velocity method for in situ dynamic modulus evaluation. Three hundred 150 by 300 mm concrete cylinders were prepared from three different mixtures with target compressive strengths of 30, 35, and 40 MPa. Static and dynamic tests were performed at 4, 7, 14, and 28 days to evaluate the compressive strength and the static and dynamic moduli of the cylinders. The results obtained from the shear-wave velocity measurements were compared with dynamic moduli obtained from standard test methods (P-wave velocity measurements according to ASTM C597/C597M-16 and fundamental longitudinal and transverse resonance tests according to ASTM C215-14). The shear-wave velocity measured from cylinders showed excellent repeatability with a coefficient of variation (COV) less than 1%, which is as good as that of the standard test methods. The relationship between the dynamic elastic modulus based on shear-wave velocity and the chord elastic modulus according to ASTM C469/C469M was established. Furthermore, the best-fit line for the shear-wave velocity was also demonstrated to be effective for estimating compressive strength using an empirical relationship between compressive strength and static elastic modulus.

1. Introduction

In the design of structures, the elastic modulus of concrete (E_c) is a fundamental parameter in estimating the deformation of a structural element under service conditions. In practice, E_c has been estimated from compressive strength based on the design code rather than on direct measurement. This practice could underestimate E_c and demand higher compressive strength to achieve a desired E_c than is actually required in structural design [1]. Furthermore, E_c has been demonstrated to be an effective parameter for condition assessment of concrete in existing structures [2, 3]. Destructive tests such as core extraction have been used widely to acquire accurate information on elastic properties of concrete. However, they are labor-intensive and time-consuming and cannot be applied ubiquitously over the entire area of the structure.

There are several nondestructive evaluation (NDE) methods to estimate elastic properties of concrete including ultrasonic pulse velocity measurements according to ASTM C597/C597M-16 [4] and resonance frequency tests according to ASTM C215-14 [5]. The elastic modulus determined by NDE methods is typically called the dynamic elastic modulus, which is generally greater than static elastic modulus measured in accordance with ASTM C469/C469M-14 [6]. Dynamic modulus has been assumed to be initial tangent modulus at zero stress determined in the standard test because only negligible stress is applied during ultrasonic pulse velocity (UPV) measurements and resonance frequency tests [7]. Philleo [8] also explained that the difference between dynamic and static moduli is based on the fact that the nonhomogeneous characteristics of concrete affect the two moduli in different ways. Moreover, the difference between the two moduli decreases as concrete strength increases: the dynamic elastic modulus is generally 20, 30, and 40% higher than the static elastic modulus for high-, medium-, and low-strength concrete, respectively [2]. Lydon and Balendran [9] proposed an empirical relationship as follows:

$$E_c = 0.83E_d \text{ (GPa)}. \quad (1)$$

The British testing standard BS8100 Part 2 [10] provides another empirical equation as follows:

$$E_c = 1.25E_d - 19 \text{ (GPa)}. \quad (2)$$

It is noteworthy that this equation does not apply to concrete containing more than 500 kg of cement per cubic meter of concrete or to lightweight aggregate concrete [7]. A more general relationship was proposed by Popovics [11] for both lightweight and normal density concrete, taking into account the effect of concrete density as follows:

$$E_c = \frac{446.09E_d^{1.4}}{\rho_c} \text{ (GPa)}, \quad (3)$$

where ρ_c is the density of the hardened concrete in a unit of kg/m^3.

The two standard NDE methods for evaluation of dynamic elastic modulus have advantages and limitations. The P-wave velocity method is convenient to use and has clear advantages over the resonance method in that the testing is not confined to regularly shaped laboratory specimens and results are not sensitive to inelastic effect [8]. P-waves move particles parallel to the direction of propagation and can propagate through media with enough lateral stiffness: they thus can propagate in solid, liquid, and gas media. Consequently, P-wave velocity in porous materials is affected by the properties of three difference phases in the media. Dynamic elastic modulus determined by P-wave velocity measurements has been known to possess the characteristics of concrete, heterogeneous, and porous material, such as type of aggregate, water content, air voids, and porosity compared with the static elastic modulus. In addition, P-wave velocity values depend on confinement conditions in solid media: the P-wave velocities in unbounded solid media (C_P), thin plates ($C_{c,\text{plate}}$), and thin rods ($C_{c,\text{rod}}$) are called constrained, partially constrained, and unconstrained compressive wave velocities, respectively, and are expressed as the following equations [12]:

$$C_P = \sqrt{\frac{E(1-\nu)}{\rho(1+\nu)(1-2\nu)}}, \quad (4)$$

$$C_{c,\text{plate}} = \sqrt{\frac{E}{\rho(1-\nu^2)}}, \quad (5)$$

$$C_{c,\text{rod}} = \sqrt{\frac{E}{\rho}}. \quad (6)$$

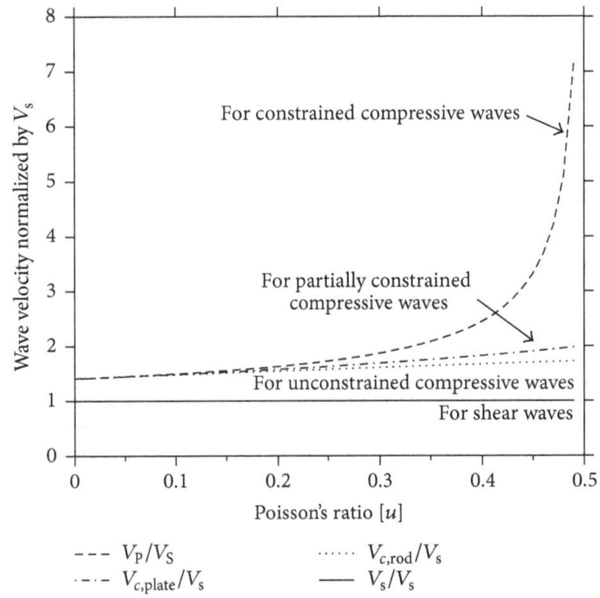

FIGURE 1: The variations of constrained, partially constrained, and unconstrained compressive waves and shear waves in elastic homogeneous and isotropic media with Poisson's ratio. The wave velocities are normalized by the S-wave velocity.

Figure 1 shows the variations of P-waves with Poisson's ratio for elastic homogeneous and isotropic media. Given Poisson's ratio, P-wave velocity is in a range between unconstrained and constrained velocities [13]. P-wave velocity increases with increasing the degree of confinement, and the confinement effect increases as Poisson's ratio increases. In the field practice, however, it is sometimes not easy to evaluate accurate dynamic elastic moduli from P-wave velocity measurements because of difficulties in estimating the confinement effect in the field and the absence of theoretical or practical formula relating P-wave velocity and dynamic elastic modulus that takes into account the confinement effect. In this sense, some of previous researchers argue that the P-wave velocity method is nonreliable and not recommended for estimating the dynamic elastic modulus of concrete [1, 7, 14]. In contrast, resonance frequency testing has been used widely to evaluate the dynamic elastic modulus of concrete, which is less sensitive than P-wave velocity. However, resonant frequency testing necessarily needs regularly shaped concrete specimens; thus, in situ application of this method is labor-intensive and time-consuming and is inappropriate for general application over the entire area of a structure.

There has been some theoretical and experimental evidence demonstrating that the shear-wave measurement has potential to be an effective in situ NDE method for estimating solid stiffness of various cementitious materials (e.g., cement paste, mortar, and concrete). Theoretically, shear waves move particles perpendicular to the direction of propagation and thus can propagate through solid media with enough shear rigidity. Consequently, S-wave propagation is less sensitive to liquids or gases in porous materials. Previous researchers have demonstrated by experiments in the laboratory that

TABLE 1: Mixture proportions of concrete.

ID	Cement type	W/B	S/A	Unit quantity (kg/m³) W	C	S	G	SCMs FA	SC	Chemical admixture AE (binder%)	SP (binder%)
Mix 1	Type I	0.45	0.46	259	121	777	934	58	69	0.9	—
Mix 2		0.35	0.47	308	166	761	886	81	85	—	1
Mix 3		0.3	0.46	357	165	714	868	94	99	—	1

Note. SCMs: Supplementary cementitious materials, W: water, C: cement, S: sand, G: gravel, FA: fly ash, SC: slag cement, AE: air-entraining agent, and SP: superplasticizer.

S-waves are less sensitive to water and/or air content and an effective parameter for evaluating stiffness (or strength) gain of early-age cementitious materials. Voigt et al. [15] showed that the reflection of the S-wave is closely related to the degree of interparticle bonding regardless of water content in cement paste. Zhu et al. [16] demonstrated experimentally that S-wave velocity is not affected by air content in a range of 0.2% to 5.2% in cement paste, while P-wave velocity is too sensitive to air content. Liu et al. [17] successfully monitored setting and hardening processes of mortar and concrete using ultrasonic S-wave velocity in the laboratory and verified that S-wave velocity correlated well with penetration resistance in mortar regardless of test setup and water-to-cement ratio. Carette and Staquet [18] observed that S-wave velocity is more sensitive to the setting process of cement-based materials than P-wave velocity. In addition, unlike P-waves, S-waves are theoretically not sensitive to confinement conditions of concrete [12]. In a homogeneous and isotropic medium, the S-wave velocity is expressed as the following equation without regard to the confinement conditions [12]:

$$C_S = \sqrt{\frac{E}{2\rho(1+\nu)}}. \tag{7}$$

Furthermore, some researchers have observed that S-wave velocity is a more stable parameter for evaluating compressive strength of hardened concrete than in P-wave velocity. An et al. [19]. observed that S-wave velocity is closely correlated with compressive strength without regard to the types of aggregate, curing age, and curing conditions. Ciancio and Helinski [20] demonstrated that the S-wave velocity measurement is effective in evaluating compressive strength of fiber reinforced shotcrete in field tests owing to its decreased sensitivity to the presence of pore water. In addition, Lencis et al. [21] observed that S-wave velocity is less sensitive to reinforcing steels in concrete than P-wave velocity. In the research, it was demonstrated that the influence of reinforcing steels is insignificant for S-wave velocity measurements in the indirect transmission method when concrete cover is at least 40 mm, which is valid in most concrete structures in the field practice. Samokrutov et al. [22] demonstrated that using S-waves compared to P-waves reduces the amount of backscattering and signal attenuation in the direction parallel to the propagating wave in hardened concrete. Recently, Lee and Oh [23] measured S-wave velocities using MIRA equipment that generates shear-wave tomography [24] and demonstrated the effectiveness of shear-wave measurements for condition assessment of reinforced and prestressed concrete slabs. In the study, it was demonstrated that S-wave velocity is statistically more stable than P-wave velocity because of lower energy attenuation and less sensitivity of confinement effects. In summary, the S-wave velocity method has potential as an in situ NDE method for evaluating the elastic modulus and compressive strength of concrete. However, it is still difficult to draw a general conclusion due to a scarcity of experimental studies on evaluating the dynamic elastic modulus with the S-wave velocity measurement.

The objectives of this study are to investigate the relationship between static and dynamic elastic moduli determined using S-wave velocity measurements and to demonstrate the practical potential of the S-wave velocity method for in situ dynamic modulus evaluation. In this study, three hundred 150 by 300 mm concrete cylinders were prepared from three different mixtures with target compressive strengths of 30, 35, and 40 MPa. Static and dynamic tests were performed at 4, 7, 14, and 28 days to evaluate the compressive strength and the static and dynamic moduli of the cylinders. The results obtained from the S-wave velocity measurements were compared with dynamic moduli obtained from standard test methods (P-wave velocity measurements in ASTM C597/C597M-16 and fundamental longitudinal and transverse resonance tests in ASTM C215-14).

2. Experimental Program

2.1. Materials and Preparation of Specimens. The concrete in this study consists of Type I Portland cement, river sand, crushed granite with a maximum size of 25 mm, and supplementary cementitious materials (fly ash and slag cement). Three different water-to-binder ratios (W/B) of 0.3, 0.35, and 0.45 were used, which results in target compressive strengths of 40, 35, and 30 MPa, respectively. The three different concrete mixtures (hereafter referred to as mix 1, mix 2, and mix 3) were prepared with ready-mixed concrete, and the proportions are summarized in Table 1.

Three hundred concrete cylinders were cast in 150 by 300 mm plastic molds in accordance with ASTM C31/C31M [25]. The cylinders were water-cured after being demolded on the next day. A series of static and dynamic modulus tests for 25 cylinders were conducted at different ages: 4, 7, 14, and 28 days.

FIGURE 2: Test setup for measuring compressive strength and static elastic modulus of a concrete cylinder.

2.2. Static Tests for Compressive Strength and Elastic Modulus. The cylinders were ground at both ends before testing to remove any surface irregularities as well as to ensure that the ends were perpendicular to the sides of the specimen. Static elastic modulus and compressive strength of the cylinders were measured using a universal testing machine (UTM) with a capacity of 1000 kN according to ASTM C469/C469M-14 [6] and ASTM C39/C39M-14 [26], respectively. Tests were performed at a loading rate of approximately 0.28 MPa/s. Deformations were measured using three sets of linear voltage differential transducers attached to two fixed rings (Figure 2). The apparatus consisted of two aluminum rings with screws for attachment to the specimen. The spacing between screws on the top and bottom rings was 150 mm. This served as a gauge length for calculating axial strain from the measured deformations. The static elastic modulus of concrete is defined as a chord modulus from the stress-strain curve, with a first point at a strain level of 0.00005 (ε_1) and the second point at 40% of maximum stress (f_c) as follows:

$$E_c = \frac{0.4f_c - \sigma(\varepsilon_1)}{\varepsilon(0.4f_c) - \varepsilon_1}. \quad (8)$$

It is of interest to compare dynamic elastic modulus and initial tangent modulus, which corresponds to a small strain and is calculated from the tangent drawn at the origin of the stress-strain curve [27]. Unfortunately, the measured stress-strain curve of cylinders in this study always showed certain fluctuations at very low strain, which posed considerable difficulties in consistent and reliable calculation of the initial tangent modulus. Instead, an initial chord modulus was defined from two points corresponding to strain levels of 0.000001 (ε_1) and 0.00005 (ε_2) as follows:

$$E_i = \frac{\sigma(\varepsilon_2) - \sigma(\varepsilon_1)}{\varepsilon_2 - \varepsilon_1}. \quad (9)$$

Note that typical stress-strain curves of cylinders in this study show that stress increases almost linearly with strain at very low values of strain, that is, approximately 0.00005. Therefore, the initial chord modulus is assumed to be equivalent to the initial tangent modulus of cylinders in this study.

2.3. P- and S-Wave Measurements for Estimating Dynamic Elastic Modulus. The S-wave velocity of concrete, C_S, was measured using the P-wave velocity procedure described in ASTM C597/C597M-16 [4] with a pair of S-wave transducers (ACS T1802) that were connected to a pulser-receiver (Ultracon 170) (see Figure 3(a)). The source transducer was driven by a 200 V square pulse, generating a transverse ultrasonic pulse. The signals generated by the source transducer and measured by the receiving transducer were digitized at a sampling frequency of 10 MHz using an NI-PXIe 6366 oscilloscope. Figure 4(a) shows typical impulse signal generated by a source transducer and an ultrasonic wave measured by a receiving transducer for the S-wave velocity measurement. The arrival of the transient stress waves through cylinders was determined from the measured ultrasonic wave using the modified threshold method [28]. In this method, an approximate arrival time was first obtained using the conventional threshold method in the literature [29]. Next, an accurate arrival time was calculated by fitting a line to the single data. The intersection of the fitting line and the calculated zero-signal level defines the S-wave travel time. Finally, the S-wave velocity was calculated by dividing the length of cylinders (L) over the travel time (t) (i.e., $C_s = L/t$). In this study, dynamic elastic modulus based on C_S is determined by rearranging (7) as follows:

$$E_{d,S} = \alpha_S \rho C_S^2, \quad (10)$$

where $E_{d,S}$ is dynamic elastic modulus of the concrete specimen using S-wave velocity measurements, α_S is constant dependent on Poisson's ratio v, that is, $2(1 + v)$, and ρ is mass density of concrete ($=m/\pi r^2 L$; m, r, and L are the mass, radius, and length of the cylinder, respectively, measured at testing day).

For comparison purposes, dynamic elastic moduli of concrete cylinders were also determined using standard NDE methods, including P-wave velocity and resonant frequency tests. In the P-wave velocity test, a longitudinal pulse of 52 kHz is transmitted through the specimen, and the P-wave velocity of concrete, C_P, is measured in accordance with ASTM C597/C597M-16 [4]. The test setup for P-wave velocity measurements was the same as that for the S-wave velocity measurements except for using a pair of P-wave transducers (MK 954). Figure 4(b) shows typical impulse signal generated by a source transducer and an ultrasonic wave measured by a receiving transducer for the P-wave velocity measurement. The dynamic elastic modulus based on C_P is determined by rearranging (4) as follows:

$$E_{d,P} = \alpha_P \rho C_P^2, \quad (11)$$

where $E_{d,P}$ is dynamic elastic modulus of the cylinder using the P-wave velocity measurements and α_P is constant dependent on Poisson's ratio υ, that is, $(1 + \upsilon)(1 - 2\upsilon)/(1 - \upsilon)$.

(a) (b)

FIGURE 3: Testing configurations for (a) ultrasonic shear-wave velocity using a pair of S-wave transducers (ACS T1802) and (b) resonance frequency of concrete specimens.

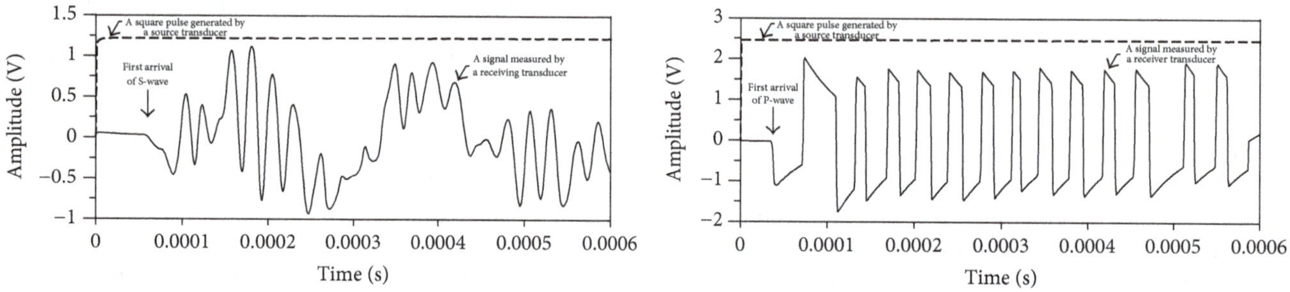

(a) (b)

FIGURE 4: Typical impulse signals generated by a source transducer and ultrasonic waves obtained from ultrasonic pulse velocity tests for (a) S-wave and (b) P-wave measurements.

2.4. Resonance Frequency Measurements for Estimating Dynamic Elastic Modulus. Fundamental longitudinal and transverse resonant frequencies of the cylinders were measured to estimate dynamic elastic moduli in accordance with ASTM C215-14 [5] (see Figure 3(b)). A steel ball having a diameter of 10 mm was used as an impact source for generating incident stress waves in concrete specimens; the steel ball was effective for generating wideband frequency signals from very low to 20 kHz, covering the frequency range of resonant frequency tests in this study. The dynamic response of the concrete cylinder was measured by an accelerometer (PCB 353B16), with ±5% frequency range of 1 to 10 kHz and resonant frequency around 70 kHz, attached to the concrete specimen according to ASTM C215-14 [5]. The acquired signals using the accelerometer were stabilized using a signal conditioner (PCB 482C16) and digitized at a sampling frequency of 1 MHz using an NI-PXIe 6366 oscilloscope. Resulting time signals were converted into the frequency domain using the FFT (fast Fourier transform) algorithm. Figure 5 shows typical spectral amplitude from the mix 2 concrete cylinders with ages of 4, 7, 14, and 28. The resonant frequencies of the cylinders are manifested as dominant amplitudes in the amplitude spectrum. The most dominant frequency was regarded as the fundamental resonant frequency for longitudinal f_L and transverse f_T modes, and the frequency values were used to calculate dynamic elastic moduli of the cylinders as follows:

$$E_{d,\mathrm{LR}} = \beta_{\mathrm{L}} m f_{\mathrm{L}}^{2}\ (\mathrm{Pa}),$$

$$E_{d,\mathrm{TR}} = \beta_{\mathrm{T}} m f_{\mathrm{T}}^{2}\ (\mathrm{Pa}), \tag{12}$$

where $E_{d,\mathrm{LR}}$ and $E_{d,\mathrm{TR}}$ are dynamic elastic moduli using longitudinal and transverse fundamental resonant frequencies, respectively; β_L is a constant depending on the dimensions of the cylinder $=5.093(L/D^2)$; L and D are the length and diameter of a cylinder, respectively; β_T is a constant depending on dimensions and Poisson's ratio of cylinder $(=1.6067(L^3T/D^4))$; and T is a correction factor obtained from ASTM C215-14[5].

3. Result and Discussion

3.1. Experimental Variability of Static and Dynamic Tests. Experimental variability of the measured static and dynamic parameters is of interest when investigating the consistency and reliability of the test methods. Uncertainty of the average value of the in-place test results is a function of the standard deviation of the results and the number of tests. In this study, the coefficient of variation (COV, the standard deviation, σ, divided by the mean value, μ, of a set of specimens) was used as a means of evaluating the experimental variability of the static and dynamic measurements of the concrete cylinders; outliers were detected by the modified Z-score method [30] and were removed in the statistical analysis. The statistical parameters (μ and COV) of the accepted static and dynamic test results are summarized in Table 2.

TABLE 2: Summary of statistical analysis.

		Day 4			Day 7			Day 14			Day 28		
		N	μ	COV	N	μ	COV	N	μ	COV	N	μ	COV
f_c [MPa]	Mix 1	25	7.9	4.36	25	10.1	4.32	25	13.9	3.42	25	19.2	4.25
	Mix 2	24	25.1	2.94	25	30.0	3.06	24	37.1	4.23	25	43.1	4.87
	Mix 3	24	23.4	2.83	25	29.3	2.62	25	39.4	4.32	25	44.8	3.63
E_c [GPa]	Mix 1	24	10.1	6.13	25	11.8	4.62	25	14.5	5.31	25	16.9	5.39
	Mix 2	25	16.5	5.35	24	17.7	4.34	25	20.4	3.86	25	23.5	4.57
	Mix 3	24	16.8	5.44	24	18.7	4.23	24	21.5	4.02	24	23.0	3.57
C_P [m/s]	Mix 1	25	3225	2.49	25	3474	1.81	25	3740	1.68	25	3958	1.34
	Mix 2	25	3948	1.51	25	4064	1.41	25	4197	1.07	25	4283	1.03
	Mix 3	25	3894	1.45	25	4055	1.07	25	4195	0.97	25	4302	0.82
C_S [m/s]	Mix 1	—	—	—	22	1664	1.15	25	1790	1.09	25	1936	1.32
	Mix 2	—	—	—	20	2015	1.00	25	2104	0.83	25	2165	0.76
	Mix 3	25	1942	0.93	25	2039	0.90	25	2138	0.90	25	2191	0.87
f_{L1} [Hz]	Mix 1	25	4596	1.56	25	4636	1.37	25	5248	1.21	25	5616	1.25
	Mix 2	25	5602	1.51	25	5752	0.98	25	5990	0.83	25	6178	0.87
	Mix 3	25	5580	1.29	25	5818	1.04	25	6064	1.22	25	6246	1.13
f_{T1} [Hz]	Mix 1	25	3346	3.28	25	3208	1.25	25	3232	1.25	25	3442	1.08
	Mix 2	25	3428	1.89	25	3490	1.49	25	3622	1.32	25	3778	1.32
	Mix 3	25	3412	1.35	25	3532	1.14	25	3678	1.52	25	3796	1.36

Note. N = the number of accepted samples used for the statistical analyses; μ = average of samples; and COV = coefficient of variable ($=\sigma/\mu \times 100$ [%]).

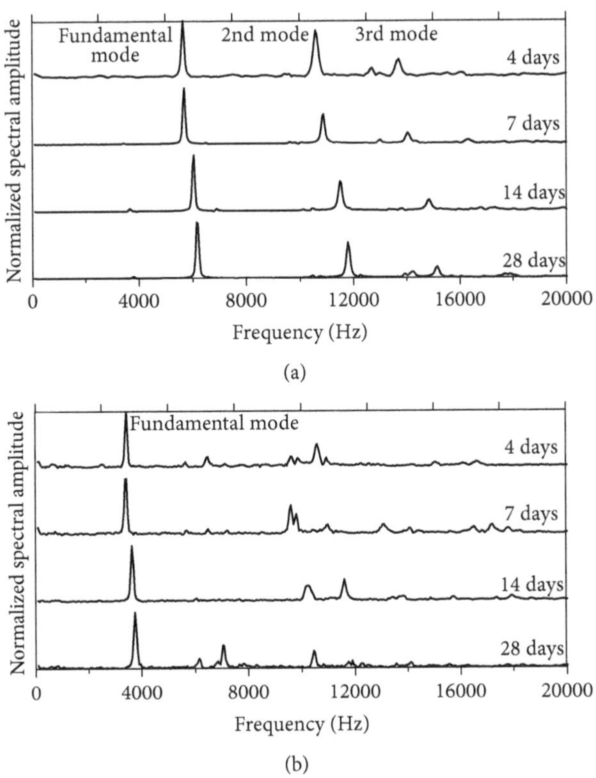

FIGURE 5: Typical spectral amplitude measured from the mix 2 concrete cylinders with ages of 4, 7, 14, and 28 days using (a) longitudinal and (b) transverse resonance frequency tests.

The COV of the compressive strength of concrete f_c ranges from 2.62% to 4.97% for concrete specimens with different mixture proportions and testing ages, which are between good (4.0 to 5.0) and excellent (<3.0) categories, according to ACI 214R [31]. The COV of the static elastic modulus E_c ranges from 3.57% to 6.13%. A slightly higher COV and more rejected observations in E_c are mainly due to the fact that the measurement of E_c involves more sources of variation than the measurement of f_c. The COVs of both f_c and E_c are reasonably consistent at the different ages of 4, 7, 14, and 28 days.

The COV of the P-wave velocity (C_P) ranges from 0.82% to 2.49%, which is consistent with observations by ACI committee 228 [32]. Furthermore, the COV of the fundamental longitudinal f_L and transverse f_T frequencies ranges from 0.83% to 1.56% and from 1.08% to 3.28%, respectively. No outlier was detected by the modified Z-score for the three data sets C_P, f_L, and f_T. The low COVs and lack of detected outliers show the excellent consistency of C_P, f_L, and f_T.

Furthermore, the COV of the accepted S-wave velocity (C_S) ranges from 0.76% to 1.32%, which is comparable to that of C_P, f_L, or f_T. This statistical result clearly shows that S-wave velocity measurements have excellent consistency and could be a consistent and reliable test method in practice. However, in early-age concrete, accurate detection of the first arrival time of S-waves is often difficult because of the interference between direct P- and S-waves; even using S-wave transducers, P-wave components with low amplitude always appear along with the S-wave components in the time

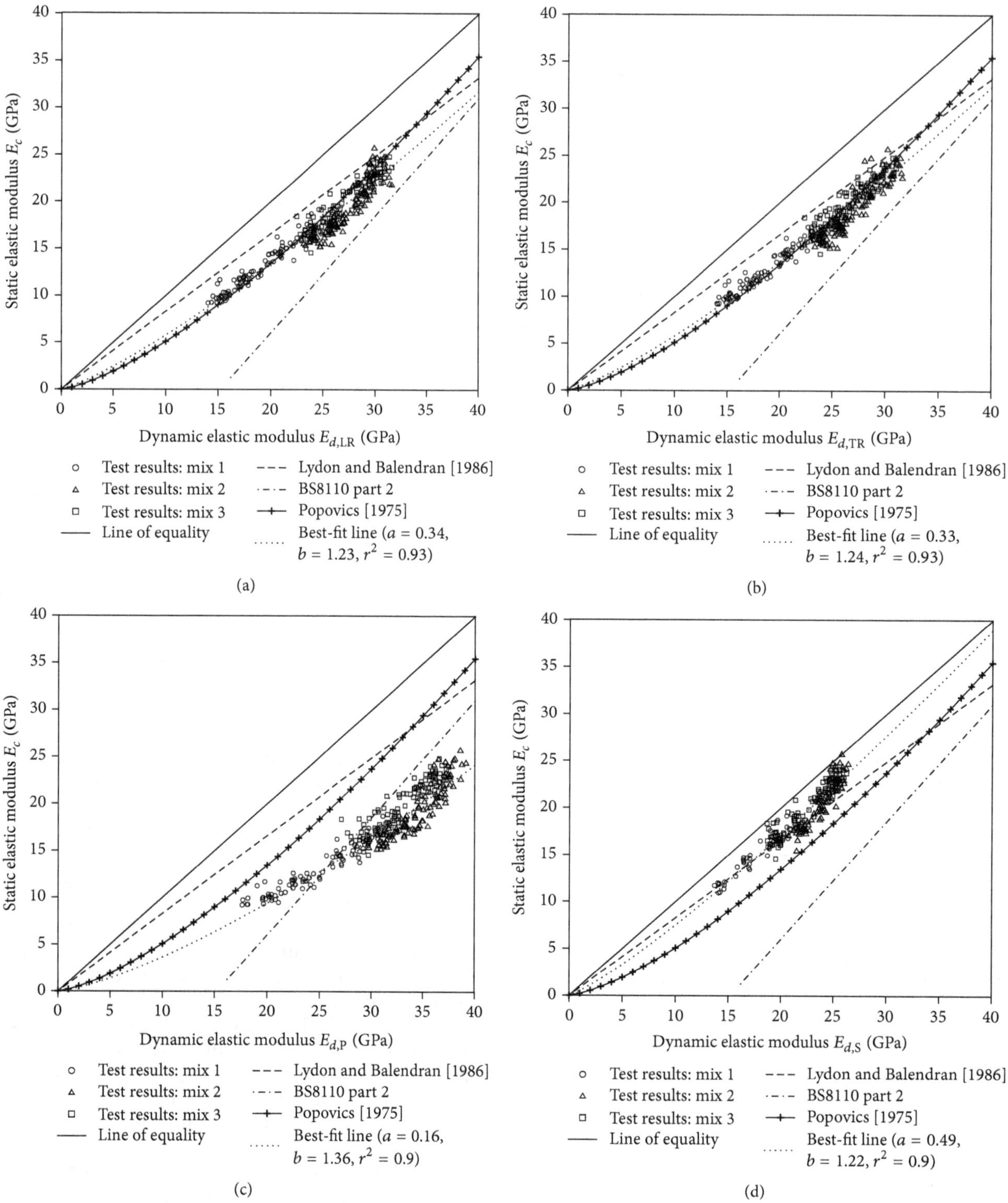

FIGURE 6: Relationship between static and dynamic elastic moduli: (a) E_c versus $E_{d,LR}$, (b) E_c versus $E_{d,TR}$, (c) E_c versus $E_{d,P}$, and (d) E_c versus $E_{d,S}$.

domain. This phenomenon was substantial in observations of cylinders with relatively low compressive strength (i.e., mixes 1 and 2 at 4 days), and measured observations did not have a normal distribution. Therefore, those data sets were not included in the statistical analysis; special care is needed to measure the S-wave velocity of concrete having relatively low compressive strength.

3.2. Relationship between Static and Dynamic Elastic Moduli. Figure 6 presents the relationship between static and dynamic

TABLE 3: Summary of constants a and b and R^2 for the best-fit lines in Figure 4.

Relationship	a	b	R^2
$E_{d,\mathrm{LR}} - E_c$	0.34	1.23	0.93
$E_{d,\mathrm{TR}} - E_c$	0.33	1.24	0.93
$E_{d,\mathrm{P}} - E_c$	0.16	1.36	0.90
$E_{d,\mathrm{S}} - E_c$	0.49	1.22	0.90

Note. The basic form of the equation is $E_c = aE_d{}^b$.

elastic modulus determined using different NDE methods, $E_{d,\mathrm{LR}}$, $E_{d,\mathrm{TR}}$, $E_{d,\mathrm{P}}$, and $E_{d,\mathrm{S}}$, with an assumed Poisson's ratio of 0.2. The use of Poisson's ratio of 0.2 is reasonable for common concrete in practice [7]. The expression relating static to dynamic elastic modulus was obtained by nonlinear regression using the following power function relationship:

$$E_c = aE_d{}^b, \tag{13}$$

where E_c is the chord elastic modulus in GPa in accordance with ASTM C469/C460M-14 [6], E_d is the dynamic elastic modulus in GPa, and a and b are best-fit constants according to the regression analysis. It has been observed that the difference between static and dynamic elastic moduli decreases with increasing dynamic elastic modulus. Therefore, the nonlinear behavior was taken into account by using a power function. Furthermore, (13) results in zero static elastic modulus for zero dynamic elastic modulus. Table 3 summarizes the constants a and b, and R^2 for the four best-fit curves, shown in Figure 6. For comparison, the same figure presents previously mentioned expressions ((1)–(3)) relating the static and dynamic elastic moduli of concrete. Table 4 summarizes mean absolute error (MAE) between the measured and the predicted static elastic modulus of concrete using the four best-fit curves proposed in this study and the three expressions ((1)–(3)).

The four dynamic moduli ($E_{d,\mathrm{LR}}$, $E_{d,\mathrm{TR}}$, $E_{d,\mathrm{P}}$, and $E_{d,\mathrm{S}}$) obtained from the resonant frequency tests and UPV method are greater than the static elastic modulus E_c (see Figure 6). However, the ratios between static and dynamic elastic modulus vary according to the measurement methods. Figures 6(a) and 6(b) show that the $E_c/E_{d,\mathrm{LR}}$ (or $E_c/E_{d,\mathrm{TR}}$) ratio, with an average of 0.72 and COV of 7.15%, was closer to the line of equality than $E_c/E_{d,\mathrm{P}}$. In the figure, the equation proposed by Lydon and Balendran [9] slightly overestimates E_c from the two resonance moduli ($E_{d,\mathrm{LR}}$ or $E_{d,\mathrm{TR}}$), with a mean absolute error (MAE) of 2.63 GPa. The equation proposed by the British Standard Institute [10] significantly underestimates E_c, with a MAE of 10.95. In contrast, the equation proposed by Popovics [11] shows very good agreement with the experimental results in this study and appears to be effective for estimating a reasonably accurate E_c from $E_{d,\mathrm{LR}}$ or $E_{d,\mathrm{TR}}$. However, as shown in Figure 6(c), the $E_c/E_{d,\mathrm{P}}$ ratio, with an average of 0.56 and COV of 9.2%, is too far away from the line of equality as well as from the three proposed equations relating static and dynamic elastic modulus.

Figure 6(d) shows that the ratio $E_c/E_{d,\mathrm{S}}$, with an average of 0.87 and COV of 6.2%, was even closer to the line of equality than $E_c/E_{d,\mathrm{LR}}$ (or $E_c/E_{d,\mathrm{TR}}$). The equations proposed by

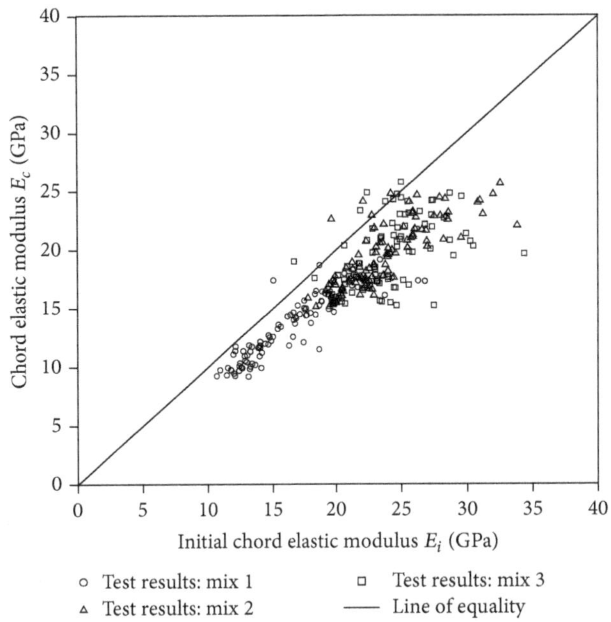

FIGURE 7: Relationship between initial chord modulus and standard chord modulus.

British Standard Institute and Popovics resulted in relatively high MAE of 6.36 GPa and 3.76 GPa, respectively. However, the best-fit curve obtained in this study and the equation suggested by Lydon and Balendran [9] shows good agreement with the experimental results in this study, with a relatively low MAE of 0.82 and 1.19 GPa, respectively. Therefore, the two equations could be a good candidate as a practical formula relating $E_{d,\mathrm{S}}$ and E_c with reasonable accuracy for normal concrete with similar mixture proportions in this study.

3.3. Relationship between Initial Chord and Dynamic Elastic Moduli. Figure 7 shows the relationship between the initial chord modulus and the standard chord modulus. The majority of the data points indicate that the initial chord modulus is greater than the standard chord modulus. This implies that the nonlinear stress-strain behavior of concrete is one of the possible reasons for differences between static and dynamic elastic modulus. It has been believed that the dynamic elastic modulus represents the tangent elastic modulus (or initial chord modulus in this study) at low strain [7]. Therefore, it is of some interest to compare dynamic elastic modulus to initial chord modulus to better understand the nonlinearity effect.

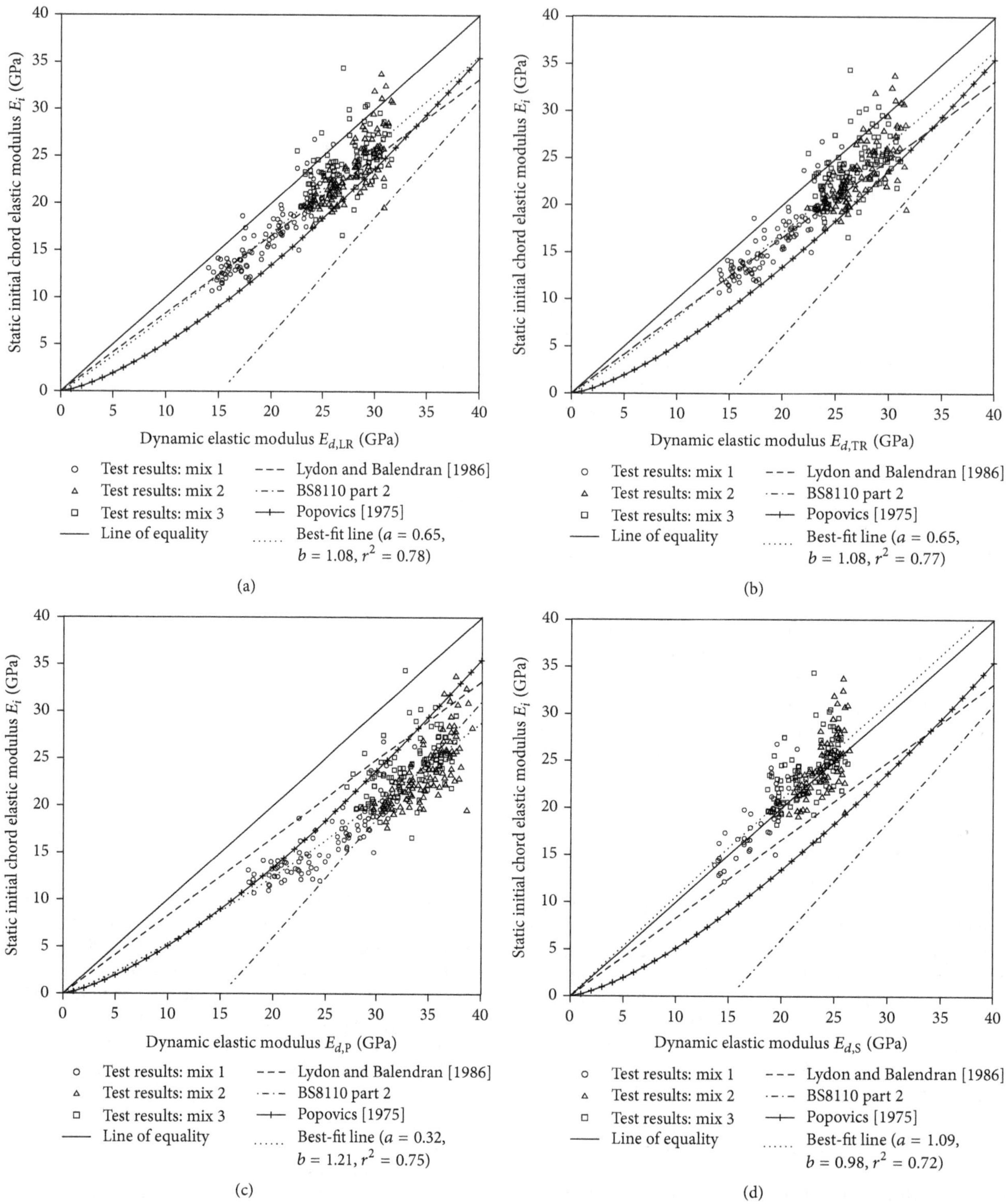

FIGURE 8: Relationship between initial chord and dynamic elastic moduli: (a) E_i versus $E_{d,\mathrm{LR}}$, (b) E_i versus $E_{d,\mathrm{TR}}$, (c) E_i versus $E_{d,\mathrm{P}}$, and (d) E_i versus $E_{d,\mathrm{S}}$.

Figures 8(a)–8(d) compare the initial chord and dynamic elastic moduli from different methods. As expected, the difference between initial chord and dynamic elastic moduli is smaller than that between standard static and dynamic moduli. Note that more scatter in Figure 8, E_i and E_d relation, than in Figure 6, E_c and E_d relation, is attributable to some fluctuating data measured in a low strain level for calculating E_i.

Figure 8(d) shows that $E_{d,\mathrm{S}}$ is close to the line of equality, with an average $E_i/E_{d,\mathrm{S}}$ ratio of 1.04 and COV of 11.0%.

TABLE 4: Mean absolute error (MAE) of expressions relating static and dynamic elastic modulus.

Equation converting E_d into E_c	MAE (GPa)			
	$E_{d,P}$	$E_{d,S}$	$E_{d,LR}$	$E_{d,TR}$
Line of equality	13.38	2.78	7.06	6.76
(1) (Lydon and Balendran [9])	8.11	1.19	2.87	2.63
(2) (British Standard Institute [10])	19.24	6.36	11.33	10.95
(3) (Popovics [11])	7.41	3.76	1.11	0.97
Best-fit line	1.05	0.82	0.84	0.85

Note. MAE is mean absolute error ($=\Sigma|Y - \widehat{Y}|/N$, where Y is the predicted values using the three proposed equations and $\widehat{Y}$ is the measured values and N is the number of specimens).

Therefore, the difference between static and dynamic elastic modulus of $E_{d,S}$ is mainly attributed to the nonlinear stress-strain curve of concrete up to 40% of compressive strength. In contrast, resonance dynamic moduli ($E_{d,LR}$ and $E_{d,TR}$), shown in Figures 8(a) and 8(b), still show considerable difference from E_i. Furthermore, $E_{d,P}$ shown in Figure 8(c) is considerably higher than E_i. It seems that there are other influential factors causing differences between static and dynamic elastic modulus for $E_{d,P}$ and the two resonance moduli ($E_{d,LR}$ and $E_{d,TR}$).

It has been suggested that differences between static and dynamic elastic moduli are attributable to the fact that the nonhomogeneous characteristics of concrete affect the two moduli in different ways [7, 8]. The static modulus of concrete has been demonstrated to be more dependent on the elastic properties of the paste, whereas, for P-wave velocity, the elastic modulus of the aggregates is the more influential factor. In normal-weight concrete, the elastic modulus of aggregates is generally greater than the paste, providing a reason for higher dynamic modulus compared to the static modulus. Similarly, the nonhomogeneity of concrete appears to affect the fundamental resonance frequencies of concrete cylinders. In contrast, the S-wave velocity method is likely the least sensitive to the heterogeneity of concrete. However, additional experimental data are needed to confirm this hypothesis due to the limited quantity of experimental data in this study and scarcity of experimental studies exploring the relationship between static modulus and dynamic elastic modulus based on S-wave measurements.

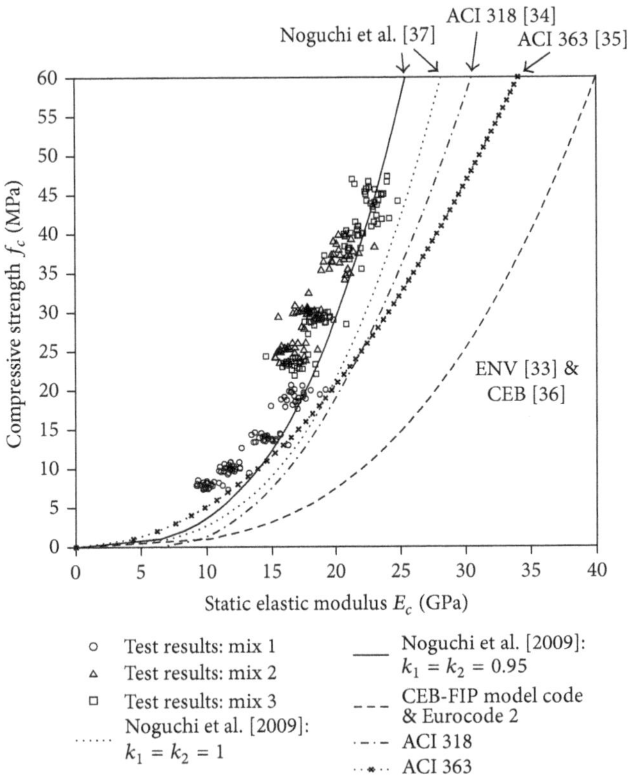

FIGURE 9: Relationship between compressive strength f_c and static elastic modulus E_c.

3.4. Relationship between Compressive Strength and Dynamic Elastic Modulus. The plot in Figure 9 represents the relationship between static elastic modulus and compressive strength measured in accordance with ASTM C469/C469M-14 [6] and ASTM C39/C39M-14 [18], respectively. Figure 9 also compares the experimental results with code equations [33–36] and a recently proposed equation by Noguchi et al. [37]. The CEB-FIP and Eurocode 2 link elastic modulus E_c to compressive strength f_c ($E = 22000(f_c/10)^{1/3}$ in MPa) for both normal-strength and high-strength concrete. ACI proposes two different equations: $E = 0.043(f_c)^{1/2}(w)^{1.5}$ MPa for ($f_c \le$ 38 MPa) by the ACI 318 committee and $E = (3320(f_c/10)^{1/2} + 6900)(w/2300)^{1.5}$ MPa for (21 MPa $< f_c <$ 83 MPa) by the ACI 363 committee. Noguchi et al. [37] proposed a practical equation based on an extensive experimental database including normal-strength and high-strength concrete: $E = k_1 k_2 33500(f_c/60)^{1/3}(w/2400)^2$, where k_1 and k_2 are correction factors for coarse aggregates and SCMs. Figure 9 shows that the three code equations significantly underestimate compressive strength, which is mainly due to the fact that some critical factors (e.g., the types of coarse aggregates and SCMs) have not been properly taken into account in the code equations. However, the Noguchi equation [37] predicts experimental results with greatly improved accuracy by using correction factors $k_1 = k_2 = 0.95$.

Figure 10 compares measured and predicted compressive strength based on different equations relating static and

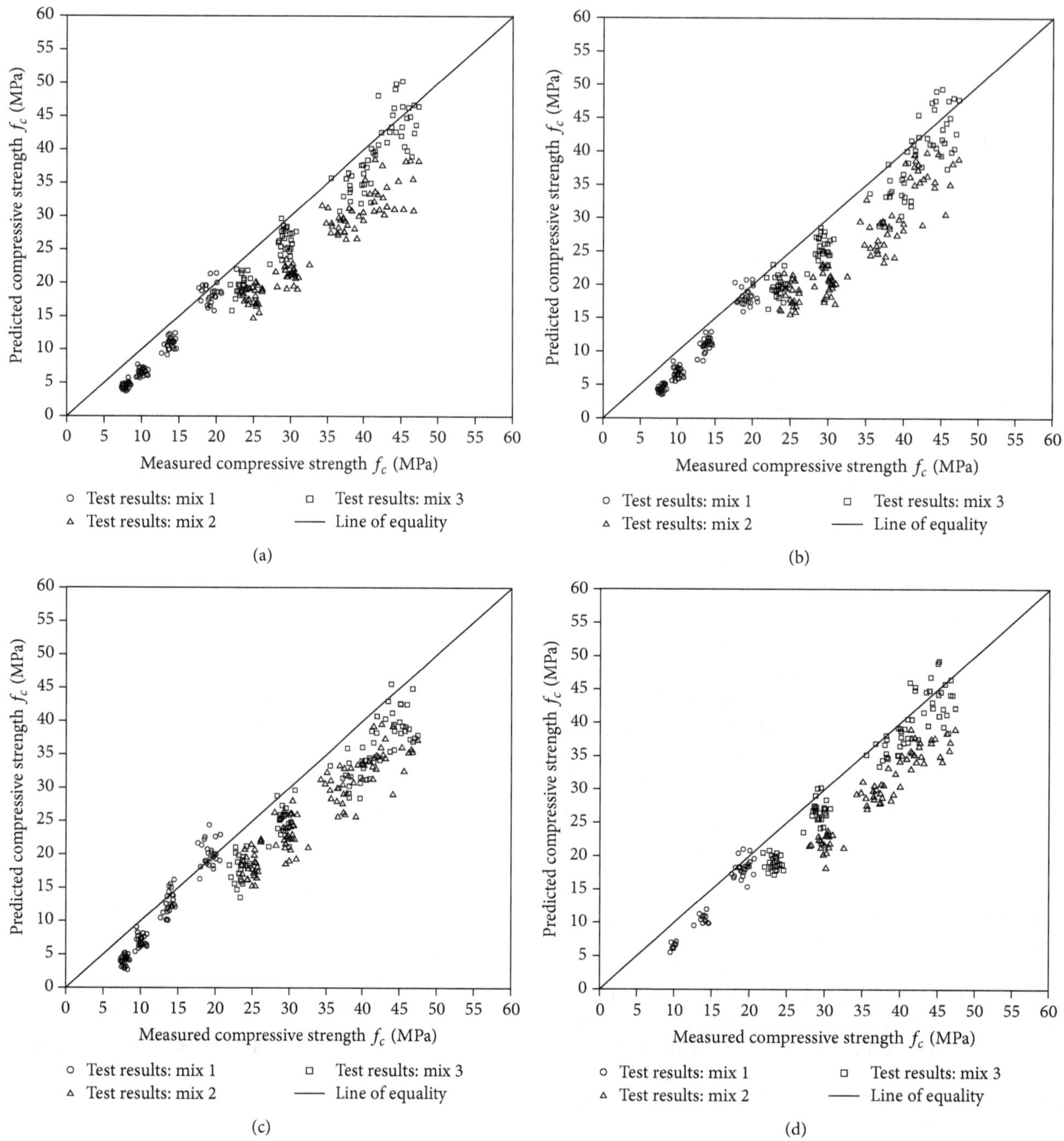

FIGURE 10: Relationship between predicted and measured compressive strength determined from various dynamic moduli: (a) $E_{d,\mathrm{LR}}$, (b) $E_{d,\mathrm{TR}}$, (c) $E_{d,\mathrm{P}}$, and (d) $E_{d,\mathrm{S}}$.

dynamic elastic modulus. The Noguchi equation [37] was used to correlate compressive strength and dynamic elastic modulus as follows:

$$f_c = \eta \left(E_c\right)^3 = \eta \left(aE_d^{\,b}\right)^3, \tag{14}$$

where f_c is estimated concrete strength, $\eta = 60(w/2400)^{-3/2}(k_1 k_2 33500)^{-3}$, and a and b are constants depending on equations relating static and dynamic elastic modulus (see Table 3). Figure 10 shows that the proposed equation (14) slightly underestimates the compressive strength of concrete when converting dynamic into static elastic modulus using the best-fit curves. The MAEs between measured and estimated values using $E_{d,\mathrm{P}}$, $E_{d,\mathrm{S}}$, $E_{d,\mathrm{LR}}$, and $E_{d,\mathrm{TR}}$ are 4.9, 4.2, 4.3, and 4.3 MPa, respectively. However, it was observed in Figure 10 that the predicted compressive strength is linearly correlated with measured compressive strength, and the plots are consistently biased toward underestimation of the compressive strength. Consequently, it was found that the proposed equation (14) could improve the accuracy of predicted compressive strength through a simple calibration process of adding a certain correction factor: MAE values would be a good candidate for a simple calibration factor. Therefore, it can be concluded that the predetermined equations relating dynamic to static elastic modulus, summarized in Table 4, are effective for estimating compressive strength within reasonable errors.

4. Conclusions

An experimental study was conducted to explore the relationship between static and dynamic elastic moduli using ultrasonic pulse velocity methods and resonance tests and to demonstrate the potential of the S-wave velocity method as an in situ dynamic elastic modulus evaluation method in practice. Conclusions are drawn as follows:

(1) It was demonstrated that the coefficient of variation (COV) of the S-wave shows very good repeatability. It is as good as P-wave velocity measurements and comparable to resonance tests. The average COV of the S-wave velocity (C_S) is 0.97%; the average COV of P-wave velocity (C_P) is 1.38%; the average COV of resonant frequency measurements are 1.19% and 1.35% for longitudinal and transverse resonance frequencies, respectively.

(2) The four dynamic moduli ($E_{d,LR}$, $E_{d,TR}$, $E_{d,P}$, and $E_{d,S}$) obtained from the resonance frequency and ultrasonic pulse velocity (UPV) tests are greater than the static elastic modulus E_c. Test results from this study and the literature show that the best-fit curve and the equation suggested by Popovics [11] ((3) in this study) are effective for predicting static elastic moduli from resonance moduli of concrete, with MAEs of 1.0 and 1.2 GPa, respectively. For velocity moduli from S-wave velocity measurements, the best-fit curve and the equation suggested by Lydon and Balendran [9] ((1) in this study) show good agreement with the experimental results in this study, with MAEs of 0.82 GPa and 1.19 GPa, respectively. However, velocity modulus from P-wave measurements is excessively greater then static elastic modulus. The empirical equations (see (1), (2), and (3)) result in significant errors between the static and dynamic moduli, with MAE in a range of 7.41 GPa to 19.24 GPa.

(3) $E_{d,S}$ was observed to be almost linearly correlated with E_i, with a mean absolute error of 2 GPa. Therefore, it can be concluded that the difference between the initial elastic modulus of concrete and $E_{d,S}$ is mainly attributable to the nonlinear stress-strain curve of concrete. In contrast, resonance dynamic moduli ($E_{d,LR}$ and $E_{d,TR}$) are slightly higher than the initial tangent modulus of concrete E_i. Moreover, $E_{d,P}$ is considerably higher than E_i. There should be other influential factors (such as heterogeneity of concrete) causing differences between static and dynamic elastic modulus for the P-wave modulus ($E_{d,P}$) and the two resonance moduli ($E_{d,LR}$ and $E_{d,TR}$).

(4) With an established equation between static and dynamic elastic modulus, the dynamic elastic modulus can be used to estimate the compressive strength of cylinders. It is demonstrated that the predetermined equations relating dynamic to static elastic modulus are effective for estimating compressive strength within reasonable errors.

Conflicts of Interest

The authors declare that there are no conflicts of interest regarding the publication of this manuscript.

Acknowledgments

This research was supported by Basic Science Research Program through the National Research Foundation of Korea (NRF) funded by the Ministry of Science, ICT & Future Planning (no. 2015R1A5A1037548).

References

[1] J. S. Popovics, "A study of static and dynamic modulus of elasticity of concrete," in *ACI-CRC Final Report*, 2008.

[2] P. K. Mehta and P. J. M. Monteiro, *Concrete-Microstructure, Properties, and Materials*, McGraw-Hill, 3rd edition, 1993.

[3] ASTM C666/C666M-15, *Standard Test Method for Resistance of Concrete to Rapid Freezing and Thawing*, West Conshohoken, Pa, USA, 2015.

[4] ASTM C597/C597M-16, *Standard Test Method for Pulse Velocity through Concrete*, West Conshohoken, Pa, USA, 2016.

[5] ASTM C215-14, *Standard Test Method for Fundamental Transverse, Longitudinal, and Torsional Resonant Frequencies of Concrete Specimens*, West Conshohoken, Pa, USA, 2016.

[6] C469 A, *Standard Test Method for Static Modulus of Elasticity and Poisson's Ratio of Concrete in Compression*, West Conshohoken, Pa, USA, 2014.

[7] A. M. Neville, *Properties of Concrete*, John Wiley & Sons, New York, NY, USA, 4th edition, 1997.

[8] R. E. Philleo, "Comparison of results of three methods for determining Young's modulus of elasticity of concrete," *Journal of the Americal Concrete Institute*, vol. 26, no. 5, pp. 461–469, 1955.

[9] F. D. Lydon and R. V. Balendran, "Some observations on elastic properties of plain concrete," *Cement and Concrete Research*, vol. 16, no. 3, pp. 314–324, 1986.

[10] British Standard Institute., *Structural Use of Concrete—Part 2: Code of Practice for Special Circumstance. BS 8110-2:1995*, BSI, London, UK, 1985.

[11] S. Popovics, "Verification of relationships between mechanical properties of concrete-like materials," *Matériaux and Constructions*, vol. 8, no. 3, pp. 183–191, 1975.

[12] K. F. Graff, *Wave Motion in Elastic Solids*, Dover publications, INC., New York, NY, USA, 4th edition, 1991.

[13] M. Cha and G.-C. Cho, "Compression wave velocity of cylindrical rock specimens: engineering modulus interpretation," *The Japan Society of Applied Physics*, vol. 46, no. 7B, pp. 4497–4499, 2007.

[14] B. J. Lee, S.-H. Kee, T. Oh, and Y.-Y. Kim, "Effect of cylinder size on the modulus of elasticity and compressive strength of

concrete from static and dynamic tests," *Advances in Materials Science and Engineering*, vol. 2015, Article ID 580638, 12 pages, 2015.

[15] T. Voigt, G. Ye, Z. Sun, S. P. Shah, and K. Van Breugel, "Early age microstructure of Portland cement mortar investigated by ultrasonic shear waves and numerical simulation," *Cement and Concrete Research*, vol. 35, no. 5, pp. 858–866, 2005.

[16] J. Zhu, S.-H. Kee, D. Han, and Y.-T. Tsai, "Effects of air voids on ultrasonic wave propagation in early age cement pasets," *Cement and Concrete Research*, vol. 41, no. 8, pp. 872–881, 2011.

[17] S. Liu, J. Zhu, S. Seraj, R. Cano, and M. Juenger, "Monitoring setting and hardening process of mortar and concrete using ultrasonic shear waves," *Construction and Building Materials*, vol. 72, pp. 248–255, 2014.

[18] J. Carette and S. Staquet, "Monitoring the setting process of mortars by ultrasonic P and S-wave transmission velocity measurement," *Construction and Building Materials*, vol. 94, pp. 196–208, 2015.

[19] J. An, J. Nam, S. Kwon, and S. Joh, "Estimation of the compressive strength of concrete using shear wave velocity," in *Proceedings of the GeoHunan International Conference*, Changsha, China, 2009.

[20] D. Ciancio and M. Helinski, "The use of shear wave velocity for assessing strength development in fibre reinforced shortcrete," in *Proceeding of the 3rd International Conference on Engineering Developments in Shotcrete*, CRC Press, Queenstown, New Zealand, 2010.

[21] U. Lencis, A. Udris, and A. Korjakins, "Decrease of the ultrasonic pulse velocity in concrete caused by reinforcement," *Journal of Materials Science and Engineering*, vol. 1, pp. 1016–1028, 2011.

[22] A. A. Samokrutov, V. N. Kozlov, and V. G. Shevaldkin, "Ultrasonic low-frequency transdcers with dry dot contact and their applications for evaluation of concrete structures," in *Proceeding of the 8th European Conference on Nondestructive Testing Conference (ECNDT '02)*, Barcelona, Spain, June 2002.

[23] Y. H. Lee and T. Oh, "The measurement of P-, S-, and R-wave velocities to evaluate the condition of reinforced and prestressed concrete slabs," *Advances in Materials Science and Engineering*, vol. 2016, Article ID 1548215, 14 pages, 2016.

[24] A. O. De La Haza, A. A. Samokrutov, and P. A. Samokrutov, "Assessment of concrete structures using the Mira and Eyecon ultrasonic shear wave devices and the SAFT-C image reconstruction technique," *Construction and Building Materials*, vol. 38, pp. 1276–1291, 2013.

[25] ASTM C31/C31M-12, *Standard Practice for Making and Curing Concrete Test Specimen in the Field*, West Conshohoken, Pa, USA, 2012.

[26] ASTM C39/C39M-14 A, *Standard Test Method for Compressive Strength of Cylindrical Concrete Specimens*, West Conshohoken, Pa, USA, 2014.

[27] A. Alam and L. Haselbach, "Estimating the modulus of elasticity of pervious concrete based on porosity," *Advances in Civil Engineering Materials*, vol. 3, no. 1, pp. 257–269, 2014.

[28] J. S. Popovics and etal., "One-sided stress wave velocity measurement in concrete," *Journal of Engineering Mechanics-Asce*, vol. 124, no. 12, pp. 1346–1353, 1998.

[29] J. Lin and M. Sansalone, "A procedure for determining P-wave speed in concrete for use in impact-echo testing using a Rayleigh wave speed measurement technique," in *Innovations in Nondestructive Testing of Concrete, SP-168*, E. S. Pessiki and L. Olson, Eds., pp. 137–165, American Concrete Institute, Farmington Hills, Mich, USA, 1997.

[30] V. Barnett and T. Lewis, *Outliers in Statistical Data*, John Wiley & Sons, New York, NY, USA, 1994.

[31] ACI committee 214, *Guide for Evaluation of Strength Test Results of Concrete (ACI 214R-11)*, Farmington Hills, Mich, USA, American Concrete Institute, 2011.

[32] ACI committee 228, "Nondestructive test methods for evaluation of concrete in structures," Report ACI 228.2R-98, American Concrete Institute, Farmington Hills, Mich, USA, 1998.

[33] ENV 1992-1-1, *Eurocode 2. Design of Concrete Structures-Part 1: General Rules and Rules for Buildings*, 2004.

[34] ACI committee 318, *Building Code Requirements for Structural Concrete (ACI 318-14) and Commentary*, 2014.

[35] ACI committee 363, "State-of-the-art report on high-strength concrete," *ACI Journal Proceedings*, vol. 84, no. 4, pp. 364–411, 1984.

[36] Comité Euro-International du Béton, "High-performance concrete, recommended extensions to the model code 90-research needs," *CEB Bulletin d'Information*, p. 46, 1995.

[37] T. Noguchi, F. Tomosawa, K. M. Nemati, B. M. Chiaia, and A. P. Fantilli, "A practical equation for elastic modulus of concrete," *ACI Materials journal*, vol. 106, no. 5, pp. 690–696, 2009.

15

Synthesis of Copper Oxide Aided by Selective Corrosion in Cu Foils

Roberto Baca Arroyo

Department of Electronics, National Polytechnic Institute, 07738 Mexico City, Mexico

Correspondence should be addressed to Roberto Baca Arroyo; rbaca02006@yahoo.com.mx

Academic Editor: Alicia E. Ares

Copper oxide films aided by selective corrosion have been synthesized. Using copper (Cu) foils of commercial printed circuit boards (PCBs), chemical attack in small areas on Cu surfaces and their oxidation with heating process in air atmosphere at low temperature have allowed to produce Cu oxide layers. Phase formation of Cu oxide samples was validated by X-ray diffraction studies, and their conduction properties were registered and evaluated by current-voltage curves. Stability analysis of the Cu oxide samples was conducted by a correlation profile between average strain and Cu oxide percentage grown, which confirms that the structure defects dependent on plastic deformation define the existence of both CuO and Cu_2O phases beneficial to mitigate corrosion and advance in Cu oxide-based devices to build electronic circuits as components on PCBs.

1. Introduction

It is well known that copper (Cu) is a nature-abundant material and is the third metal more employed in the world, after iron and aluminum. It has an important biological role in photosynthesis, and it is essential for human life. Despite the standard negative electrode potential with −0.345 V, its corrosion can occur at a significant rate under strenuous conditions [1]. Both cupric oxide (CuO) and cuprous oxide (Cu_2O) are the two chemical compounds formed by oxygen and copper in the nature only. Such oxides are semiconductors, where Cu_2O has a direct bandgap of 2.1 eV and reddish appearance, while CuO with a bandgap of 1.3 eV is black in color [2].

Due to the miniaturization in the integrated circuits (ICs), Cu has been widely used for microelectronic device industry, including wiring technology, electromagnetic interference shielding, and electrostatic dissipation technology [3]. However, the reduced spacing between IC components on a printed circuit board (PCB) can produce various types of corrosions, although the underlying mechanism in all cases has been electrochemical migration (EM) as one of the most severe problems, where events such as temperature cycles during soldering at temperature above 150°C, degradation of the thermal interface in contacts, existence of dust particles between contacts, and application of high voltage for operation of the electronic devices have made it easy for corrosive environments [4, 5]. The high-temperature corrosion occurs when gradual EM by Cu ions along an interconnection under high current density (current crowding) increases the resistance of the Cu lines with reduction in their cross-sectional area [6]. High-temperature corrosion can be able to assist the oxidation in the Cu lines where their surface will remain safe, and small fluctuations will degrade the oxide in few critical points due to the pit formation under humidity conditions. The interior of a pit is naturally poorer in oxygen, and pH locally decreases to very low values, thus pitting corrosion increase due to the catalytic process where dendrite growth from cathode to anode produces an electric short and premature failure. Figure 1 shows pitting corrosion and dendrite growth on a PCB when high electric field was applied [7].

Previously, studies have shown that for maximizing reliability in IC design and PCB fabrication, width scaling of Cu lines must be investigated by EM phenomena where to reduce failures by corrosion environments, two scenarios

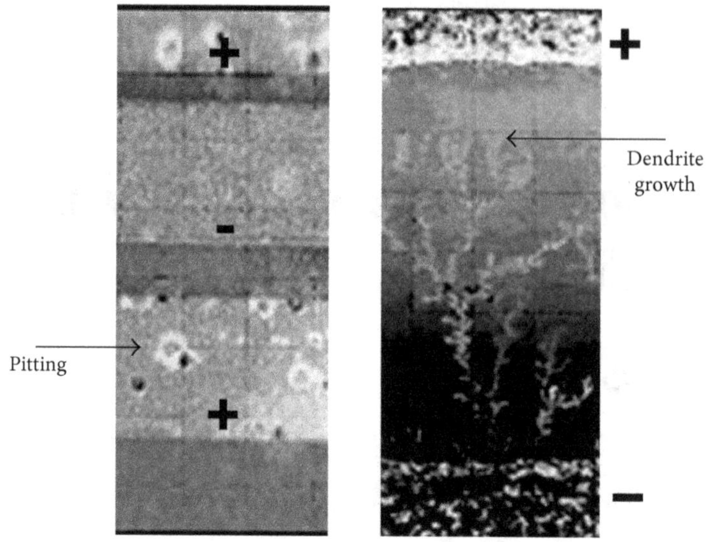

Figure 1: Corrosion phenomena on a PCB based on EM under high electric field conditions.

have been recommended with respect to the mechanisms governing the Cu ion migration with the suggested spacing between IC components satisfying a width $<1\,\mu$m, whereas a width $>1\,\mu$m is for components on PCBs [6, 8].

Nowadays, researchers have synthesized p-channel Cu_2O thin-film transistors on flexible polyethylene terephthalate (PET) substrates by using magnetron sputtering and vacuum annealing techniques [9, 10]. Such novel electronic devices provide modest transport properties with hole mobility (μ_h) in the range of 1 to 10 cm^2/Vs and hole concentration (n_h) from 10^{16} to $10^{17}\,cm^{-3}$. Its conduction mechanism is attributed to the electronic structure of intrinsic defects such as Cu vacancies $[V_{Cu}]$, which create trap states [11]. Conversely, the operation characteristics of the Cu_2O thin-film transistors are of no use today to engineering applications, and the market for Cu_2O-based devices has been delayed by setback in its scalability, degradation of their electronic properties at high-temperature conditions, and requirements for high quality of Cu_2O with a minimum in structure defects. Consequently, to achieve useful engineering applications using Cu oxide-based devices compatible with silicon technology and controllable operation characteristics, the formation and growth of CuO on the Cu surface can increase the chemical bonding and path of the specific thermal resistance in Cu_2O-based devices for electronic circuits designed as components on PCBs [5]. The layer CuO is accompanied by defects such as dislocations and grain boundaries where reactivity of these is dependent on intrinsic structure disorder (elastic strain effects) [12]. Thus, transformation in the microstructure of the Cu surface (process known as plastic deformation) can be a strategy to adjust defect density and mitigate corrosion issues [3].

In the past, oxidation procedures at high temperature around 1000°C in air atmosphere have been developed to oxidize Cu foils in order to obtain polycrystalline Cu_2O crystals [11], where Cu oxidation has taken place by diffusion process of the Cu ions which migrate because there is an opposite diffusion of $[V_{Cu}]$ from oxide layer to the Cu foil [1]. Nevertheless, as CuO is less stable than Cu_2O and its formation energy is lower, the presence of CuO appearing together with Cu_2O can be attained by plastic deformation on the Cu surface at low temperatures [2, 4]. Thereby, the work presents the synthesis of Cu oxide films aided by selective corrosion of Cu foils with the subsequent heating process in air atmosphere at low temperatures. Such experimental procedure will be discussed in Section 2. Phase formation and conduction properties in Cu oxide films and stability analysis as a function of average elastic strain originated by selective corrosion will be covered in Section 3. Conclusions about this research are presented in Section 4.

2. Experimental Procedure

Cu foils with an average thickness of 15 μm were obtained from commercial PCBs with copper joined in a fiberglass epoxy polymer. Before, Cu foils were cut in a cross-sectional area of 1 cm^2; after, cleaning process to remove nonpolar substances was developed with acetone; finally, washing process was completed in deionized water and successively dried with nitrogen. In the synthesis of Cu oxide aided by selective corrosion, the following four stages are involved: (a) choosing the area for chemical attack, (b) immersing Cu foils in $FeCl_3/H_2O$ solution, (c) removing residual $FeCl_3/H_2O$ by soaking the chemically attacked Cu foils, and (d) heating the Cu foils in air atmosphere inside of quartz tube furnace. Figure 2 shows the stages in the synthesis to obtain Cu oxide films.

A nontoxic permanent marker typically engaged in the design of PCBs was used to synthesize Cu oxide samples. An area of 4 mm × 4 mm was defined on each Cu foil under study. A permanent marker is resistant at $FeCl_3/H_2O$ solution; therefore, during corrosion, the outside area was protected as shown in Figure 2. Iron chloride ($FeCl_3$) solution activated with H_2O was employed as the electrolyte.

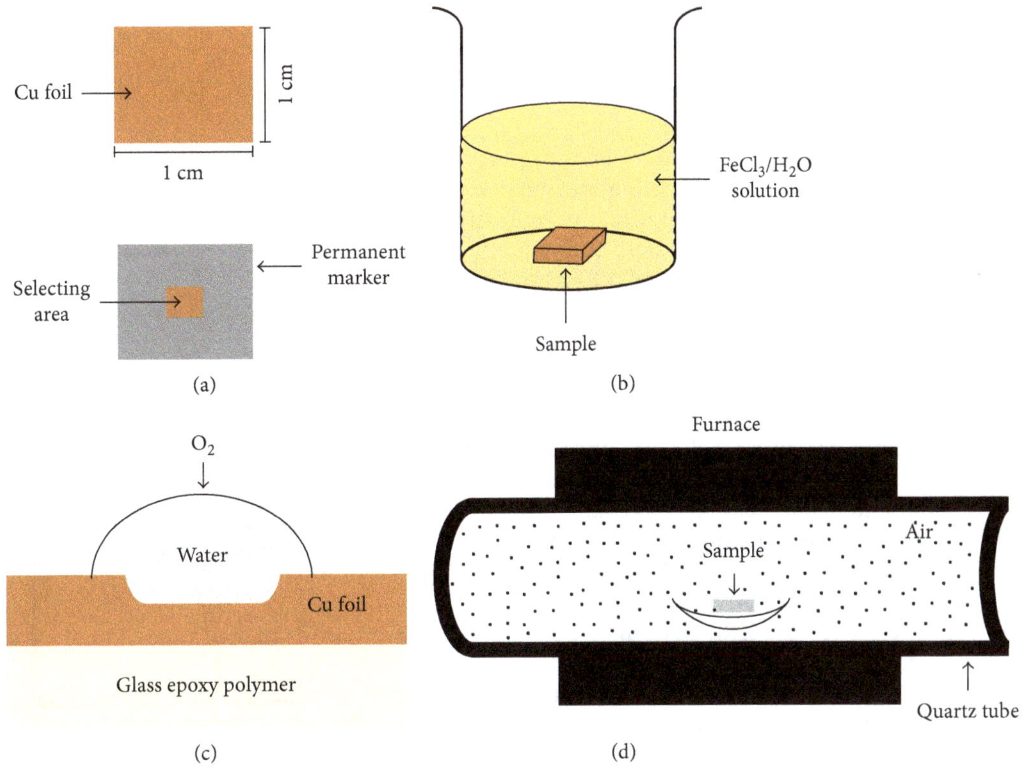

FIGURE 2: Stages employed in the synthesis of Cu oxide films.

The corrosion behavior by chemical attack was studied at different $FeCl_3 : H_2O$ rates and immersion times as indicated in Table 1.

The reaction mechanism for Cu corrosion can be written in two stages as

$$FeCl_3/H_2O\,(aq) + Cu\,(s) \rightarrow FeCl_2\,(aq) + CuCl\,(s) \quad (1)$$

$$FeCl_3/H_2O\,(aq) + CuCl\,(s) \rightarrow FeCl_2\,(aq) + CuCl_2\,(s) \quad (2)$$

The species dissolved in $FeCl_3$ activated with H_2O are $Fe^{2+}Cl_2^-$, $Fe^+(OH)^-$, and H^+Cl^- which indicate that corrosion in Cu foils occur because initially Cu^+ ion in the presence of H^+ ion produces cupric chloride (CuCl), and later cuprous chloride ($CuCl_2$) is fully formed by the reaction of CuCl with Cu^{2+} ion [1, 13, 14].

Once Cu foils were immersed in the $FeCl_3/H_2O$ solution to obtain $CuCl_2$ and soaked with water, heating process at 35°C with a duration of 2 min at atmospheric pressure was completed. In accordance with the following reaction mechanism, the formation of Cu oxide films took place in two stages:

$$2CuCl_2 + O_2 + 2H_2O \rightarrow 2Cu\,(OH)\,(aq) + 2HCl\,(aq) + O_2 \quad (3)$$

$$Cu^+(OH)^- + H^+Cl^- + \frac{1}{2}O_2^{2-} \xrightarrow{heat} Cu_xO_{1-x}\,(oxide) + H_2O \quad (4)$$

where x is the chemical composition in Cu oxide samples.

TABLE 1: Parameters for selective corrosion in Cu foils.

Sample	Volume (mL)		Immersion time (min)
	$FeCl_3$	H_2O	
CS-1	5	2.5	2
CS-2	2.5	2.5	4
CS-3	2.5	5	6

The heating temperature for the Cu samples was chosen as a function of the boiling point for HCl of 48°C, where Cl^- ions and residual water are fully evaporated. The surface of Cu oxide samples resulted with opaque gray color after thermal process. The presence of ionic species such as copper hydroxide (CuOH) and acidic chloride (HCl) determines the oxidation degree of the $CuCl_2$ surfaces because displacement interactions between OH^- and H^+ ions initially synthetize CuO by surface migration of such ionic species and after Cu_2O by diffusion of Cu^+ ions and reaction with atmospheric oxygen [3, 14]. Nevertheless, as the activity of these species is poorer under heating process within humidity, the Cu oxide films were nonstoichiometric.

Using X-ray diffraction with PANalytical diffractometer of CuK_α radiation ($\lambda = 0.1541$ nm), phase formation of Cu oxide films was validated. Conduction behavior in the samples was registered and evaluated by current-voltage curves at room temperature. The electrical parameters of the samples were collected by employing a digital storage oscilloscope (Tektronix, TDS1012C). To know the performance of the Cu oxide samples as a function of structure defects involved, a stability analysis was done by a correlation profile between elastic strain and Cu oxide percentage.

3. Results and Discussion

This section discusses phase formation and conduction properties in the Cu oxide samples. Here are the revised XRD patterns and electrical characterization to demonstrate the potential of the selective corrosion technique with emphasis in the oxidation degree of the Cu foils dependent on their plastic deformation.

3.1. X-Ray Diffraction in Cu Oxide Samples. Figure 3 shows XRD patterns of Cu oxide films synthesized with the heating process. The samples labeled as CS-1, CS-2, and CS-3 exhibit monoclinic phase (CuO) with peaks located at 33.9° and plane (110), 58.1° and plane (202), and 65.2° and plane (022) according to PANalytical Card number 00-041-0254. Also, peak positioned at 69.7° and plane (310) corresponds to cubic phase (Cu_2O) according to PANalytical Card number 00-005-0667. As a reference, the XRD pattern of Cu foil also is shown. The larger peak-position displacement between XRD patterns of the Cu oxide films and XRD pattern of the Cu foil was observed which ensures presence of the elastic strain in Cu oxide surface induced by high density of the defects in chemically attacked Cu foils.

3.2. Conduction Properties in Cu Oxide Samples. The performance of Cu oxide samples has been confirmed with the schematic diagram of Figure 4(a). To ensure linear response, a function generator (Matrix, MFG-8250A) was used to produce a linear ramp signal at low frequency ($f = 100$ Hz) with voltage ranging from −2 V to 2 V to conserve low-voltage operation across the aluminum (Al) contacts. Also, a resistor of $R = 1\,\mathrm{k}\Omega$ was chosen to monitor the current signal. The log current-voltage plots for samples labeled as CS-1, CS-2, and CS-3 are shown in Figure 4(b), where such current-voltage curves have shown ohmic behavior.

To investigate conduction parameters, a model based on the point form of Ohm's law [15] has been proposed by

$$\frac{\Delta V}{\Delta I} = \frac{l_c}{\sigma_o \exp\left(-(E_a/k_B T)\right) \cdot S}, \tag{5}$$

where $\sigma_o = 5.8 \times 10^5 \Omega^{-1}\mathrm{cm}^{-1}$ is the Cu conductivity at room temperature, E_a is the activation energy of Cu oxide, k_B is the Boltzmann constant, T is the temperature in samples under test, $l_c = 0.77$ nm is the interatomic distance into CuO crystalline lattice associated with minimal free path of charge carries [16], and S is the cross-sectional area ($S = t \cdot d$), where $d = 0.5$ mm is the diameter of aluminum (Al) contact and t the thickness of the samples.

The $\Delta V/\Delta I$ rate was extracted from Figure 4(b), with ΔV as the differential voltage and ΔI as the differential current, respectively, while E_a was estimated with the slope of the dotted line in Figure 4(b) which crosses at the voltage axis. Using MATLAB program, the cross-sectional area $S = t \cdot d$ is computed by solving (5) and by the Arrhenius conductivity dependent on temperature [17]. $[V_{Cu}]$ was estimated by

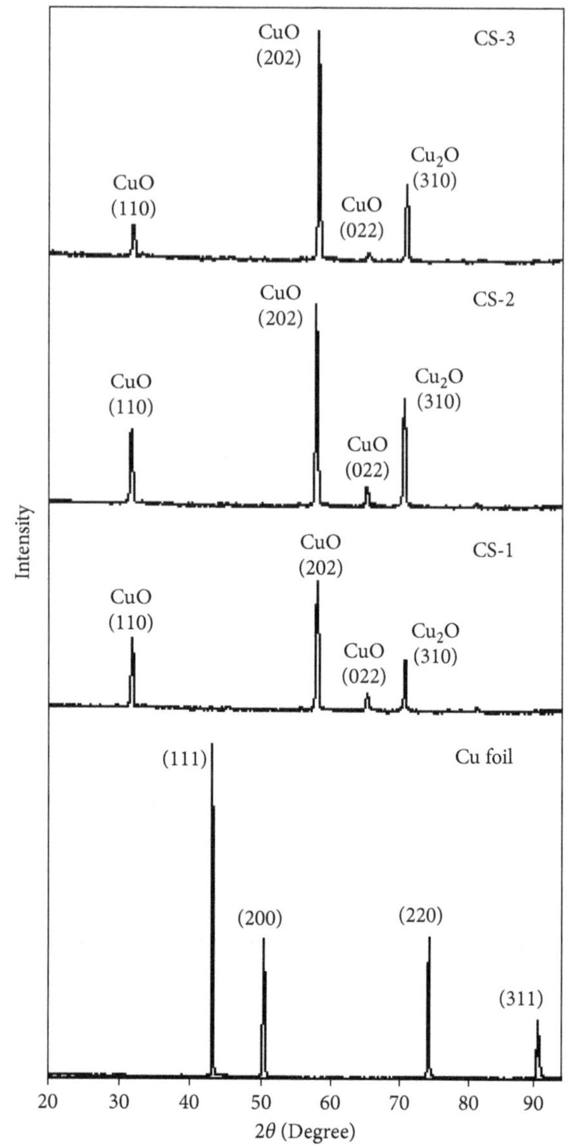

Figure 3: XRD patterns in Cu oxide films grown on Cu foils.

$$[V_{Cu}] \cdot e \cdot \mu_{Cu} = \sigma_o \exp\left(-\frac{E_a}{k_B T}\right), \tag{6}$$

where $\sigma_s(E_a) = [V_{Cu}] \cdot e \cdot \mu_{Cu}$ is the conductivity for the Cu oxide layer, while μ_{Cu} corresponds to the mobility of Cu vacancies. The quantity $[V_{Cu}]\mu_{Cu}$ from (6) was computed, confirming that $[V_{Cu}]$ is attained in the range of 10^{14} to $10^{15}\,\mathrm{cm}^{-3}$ and mobility from 1 to 40 cm^2/Vs, which confirms that the operation characteristics in Cu oxide samples would be useful in engineering applications because their specific thermal resistance chemically controlled by CuO and electronic properties driven by Cu_2O allow advances in design of Cu oxide-based devices whose principle of operation similarly to previous studies by other authors is dependent on strain effects [18, 19]. Table 2 summarizes the conduction parameters found.

3.3. Stability in Cu Oxide Samples. To evaluate the performance of the Cu oxide samples as a function of the plastic

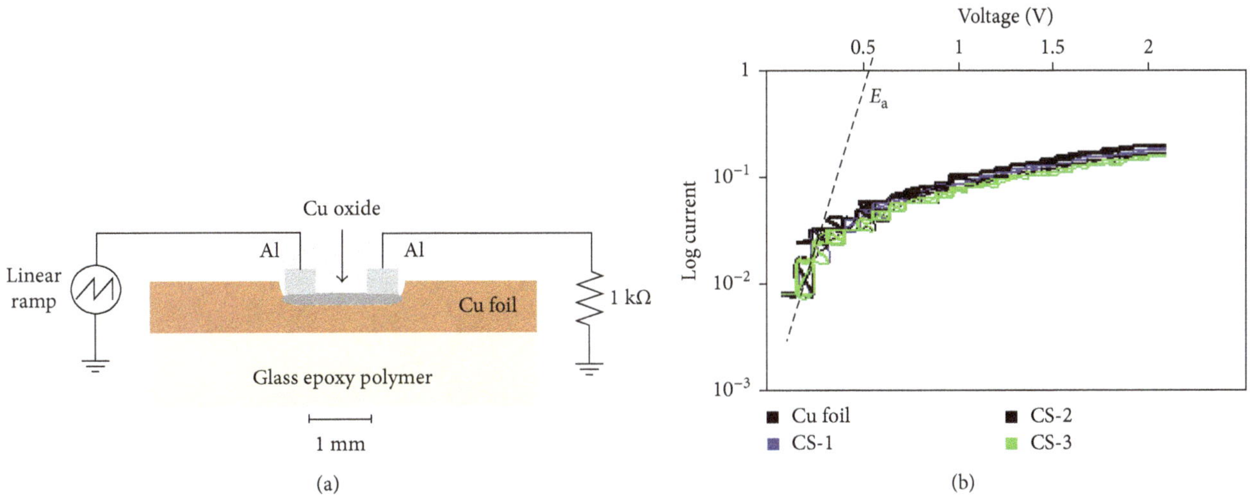

FIGURE 4: (a) Schematic diagram of practical circuit implemented to measure the current-voltage curves in Cu oxide samples. (b) Log current-voltage plots of each sample.

TABLE 2: Conduction parameters in Cu oxide samples.

Sample	E_a (eV)	$\Delta V/\Delta I(\Omega)$	$\sigma_s(E_a)\cdot(\Omega^{-1}\cdot cm^{-1})$	t (nm)	$\mu\cdot[V_{Cu}]\cdot(cm^{-1}\cdot V^{-1}\cdot s^{-1})$
CS-1	0.54	130.43	5.55×10^{-4}	212.1	3.5×10^{15}
CS-2	0.56	190.47	2.57×10^{-4}	314.6	1.6×10^{15}
CS-3	0.56	315.78	2.57×10^{-4}	189.7	1.6×10^{15}

deformation in Cu foils, average strain along each crystallographic plane on the oxidized Cu surface with respect to the Cu foil as substrate is approximately estimated by $(\Delta d/d) = \Delta\theta/\tan\phi$ [12], where $\Delta\theta$ is the difference between measured incident angle θ_m and reference angle θ_r, while the angle of each Cu foil plane corresponds to ϕ. The parameter $(x \approx \Delta d/d)$ can be linked with the chemical composition in Cu_xO_{1-x} which defines its nonstoichiometry [20].

The elastic strain must be analyzed as a function of Cu oxide grown, being it proportional to the intensity in percentage of each peak [12, 20]. Figure 3 allows evaluating $\Delta\theta$ at each crystallographic plane of both oxides (Cu_2O and CuO), as well as $\tan\phi$ at each crystallographic plane in the Cu foil; thus, a correlation profile between x and Cu oxide percentage is built in Figure 5.

The structure defects can be validated by stability analysis, which determines the ability of the Cu oxide layers to remain structurally stable during electrical conduction. The stability in the Cu oxide samples is a function of both tensile and compressive stress into a limit where Cu oxide (Cu_xO_{1-x}) is fully grown into two phases. In Figure 5, both oxides (Cu_2O and CuO) give a slight deviation from their nonstoichiometry into the range of $-1 < x < 1$, which corresponds to the limit where the Cu oxide percentage from 30 to 100% is stable. In contrast, anomalous Cu oxide is lower than30%, because its severe nonstoichiometry is associated with intergranular local attack influenced by poor quality of the initial Cu surface.

Selective corrosion followed with the heating process at 35°C had allowed obtaining both Cu_2O and CuO phases in accordance with the experimental procedure detailed in Section 2, where oxidation in Cu foils occurs beginning with a first layer of CuO grown on Cu, and subsequently a layer of Cu_2O grows between Cu and CuO. These two layers increase with time because the Cu_2O has a very small expansion coefficient at lower temperatures of 300°C where the lattice constant changes to less than 0.5% and the dissociation pressure of CuO is lower than the atmospheric pressure [11, 13]. As Cu vacancies agglomeration or trapped oxygen happen into grain boundaries, intrinsic concentration of defects corresponding with $[V_{Cu}]$ and the opaque gray color in Cu oxide samples explicate the stability of the CuO phase.

4. Conclusions

Cu oxide samples synthesized by corrosion and heating processes at low temperature have been studied. Selective corrosion was achieved by chemical attack in smaller areas from Cu foils of commercial PCBs. Both CuO and Cu_2O phases were formed under air atmosphere. The study conducted by X-ray diffraction and electrical characterization found that phase formation and conduction properties are linked with structure defects into Cu foils. Stability analysis has confirmed that the structure behavior in Cu oxide samples is strongly dependent on its intrinsic disorder. Thus, the selective corrosion technique has significant benefits as physical and mechanical properties are influenced by plastic deformation, which has resulted in stable Cu oxide samples with next electronic engineering applications for Cu oxide-based devices.

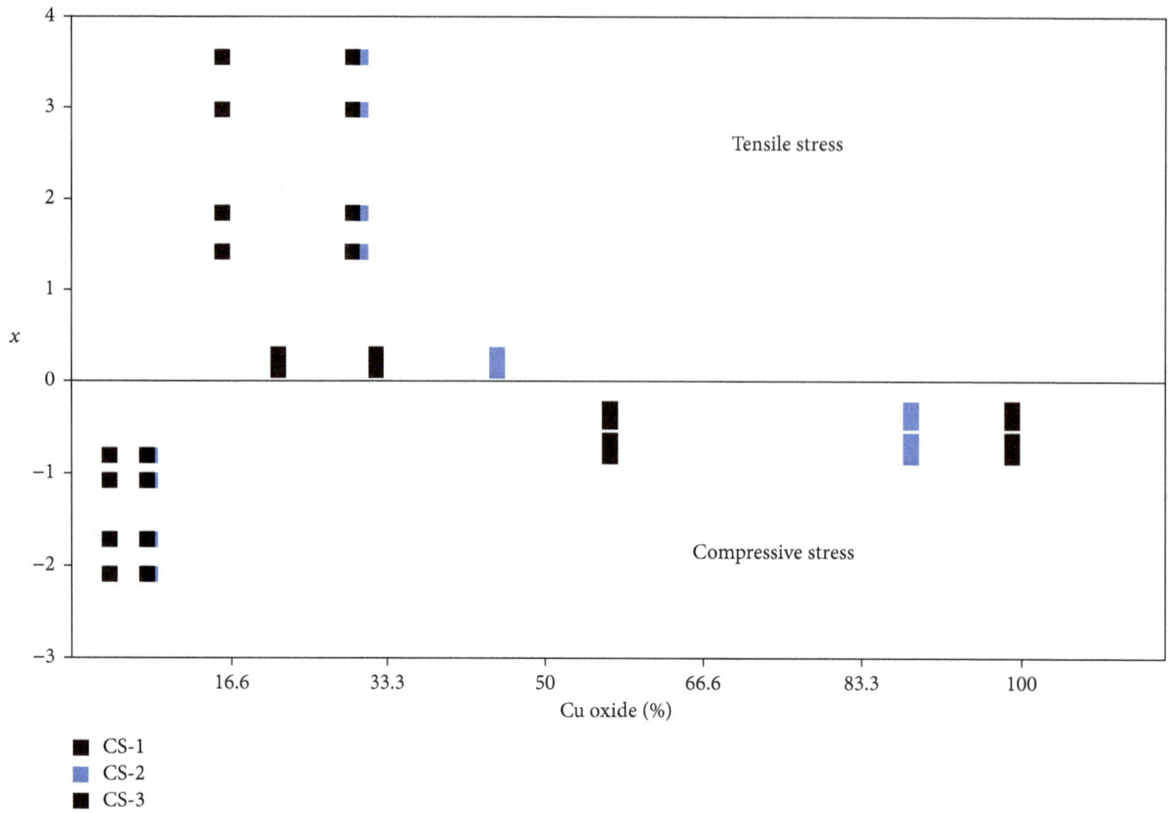

FIGURE 5: Correlation profile for different Cu oxide samples.

Conflicts of Interest

The author declares that there are no conflicts of interest, but there is an interest regarding the publication of this manuscript to share knowledge about green material processing and their applications.

Acknowledgments

The structural data collected from XRD studies have been supported thanks to Dr. José Alberto Andraca Adame from CNMN belonging at the National Polytechnic Institute, Mexico City, Mexico.

References

[1] A. C. Keyser, *Materials Science in Engineering*, Prentice Hall, Inc., México City, Mexico, 6th edition, 1990.

[2] L. Y. Isseroff and E. A. Carter, "Electronic structure of pure and doped cuprous oxide with copper vacancies: suppression of trap states," *Chemistry of Materials*, vol. 25, no. 3, pp. 253–265, 2013.

[3] W. Deng, P. Lin, Q. Li, and G. Mo, "Ultrafine-grained copper produced by machining and its unusual electrochemical corrosion resistance in acidic chloride pickling solutions," *Corrosion Science*, vol. 74, pp. 44–49, 2013.

[4] J.-M. Song, D.-S. Wang, C.-H. Yeh, W.-C. Lu, Y.-S. Tsou, and S.-C. Lin, "Texture and temperature dependence on the mechanical characteristics of copper electrodeposits," *Materials Science and Engineering A*, vol. 559, pp. 625–644, 2013.

[5] R. Skuriat, J. F. Li, P. A. Agyakwa, N. Mattey, P. Evans, and C. M. Johnson, "Degradation of thermal interface materials for high-temperature power electronics applications," *Microelectronics Reliability*, vol. 53, no. 12, pp. 1933–1942, 2013.

[6] Md. K. Rahman, A. M. M. Musa, B. Neher, K. A. Patwary, M. A. Rahman, and Md. S. Islam, "A review of the study on the electromigration and power electronics," *Journal of Electronics Cooling and Thermal Control*, vol. 6, no. 1, pp. 19–31, 2016.

[7] G. S. Frankel, *Corrosion Mechanisms in Theory and Practice*, Marcel Dekker, New York, NY, USA, 1995.

[8] M. H. Lin, K. P. Chang, K. C. Su, and T. Wang, "Effects of width scaling and layout variation on dual damascene copper interconnect electromigration," *Microelectronics Reliability*, vol. 47, no. 12, pp. 2100–2108, 2007.

[9] Z. Q. Yao, S. L. Liu, L. Zhang et al., "Room temperature fabrication of p-channel Cu_2O thin-film transistors on flexible polyethylene terephthalate substrates," *Applied Physics Letters*, vol. 101, no. 4, p. 042114, 2012.

[10] J. Yu, G. Liu, A. Liu, Y. Meng, B. Shin, and F. Shan, "Solution-processed p-type copper oxide thin-film transistors fabricated by using a one-step vacuum annealing technique," *Journal of Materials Chemistry C*, vol. 3, no. 37, pp. 9509–9513, 2015.

[11] N. Cabrera and N. F. Mott, "Theory of the oxidation of metals," *Reports on Progress in Physics*, vol. 12, no. 1, pp. 163–184, 1949.

[12] P. F. Fewster, *X-Ray Scattering from Semiconductors*, Imperial College Press, Singapore, 2nd edition, 2003.

[13] A. Cotton and G. Wilkinson, *Advanced Inorganic Chemistry*, John Wiley & Sons, New York, NY, USA, 4th edition, 2008.

[14] M. Ottosson, M. Boman, P. Berastegui et al., "Copper in ultrapure water, a scientific issue under debate," *Corrosion Science*, vol. 122, pp. 53–60, 2017.

[15] A. S. Grove, *Physics and Technology of Semiconductor Devices*, John Wiley & Sons, New York, NY, USA, 1st edition, 1967.
[16] M. Lundstrom, *Fundamentals of Carrier Transport*, Cambridge University Press, New York, NY, USA, 2nd edition, 2000.
[17] J. J. O'Dwyer, *The Theory of Electrical Conduction and Breakdown in Solid Dielectrics*, Clarendon Press, Oxford, UK, 1st edition, 1973.
[18] R. K. Zheng, C. Chao, H. L. W. Chan, C. L. Choy, and H. S. Luo, "Converse piezoelectric control of the lattice strain and resistance in $Pr_{0.5}Ca_{0.5}MnO_3$/PMN-PT structures," *Physical Review B*, vol. 75, no. 2, p. 024110, 2007.
[19] V. P. Malinenko, L. A. Aleshina, A. L. Pergament, and G. V. Germak, "Switching Effects and Metal-Insulator Transition in Manganese Oxide," *Journal on Selected Topics in Nano Electronics and Computing*, vol. 1, no. 1, pp. 44–50, 2013.
[20] R. Baca-Arroyo, "Structural characterization of the chromium-iron alloy formed by thermal diffusion processes," *Advanced Materials Research*, vol. 651, pp. 284–288, 2013.

16

Investigation of Erosion of Cement-Bentonite via Piping

Zijun Wang,[1] Boming Zhao,[1] and A. C. D. Royal[2]

[1]*School of Civil Engineering, Beijing Jiaotong University, Beijing, China*
[2]*School of Civil Engineering, University of Birmingham, Edgbaston, Birmingham, UK*

Correspondence should be addressed to Boming Zhao; bmzhao@bjtu.edu.cn

Academic Editor: Francesco Ruffino

Cement-bentonite is one of the main materials used in the seepage barriers to protect earth dams and levees from water erosion. However, the current understanding of the erodibility of the cementitious materials and the interactions between cracked seepage barriers and the water flow is inadequate. Based on the laboratory pinhole erosion test, we first investigated the impacts of cement-bentonite treatments by using the ground granulated blast-furnace slag (GGBS) as replacement on the erosion characteristics, compared with the original mixtures; the inclusion of GGBS highlighted a potential advantage against water erosion. In addition, we proposed to calculate the erosion percentage and establish the mathematical relationships between the erosion percentage and different regimes, that is, different curing period, erosion time, and sizes of initial holes. Results showed that enough curing period was critical to avoid the increases of hydraulic conductivity in the macrofabric of the barrier; meanwhile, the materials were eroded quickly at the beginning and slowed down with the erosion time, where the enlargement of the initial creaks would be stabilised at some point in time. Moreover, the sizes of initial holes may affect the erosion situation varying from the sample curing periods.

1. Introduction

Seepage barriers are often used in earth dams and levees to reduce the water flow through their foundations or embankments [1, 2]. However, internal erosion is known to occur in these seepage barriers and the ability of predicting this kind of piping has been inadequate and leads to many failures [3, 4]. Govindaraju et al. [5] concluded that the internal erosion and piping process consists of the particle detachment and subsequent migration. The initiation of a pipe is generally considered to originate from a concentrated leak and the pipe formation often progresses from the downstream to upstream face [6]. There are two parameters affecting the pipe development: the strength of soil structures and the sufficient seepage flow to overcome the critical shear strength between particles for erosion and transportation [7].

Cement-bentonite is one of the main materials used in seepage barriers [8, 9]; in addition, to achieve the recommended hydraulic conductivity and maintain a slurry that is pumpable and economic, the cement replacement materials such as pulverised fuel ash (PFA) and ground granulated blast-furnace slag (GGBS) are commonly required [10, 11]. However, the current understanding of the erodibility of these materials and the interactions between cracked seepage barriers and the water flow is inadequate, where limited data was available to provide the sufficient information regarding the impacts of cement replacements, curing period, erosion time, and sizes of initial cracks.

Based on the laboratory pinhole erosion test [12], which can be a convenient method to quantify the material erodibility, this paper investigated the impact of cement-bentonite treatments on erosion characteristics. In addition, the hole erosion test [13, 14] was also referred to simulate the erosion phenomenon in open cracks. In the following sections, the testing procedures and theoretical models were presented; then we proposed to calculate the erosion percentage and establish the mathematical relationships between the erosion percentage and different regimes, that is, different cementitious replacement, curing period, erosion time, and sizes of initial holes to aid in assessing the internal erosion propagating from cracks in cement-bentonite seepage barriers.

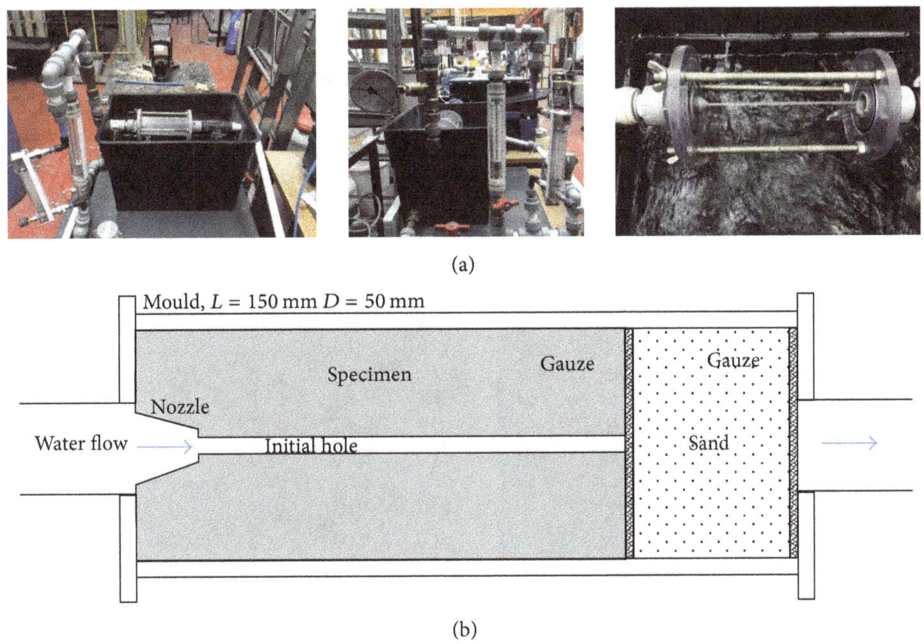

FIGURE 1: (a) Internal erosion apparatus (left); measurements of water flow and pressure (middle); the controlled water jet (right); (b) detailed erosion device setup schematic.

2. Experimental Methodology

2.1. Test Apparatus. The internal erosion test measures the water erosions that take place in the predrilled samples. The test apparatus (Figure 1) is designed to set the specimen mould with an inner diameter of 50 mm and a length of 150 mm. The pressurized water is provided by a pump using the water inside the black box circularly while the pressure is controlled by a regulator, where the maximum value is limited by about 1.0 bar. Water flow is measured upstream throughout the test using a flow meter, which is accurate to a tenth of a litter per minute. A nozzle is installed at the inlet to control the water jet only passing through the initial hole within the specimen, avoiding leaks along the edge.

For the fills to be used at the exit to fix the sample in the correct position of the mould, rubber rings and plastic foam rings were tried once, but the water impulse was so strong that these fills may cut into the samples. Therefore, sand was chosen to be used at the exit to fix the specimen in the correct position and two gauze screens were used as filters to hold the sand. Trials illustrated that they worked very well since the specimen could be positioned stably without any undesired damage. The detailed device setup is shown in Figure 1(b).

2.2. Sample Mixture Designs. Two types of sample mixtures were used in this study. The first one was the cement-bentonite mixture and the second one was the cement-bentonite mixture produced by using GGBS (BS EN 15167-1) as the cement replacement material, which can bring benefits to strength, setting ability, and reduction in permeability [15]. Because the substitution levels commonly used are relatively high (e.g., Garvin and Hayles [16] recommend 70–80% GGBS as cement replacement in cement-bentonite slurry) for achieving a given permeability, the proportion of GGBS was adopted as 80% by weight of the cement component in this study. The cement used (Rugby PFA cement, CEM II/B-V, 32.5N, supplied by Cemex) contained 20% PFA (Class F, in accordance with BS EN197-1:2011; BSi 2011 [17]), which is considered to be pozzolanic [18]. The bentonite used was the sodium carbonate activated type (Berkbent 163 Bentonite, supplied by Tolsa UK Ltd.). Therefore, two mixtures were created to represent the mix designs of real seepage barriers in dams that have experienced seepage problems. We defined them as type 1 CB-PFA samples and type 2 CB-GGBS samples. The proportion of bentonite and water remained constant and the two mixtures are specified in Table 1.

2.3. Sample Creation and Preparation. Specimens for erosion tests were formed from slurry to ensure that they were homogenous and fully saturated [19]. The CB-PFA samples and CB-GGBS samples were manufactured broadly according to the procedures of Garvin and Hayles [16]. The bentonite and laboratory-purified water (produced by reverse osmosis) were shaken in a mixer for 30 minutes and the mixture was left to hydrate for a period of 24 hours; the cementitious component was subsequently added to the slurry and it was mixed once again; the slurry was then poured into the cylindrical moulds with an inner diameter of 50 mm and a length of 150 mm for consolidation.

In order to create a hole through the sample, we introduced a wire running vertically and centrally from the top cap to the bottom cap throughout the mould. Two weights were attached to the wire at both ends to ensure the wire to be tight during the consolidation process. Once the pouring was finished the decanted slurry was agitated on a vibrating table to de-air it; the final assembled setup is shown in Figure 2(a). Creating the hole within samples in place can reduce the disturbance of the hole wall to a large extent compared to

TABLE 1: Proportion of the constituent components within the two cement-bentonite mixtures.

Mass of the constituent components of the cement-bentonite samples		Mixture	Proportion of PFA within cementitious replacement (%)	Proportion of GGBS within cementitious replacement (%)
Component	Mass (g)			
Water	1000	No. 1	20	/
Cement and cement replacement material	200	No. 2	4	80
Bentonite	40			

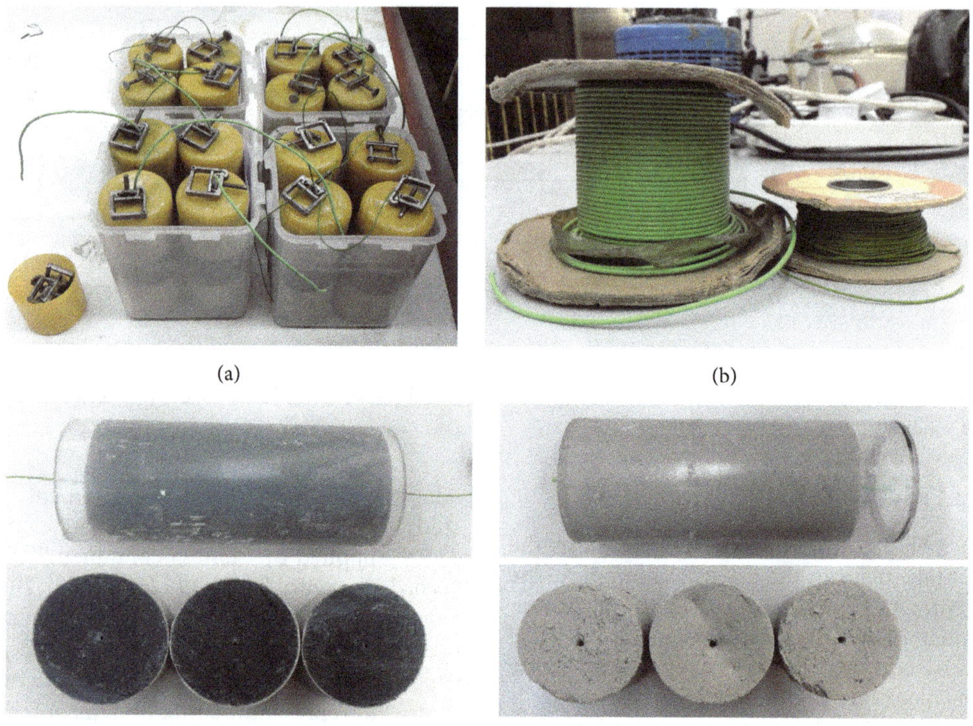

FIGURE 2: Illustration of the handling of samples; (a) decanted slurry curing within moulds stored in a constant temperature room at a temperature of 5°C; (b) two sizes of wires were used to create initial holes; (c) CB-GGBS sample with 1 mm initial hole; (d) CB-PFA sample with 2 mm initial hole.

the practice of inserting the wire afterwards [12, 20]; thus our method can help to prevent the soil around the hole from the extensive disturbance. In this study, we used two sizes of wires (1 mm and 2 mm) to create the initial holes to investigate how different conduits affect the sample erosion.

The samples were cured in a constant temperature room at a temperature of 5°C for different curing times, where test ages of 7, 14, 28, and 60 days were chosen to assess the performance of the investigated samples since these regimes were commonly used. However, trials showed that the curing process was incomplete for CB-PFA samples of 7 days and the physical properties were still in flux; even the strength was not strong enough to maintain the initial hole to be stable when taking out the wire; thus these samples were eliminated.

When the curing was completed, the wire could be smoothly removed by hand without any rotation since the wire surface did not adhere to the initial hole wall. For checking purpose, the prepared extra samples were cut into sections to confirm that the initial holes made by the wire could be straight, horizontal, and intact along the middle of the samples (Figures 2(c) and 2(d)).

A countersinker of 16.5 mm diameter was hand drilled into the sample approximately 12 mm to provide a pilot hole on the lower surface, which was regarded as the inlet face and therefore the nozzle can sit flush towards the initial conduit. When drilling the pilot hole, the countersinker should be inserted slowly and carefully to reduce the disturbance of the sample and avoid the drilled material blocking the hole. Since bleeding could cause the upper face of poured slurry to be varying in physical properties, the cured samples may have wrong length and uneven upper surfaces. Method of casting and trim was selected in preference to casting a sample of the required length; while cutting the sample is not ideal, it was considered to be a preferable alternative [10]. Thus samples were cut to size by a pallet knife to produce the samples with a length of 100 mm exclusive from the pilot hole.

Because the creation was from batched slurry, individual samples were of varying physical properties, three samples were prepared for each batch, and the mean behaviour of these samples was used to evaluate the characters. Therefore, a total of 126 tests were conducted and Table 2 presents the groups of the tested samples, where the different material

TABLE 2: Summary of experiments of different regimes.

Sample types	Initial holes	Number of samples investigated											
		Eroded for 1 hour				Eroded for 2 hours				Eroded for 3 hours			
		7 days	14 days	28 days	60 days	7 days	14 days	28 days	60 days	7 days	14 days	28 days	60 days
CB-PFA	1 mm	/	3	3	3	/	3	3	3	/	3	3	3
	2 mm	/	3	3	3	/	3	3	3	/	3	3	3
CB-GGBS	1 mm	3	3	3	3	3	3	3	3	3	3	3	3
	2 mm	3	3	3	3	3	3	3	3	3	3	3	3

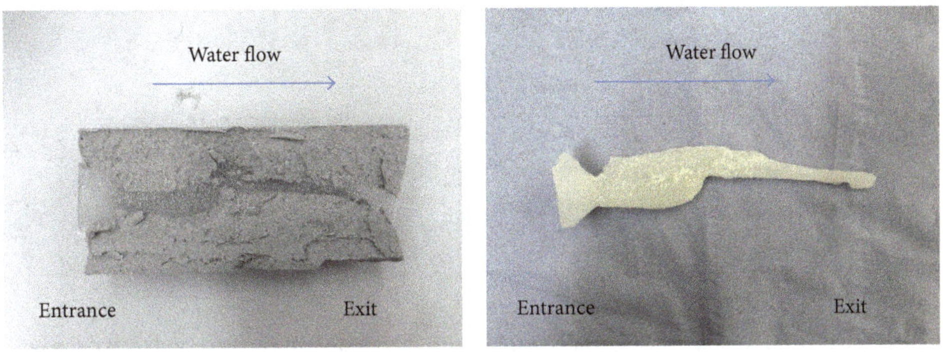

FIGURE 3: Cross-section showing the cast of an eroded cavity.

component, curing period, erosion time, and size of initial holes are investigated.

2.4. Data Collection and Erosion Calculation. We set the given flow rate to be 3 L/min for the designed erosion durations of 60, 120, and 180 minutes, respectively. Based on the law of volumetric flow rate, the flow velocity (v) can be obtained as the quotient of flow rate (Q) to the cross-area (A) that fluid travels through, as specified $v = Q/A$. Thus the flow velocity of the 1 mm initial hole is 4 times larger than that of the 2 mm initial hole at the beginning. After the samples were subjected to the pinhole erosion test, the liquid wax was chosen to cast the eroded conduits to provide undisturbed images of the erosion cavity. The samples were positioned within the moulds for pouring the liquid wax and allowing the wax to flow from the entrance and fully occupy the eroded holes. It was required to give some time to ensure the exit to be sealed before casting the whole conduit. Then the samples were disassembled to extract the cast to be inspected and photographed for the erosion patterns. Figure 3 shows a typical example of the cross-section of an eroded cavity after test.

The cast was washed carefully to remove the sticking material and dried before measuring the weight. In this process, the cast of nozzle was rejected from this measurement. It was straightforward to work out the erosion volume of the sample via dividing the cast weight by the wax density (tested with a value of $0.8\ g/cm^3$).

3. Results

One example of the 14-day CB-PFA sample being eroded for 2 hours is illustrated in Figure 4(a), which shows the cross-section of the initial hole before and after test for both entrance and exit where the 2 mm diameter hole became 9.5 mm at the entrance. The cast of the eroded cavity (Figure 4(b)) was extracted to the black silhouette (Figure 4(c)) to clearly illustrate the shape changes of the initial conduit. In addition, a part of the investigated samples, regardless of the material types, exhibited brittle transverse microcracks during eroding (Figure 4(d)), possibly because the water jet was so intense, thereby resulting in the formation of these microcracks through the sample.

For CB-PFA samples, if the curing period was 14 and 28 days, microcracks occurred in all test specimens. If curing period reached 60 days, test specimens indicated basically the same performance; when erosion time lasted for 1 hour and 2 hours, for initial hole of 1 mm, only one out of six test specimens indicated microcracks; for initial hole of 2 mm, two out of six test specimens indicated microcracks; when erosion time lasted for 3 hours, all test specimens indicated microcracks. For CB-GGBS samples, if the curing period was 7 days, microcracks occurred in all test specimens. If curing period was 14 to 60 days, test specimens indicated basically the same performance. In other words, when erosion time lasted for 1 hour and 2 hours, for initial hole of 1 mm, 83.3% of test specimens did not indicate microcracks; for initial hole of 2 mm, 72.2% of test specimens did not indicate microcracks.

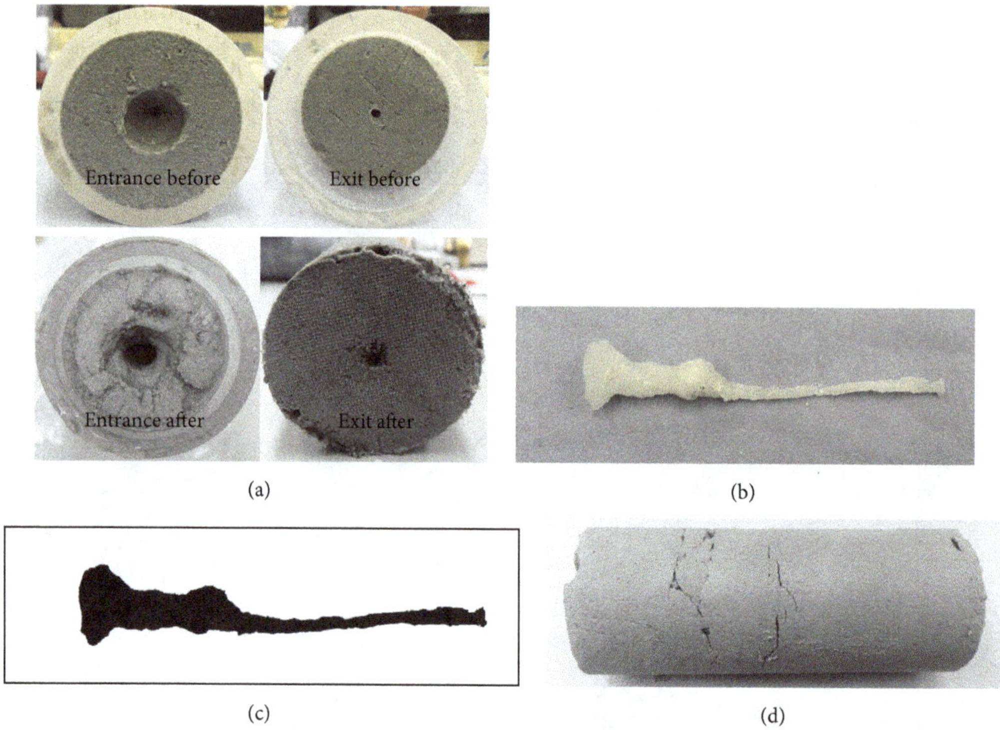

FIGURE 4: Example of the 14 days CB-PFA sample being eroded for 2 hours. (a) Surfaces of entrance and exit of the hole before and after test; (b) cast of the eroded cavity; (c) extracted black silhouette of the cast; (d) brittle transverse microcracks appeared during eroding.

When erosion time increased to 3 hours, all test specimens indicated microcracks.

To sum up, test specimens with smaller initial holes were less likely to indicate microcracks than test specimens with larger initial holes. But their differences were not significant. However, there was an apparent trend of decreasing microcrack widths with curing duration but increasing crack widths with erosion time. With the increase of the curing, both types of samples became stiffer and stronger as the microcrack widths reduced; for instance, there were not any microcracks for both samples cured for 60 days after 1-hour erosion; however, as the erosion time increased to 3 hours, the brittle microcracks appeared again. Although there was water in the microcracks, the water flow passes mainly through the initial holes that could guarantee the test purpose.

We defined the erosion percentage (%) as the ratio of the eroded material volume to the original volume of the specimen. The original volume was calculated according to the sample geometry and the eroded volume was obtained via dividing the cast weight by the wax density. The erosion percentage is specified in the following:

$$P_{\text{erosion}} = \frac{V_{\text{eroded}}}{V_{\text{original}}} = \frac{M_{\text{eroded}}/\rho_{\text{wax}}}{V_{\text{original}}}. \quad (1)$$

When we calculate the original volumes, the volume of the initial holes should be excluded. The summary of the erosion percentages for the CB-PFA and CB-GGBS samples is listed in Table 3.

The relationships between erosion percentage of the investigated CB-PFA samples and CB-GGBS samples against erosion time are shown in Figure 5. Each point corresponds to the mean value out of three tests; that is, Figure 5(a) shows the results of 54 tests for CB-PFA samples and Figure 5(b) shows the results of 72 tests for CB-GGBS samples. According to the erosion percentage and the shape of the conduit cast, there was an overall increase in erosion with increased erosion time for both types of samples. Meanwhile, the trend also showed that material was eroded quickly at the beginning of the test and then slowed down with time.

With respect to the CB-PFA samples cured for 60 days, there was no obvious difference in the erosion percentage between 1-hour erosion and 3-hour erosion. As for the CB-GGBS samples, the conduits of 7-day specimens were witnessed to widen when the erosion time approached/exceeded 2 hours. Conversely, samples cured for more than 14 days exhibited similar behaviour that the impact of erosion time on the erosion percentage reduced to a certain degree.

From Figure 6, it is apparent that if the curing time increases, the erosion percentages for all cases drop down. Specifically, the difference of erosion percentage between each curing interval reduced, meaning that the erosion rate was reduced as the curing increased for the given barrier materials. However, the downward trend of the CB-GGBS samples was more obvious than that of the CB-PFA samples. In addition, the erosion percentage of the CB-PFA was approximately two times higher than that of the CB-GGBS for the same curing period. Even CB-PFA samples cured for 28 days did exceed the erosion percentage for CB-GGBS samples cured for only 14 days. As the curing period increased to 60 days, the difference in the erodibility between these two types reduced that the erosion percentage of the former was as approximately 1.3 to 1.7 times as that of the latter. This

TABLE 3: Summary of the erosion percentages for the CB-PFA samples and the CB-GGBS samples.

Number	CB-PFA	Erosion percentage (%)	CB-GGBS	Erosion percentage (%)
(1)	/	/	1 mm-7 days-1 h	2.331
(2)	/	/	1 mm-7 days-2 h	3.675
(3)	/	/	1 mm-7 days-3 h	4.671
(4)	/	/	2 mm-7 days-1 h	1.639
(5)	/	/	2 mm-7 days-2 h	1.900
(6)	/	/	2 mm-7 days-3 h	1.917
(7)	1 mm-14 days-1 h	1.161	1 mm-14 days-1 h	0.197
(8)	1 mm-14 days-2 h	1.348	1 mm-14 days-2 h	0.426
(9)	1 mm-14 days-3 h	1.470	1 mm-14 days-3 h	0.625
(10)	2 mm-14 days-1 h	0.947	2 mm-14 days-1 h	0.725
(11)	2 mm-14 days-2 h	1.705	2 mm-14 days-2 h	0.842
(12)	2 mm-14 days-3 h	2.333	2 mm-14 days-3 h	0.973
(13)	1 mm-28 days-1 h	0.790	1 mm-28 days-1 h	0.121
(14)	1 mm-28 days-2 h	1.197	1 mm-28 days-2 h	0.129
(15)	1 mm-28 days-3 h	1.227	1 mm-28 days-3 h	0.460
(16)	2 mm-28 days-1 h	0.421	2 mm-28 days-1 h	0.254
(17)	2 mm-28 days-2 h	0.587	2 mm-28 days-2 h	0.540
(18)	2 mm-28 days-3 h	1.007	2 mm-28 days-3 h	0.653
(19)	1 mm-60 days-1 h	0.144	1 mm-60 days-1 h	0.067
(20)	1 mm-60 days-2 h	0.147	1 mm-60 days-2 h	0.087
(21)	1 mm-60 days-3 h	0.152	1 mm-60 days-3 h	0.124
(22)	2 mm-60 days-1 h	0.343	2 mm-60 days-1 h	0.215
(23)	2 mm-60 days-2 h	0.541	2 mm-60 days-2 h	0.457
(24)	2 mm-60 days-3 h	0.562	2 mm-60 days-3 h	0.547

indicated that the inclusion of cementitious materials within the slurry will clearly result in changing physical properties; samples containing GGBS have a higher rate of strength increase with curing; that is, in order to achieve a certain performance in resisting water, the CB-GGBS samples require a shorter curing period.

The sizes of initial holes may affect the erosion situation varying from the sample curing regardless of the material types. For juvenile samples, the erosion percentage of samples with 1 mm initial hole was approximately twice higher than that of the one with 2 mm initial hole. By contrast, when the curing duration was enough, samples with larger initial holes would experience more severe erosion.

4. Discussion and Conclusion

Normally before the construction of a seepage barrier, material that is to be used should be subjected to a variety of tests to determine its properties to assess its suitability. The erodibility of material is controlled by two factors, the strength of material, or its erosion resistance, and the flow regime of the eroding fluid, that is, sufficient seepage flow to overcome a critical shear strength between particles for erosion and transportation [7, 21–24]. The strength is correlated to the cementitious properties of replacement and curing period while the seepage flow is corresponding to the erosion time.

The cement-bentonite mixtures containing GGBS have a higher strength and stiffness in comparison with the behaviour of CB-PFA samples, highlighting the potential advantage with the inclusion of GGBS in the cement-bentonite mixtures, which could bring great benefits to slurry loss and in-trench mix curing. Curing period is significant to the sample performance. Juvenile samples had significant erosion percentage; however, an apparent downward trend of the erosion rate was obtained for the samples cured for 28 days or longer; when the samples were cured for 60 days, there were even no obvious changes in the shape of the initial conduits. This suggests that enough curing period for a barrier could improve the ability to avoid the increases of hydraulic conductivity in the macrofabric of the barrier that may cause the fail in its primary function.

Since the flow velocity controls the shear stress [23] and for a constant flow rate (Q), initially the velocity within a smaller crack is higher and more erosive to the barrier material to be developed into pipes. The continued flow along the pipes will further propagate the erosion, which is considered to be the mechanism of the concern in several existing seepage barriers [25]. With the increase of the cross-section area of the initial conduit during erosion, the flow velocity reduces; this reduced velocity will result in a decrease of the shear stress and thus a decline in erosion rate as well. Therefore, even if a crack on a barrier could be eroded under a certain gradient, the crack width will initially increase, but,

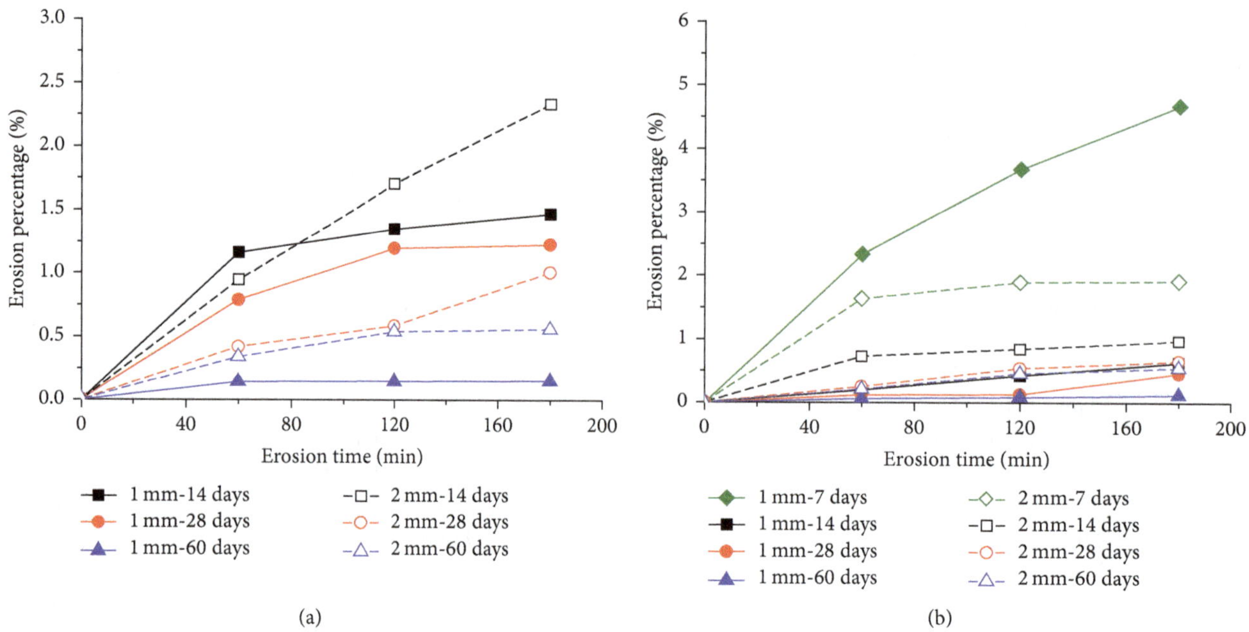

FIGURE 5: Erosion percentage against erosion time for (a) CB-PFA samples and (b) CB-GGBS samples.

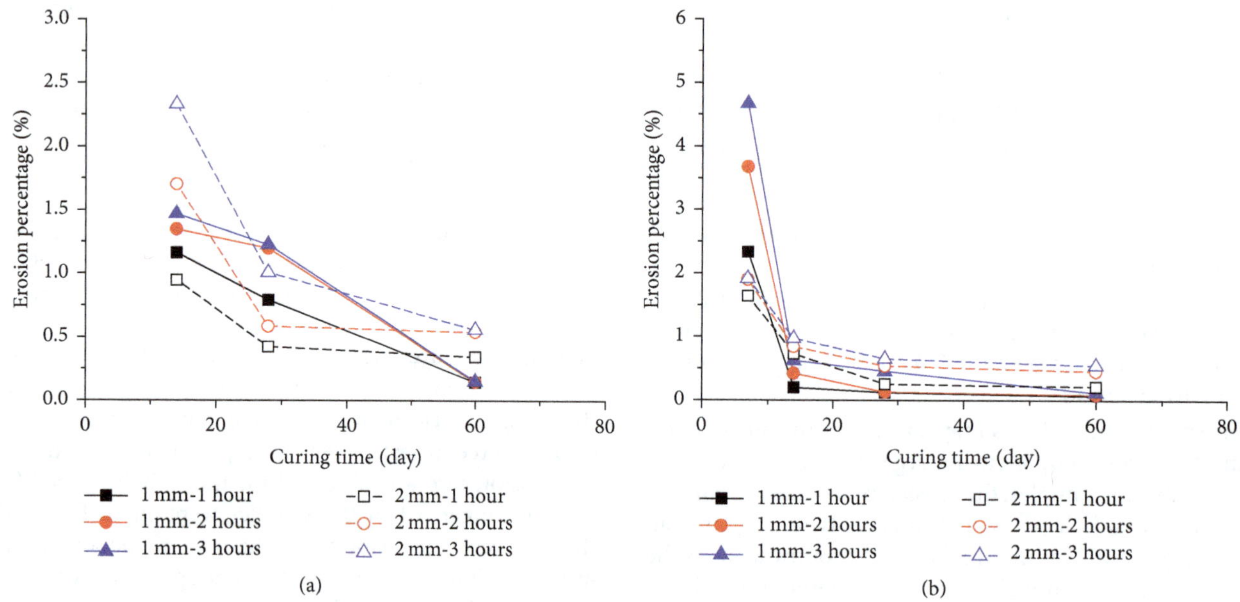

FIGURE 6: Erosion percentage against curing time for (a) CB-PFA samples and (b) CB-GGBS samples.

at some point in time, the width will be stabilised as the water velocity decreases in the crack.

With respect to the issue of how different sizes of initial cracks affect the barrier behaviours against eroding, the material erodibility and the water velocity inside the mass should be considered concurrently. When the curing duration is incomplete, the flow velocity dominates the erosion of the barrier that smaller cracks can cause higher erosion percentage. However, with the increase of the critical shear stress in correlation with the curing, the difference between the shear stress caused by velocity and the critical stress reduced, meaning that the dominant factor affecting the erosion is changed. On the other hand, the generated pipes with larger diameters can provide more particles to be detached from the conduit wall compared with pipes of smaller diameters. The more particles having the potential to be detached, the higher erosion percentage to be caused. However, to what extent the larger initial hole may affect the erosion situation is not easy to ascertain, while, according to Hubbe [23], no matter how small is the applied force, the particles will be released after a characteristic length of time; thus, this might be a time-related issue which should be studied further.

Conflicts of Interest

The authors declare that they have no conflicts of interest.

Acknowledgments

This research has been supported by the State Key Program of National Natural Science Foundation of China (Grant no. U1434210), the National Natural Science Foundation of China (Grant no. 51778046), the National Natural Science Foundation of China (Grant no. 51278045), and the Project of Shenzhen Metro Group Co., Ltd. (no. SZ-CGM-KY001/2014).

References

[1] S. A. Jefferis, "The origins of the slurry trench cut-off and a review of cement-bentonite cut-off walls in the UK," in *Proceedings of the International Containment Technology Conference and Exhibition*, St. Petersburg, Fla, USA, 1997.

[2] S. M. Opdyke and J. C. Evans, "Slag-cement-bentonite slurry walls," *Journal of Geotechnical and Geoenvironmental Engineering*, vol. 131, no. 6, pp. 673–681, 2005.

[3] E. Sinco, M. Xiao, L. M. Owaidat, and L. Smith, "Slurry induced piping progression of a sand," in *Proceedings of the 5th International Conference on Scour and Erosion*, pp. 408–417, San Francisco, Calif, USA, November 2010.

[4] C. F. Wan and R. Fell, "Investigation of rate of erosion of soils in embankment dams," *Journal of Geotechnical and Geoenvironmental Engineering*, vol. 130, no. 4, pp. 373–380, 2004.

[5] R. S. Govindaraju, L. N. Reddi, and S. K. Kasavaraju, "A physically based model for mobilization of kaolinite particles under hydraulic gradients," *Journal of Hydrology*, vol. 172, no. 1, pp. 331–350, 1995.

[6] J. A. A. Jones, J. M. Richardson, and H. J. Jacob, "Factors controlling the distribution of piping in Britain: A reconnaissance," *Geomorphology*, vol. 20, no. 3-4, pp. 289–306, 1997.

[7] L. N. Reddi and M. V. Bonala, "Analytical solution for fine particle accumulation in soil filters," *Journal of Geotechnical and Geoenvironmental Engineering*, vol. 123, no. 12, pp. 1143–1152, 1997.

[8] J. C. Evans, A. R. Dawson, and S. Opdyke, "Slurry walls for groundwater control: a comparison of UK and US practice," in *Proceedings of the ASCE/PENNDOT Central PA Geotechnical Conference*, 2002.

[9] Institution of Civil Engineers (ICE), *Specification for the Construction of Slurry Trench Cut-off Walls*, Thomas Telford, London, UK, 1999.

[10] A. C. D. Royal, Y. Makhover, S. Moshirian, and D. Hesami, "Investigation of cement-bentonite slurry samples containing PFA in the UCS and triaxial apparatus," *Geotechnical and Geological Engineering*, vol. 31, no. 2, pp. 767–781, 2013.

[11] L. K. Sear, *Properties and Use of Coal Fly Ash: A Valuable Industrial by-Product*, Thomas Telford, London, UK, 2001.

[12] J. L. Sherard, E. F. Steele, R. S. Decker, and E. F. Steele, "Pinhole test for identifying dispersive soils," *Journal of the Geotechnical Engineering Division*, vol. 102, no. 1, pp. 69–85, 1976.

[13] C. F. Wan and R. Fell, "Investigation of internal erosion and piping of soils in embankment dams by the slot erosion test and the hole erosion test," UNICIV Report No R-412, University of New South Wales, New South Wales, Australia, 2002.

[14] N. Benahmed and S. Bonelli, "Investigating concentrated leak erosion behaviour of cohesive soils by performing hole erosion tests," *European Journal of Environmental and Civil Engineering*, vol. 16, no. 1, pp. 43–58, 2012.

[15] M. Williams and G. S. Ghataora, "Effect of fibre reinforcement on the properties of ground granulated blast furnace slag-cement-bentonite slurry," *Studia Geotechnica et Mechanica*, vol. 33, no. 4, pp. 63–83, 2011.

[16] S. L. Garvin and C. S. Hayles, "Chemical compatibility of cement-bentonite cut-off wall material," *Construction and Building Materials*, vol. 13, no. 6, pp. 329–341, 1999.

[17] BSI, *BS EN 197-1:2011. Cement Part 1: Composition, Specifications and Conformity Criteria for Common Cements*, The British Standards Institution, 2011.

[18] S. H. Gebler and P. Klieger, "Effect of fly ash on physical properties of concrete," *Journal of the American Concrete Institute*, vol. 83, no. 2, p. 328, 1986.

[19] J. H. Atkinson, J. A. Charles, and H. K. Mhach, "Examination of erosion resistance of clays in embankment dams," *Quarterly Journal of Engineering Geology and Hydrogeology*, vol. 23, no. 2, pp. 103–108, 1990.

[20] C. C. Hird and V. J. Moseley, "Model study of seepage in smear zones around vertical drains in layered soil," *Geotechnique*, vol. 50, no. 1, pp. 89–97, 2000.

[21] R. Flores-Berrones and N. P. Lopez-Acosta, "Internal erosion due to water flow through earth dams and earth structures," in *Soil Erosion Studies*, D. Godone, Ed., InTech, 2011.

[22] L. A. Clark and T. M. Wynn, "Methods for determining streambank critical shear stress and soil erodibility: Implications for erosion rate predictions," *Transactions of the ASABE*, vol. 50, no. 1, pp. 95–106, 2007.

[23] M. A. Hubbe, "Theory of detachment of colloidal particles from flat surfaces exposed to flow," *Colloids and Surfaces*, vol. 12, no. C, pp. 151–178, 1984.

[24] J. R. Talbot, "The mechanics of cracking in embankment dams," in *Fracture Mechanics Applied to Geotechnical Engineering: Proceedings of Sessions Sponsored by the Geotechnical Engineering Division of the American Society of Civil Engineers in Conjunction with the ASCE National Convention*, E. Vallejo and R. Y. Liang, Eds., American Society of Civil Engineers, pp. 118–130, Atlanta, Ga, USA, 1994.

[25] J. D. Rice and J. M. Duncan, "Deformation and cracking of seepage barriers in dams due to changes in the pore pressure regime," *Journal of Geotechnical and Geoenvironmental Engineering*, vol. 136, no. 1, pp. 16–25, 2010.

17

Analysis of Titania Nanosheet Adsorption Behavior Using a Quartz Crystal Microbalance Sensor

Yuichiro Tashiro,[1] Satoshi Komasa,[1] Akiko Miyake,[2] Hiroshi Nishizaki,[1] and Joji Okazaki[1]

[1]*Department of Removable Prosthodontics and Occlusion, Osaka Dental University, 8-1 Kuzuhahanazono-cho, Hirakata-shi, Osaka, Japan*

[2]*Department of Oral Health Engineering, Faculty of Health Sciences, Osaka Dental University, 1-4-4 Makino-honmachi, Hirakata-shi, Osaka, Japan*

Correspondence should be addressed to Satoshi Komasa; komasa-s@cc.osaka-dent.ac.jp

Academic Editor: Marco Cannas

We investigated the adsorption of albumin and fibronectin on a titania nanosheet- (TNS-) modified quartz crystal microbalance (QCM) sensor. A Ti QCM sensor was fabricated by reactive magnetron sputtering. A thin layer of Ti was deposited on the QCM sensor. This sensor was then alkali-modified by treatment with NaOH at room temperature to fabricate the titania nanosheets. Scanning probe microscopy, X-ray photoelectron spectroscopy, and scanning electron microscopy were performed to investigate the surface topology and chemical components of each sensor. The TNS had a titanium oxide film exhibiting a nodular structure and a thickness of 13 nm on the QCM sensor. Furthermore, QCM measurements showed significantly greater amounts of albumin and fibronectin adsorbed on the TNS than on titanium. The NaOH treatment of titanium modified the sensor surface and improved the adsorption behaviors of proteins related to the initial adhesion of bone marrow cells. Therefore, we concluded that TNS improves the initial adhesion between the implant materials and the surrounding tissues.

1. Introduction

Titanium is a biocompatible material that is commonly used in dentistry and orthopedic reconstruction. The surface features of implant materials have important functions in cell or extracellular matrix interrelationships and eventual osseointegration. Therefore, interrelationships between cells and microtopography have been intensively studied.

An important contemporary advance in dental implant research is the ability to modify implant surface materials at the nanoscale [1, 2]. Materials with an expanded surface region and a better surface roughness may yield better mechanical interlocking between tissues and titanium [3]. However, more importantly, such nanoscale features are also believed to directly affect osteogenic cell behaviors around implant fixtures with nonconventional surfaces, creating a biomimetic relationship between alloplastic surfaces and host tissues by the replication of the natural cellular environment at the nanometer level [1, 2, 4].

Low-dimensional TiO_2 nanostructures have attracted recent attention because these materials can take the forms of nanotubes [5], nanofibers [6], and nanowires [7]. Compared with mass materials or nanoparticles, TiO_2 nanotubes have high particular surface zones accessible for the adsorption of color sensitizers, and they provide channels to improve electron exchange, thus expanding the effectiveness of solar cells [5]. Titania nanosheets (TNSs) are similar to TiO_2 nanotubes, created by titania deposition using the sputtering process [5].

In our previous study [8], we showed that TNSs created by compound processing enhance the osteogenic separation of rodent bone marrow (RBM) cells. The surface properties and structural characteristics of materials play an important role in protein and cell adsorption behaviors. The initial amount adsorbed and the conformation of proteins contained in the serum could alter the bioactivity of stem cells on TNSs. However, the bioactivities of TNS materials, including their roles in osteogenic differentiation and the biointegration of dental implants into the alveolar bone, have not been elucidated.

A quartz crystal microbalance (QCM) sensor is a profoundly delicate and handy device that is used to observe protein adsorption and cell behavior in situ. A QCM-based sensor comprises a quartz crystal and a detection material. A 27 MHz QCM can provide highly sensitive measurements of mass in aqueous solutions; the resonance frequency decreases in relation to the mass of the protein bound on the QCM electrode surface. We previously fabricated several types of QCM sensors by coating the gold electrode of the quartz crystal with a thin film of a biomaterial [9]; this QCM sensor achieved an increase in sensitivity approximately 24 times that of a conventional 5 MHz QCM. PMMA (poly (methyl methacrylate)), Au, and Ti have been used as QCM electrode surface materials to imitate denture materials and evaluate the adsorption behaviors of various bovine salivary proteins [10]. These previous findings support the potential utility of the QCM method for the evaluation of protein adsorption behaviors on implant surfaces.

In this study, we evaluated the effects of modified surfaces on the adsorption of albumin and fibronectin in RBM cells and simulated body fluid (SBF) in QCM analyses.

2. Materials and Methods

2.1. Sputtering Procedure. A thin layer of Ti was deposited on quartz discs (diameter: 8 mm, area: 4.9 mm^2) by reactive magnetron sputtering using a radio-frequency magnetron sputtering system (CFS-4ES-231; Shibaura Mechatronics Co., Ltd., Kanagawa, Japan). The QCM crystal was cleaned using piranha solution (H_2SO_4/30% H_2O_2 (v/v): 7 : 3). Before deposition was conducted, the quartz surfaces were ultrasonically cleaned in high-purity acetone (99.999%). Pure Ti powder was used to prepare the target, which was formed by pressing the powder into a disc with a diameter of 75 mm. The quartz substrates were positioned 85 mm above the target, and the magnetron sputtering chamber was evacuated to a pressure of 3×10^{-3} Pa. Argon was used as the working gas, and its pressure was kept constant at 6.7×10^{-1} Pa. All of the films were fabricated using a constant radio-frequency discharge power of 480 W, and the Ti thin films were deposited at room temperature at a deposition rate of 20.0 nm/min, yielding a film thickness of approximately 240 nm. The crystals were washed and cleaned with both sodium dodecyl sulfate and UV-Ozone Cleaner (PC450; Meiwafosis Co., Ltd., Osaka, Japan) prior to QCM measurements.

2.2. Sample Preparation. In the TNS group, Ti sensors were treated to produce TNS on their surfaces. An unprocessed QCM sensor was used as the Ti sensor. These sensors were immersed in 10 M NaOH (aq) and were then placed in an oil bath, which was kept at a temperature of 30°C for 24 h. The solution in each flask was replaced and treated with distilled water (200 mL), and this procedure was repeated until the point that a conductivity of 5 μS/cm^3 was reached. The specimens were then dried at room temperature.

2.3. Characterization of Materials. Scanning electron microscopy (S-4800; Shimadzu, Kyoto, Japan) and scanning probe microscopy (SPM-9600; Shimadzu) over a surface area of 2.0 μm × 2.0 μm were conducted to observe the surface topology and roughness of the fabricated TNS and Ti sensors. The composition of the coating was analyzed by X-ray photoelectron spectroscopy (XPS; ESCA 5600, Ulvac-Phi Inc., Kanagawa, Japan) using surface etching with ionized argon. In addition, the surfaces of the fabricated Ti sensors were subjected to XPS analysis with an Al Kα line (15 kV, 300 W) as an X-ray source. During XPS, argon ion sputtering was applied to determine the thickness and structure of the surface layers.

2.4. Contact Angle Measurements. Contact angles were measured for the TNS and Ti sensors using a video contact angle measurement system (model VSA 2500 XE; AST Products Inc., Billerica, MA, USA). A small droplet of a deionized water solution with Hanks' Balanced Salt Solution and bovine serum albumin (BSA; approximately 3 mg) was placed on the TNS to measure the static contact angle. Estimation of the contact edge is a straightforward strategy for breaking down the vitality and hydrophilic nature of a surface.

2.5. Proteins. BSA (Wako Pure Chemical Industries Ltd., Osaka, Japan) was dissolved in phosphate-buffered saline (PBS; pH 7.4) at 200 μg/mL. Human plasma fibronectin (HFN; Nacalai Tesque, Inc., Kyoto, Japan) was dissolved in PBS (pH 7.4) at 500 μg/mL.

2.6. Cell Culture. Since most bone embed materials are embedded in adult bone that is in direct contact with bone marrow tissue, the effects and success of new embed materials can be investigated by examining bone marrow cell cultures from adult rats. RBM cells multiply and separate into a phenotype that expresses bone cell markers in vitro. RBM cells were extracted from the femurs of 7-week-old Sprague–Dawley rats. The rats were humanely sacrificed utilizing 4% isoflurane, and the bones were aseptically extracted from the hind limbs. The external soft tissues were discarded, and the extracted bone samples were immersed in 50 mL of Eagle's minimal essential medium (EMEM; Wako) supplemented with 20% fetal bovine serum (lot number 1,412,447; Invitrogen, Life Technologies Corp., Carlsbad, CA, USA) and penicillin (850 U/mL) for approximately 15 min.

The proximal end of the femur and the distal end of the tibia were cut. An 18-gauge needle (TERMO, Japan) was introduced into the opening at the knee-joint end of each bone, and the marrow was washed out of the bone shaft by EMEM. The obtained marrow pellet was separated by trituration, and the cell suspensions obtained from all the bones were combined by centrifugation. RBM cells were cultured in 75 cm^2 culture flasks (TD75; Falcon) in EMEM. At confluence, the cells were removed by trypsinization, washed twice in EMEM, resuspended in culture medium, and seeded on test and control titanium disks at a concentration of 4×10^4 cells/cm^2 in 24-well tissue culture plates. The cells were incubated for 3 days in a CO_2 incubator at 37°C. This investigation was conducted in accordance with the Guidelines for Animal Experimentation of Osaka Dental University (Approval no. 16-08001).

2.7. QCM Measurements. The amounts of proteins (BSA and HFN) and RBM cells were determined by QCM measurements (Affinix QN μ; Initium Co., Ltd., Tokyo, Japan). Affinix QN μ had a 550 μL cell outfitted with a 27 MHz QCM plate at the base of the cell. The diameter of the quartz plate was 8 mm, and the area of the gold-plated quartz was 4.9 mm^2. The unit also included a mixing bar and a temperature controller. The adjustment in recurrence was checked utilizing a universal frequency counter connected to a microcomputer.

The Ti QCM sensors and TNS were immersed in 500 μL of PBS (0.01 M PBS at pH 7.4). Changes in the QCM frequency were measured as a function of time; recording started immediately after the infusion of 5 μL (20 μg/mL) of BSA, HFN, and RBM cells. The solution was mixed to avoid any influence of protein dispersion on the measured results. Stirring did not influence the soundness of the frequency or the degree of frequency adjustment. The frequency change relied upon the adsorbed mass in accordance with the Sauerbrey equation:

$$\Delta F = -\frac{2F_0^2 \Delta m}{A\sqrt{\rho_q \mu_q}}. \quad (1)$$

As per this condition, at 27 MHz, a frequency shift of 1 Hz relates to a mass difference in roughly 0.62 ng·cm^{-2}. In the Sauerbrey equation, F_0 is the fundamental frequency of the quartz crystal (27×10^6 Hz), ΔF is the measured frequency shift (Hz), ρ_q is the density of quartz (2.65 g·cm^{-3}), Δm is the mass change (g), A is the electrode area (0.049 cm^2), and μ_q is the shear modulus of quartz (2.95×1011 dyn·cm^{-2}). QCM observation was performed at 25°C, and the test was repeated four times. Results are depicted as mean ± standard deviation.

2.8. XPS Analysis after Measuring the Adsorption of BSA and HFN Using Ti and TNS QCM Sensors. The biochemical constituents of the adsorbed protein films on the QCM sensors were investigated by XPS on an AXIS Ultra DLD spectrometer (Kratos Instruments, Manchester, UK) equipped with a monochromated Al Kα X-ray source (hv = 1486.6 eV) operated at 75 W. XPS was utilized to investigate the proteinaceous carbon (C1s) and nitrogen (N1s) signals produced by the protein. Evaluation of the C1s and N1s signals emerging from the peptide bonds of the protein was conducted to determine the relative amount of protein adsorbed on various surfaces.

2.9. Preparation of SBF. Since SBF is supersaturated with respect to apatite, improper planning can prompt the precipitation of apatite in the solution. Therefore, the solution should remain colorless and transparent, and there should not be any deposition on the inner surface of the container. If precipitation is observed, the preparation of SBF should be halted, and the procedure should be restarted at the step of washing the apparatus.

For preparation of 1 L of SBF, 700 mL of ion-exchanged and distilled water was added to a 1 L plastic beaker with a stir bar. The beaker was covered with plastic wrap, and the water was heated to 36.5 ± 1.5°C under stirring. Reagents 1–8 were then dissolved into the solution in order (given in Table 1) at 36.5 ± 1.5°C, and reagents 9 (Tris) and 10 (HCl) were added after pH adjustment.

Note that during this procedure, instead of glass containers, plastic containers with smooth, unscratched surfaces were used because apatite nucleation can be induced on the surfaces of glass containers or the edges of scratches. Additionally, reagents were dissolved completely before addition of the next reagent. The volume of 1 M HCl was measured using a cylinder after the cylinder had been washed with 1 M HCl. Finally, hygroscopic reagents, such as KCl, $K_2HPO_4 \cdot 3H_2O$, $MgCl_2 \cdot 6H_2O$, $CaCl_2$, and Na_2SO_4, were measured as quickly as possible.

The temperature of the solution was set to 36.5 ± 1.5°C. If the amount of solution was less than 900 mL, ion-exchanged and distilled water was added to increase the volume to 900 mL. The pH of the solution was then determined. Before Tris was added to the solution and dissolved, the pH of the solution was 2.0 ± 1.0. The solution temperature was kept within the range of 35–38°C (optimal: 36.5 ± 0.5°C), and the added Tris was dissolved slowly, while changes in pH were noted. Tris was added until the pH reached approximately 7.45.

3. Results

3.1. Scanning Electron Microscopy (SEM) and SPM Analysis. SEM images are shown in Figure 1. After modification in NaOH at 30°C, the TNS sensor surfaces showed a nanoscale network structure. SPM was utilized to gauge the depth of the surface characteristics of the specimens, and the surface morphologies of the TNS and Ti sensors are shown in Figure 2. Many nanonodules were detected on the TNS surface; these formations had horizontal dimensions of approximately 300 nm. The surface roughness (Ra) values were 3.9 and 18.9 nm for the Ti and TNS sensors, respectively.

3.2. XPS Analysis. Figure 3 shows the results of wide-scan XPS surface chemical analyses of Ti and TNS QCM sensors. The presence of Ti, O, C, and N was confirmed on the surfaces of Ti and TNS QCM sensors. In addition, the presence of Na was confirmed on the surface of the TNS QCM sensor. Moreover, the Ti and O concentrations of TNS QCM sensors were higher than those on the surface of TNS QCM sensors.

3.3. Surface Wettability. The contact angle for a water drop on the Ti sensor was 42.5°. However, the water drops spread very rapidly as they reached the test sensor, and the contact angle could not be determined. However, the behavior on

TABLE 1: Order, amounts, weighing containers, purities, and formula weights of reagents for the preparation of 1 L of SBF.

Order	Reagent	Amount	Container	Purity (%)	Formula weight
1	NaCl	8.035 g	Weighing paper	99.5	58.4430
2	$NaHCO_3$	0.355 g	Weighing paper	99.5	84.0068
3	KCl	0.225 g	Weighing bottle	99.5	74.5515
4	$K_2HPO_4{\cdot}3H_2O$	0.231 g	Weighing bottle	99.0	228.2220
5	$MgCl_2{\cdot}6H_2O$	0.311 g	Weighing bottle	98.0	203.3034
6	1.0 M HCl	39 mL	Graduated cylinder	—	—
7	$CaCl_2$	0.292 g	Weighing bottle	95.0	110.9848
8	Na_2SO_4	0.072 g	Weighing bottle	99.0	142.0428
9	Tris	6.118 g	Weighing paper	99.0	121.1356
10	1.0 M HCl	0–5 mL	Syringe	—	—

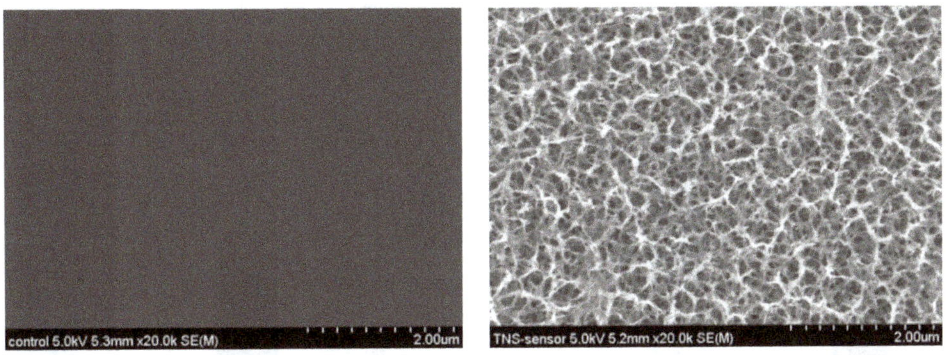

(a) (b)

FIGURE 1: SEM images of (a) Ti and (b) TNS QCM sensors.

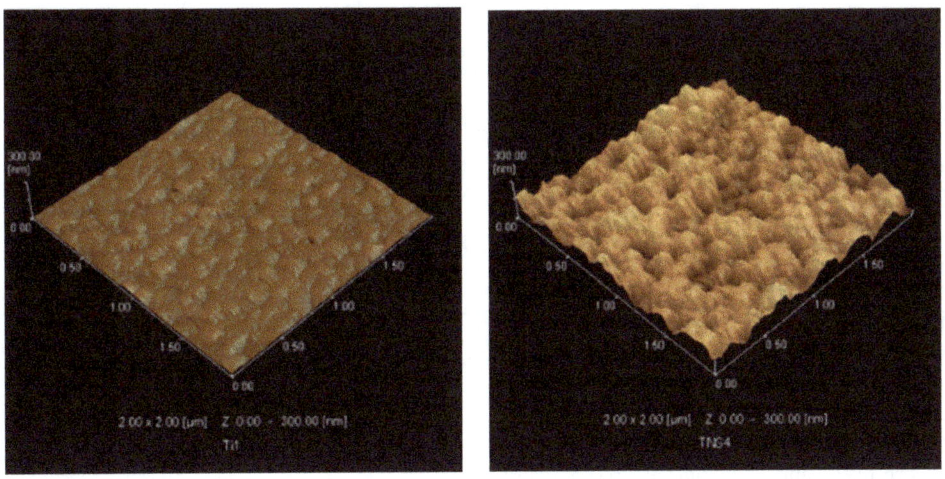

(a) (b)

FIGURE 2: SPM images of (a) Ti and (b) TNS QCM sensors.

the TNS sensor indicated that the sensor was superhydrophilic (contact angle of less than 5°).

3.4. QCM Measurement of Proteins. Figure 4 shows the adsorption of albumin and fibronectin based on QCM measurements. An immediate decrease in frequency was observed after the injection of albumin and fibronectin. This decrease in frequency was identified with the adsorption of albumin and fibronectin. The adsorption of albumin and fibronectin on the TNS sensor produced a decrease in frequency that was greater than that measured for the Ti sensor. According to Sauerbrey's equation, at 27 MHz, a frequency decrease of 1 Hz compares to a mass difference of around 0.62 ng/cm^2 [10]. After 30 min, the amounts of albumin adsorbed on the TNS and Ti sensors were 187.8 and

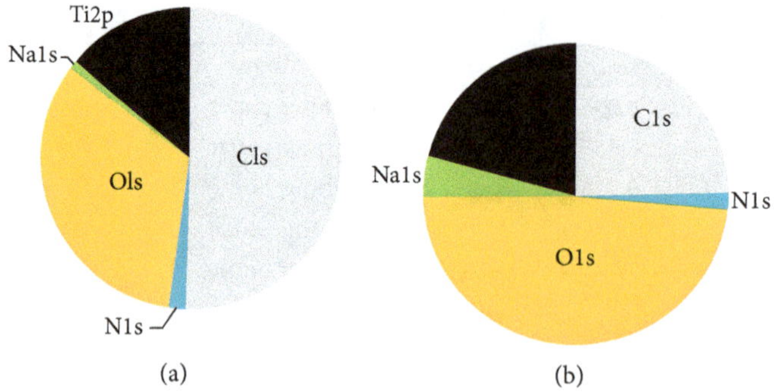

FIGURE 3: XPS analysis of (a) Ti and (b) TNS QCM sensors.

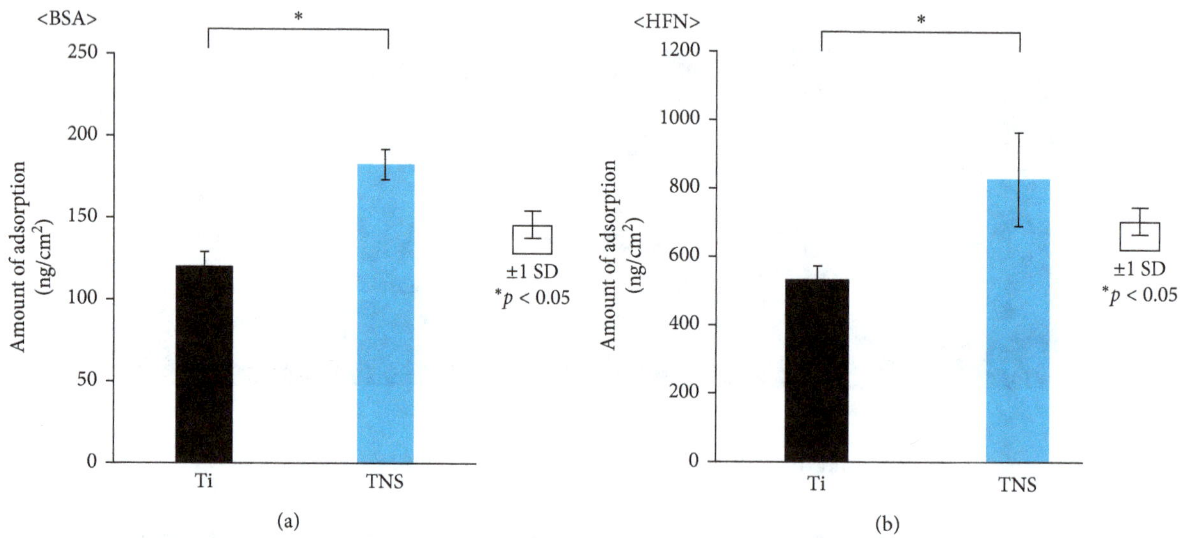

FIGURE 4: Adsorption of two proteins on Ti and TNS QCM sensors. (a) BSA. (b) HFN.

117.5 ng/cm^2, respectively, and the amounts of fibronectin adsorbed on the TNS and Ti sensors were 827.6 and 533.7 ng/cm^2, respectively.

3.5. QCM Measurement of RBM Cells. Figure 5 shows the adsorption of rat bone marrow cells based on QCM measurements. An immediate decrease in frequency was observed after the injection of RBM cells. The adsorption of RBM cells on the TNS sensor produced a decrease in frequency that was greater than that measured for the Ti sensor. Using Sauerbrey's equation, after 30 min, the amounts of albumin adsorbed on the TNS and Ti sensors were 1491.1 and 872.5 ng/cm^2, respectively.

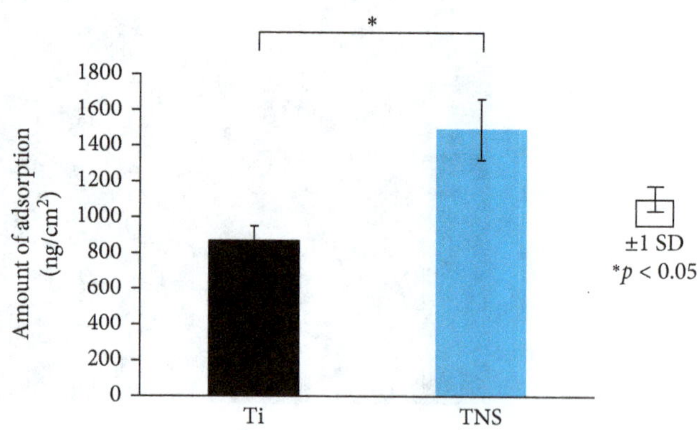

FIGURE 5: Adsorption of rat bone marrow cells on Ti and TNS QCM sensors.

3.6. XPS Analysis after Adsorption of Proteins. The XPS spectra of the Ti and TNS QCM sensors after immersion in BSA are summarized in Figures 6 and 7, and those of the Ti and TNS QCM sensors after immersion in HFN are shown in Figures 8 and 9. For XPS analysis, the N1s and C1s spectra were measured. The coupling energies (BEs) of the C1s range for adsorbed BSA were 284.8 eV (C-C/C-H), 286.4 eV (C-O/C-N), and 288.2 eV (O=C-O) for the Ti QCM sensors and 284.8 eV (C-C/C-H), 286.4 eV (C-O/C-N), and 288.6 eV (O=C-O) for the TNS QCM sensors. There were N1s peaks for the Ti and TNS QCM sensors after the adsorption of BSA. The BEs of the C1s spectrum for adsorbed HFN were 284.8 eV (C-C/C-H), 286.4 eV (C-O/C-N), and 288.0 eV (O=C-O) in the Ti QCM sensors and 284.7 eV (C-C/C-H), 286.4 eV (C-O/C-N), and 288.0 eV (O=C-O) for the TNS QCM sensors. After adsorption of the HFN, there were also N1s center-level spectra for the Ti and TNS QCM sensors.

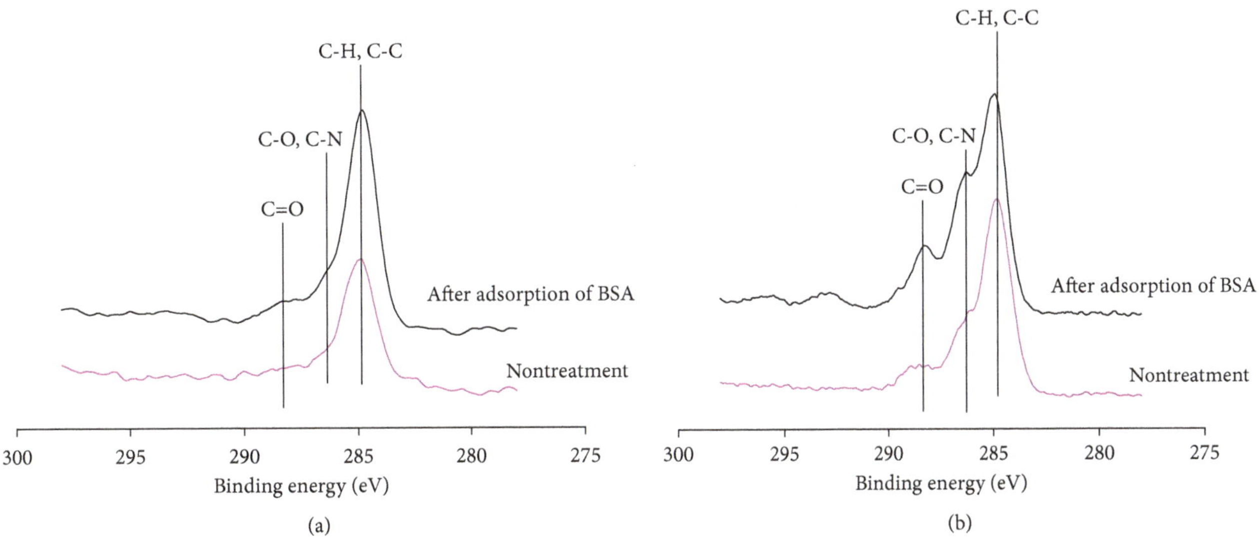

FIGURE 6: C1s XPS spectra of (a) Ti and (b) TNS QCM sensors after immersion in BSA.

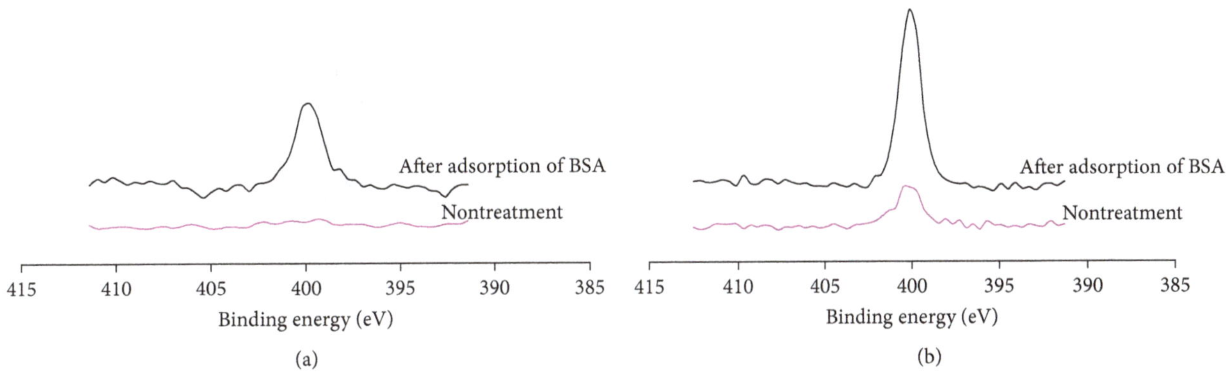

FIGURE 7: N1s XPS spectra of (a) Ti and (b) TNS QCM sensors after immersion in BSA.

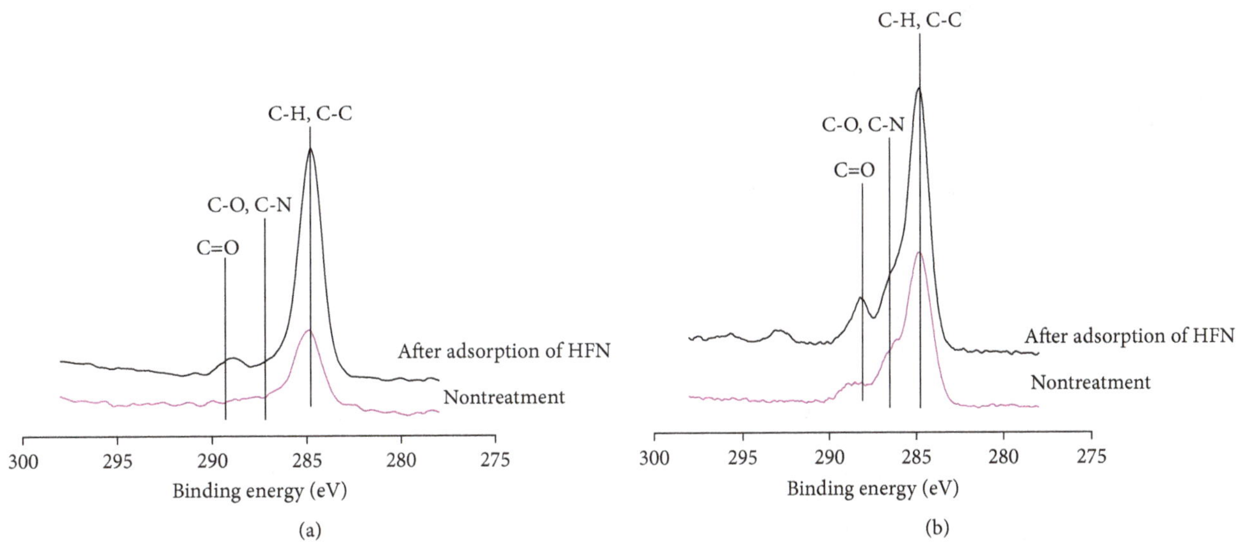

FIGURE 8: C1s XPS spectra of (a) Ti and (b) TNS QCM sensors after immersion in HFN.

3.7. Characterization after Immersion in SBF. Figure 10 shows the adsorption of apatite after exposure to the SBF solution based on QCM measurements. A quick frequency reduction was seen after the infusion of SBF. The adsorption of SBF on the TNS sensor produced a decrease in the frequency that was greater than that measured for the Ti sensor.

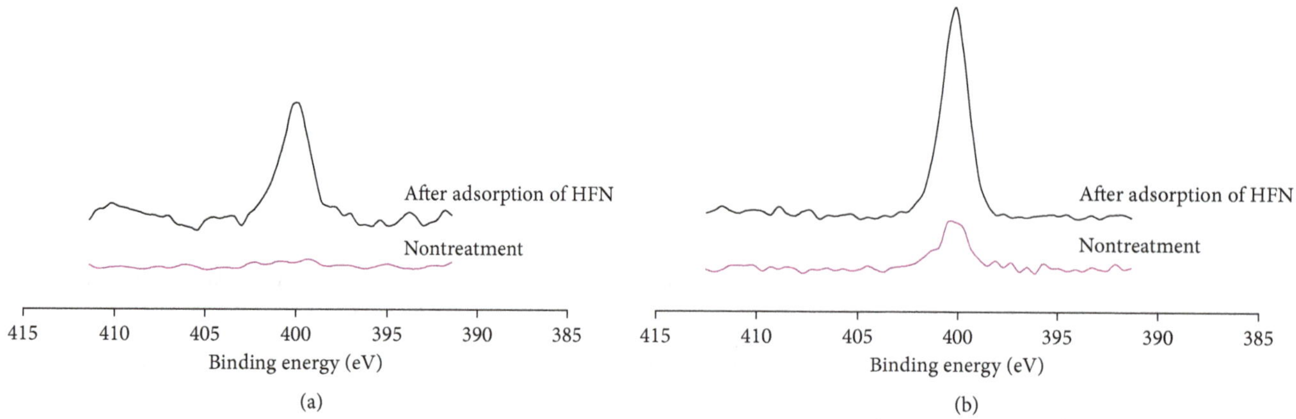

FIGURE 9: N1s XPS spectra of (a) Ti and (b) TNS QCM sensors after immersion in HFN.

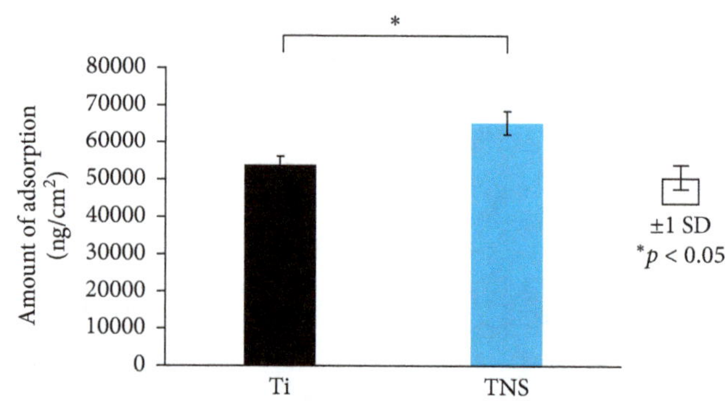

FIGURE 10: Adsorption of apatite on Ti and TNS QCM sensors.

Figure 11 shows the surface morphology of the Ti and TNS QCM sensors after immersion in SBF for 24 h. After immersion, a recently framed layer was seen on the Ti and TNS surfaces. The surface was covered by single and grouped ball-like particles with a size of around 1.5 μm. The number of particles on the surface of the TNS QCM sensor was clearly greater than that on the surface of the Ti QCM sensor.

4. Discussion

In this study, Ti sensors were realized by depositing a thin Ti film on a QCM electrode using a reactive DC magnetron sputtering technique; this sensor was further alkali modified using NaOH treatment to fabricate the TNS. SPM and XPS analyses were carried out to characterize the nanostructure of the TNSs. We evaluated the quality of the TNSs and the potential application of the high-frequency TNS sensors by investigating the in situ binding behaviors for TNS sensors, two proteins, RBM cells, and SBF to determine the effects of these parameters on biological reactions in solution.

Several studies have demonstrated that implant surfaces affect nanoscale topography and thereby alter cell behaviors or change the nanofeatures of structures to improve the osseointegration process [4, 11, 12]. The embed surface can be adjusted by various approaches to add nanoscale features to the surfaces in specific combinations. The most well-known techniques are chemical processes, such as alkaline hydrothermal [13, 14] or acid [15, 16] oxidation, on titanium surfaces to produce diverse nanoscale topographies. Kasuga et al. [5] demonstrated that TiO_2 nanotubes with a diameter of about 8 nm and a length of about 100 nm could be formed by Ti treatment with a 10 M NaOH aqueous solution for 20 h at 110°C, without the need for templates or replication. Essential factors in regulating cell reactions at the implant-tissue interface can dramatically affect tissue coordination [17]. In a recent work, we demonstrated that TiO_2 nanotubes and TNSs could be formed on titanium metal surfaces by treatment with a 10 M NaOH aqueous solution at 30°C, and we used this method to prepare TNS-modified disks. Komasa et al. [8] suggested that TNSs on titanium surfaces can be applied to control the osteogenic differentiation of bone marrow cells and enhance mineralization. Our results demonstrated that TNS-modified titanium disks were more hydrophilic and showed uniquely enhanced wettability in comparison with unmodified disks. Further studies of the surface roughness and topography of modified titanium alloy surfaces are needed to assess their wettability. Ra is a commonly used height parameter to describe implant surface roughness; Ra of the TNS-modified titanium surface was 19 nm, which was greater than that of the untreated titanium surface. The contact angles of the alkali-treated titanium disks gradually decreased in comparison with those of the control group, indicating that the wettability of the surface of the test group was reduced by NaOH treatment. In previous studies, a surface roughness of between 13 and 16 nm was found to be optimal for RBM cell culture [18, 19]. The nanonetwork structure framed on the titanium disks here was like the hierarchical structure outlined by Zhao et al. [12]. In their work, hierarchical nanotextured titanium surface topographies with TNS structures that mirrored the hierarchical structures of bone tissues were created by etching followed by anodization. Natural tissues are hierarchical structures of nanoscale building blocks organized in a structured way. Hierarchical structures composed of nanocomponents may give a more reasonable surface topography for bone marrow cell functions than simpler structures because they can better copy the structures of natural tissues. Our research revealed that NaOH treatment prompts the development of a Ti-O-Na titanate layer on the titanium surface. Thus, we expect that NaOH treatment

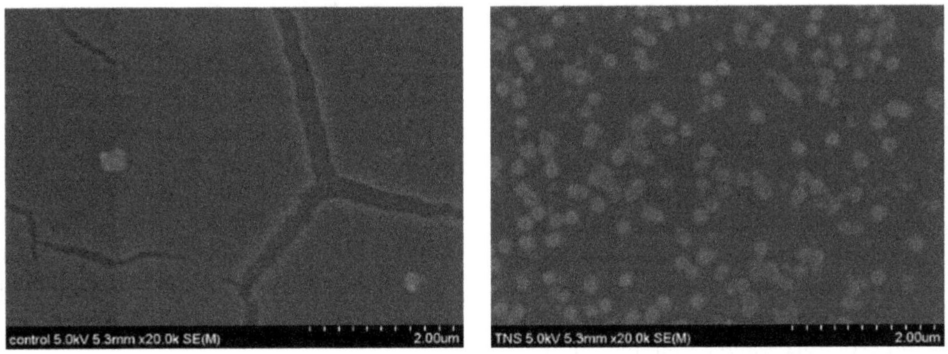

(a) (b)

FIGURE 11: SEM images of exposure to SBF solution on (a) Ti and (b) TNS QCM sensors.

results in the formation of a thick oxide film on the TiO_2 layer on the titanium surface. The deconvolution procedure suggested that this may have been due to surface contamination resulting from the binding of O to C [20].

Before argon ion etching was performed, the XPS depth profiles perpendicular to the surface of the sample showed a carbon contamination overlayer on the surface. This carbon contaminant was removed after the initial etching cycles. The oxygen and Ti concentrations gradually decreased and increased, respectively, as the number of etching cycles increased. Kim et al. [21] demonstrated that NaOH and heat treatment resulted in the covering of the Ti substrate with a titanium oxide layer with a thickness of approximately 1000 nm; AES depth profiles of the Ti oxide layer indicated the presence of an amorphous sodium titanate hydrogel layer. Kasuga et al. [5] likewise showed that NaOH treatment prompted the formation of a Ti-O-Na titanate layer on the Ti surface. At a greater depth of approximately 250 nm, the oxygen concentrations remained almost constant at 5 atom%. In addition, the remaining oxygen content was approximately 5 atom% inside the titanium film, which was attributed to the presence of residual oxygen in the chamber after preparation of the Ti sample.

All embed surfaces are promptly covered with a layer of protein from the in vitro culture medium or in vivo biological fluids, and this interface regulates the course of cell reactions and behaviors [22]. To elucidate the connection between the Ti implant surface properties, opsonization, and phagocytosis under in vivo conditions, phagocytic experiments were previously conducted using a cell culture medium supplemented with serum albumin and human opsonizing serum factors [23]. Fibronectin plays a crucial role in the progressive differentiation of osteoblasts [24]. Additionally, fibronectin has RGD sequences and is a large extracellular matrix dimer glycoprotein [25] with a molecular weight of approximately 440 kDa [26]. Albumin is the most abundant plasma protein; it suppresses the adsorption of other proteins that may empower aggravation and bacterial colonization [27]. The molecular weight of BSA is approximately 6-7 kDa [28]. In this study, the amount of fibronectin adsorbed was greater than the amount of albumin adsorbed, suggesting that adsorption quantity was related to the molecular weight of the injection solutions. After adsorption of BSA and BSF, N1s core-level spectra were obtained for both QCM sensors. Endo [29] detected C, O, and N obtained from the organic material or metal oxide on the titanium surface by XPS investigation. In the present study, based on XPS, the C1s and N1s peaks derived from BSA and BSF were detected on both QCM sensor surfaces tested. Evaluation of the N1s peak emerging from the peptide bonds of the implant-bone bonding protein may be a successful measure of the relative amount of protein adsorbed onto implant surfaces on the TNS QCM sensor.

Our results are the first comparison of RBM cell proliferation on TNS-modified titanium surfaces and unprocessed controls. Surfaces with nanostructures have higher surface areas than those of surfaces without such features [11, 30]. This expanded surface territory permits increased adhesion of cells, such as osteoblasts and fibroblasts. Thus, the altered surface energies of materials with nanostructures may promote tissue growth via increased adsorption of select proteins compared with materials with microscale features [31]. The adsorption of select proteins can in this manner guide the adhesion of cells on the implant material surface, among other capabilities. Various investigations have exhibited improved cell adhesion and multiplication on nanostructured surfaces, with various potential tissue applications, including applications in the bladder, bone, vasculature, and nervous system [32–34]. A prior study suggested that TNSs on titanium surfaces facilitate the regulation of osteoblastic differentiation of bone marrow cells and enhance mineralization. The current study showed that the TNSs formed nanonodules with a diameter of about 19 nm on the titanium surfaces, and these structures promoted the adhesion and/or multiplication of cells. The network structure of TNSs on titanium alloy facilitated rapid cell adhesion, spreading, and multiplication due to the mechanics of the TNS structure and chemical nature of the Ti-O-Na layer.

To enhance bone-titanium bonding, Kokubo et al. recently demonstrated that a blend of alkalis resulted in the formation of bone-like apatite on the surface of titanium in SBF with an ion concentration almost equivalent to that of human blood [35]. Apatite development on the material surface is accepted to be essential for bioactivity, that is, direct bone bonding. In our study, there was an apatite layer on the TNS QCM sensor after 24 h. Thus, the increased adhesion of RBM cells and SBF on the TNS QCM sensor suggested that TNS induced bone differentiation.

In the TNS sensor, the concentration of Ti exceeded that of oxygen at a depth of approximately 170 nm. Subsequently, the NaOH treatment was thought to produce a thick oxide film. Nonetheless, the chemical structure of the treated surfaces did not vary fundamentally; Ti oxides (mostly TiO_2) formed at the surface. Albumin and fibronectin were readily adsorbed, to a greater extent on the TNS sensor than on the reference Ti sensor. The TNS surface seemed to adsorb more protein for a given geometric surface area than the reference Ti sensor; this may have been a result of its rough morphology, as evidenced by the SEM and SPM results. Several studies [36, 37] have shown that nanostructured topographies can act as good mimics of natural extracellular matrixes. Advancement of the surface topography could be indirect; the adsorption of proteins or ions may function as an extension between the nanosurface structure and cells [38]. Webster et al. [39] observed increases in the adsorption of vitronectin on nanostructured surfaces (compared with conventional surfaces), which resulted in preferable adhesion of osteoblasts. In addition, here, the protein adsorption rate on the TNSs was correlated with the contact angle, suggesting that the hydrophilicity of titanium greatly affected its protein adsorption ability. The contact angle of the TNS and Ti sensors fluctuated in the hydrophilic range; the hydrophilicity of the surfaces expanded after treatment, corresponding to the formation of the TNSs. Thus, the growth of the titanium oxide layer increased the surface energy, resulting in a more hydrophilic surface [40].

5. Conclusions

In conclusion, TNS structures were obtained on a titanium surface via treatment with a NaOH aqueous solution at room temperature. Nanoscale network structures and a large number of nanoscale nodules were observed by SEM and SPM. In addition, the chemical composition of the TNS structures was estimated by XPS. The results confirmed the presence of a combined titanium and oxide titanate layer, which induced the adsorption of albumin and fibronectin. In the fields of tissue engineering and biomaterials, nanostructuring technologies are expected to yield novel, biologically optimized surfaces.

Conflicts of Interest

The authors declare that they have no conflicts of interest regarding the publication of this article.

Authors' Contributions

Satoshi Komasa conceived and designed the experiments; Yuichiro Tashiro performed the experiments; Yuichiro Tashiro, Satoshi Komasa, and Akiko Miyake analyzed the data; Hiroshi Nishizaki and Joji Okazaki contributed reagents, materials, and analysis tools; Yuichiro Tashiro and Satoshi Komasa wrote the paper.

Acknowledgments

The authors wish to express their thanks to Tohru Sekino from Osaka University for setting up the nanosheets and for his supportive recommendations. The authors are also grateful to the members of the Department of Removable Prosthodontics and Occlusion and the Department of Oral Health Engineering for their advice and assistance. The authors also thank Toshio Tamaki and Hirokazu Hojyo. This work was supported by grants from the Japan Society for the Promotion of Science (16K20524).

References

[1] M. Annunziata, A. Oliva, A. Buosciolo, M. Giordano, A. Guida, and L. Guida, "Bone marrow mesenchymal stem cell response to nano-structured oxidized and turned titanium surfaces," *Clinical Oral Implants Research*, vol. 23, no. 6, pp. 733–740, 2012.

[2] G. Mendonca, D. Mendonca, F. J. L. Aragao, and L. F. Cooper, "Advancing dental implant surface technology–From micron- to nanotopography," *Biomaterials*, vol. 29, no. 28, pp. 3822–3835, 2008.

[3] L. Meirelles, F. Currie, M. Jacobsson, T. Albrektsson, and A. Wennerberg, "The effect of chemical and nanotopographical modifications on the early stages of osseointegration," *International Journal of Oral and Maxillofacial Implants*, vol. 23, no. 4, p. 641, 2008.

[4] K. Kubo, N. Tsukimura, F. Iwasa et al., "Cellular behavior on TiO_2 nanonodular structures in a micro-to-nanoscale hierarchy model," *Biomaterials*, vol. 30, no. 29, pp. 5319–5329, 2009.

[5] T. Kasuga, M. Hiramatsu, A. Hoson, T. Sekino, and K. Niihara, "Titania nanotubes prepared by chemical processing," *Advanced Materials*, vol. 11, no. 15, pp. 1307–1311, 1999.

[6] X. Gao, H. Zhu, G. Pan et al., "Preparation and electrochemical characterization of anatase nanorods for lithium-inserting electrode material," *Journal of Physical Chemistry B*, vol. 108, no. 9, pp. 2868–2872, 2004.

[7] A. R. Armstrong, G. Armstrong, J. Canales, and P. G. Bruce, "TiO_2 nanowires," *Angewandte Chemie International*, vol. 43, no. 17, pp. 2286–2288, 2004.

[8] S. Komasa, Y. Taguchi, H. Nishida, M. Tanaka, and T. Kawazoe, "Bioactivity of nanostructure on titanium surface modified by chemical processing at room temperature," *Journal of Prosthodontic Research*, vol. 56, no. 3, pp. 170–177, 2012.

[9] Y. Hashimoto, S. Minoura, A. Nishiura et al., "Development of titanium quartz crystal microbalance sensor by magnetron sputtering," *Journal of Oral Tissue Engineering*, vol. 8, pp. 52–59, 2010.

[10] A. Miyake, S. Komasa, Y. Hashimoto, Y. Komasa, and J. Okazaki, "Adsorption of saliva related protein on denture materials: an X-ray photoelectron spectroscopy and quartz crystal microbalance study," *Advances in Materials Science and Engineering*, vol. 2016, Article ID 5478326, 9 pages, 2016.

[11] G. Mendonca, D. B. Mendonca, F. J. Aragao, and L. F. Cooper, "The combination of micron and nanotopography by H(2)SO(4)/H(2)O(2) treatment and its effects on osteoblast-specific gene expression of hMSCs," *Journal of Biomedical Materials Research A*, vol. 94, no. 1, pp. 169–179, 2010.

[12] L. Zhao, S. Mei, P. K. Chu, Y. Zhang, and Z. Wu, "The influence of hierarchical hybrid micro/nano-textured titanium surface with titania nanotubes on osteoblast functions," *Biomaterials*, vol. 31, no. 19, pp. 5072–5082, 2010.

[13] W. Zhang, Z. Li, Y. Liu et al., "Biofunctionalization of a titanium surface with a nano-sawtooth structure regulates the behavior of rat bone marrow mesenchymal stem cells," *International Journal of Nanomedicine*, vol. 7, pp. 4459–4472, 2012.

[14] W. Dong, T. Zhang, J. Epstein et al., "Multifunctional nanowire bioscaffolds on titanium," *Chemistry of Materials*, vol. 19, no. 18, pp. 4454–4459, 2007.

[15] S. Bauer, J. Park, K. von der Mark, and P. Schmuki, "Improved attachment of mesenchymal stem cells on super-hydrophobic TiO_2 nanotubes," *Acta Biomaterials*, vol. 4, no. 5, pp. 1576–1582, 2008.

[16] E. Hosono, H. Matsuda, I. Honma, M. Ichihara, and H. Zhou, "Synthesis of a perpendicular TiO_2 nanosheet film with the superhydrophilic property without UV irradiation," *Langmuir*, vol. 23, no. 14, pp. 7447–7450, 2007.

[17] T. Albrektsson, P. I. Brånemark, H. A. Hansson, and J. Lindström, "Osseointegrated titanium implants: requirements for ensuring a long-lasting, direct bone-to-implant anchorage in man," *Acta Orthopaedica*, vol. 52, no. 2, pp. 155–170, 1981.

[18] H. Xing, S. Komasa, Y. Taguchi, T. Sekino, and J. Okazaki, "Osteogenic activity of titanium surfaces with nanonetwork structures," *International Journal of Nanomedicine*, vol. 9, p. 1741, 2014.

[19] T. Fujino, Y. Taguchi, S. Komasa, T. Sekino, and M. Tanaka, "Cell differentiation on nanoscale features of a titanium surface: effects of deposition time in NaOH solution," *Journal of Hard Tissue Biology*, vol. 23, no. 1, pp. 63–70, 2014.

[20] M. Pisarek, A. Roguska, M. Andrzejczuk et al., "Effect of two-step functionalization of Ti by chemical processes on protein adsorption," *Applied Surface Science*, vol. 257, no. 19, pp. 8196–8204, 2011.

[21] H. M. Kim, F. Miyaji, T. Kokubo, S. Nishiguchi, and T. Nakamura, "Graded surface structure of bioactive titanium prepared by chemical treatment," *Journal of Biomedical Materials Research*, vol. 45, pp. 100–107, 1999.

[22] M. Roser, D. Fischer, and T. Kissel, "Surface-modified biodegradable albumin nano-and microspheres. II: effect of surface charges on in vitro phagocytosis and biodistribution in rats," *European Journal of Pharmaceutics and Biopharmaceutics*, vol. 46, no. 3, pp. 255–263, 1998.

[23] P. Roach, D. Farrar, and C. C. Perry, "Interpretation of protein adsorption: surface-induced conformational changes," *Journal of the American Chemical Society*, vol. 127, no. 22, pp. 8168–8173, 2005.

[24] A. M. Moursi, C. H. Damsky, J. Lull et al., "Fibronectin regulates calvarial osteoblast differentiation," *Journal of Cell Science*, vol. 109, pp. 1369–1380, 1996.

[25] J. P. Quigley, L. I. Gold, R. Schwimmer, and L. M. Sullivan, "Limited cleavage of cellular fibronectin by plasminogen activator purified from transformed cells," *Proceedings of the National Academy of Sciences U. S. A.*, vol. 84, no. 9, pp. 2776–2780, 1987.

[26] D. Khang, S. Y. Kim, P. Liu-Snyder, G. T. R. Palmore, S. M. Durbin, and T. J. Webster, "Enhanced fibronectin adsorption on carbon nanotube/poly (carbonate) urethane: independent role of surface nano-roughness and associated surface energy," *Biomaterials*, vol. 28, no. 32, pp. 4756–4768, 2007.

[27] C. McFarland, C. De Filippis, M. Jenkins et al., "Albumin-binding surfaces: in vitro activity," *Journal of Biomaterials Science Polymer Edition*, vol. 9, no. 11, pp. 1227–1239, 1998.

[28] A. Amaral, N. Alvarado, I. Marigomez, R. Cunha, K. Hylland, and M. Soto, "Autometallography and metallothionein immunohistochemistry in hepatocytes of turbot (*Scophthalmus maximus* L.) after exposure to cadmium and depuration treatment," *Biomarkers*, vol. 7, no. 6, pp. 491–500, 2002.

[29] K. Endo, Y. Araki, H. Ohno, and K. Matsuda, "ESCA analysis of tarnish films on dental alloys removed from the oral cavities (Part 1) Ag-In alloys," *Journal of Dental Materials*, vol. 7, pp. 184–191, 1988.

[30] K. Anselme, "Osteoblast adhesion on biomaterials," *Biomaterials*, vol. 21, no. 7, pp. 667–681, 2000.

[31] G. Balasundaram and T. J. Webster, "An overview of nano-polymers for orthopedic applications," *Macromolecular Bioscience*, vol. 7, no. 5, pp. 635–642, 2007.

[32] R. L. Price, K. Ellison, K. M. Haberstroh, and T. J. Webster, "Nanometer surface roughness increases select osteoblast adhesion on carbon nanofiber compacts," *Journal of Biomedical Materials Research Part A*, vol. 70, no. 1, pp. 129–138, 2004.

[33] R. Langer and D. A. Tirrell, "Designing materials for biology and medicine," *Nature*, vol. 428, no. 6982, pp. 487–492, 2004.

[34] L. Zhang and T. J. Webster, "Nanotechnology and nanomaterials: promises for improved tissue regeneration," *Nano Today*, vol. 4, no. 1, pp. 66–80, 2009.

[35] T. Kokubo, F. Miyaji, H. M. Kim, and T. Nakamura, "Spontaneous apatite formation on chemically surface treated Ti," *Journal of the American Ceramic Society*, vol. 79, no. 4, pp. 1127–1129, 1996.

[36] K. Woo, G. Wei, and P. Ma, "Enhancement of fibronectin-and vitronectin-adsorption to polymer/hydroxyapatite scaffolds suppresses the apoptosis of osteoblasts," *Journal of Bone Mineral Research*, vol. 17, p. 49, 2002.

[37] K. Woo, R. Zhang, H. Deng, and P. Ma, "Protein-mediated osteoblast survival and migration on biodegradable polymer/hydroxyapatite composite scaffolds," in *Proceedings of Transactions of the 28th Annual Meeting of the Society for Biomaterials*, Tampa, FL, USA, April 2002.

[38] V. Bucci-Sabattini, C. Cassinelli, P. G Coelho, A. Minnici, A. Trani, and D. M. D. Ehrenfest, "Effect of titanium implant surface nanoroughness and calcium phosphate low impregnation on bone cell activity in vitro," *Oral Surgery, Oral Medicine, Oral Pathology, Oral Radiology, Endodontology*, vol. 109, no. 2, pp. 217–224, 2010.

[39] T. J. Webster, C. Ergun, R. H. Doremus, R. W Siegel, and R. Bizios, "Specific proteins mediate enhanced osteoblast adhesion on nanophase ceramics," *Journal of Biomedical Materials Research*, vol. 51, no. 3, pp. 475–483, 2000.

[40] Y. Shibata, D. Suzuki, S. Omori et al., "The characteristics of in vitro biological activity of titanium surfaces anodically oxidized in chloride solutions," *Biomaterials*, vol. 31, no. 33, pp. 8546–8555, 2010.

18

Simulation Calculation and Analysis of Electronic Structure and Electrical Properties of Metal-Doped SnO_2

Jing-Qin Wang, Hui-Ling Kang, and Ying Zhang

Province-Ministry Joint State Key Laboratory of Reliability and Intelligence of Electrical Equipment, Hebei University of Technology, Tianjin 300130, China

Correspondence should be addressed to Hui-Ling Kang; 15222159859@163.com

Academic Editor: Fernando Rubio-Marcos

SnO_2 is an almost insulated semiconductor material, which increases the contact resistance of the $AgSnO_2$ electrical contact material. Therefore, by improving the electrical performance of SnO_2, the electrical properties of the $AgSnO_2$ can be optimized. The first principle method based on density functional theory is used to calculate the electronic structure, formation energy, band structure, density of states, and differential charge density of SnO_2 doped with the metals Ti, Sr, Ge, Sb, and Ga. The results show that metal-doped SnO_2 materials are still direct bandgap semiconductor materials, and the effect of the electronic states of the metallic elements enhances the localization of the energy band, decreases the bandgap, increases the carrier concentration at the Fermi level, and enhances the electrical performance of the materials, and the bandgap of Ga-doped SnO_2 is the smallest, 0.041 eV. And the charge transfer between Sb, Sr, Ga, Ti, and Ge metal atoms and O atoms increases, especially between Ga atom and O atom; that is, the electrical performance of Ga doping is better.

1. Introduction

As a new type of pollution-free contact material, $AgSnO_2$ is the most promising to replace the toxic AgCdO, and it is widely used in low-voltage electrical appliances such as relay and circuit breaker because of its good resistance to welding and erosion [1]. And the material is a mixture of Ag and SnO_2, with Ag as the main component and SnO_2 as the auxiliary material to enhance the viscosity of the silver liquid and prevent the splash of Ag droplets. However, SnO_2 is a kind of the wide bandgap semiconductor material with a bandgap value of 3.6 eV [2, 3], which is the main reason for increasing the contact resistance of $AgSnO_2$ material; therefore, improving the electrical performance of SnO_2 has become an urgent problem to be solved. Studies have shown that the doping of metals, metal oxides, or rare earth elements in SnO_2 can enhance the electrical performance of SnO_2 [4, 5], thereby achieving the purpose of optimizing the electrical properties of $AgSnO_2$.

In recent years, scholars have done a lot of research on the doping modification of SnO_2. Shan Lin-ting et al. prepared Ce-Cu codoped SnO_2 nanopowders by the sol-gel method and studied the influence of doping on the electronic structure and photoelectric properties of SnO_2 by first principles; the results showed that the bandgap was reduced by Ce-Cu codoping, and the electrical properties of SnO_2 were improved [6]; Liu calculated the band structure and state density of Bi-, Cu-, Ni-, Co-, and Ti-doped SnO_2 by CASTEP software and showed that several doped elements can improve the electrical performance of SnO_2, and the best doping element is Bi [7]; Liu et al. prepared SnO_2-TiO_2 powder by the sol-gel method and prepared Ag-SnO_2-TiO_2 contact material by electroplating and showed that the conductivity of the Ag-SnO_2-TiO_2 contact material was 66.9% IACS and the density was 9.63 g/cm^3 which were higher than the national standard, indicating that TiO_2 doping improved the electrical performance of $AgSnO_2$ [8]; Wang et al. studied the electronic structure, band structure, and optical properties of Fe and Mn codoped SnO_2 by first principles and concluded that after the Fe and Mn codoping, the material exhibited semimetallic properties [9]. However, at present, the theoretical calculation and analysis of Sr-, Ti-, Sb-, Ge-, and Ga-doped SnO_2 are rarely reported.

In this research, a $2 \times 2 \times 2$ SnO_2 supercell model was built and the electronic structure, energy band structure, state density, differential charge density of Sr-, Ti-, Sb-, Ge-, and Ga-doped SnO_2 were calculated by the supersoft pseudopotential method in the first principle, and the best doping element to improve the electrical performance of SnO_2 was also analyzed. Then, the metal-doped SnO_2 powder materials were prepared by the sol-gel method, and the metal-doped Ag-SnO_2 contact materials were prepared by powder metallurgy. Finally, the conductivity of metal-doped Ag-SnO_2 contact materials was tested. The research provides theory and data support for further study, and a feasibility method for the study of $AgSnO_2$ contact material.

2. Crystal Cells and Calculation Method of Metal-Doped SnO_2 Materials

Figure 1 shows the crystal structures of intrinsic SnO_2. The intrinsic SnO_2 derived directly from the structure library of Materials Studio is a tetragonal rutile structure with the lattice constants of $a = b = 4.737$ Å, $c = 3.816$ Å, and $\alpha = \beta = \gamma = 90°$ [10–12], and it contains two Sn atoms and four O atoms. The two Sn atoms occupy the center and vertex of the tetrahedron. The four O atoms are located in the tetrahedron and the surface, respectively. In this research, a supercell model of $2 \times 2 \times 2$ SnO_2 was established, and in order to analyze the influence of Sb, Sr, Ga, Ti, and Ge doping on the electric performance of SnO_2, the $2 \times 2 \times 2$ doped models were established by using the Sb, Sr, Ga, Ti, and Ge atoms to replace a Sn atom so that the doping ratio is 6.25%, and this ratio has reached a relatively good doping ratio verified by the research group. Therefore, this doping ratio is directly adopted herein.

The electronic structure, formation energy, band structure, density of states, and differential charge densities of Sb-, Sr-, Ga-, Ti-, and Ge-doped SnO_2 are calculated by using the CASTEP module in Materials Studio 7.0 software based on the density functional theory. And the calculation process is divided into two parts. Firstly, the supersoft pseudopotential algorithm is used to optimize the models of Sb-, Sr-, Ga-, Ti-, and Ge-doped SnO_2 to obtain the structural parameters and the stable structure with the lowest energy. Then, the energy of the optimized structure is calculated by using the generalized gradient approximation (GGA), the interaction between the valence electron and the real ion is approximately described by the supersoft pseudopotential, and the band structure, density of states, and differential charge density are obtained.

In order to make Sb-, Sr-, Ga-, Ti-, and Ge-doped SnO_2 calculation results comparable, the parameters used for the calculation are set to be consistent. The calculation parameters are set as follows: the energy cutoff of the plane wave is selected as 340 eV, the self-consistent field (SCF) convergence rate is 2.0×10^{-6} eV/atom, the number of self-consistent convergence is 200 times, and the Brillouin region k-grid is selected as $3 \times 3 \times 4$, and the electron energy as 1.0×10^{-5} eV/atom. The calculation process is carried out in the reciprocal space [13, 14], and the valence-electron configurations are chosen as Sn:$5s^2 5p^2$, O:$2s^2 2p^4$, Sb: $5s^2 5p^3$, Sr:$4p^6 5s^2$, Ga:$3d^{10} 4s^2 4p^1$, Ti:$3p^6 3d^2 4s^2$, and Ge: $4s^2 4p^2$.

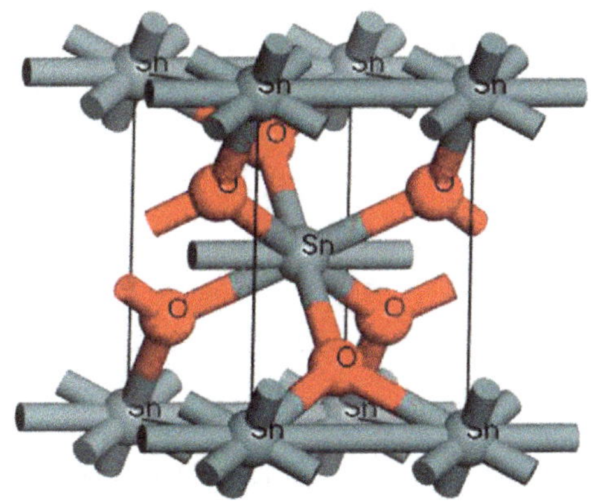

Figure 1: Intrinsic SnO_2 model.

3. Analysis of the Calculation Results of Metal-Doped SnO_2 Materials

3.1. Analysis of Crystal Structure and Formation Energy. Table 1 shows the crystal cell after the optimization of Sb-, Sr-, Ga-, Ti-, and Ge-doped SnO_2. It can be seen from the optimized data that after Sb, Sr, Ga, Ti, and Ge doping, the lattice parameters a and c and the volume V are bigger than that of intrinsic SnO_2, and this is because the ionic valence radii of the metals Sb, Sr, and Ti are both larger than that of the Sn ions [15, 16]. The common valence states and ionic radius of each metal are shown in Table 2. Although the ionic radii of the metals Ge and Ga are smaller than those of Sn ions, it can be seen from the change of the bond lengths between the atoms after doping the metal (x = Sb, Sr, Ga, Ti, and Ge), as shown in Figure 2(a), that the Ge-Sn bond and the Ga-Sn bond increase compared with the Sn-Sn bond, and the Sn-O bond is also increased compared with the Sn-O bond of pure SnO_2, which is the primary cause for the growth of lattice parameters and volumes.

The Sr ionic radius is larger than that of Sn, and the ratio of bond length between atoms after Sr doping, as shown in Figure 2(b), is larger; therefore, the cell volume expansion is larger after doping the Sr metal element, and the cell volume expansion is relatively smaller after doping the remaining metal elements. The calculation results conform to the theoretical basis, indicating that the optimization results are effective, and the structure and method used to calculate are reasonable.

The formation energy E_f can be used to judge the difficulty of metal doping and the structural stability of metal-doped SnO_2. And the larger the formation energy of metal doping is, the more difficult the metal doping is and the lower the stability is. E_f can be calculated by [17] the following equation:

$$E_f = E_{SnO_2:x} - E_{SnO_2} - E_x + E_{Sn}, \quad (x = \text{Sb, Sr, Ga, Ti, Ge}), \tag{1}$$

where $E_{SnO_2:x}$ is the total energy of the metal-doped SnO_2 system, E_{SnO_2} is the total energy of the pure SnO_2 supercell

TABLE 1: Optimized unit cell parameters.

SnO_2 model	Undoped	Sb doped	Sr doped	Ga doped	Ti doped	Ge doped
a (Å)	4.7373	4.9305	4.9733	4.9124	4.9082	4.8995
Growth ratio (%)	—	4.0783	4.9817	3.5644	3.6075	3.4239
c (Å)	3.1864	3.3053	3.3276	3.2844	3.2807	3.2780
Growth ratio (%)	—	3.7315	4.4313	3.0756	2.9595	2.8747
Volume (Å^3)	572.07	642.82	658.42	634.05	632.26	629.52
Growth ratio (%)	—	12.367	15.094	10.834	10.521	10.042
Energy (eV)	−1937.9	−15562	−16245	−17463	−17019	−15519

TABLE 2: Common valence states and ionic radii of doped metals.

	Sn	Sb	Sr	Ga	Ti	Ge
Common valence states	+4	+3	+2	+3	+2	+4
Ionic radii (Å)	0.690	0.760	1.18	0.620	0.860	0.530

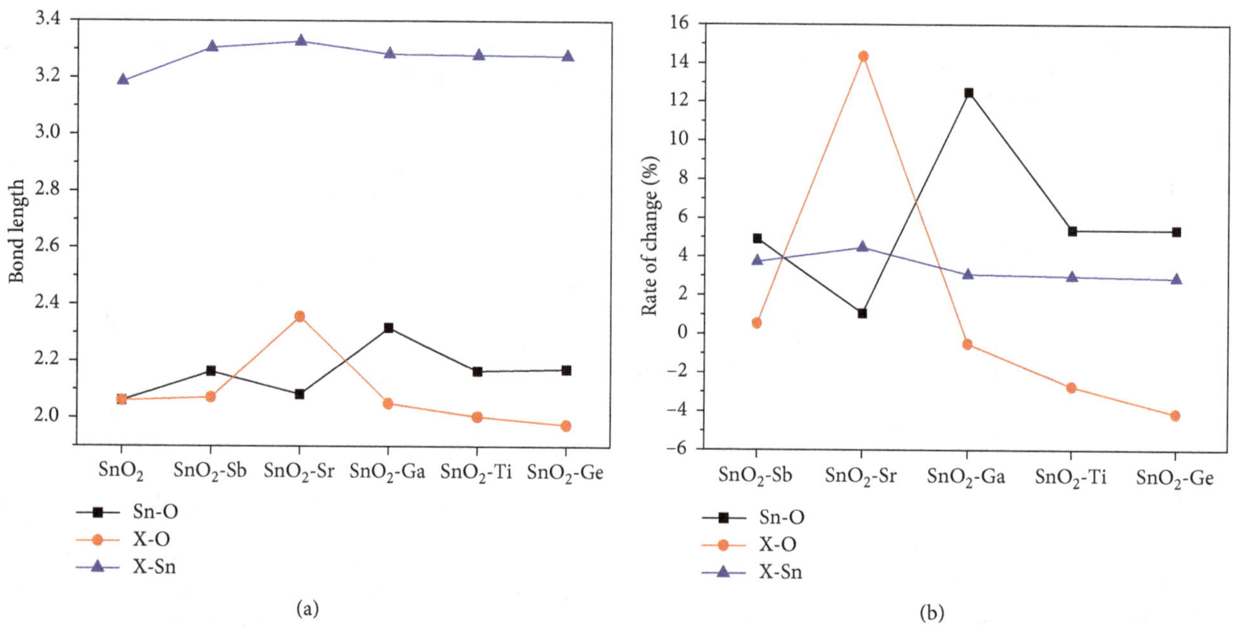

FIGURE 2: (a) Bond length and (b) rate of change.

system of the same size with the doping system, and E_x and E_{Sn} are the energies of each molecule of the stable metal phase; the calculation results of the formation energy E_f are shown in Figure 3. It can be seen that the formation energy E_f of Ti doping is the smallest (i.e., the stability is the best), followed by Ge, Sb, Ga, and Sr.

3.2. Analysis of the Energy Band Structure. Figures 4(a)–4(e) show the band structure of Sb-, Sr-, Ga-, Ti-, and Ge-doped SnO_2. The energy level above the Fermi surface (0 eV) is the conduction band, and the energy level below the Fermi surface is the valence band. It can be seen from the energy band structure of pure SnO_2 in [18] that SnO_2 belongs to the direct bandgap semiconductor material and the bandgap value of pure SnO_2 is 1.003 eV. It is known from the semiconductor theory that the physical properties of materials are mainly determined by the energy bands near the Fermi surface; therefore, in this study, the energy bands near the Fermi surface are mainly analyzed. It can be seen from Figures 4(a)–4(e) that after SnO_2 is doped with Sb, Sr, Ga, Ti, and Ge metal elements, the top of the valence band and the bottom of the conduction band are still at point G; that is, Sb, Sr, Ga, Ti, and Ge metal element-doped SnO_2 materials are still direct bandgap semiconductor materials. And compared with pure SnO_2, the valence band near the Fermi surface is remarkably refined, and the number of energy levels in the valence band increases significantly, which increases the number of electrons that may transition in the valence band; the interaction between electrons is enhanced, and the electrical performance is enhanced. According to the studies of Jiang et al. [19], Lu et al. [20], and other scholars, we can see that the smaller the bandgap is, the better the electrical performance is; therefore, the electrical

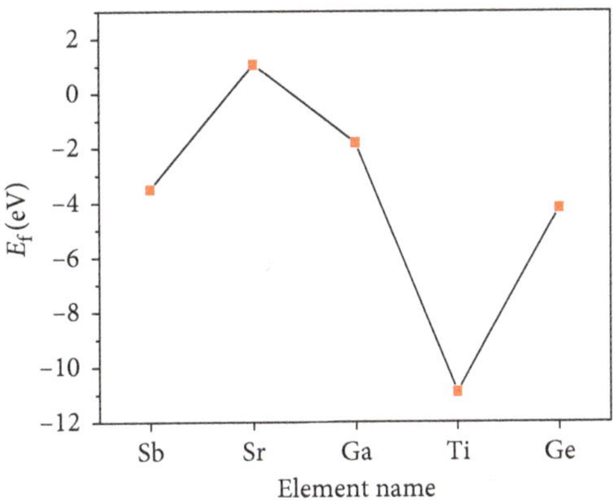

FIGURE 3: Formation energy of the metal-doped SnO_2 system.

properties of Sb-, Sr-, Ga-, Ti-, and Ge-doped SnO_2 are analyzed in detail from the bandgap.

As can be seen from Figure 4(a), after the Sb doping, the energy level of the conduction band shifts to the lower level and becomes denser than that of pure SnO_2, mainly concentrated in the range of (0 eV~5 eV) and the locality is enhanced, and the bottom of the conduction band moves downward crossing the Fermi surface. And the energy level of the valence band also becomes denser after Sb doping and moves slightly to the lower energy level. Therefore, after Sb doping, the bandgap between the valence band and the conduction band is reduced; that is, the bandgap value is smaller than that of pure SnO_2. From Figures 4(b)–4(e), it can be seen that after Sr, Ga, Ti, and Ge doping, the energy levels of the conduction band all move to the lower energy level, mainly concentrated in the range of (0 eV~5 eV), and Ga doping causes the bottom of the conduction band cross the Fermi surface. And the valence band varies with the doping metal. After doping with Sr and Ga, as shown in Figures 4(b) and 4(c), respectively, the new energy levels appear in the valence band region of −15 eV~−10 eV, and the Sr doping causes another new energy level to appear at −35 eV~−30 eV, which makes the valence band broadened. After Ti doping, as shown in Figure 4(e), the new energy levels appear at −35 eV~−30 eV and −60 eV~−55 eV, making the valence band further broadened.

After the analysis of Sb, Sr, Ga, Ti, and Ge doping, the valence band levels become denser near the Fermi surface so that the locality is strengthened, and after Ti doping, the top of the valence band crosses the Fermi surface. Therefore, when SnO_2 is doped with Sb, Sr, Ga, Ti, and Ge, the bandgap between the valence band and the conduction band is different; that is, the bandgap values are different. The calculated bandgap values are listed in Table 3.

It can be seen from Table 3 that, in the given doped metal elements, the bandgap value after Ga doping is relatively smaller (i.e., the energy required for the electrons to be excited from the valence band to the conduction band is also smaller, and the electrical performance is relatively better), followed by Ge, Sr, Ti, and Sb.

3.3. Analysis of State Density. Figures 5(a)–5(e) show the total state density and partial state density of Sb-, Sr-, Ga-, Ti-, and Ge-doped SnO_2. As can be seen from the total state density of Figures 5(a)–5(e), after Sb, Sr, Ga, Ti, and Ge doping, the peak value of the total state density becomes larger than that of pure SnO_2 proposed in [18], indicating the localization of the energy band enhanced. And the energy levels of the conduction band move closer to the Fermi level, mainly concentrated in the region of 0 eV~5 eV, and because of the doping of Sr and Ti, the new energy levels are formed; thus, the valence band is broadened.

It can be seen from Figure 5(a) that after the Sb doping, the contribution of the 2p electronic state of O and the 5p electronic state of Sn is reduced compared with pure SnO_2; however, because of the contribution of the 5s and 5p electronic states of Sb, the bottom of the conduction band crosses the Fermi surface, and the 5s electronic state of Sb and the 5s electronic state of Sn occur orbital hybridization at the Fermi surface. And in the region of the valence band, because of the contribution of 5s and 5p electronic states of Sb, the peak value of the total state density increases; therefore, the localization of the energy band is enhanced and the energy band becomes denser, especially the energy band near the Fermi surface. Therefore, after Sb doping, the movement of the valence band and the conduction band makes the bandgap value decrease. It can be seen from Figure 5(b) that after the Sr doping, the conduction band moves correspondingly closer to the Fermi surface because of the contribution of the 5s electronic state of Sr, and in the region of −1.8 eV~0.2 eV, the 2p orbital of O hybridizes with the 5s, 4p orbitals of Sr; as a result, the concentration of carriers here increases. And because of the contribution of the 5s electronic state of Sr, the valence band region (−30 eV~−31 eV) has a new energy level, which makes the valence band to broaden, and another new energy band appears in the region of −11.6 eV~−13.2 eV because of the contribution of the 4p electronic state of Sr; the valence bands near the Fermi surface become denser, and the top of the valence band is closer to the Fermi surface because of the contribution of 5s and 4p electronic states of Sr; as a result, the bandgap is decreased. It can be seen from Figure 5(c) that after the Ga doping, the conduction band also moves closer to the Fermi surface because of the 4s and 4p electronic states contribution of Ga, and the movement is relatively larger such that the bottom of the conduction band crosses the Fermi surface; the hybridization occurs between the 2p electronic state of O and the 4s and 4p electronic states of Ga in the conduction band region so that the concentration of carriers increases. And in the valence band, a new energy band appears in the region of −13.8 eV~−12.3 eV because of the main contribution of the 3d electronic state of Ga, and the other changes are similar to those in Figure 5(a); in general, the doping of Ga enhances the localization of the energy band. It can be seen from Figure 5(d) that after the Ti doping, the conduction band also moves closer to the Fermi surface because of the contribution of the 3d electronic state of Ti, while the valence band has two new energy bands due to the 3p and 4s electronic states of Ti, which causes the valence band to broaden, and the valence band becomes denser near the Fermi surface because of the 3d electronic state of Ti; thus, the bandgap decreases. The total state

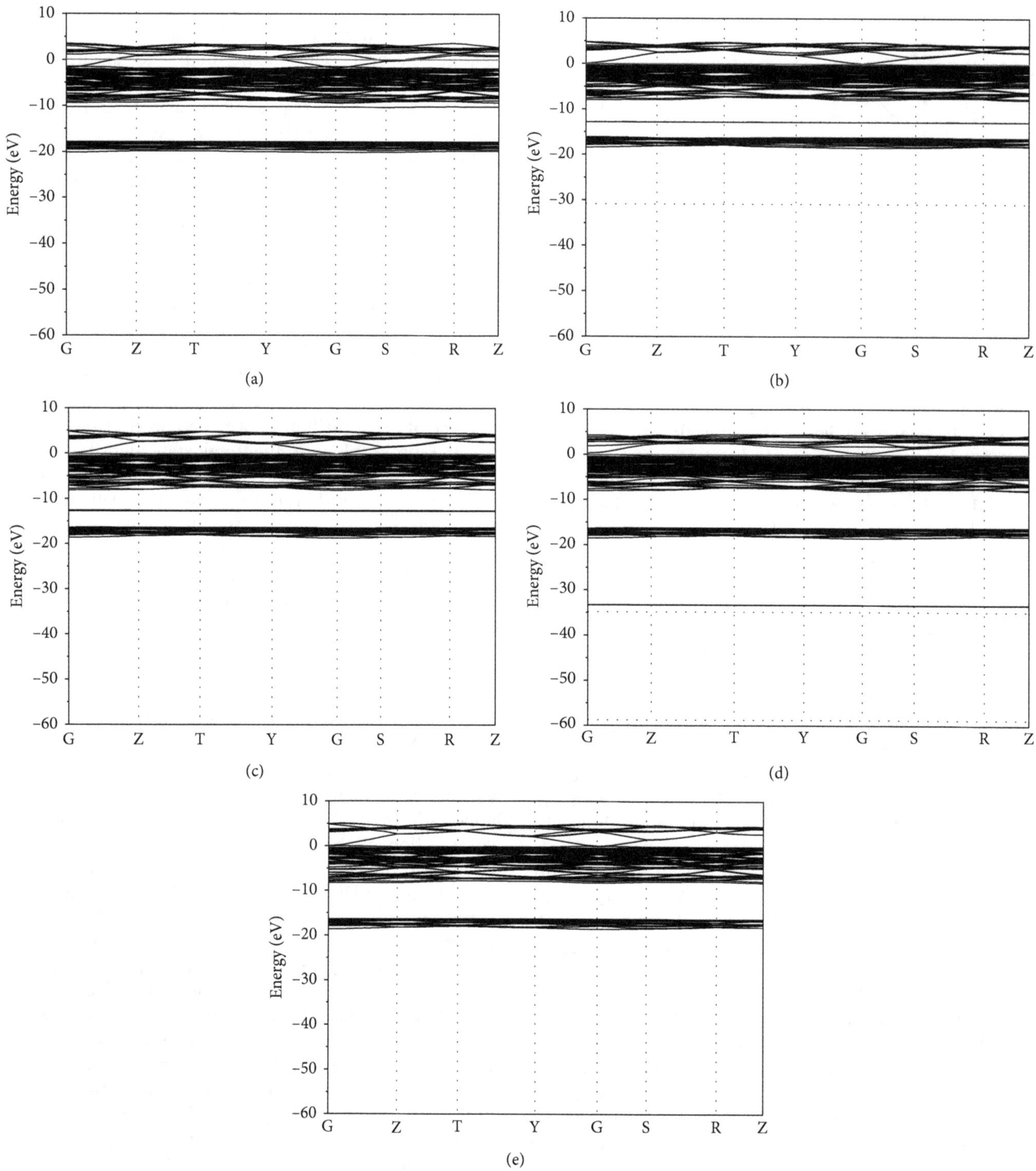

FIGURE 4: Band structure of pure SnO_2 and metal-doped SnO_2. (a) Sb-doped SnO_2. (b) Sr-doped SnO_2. (c) Ga-doped SnO_2. (d) Ti-doped SnO_2. (e) Ge-doped SnO_2.

density of the Ge doping is similar to that of the Sb doping, as shown in Figure 5(e); however, the bottom of the conduction band does not cross the Fermi surface after the doping of Ge, and the 4s and 4p electronic states of Ge are hybridized with the 2p electronic states of O in the conduction band region, so the concentration of carriers also increases accordingly.

From the above analysis of the state density, it is found that after SnO_2 is doped with Sb, Sr, Ga, Ti, and Ge, the contribution of the doped metals electronic states makes the movement degree of the conduction band and valence band different; therefore, the bandgap value is decreased to a different degree, which is in accordance with the energy band analysis results.

TABLE 3: Metal-doped SnO_2 bandgap parameters.

Doping model	Bandgap value (eV)	Change ratio (%)
SnO_2	1.003	—
Sb-SnO_2	0.2814	−71.94
Sr-SnO_2	0.1646	−83.59
Ga-SnO_2	0.0410	−95.91
Ti-SnO_2	0.2755	−72.53
Ge-SnO_2	0.0925	−90.78

3.4. Analysis of Differential Charge Density. Figures 6(a)–6(f) show the differential charge density of pure SnO_2 and Sb-, Sr-, Ga-, Ti-, and Ge-doped SnO_2. In this study, the differential charge density of the (0.7, 0.7, 0) cross section is taken. Figure 6(a) shows the differential charge density of pure SnO_2, and the red region indicates the increases in the charge density, and the blue, green, and yellow regions indicate the decreases in the charge density. The distribution of the differential charge density around the O atom appears directionality, and this is because that the coupling of the 5s orbit of Sn and the 2p orbit of O; the Sn atom and O atom form a semiconductor by hybrid mode, and the isolated electrons of Sn atom enter the hybrid orbit to form the coordination bond, which strengthens the combination of the Sn atom and O atom; the gains and losses of electron between Sn atom and O atom are more compact; therefore, the distribution of the differential charge density appears directionality. From the differential charge density of pure SnO_2, it can be seen that surrounding the Sn atom mainly gains electron, while the center of the Sn atom slightly losses electrons, and the center of the O atom losses electrons seriously, and surrounding the O atom gains electron, indicating that between the Sn atom and O atom are mainly ionic bonds with a certain covalent bond.

Figure 6(b) shows the differential charge density of Sb-doped SnO_2, and the red-yellow region indicates the increases of the charge density, and the blue-green region indicates the decreases of the charge density. Compared with the differential charge density of pure SnO_2, the distribution of the differential charge around the O atom and the Sn atom appears directionality after SnO_2 is doped with Sb, and the charge density in the Sn-O bond direction decreases slightly, while the charge density in the Sb-O bond direction increases, indicating that the doping of Sb has a certain influence on the electronic structure of SnO_2. Figure 6(c) shows the differential charge density of Sr-doped SnO_2. The charge transfer between the Sn and O atoms around the Sr atom is not obvious; the Sr atom gains electron, and the O atom around the Sr atom losses electron, indicating that the doping of Sr also has a certain influence on the electronic structure of SnO_2 and making it a shallow acceptor. Figure 6(d) shows the differential charge density of Ga-doped SnO_2. The distribution of the differential charge density of the Ga atom and its surrounding O atoms all appears directionality, and this is because of the hybridization between the 4s and 4p electronic states of Ga and the 2p electronic states of O, making the isolated electrons of the O atom to enter the hybrid orbital to form the coordination bond; thus the combination and the gains and losses of electron between the Ga atom and the O atom are more compact, so the distribution of the differential charge density appears directionality. However, the distribution of the differential charge density of the Sn atom does not appear the directionality, which indicates that the charge density in the direction of Ga-O bonds increases and the charge density in the direction of the Sn-O bonds decreases; that is, the charge transfer degree on the Ga-O bonds is larger, so the conductive property of SnO_2 is enhanced. Figure 6(e) shows the differential charge density of Ti-doped SnO_2. The distribution of the differential charge density of the Ti atom also appears directionality, and the center of Ti atom gains electron, while the center of O atom losses electron. And comparing the differential charge density around the Ti atom and the Sn atom, it is known that the charge density around the Ti atom is larger than that of the Sn atom, indicating that the charge transfer on the Ti-O bonds is stronger than that on the Sn-O bonds; the electrical performance is improved. Figure 6(f) shows the differential charge density of Ge-doped SnO_2. The center of the O atom losses electrons, and its surrounding gains electron, while the center of the Ge atom gains electrons, and its surrounding losses electron, comparing the differential charge density around Ge and Sn atoms. The number of the Ge atom losing electron is more than that of the Sn atom, so the charge transfer between the Ge atom and the O atom is more intense, thus improving the electrical performance of SnO_2.

From the above analysis, it can be seen that the electrical performance can be improved after SnO_2 is doped with Sb, Sr, Ga, Ti, and Ge , and the degree of charge interaction between Ga and O atoms is the strongest, followed by Ge, Sr, Ti, and Sb.

4. Experimental Verification of Metal-Doped $AgSnO_2$

In this study, firstly, the sol-gel method was used to prepare the powder materials of Sb-, Sr-, Ga-, Ti-, and Ge-doped SnO_2; therefore, Sb-, Sr-, Ga-, Ti-, and Ge-doped SnO_2 materials are more proximate to the built structures used in the simulation calculation. And the X-ray diffraction experiment of metal-doped SnO_2 materials was carried out, and the Sr-doped SnO_2 powder material prepared by the sol-gel method is taken as an example to carry out the XRD phase analysis and to illustrate that the sol-gel method can make the doping element into SnO_2 structure. The XRD diagrams are shown in Figure 7. It can be seen that the diffraction peak is basically similar to that of pure SnO_2, and the intensity of the diffraction peak is reduced, which indicates that the Sr^{2+} infiltrates into SnO_2 structure. And the diffraction angle of the three diffraction peaks marked in the figure is basically similar to that of pure SnO_2, indicating that Sr doping has almost no effect on the SnO_2 crystal structure and achieves better gap doping.

Then, the metal-doped Ag-SnO_2 contact materials were obtained by the powder metallurgy technology, and this technology will not change the phase structure; the material ratio is $Ag : Sn_{0.9375}X_{0.0625}O_2 = 88 : 12 (x = Sb, Sr, Ga, Ti, Ge)$.

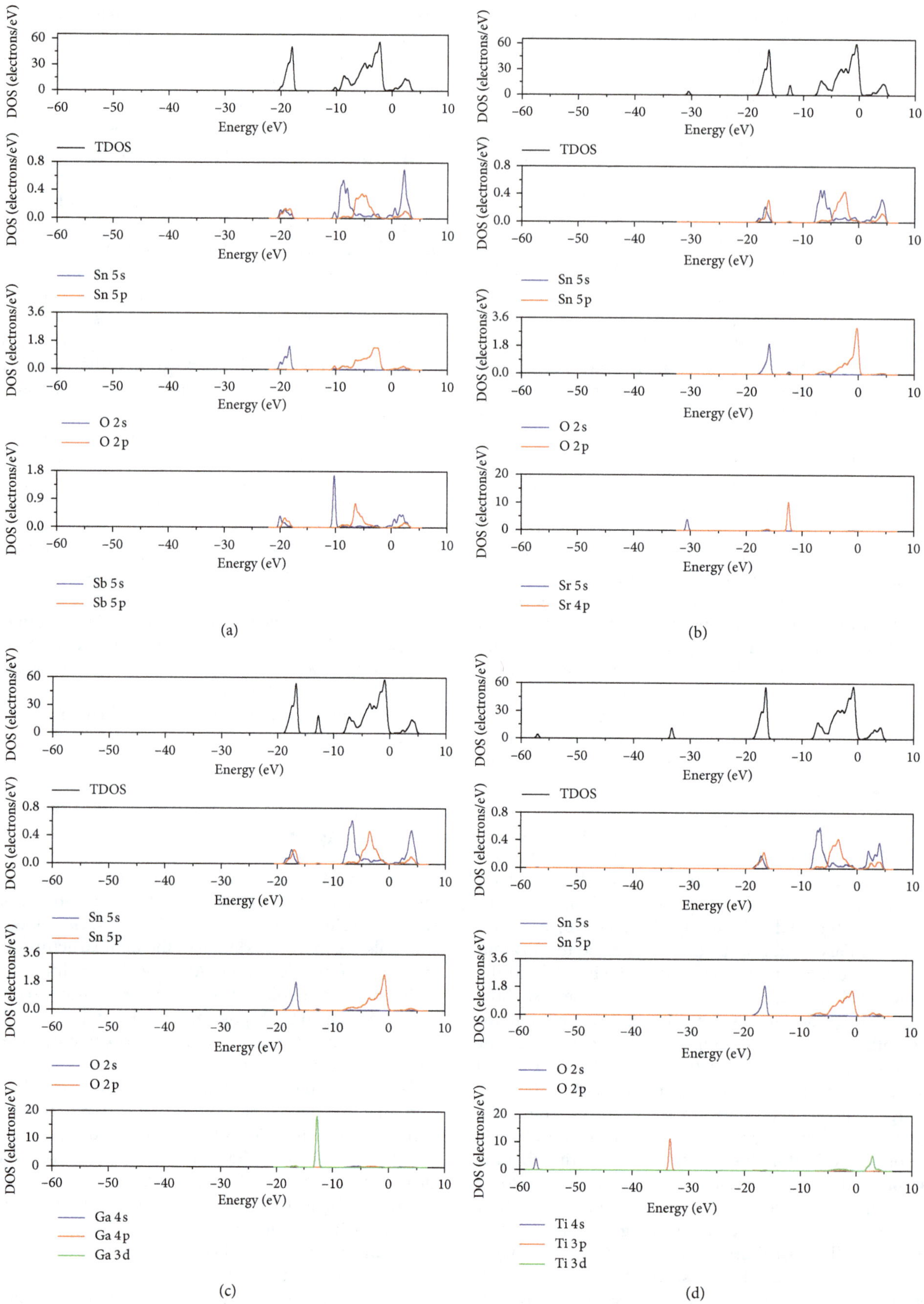

FIGURE 5: Continued.

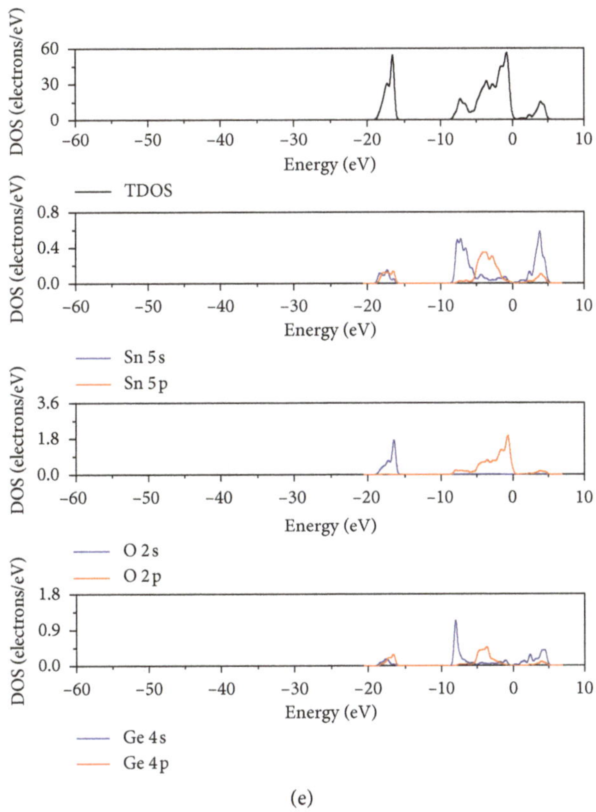

FIGURE 5: State density of pure SnO_2 and metal-doped SnO_2. (a) Sb-doped SnO_2. (b) Sr-doped SnO_2. (c) Ga-doped SnO_2. (d) Ti-doped SnO_2. (e) Ge-doped SnO_2.

And the preparation process is mainly divided into the following steps: mixing Ag powder and metal-doped SnO_2 powder, initial pressing the mixed powder, initial sintering, second pressing, second sintering, and polishing treatment. Finally, the electrical conductivities of pure $AgSnO_2$ and Sb-, Sr-, Ga-, Ti-, and Ge-doped $AgSnO_2$ contact materials were measured by the conductivity tester, and the conductivity of each metal-doped system is tested five times, and then the average value is obtained. The data are shown in Table 4.

It can be seen from the data that the conductivity of $AgSnO_2$ contact material is 50.36%IACS, which is basically consistent with the conductivity data 50.84 %IACS of $AgSnO_2$ contact material measured by Wang et al. [21]. And comparing the above data with the national standard value 54% IACS of $AgSnO_2$ contact material, it can be seen that Sb, Sr, Ga, Ti, and Ge doping can improve the electrical performance of $AgSnO_2$ contact material, and the metal Ga doping has the best improvement effect, followed by Ge, Sr, Ti, and Sb, which further verified the accuracy of the simulation calculation. This study initially proves that the first-principles calculation method can be used to calculate the electrical performance of the materials and to guide the test results.

5. Conclusion

In this study, the first principles based on the density functional theory is used to calculate the energy band structure, density of states, and the differential charge density of pure SnO_2 and Sb-, Sr-, Ga-, Ti-, and Ge-doped SnO_2. And Sb-, Sr-, Ga-, Ti-, and Ge-doped SnO_2 powders were prepared by the sol-gel method, and then the metal-doped $AgSnO_2$ contact materials were prepared by the powder metallurgy technology; the conductivity was tested. The conclusion reveals that after SnO_2 doping with the metals Sb, Sr, Ga, Ti, and Ge, the bandgap decreases and the energy bands are concentrated near the Fermi surface, thus the locality is enhanced. The number of electrons that may transition on the valence band increases, and the interaction between electrons is enhanced; that is, the energy required for the carriers to transition from the valence band to the conduction band is reduced, and the electrical performance is improved. The bandgap of Ga-doped SnO_2 is the smallest, that is, 0.041 eV. And after the doping of the metals Sb, Sr, Ga, Ti, and Ge, the charge transfer between Sn and O atoms decreases, while the charge transfer between the metal atoms Sb, Sr, Ga, Ti, Ge, and O increases, especially between Ga and O atoms, so the electrical performance of Ga doping is relatively better. And the conductivity of metal-doped $AgSnO_2$ is higher than the national standard, and the conductivity of Ga doped is the largest, that is, 59.30% IACS, which indicates that Sb, Sr, Ga, Ti, and Ge doping improves the electrical performance of $AgSnO_2$, thus verifying the correctness of the simulation results. This research provides a feasible method and a reference for the study of $AgSnO_2$ contact materials.

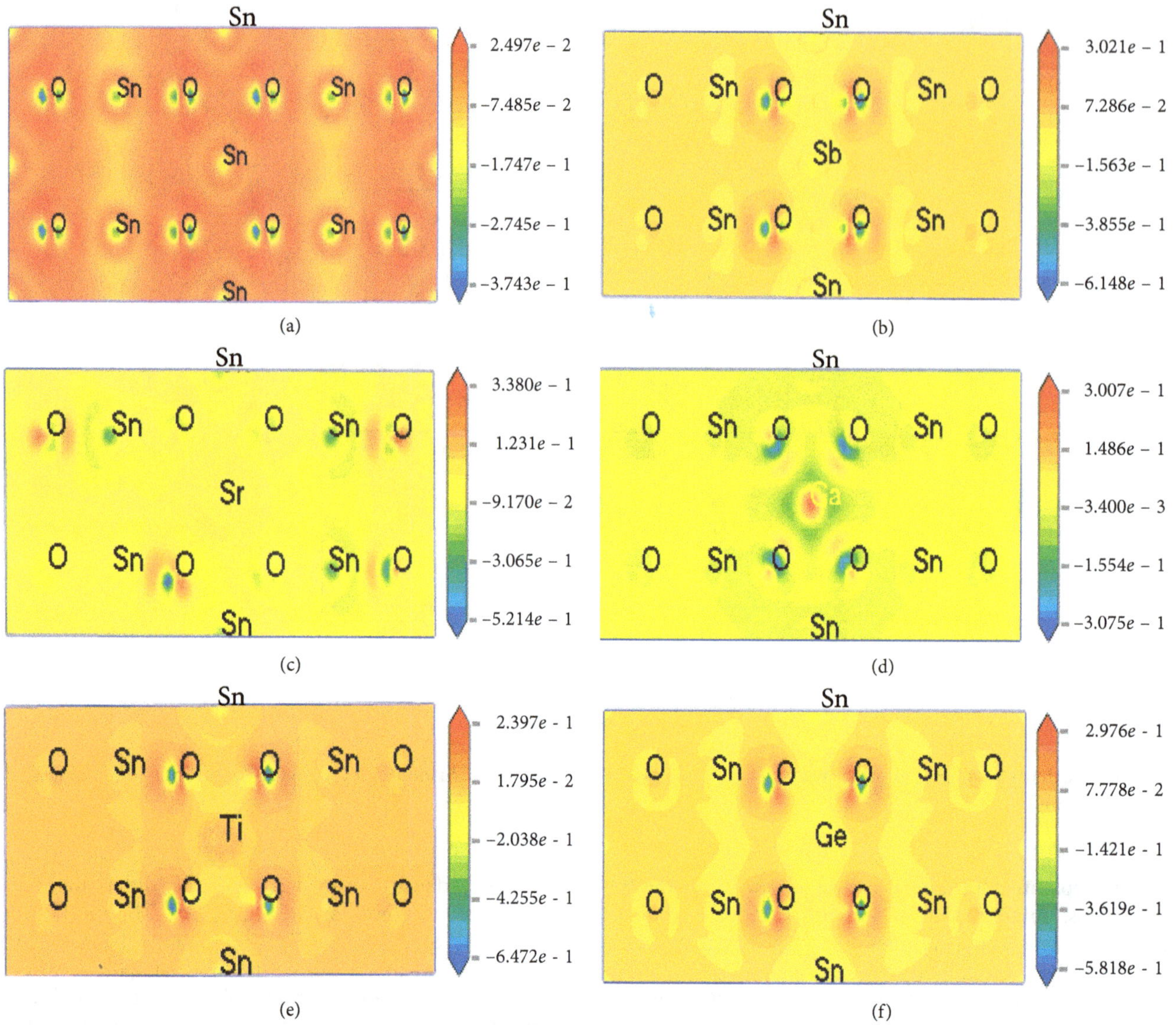

Figure 6: Differential charge density of pure SnO_2 and metal-doped SnO_2. (a) SnO_2. (b) Sb-doped SnO_2. (c) Sr-doped SnO_2. (d) Ga-doped SnO_2. (e) Ti-doped SnO_2. (f) Ge-doped SnO_2.

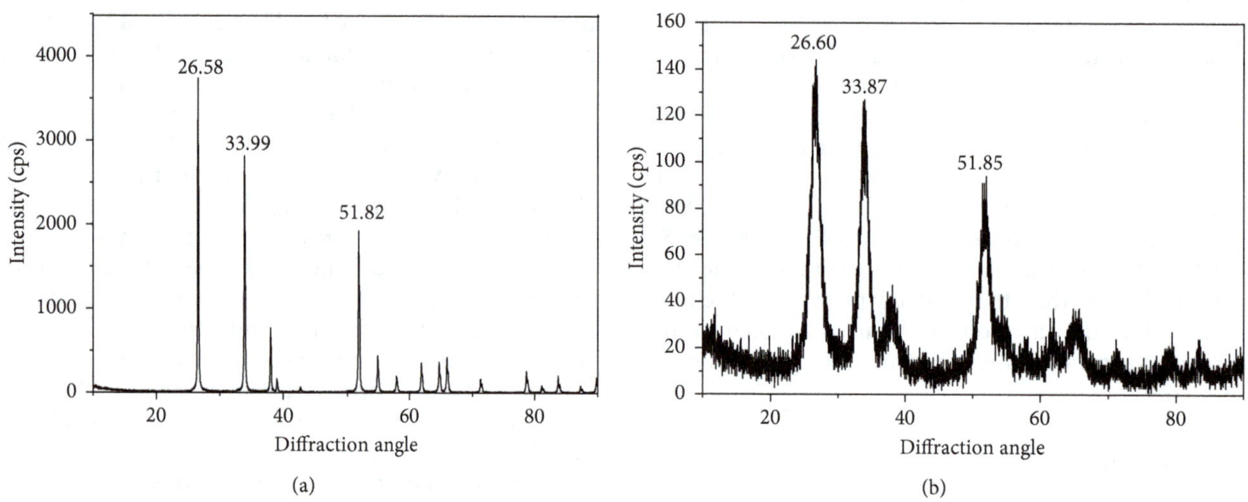

Figure 7: X-ray diffraction experimental diagram of pure SnO_2 and Sr-doped SnO_2 powder.

TABLE 4: Conductivity.

Doping model	Conductivity (%IACS)
$AgSnO_2$	50.36
$AgSnO_2$-Sb	54.16
$AgSnO_2$-Sr	55.91
$AgSnO_2$-Ga	59.30
$AgSnO_2$-Ti	54.21
$AgSnO_2$-Ge	58.09

Conflicts of Interest

The authors declare that they have no conflicts of interest.

References

[1] C. T. Zhao and J. Q. Wang, "Electrical and mechanical properties of Cu-doped SnO_2 with different concentrations," *Chinese Journal of Materials Science and Technology*, vol. 25, no. 6, p. 34, 2017.

[2] Y. P. Wang and H. Y. Li, "Improved workability of the nano-composited $AgSnO_2$ contact material and its microstructure control during the arcing process," *Metallurgical and Materials Transactions A*, vol. 48, no. 2, pp. 609–616, 2017.

[3] S. Hassan and S. Narges, "First principle DFT study of electric field effects on the characteristics of bilayer graphene," *Zeitschrift für Naturforschung A*, vol. 72, no. 1, pp. 1–7, 2017.

[4] H. T. Wang, P. L. Wen, L. Liang, Y. C. Zhu, and X. W. Wang, "Effect of wettability on properties of $AgSnO_2TiO_2$ contact materials," *Electronic Components and Materials*, vol. 35, no. 7, p. 80, 2016.

[5] B. Elodie, C. Patrick, D. Henri, R. Philippe, K. Vincent, and B. Alexandre, "Effective thermal and electrical conductivities of $AgSnO_2$ during sintering. part I: experimental characterization and mechanisms," *Metallurgical and Materials Transactions A*, vol. 47, no. 12, pp. 6304–6318, 2016.

[6] L. T. Dan, D. C. Ba, Q. Cao, X. Y. Hou, and J. C. Li, "Effect of Ce-Cu codoping on optoelectronic property of SnO_2 film," *Acta Metallurgica Sinica*, vol. 50, no. 1, p. 95, 2014.

[7] Z. Y. Liu, *Calculation and Study of Nano-doping $AgSnO_2$ Electrical Contact Materials*, Tianjin University, Tianjin, China, 2007.

[8] X. M. Liu, J. Zheng, S. L. Li, Q. Y. Li, Y. T. Li, and K. T. Lv, "Study on preparation and performance of nano Ag-SnO_2-TiO_2 contact," *Material Development and Application of Materials*, vol. 18, no. 3, p. 4, 2003.

[9] Z. Wang, Y. Lu, and P. J. Wang, "Study on electronic structure and optical properties of Fe and Mn codoped SnO_2 superlattice," *Journal of Synthetic Crystals*, vol. 42, no. 7, p. 1429, 2013.

[10] D. L. Chen and X. Gao, "Preparation of (Ce, Cu)-SnO_2 nanopowders by polymer precursor method and its sintering behavior," *Journal of Inorganic Materials*, vol. 19, no. 1, p. 58, 2004.

[11] J. Z. Wang, Y. P. Wang, Z. M. Yang, W. Wang, and B. J. Ding, "Influences of CuO additives on wettability and interface properties of Ag/SnO_2," *Rare Metal Materials and Engineering*, vol. 34, no. 3, p. 405, 2005.

[12] H. Ali, M. Behrang, H. Hossein, N. Mohsen, G. Mir, and S. Hassan, "Synthesis of SnO_2 nano-powders by a sol-gel process using propanol-isopropanol mixture," *Open Chemistry*, vol. 6, no. 4, p. 651, 2008.

[13] X. Li, R. Deng, Y. F. Li et al., "Effect of Mg doping on optical and electrical properties of SnO_2 thin films: an experiment and first-principles study," *Ceramics International*, vol. 42, no. 4, pp. 5299–5303, 2016.

[14] J. Q. Jia, X. J. Xie, Z. H. Liang, X. C. Zhang, C. M. Fan, and P. D. Han, "First-principle study of Ti-doped SnO_2 semiconductor solid solution," *Chemical Journal of Chinese Universities*, vol. 33, no. 5, pp. 1050–1056, 2012.

[15] J. Q. Jia, *Study on Structure and Properties of Doped SnO_2 Semiconductor Solid Solution Electrodes*, Taiyuan University of Technology, Taiyuan, China, 2012.

[16] J. Zheng, S. L. Li, and J. Guo, "The influence of rare earth oxide on the structure and properties of $AgSnO_2$ electrical contact materials," *Advanced Materials Research*, vol. 479–481, pp. 1986–1989, 2012.

[17] Q. Y. Hou, Z. Y. Lv, and C. W. Zhao, "Effect of V-heavy-doped ZnO on Electric conductivity performance and absorption spectrum," *Acta Physica Sinica*, vol. 63, no. 19, p. 296, 2014.

[18] J. Q. Wang, H. L. Kang, and Y. Zhang, "Analysis of the influence of Ti doping on the electrical properties of SnO_2," *Materials Research Express*, vol. 5, no. 7, article 075902, 2018.

[19] L. Jiang, P. J. Wang, C. W. Zhang, X. Y. Feng, Y. Lu, and G. L. Zhang, "Electronic structure and optical properties of Cr doped SnO_2 superlattice," *Acta Physica Sinica*, vol. 60, no. 9, p. 227, 2011.

[20] Y. Lu, P. J. Wang, C. W. Zhang, L. Jiang, G. L. Zhang, and P. Song, "Material opto-electronic properties of in, N Co-doped SnO_2 studied by first principles," *Acta Physica Sinica*, vol. 60, no. 6, p. 223, 2011.

[21] H. T. Wang, L. Liang, P. L. Wen, Y. C. Zhu, X. W. Wang, and J. T. Huo, "Effect of adding Bi_2O_3 on physical contact properties of $AgSnO_2$ contacts," *Precious Metals*, vol. 37, no. 4, p. 21, 2016.

19

Determination of Anisotropic Mechanical Properties for Materials Processed by Laser Powder Bed Fusion

Even W. Hovig,[1] Amin S. Azar,[2] Frode Grytten,[2] Knut Sørby,[1] and Erik Andreassen[2]

[1]*Norwegian University of Science and Technology (NTNU), Department of Production and Quality Engineering, Trondheim, Norway*
[2]*SINTEF Industry, Oslo, Norway*

Correspondence should be addressed to Even W. Hovig; even.w.hovig@ntnu.no

Academic Editor: Marco Rossi

Improving the success rate in additive manufacturing and designing highly optimized structures require proper understanding of material behaviour. This study proposes a novel experimental method by which anisotropic mechanical properties of additively manufactured materials can be assessed. The procedure is based on tensile testing of flat specimens, manufactured by laser powder bed fusion (LPBF) at different orientations relative to the build plate. In this study, the procedure was applied to the Inconel 718 alloy. Three identical specimen sets were built, each of which received complementary postprocessing treatments. The tensile tests were carried out on specimens with as-built surface finish. Digital image correlation was used to record the strain field evolution on two perpendicular surfaces of the tensile specimens under loading. An optimization algorithm is also proposed for determining the anisotropic elastic constants using only a few tensile test results. It was observed that both build orientation and postprocessing have strong influence on the anisotropic mechanical properties of the material. The effect of microstructure was also investigated and characterised. Consequently, three transversely isotropic compliance matrices were constructed, representing the effect of the different processing conditions.

1. Introduction

Most of the commercial additive manufacturing (AM) technologies can produce materials with relatively high density and decent mechanical properties, close to that of classical processing methods. Capability of manufacturing complex geometries by AM brought about new opportunities such as "topology optimization" and "design for additive manufacturing (DfAM)" and caused commencement of a new chapter in performance-oriented design of components. Several challenges, such as anisotropic behaviour [1–6] and effects of postprocessing steps [7–10], need to be addressed in the design phase to release the full potential of AM. Despite these facts, most of the current algorithms for generative design use isotropic material models or globally uniform and defect-free properties [11–13]. Thus, we propose a methodology for calibrating anisotropic constitutive material models based on systematic mechanical tests and as-built surface conditions. Inconel 718 (IN718) processed by laser powder bed fusion (LPBF) are the material and process of choice in this study.

1.1. Determination of Elastic Constants—Transversely Isotropic Elastic Materials. Anisotropic elasticity can be modelled using the generalized Hooke's law. If the material is assumed to be transversely isotropic, that is, isotropic in the *XY* plane, Hooke's law can be expressed as [14]

$$\begin{bmatrix} \varepsilon_{xx} \\ \varepsilon_{yy} \\ \varepsilon_{zz} \\ \varepsilon_{yz} \\ \varepsilon_{xz} \\ \varepsilon_{xy} \end{bmatrix} = \begin{bmatrix} \frac{1}{E_{xx}} & -\frac{\nu_{xy}}{E_{xx}} & -\frac{\nu_{zx}}{E_{zz}} & 0 & 0 & 0 \\ -\frac{\nu_{xy}}{E_{xx}} & \frac{1}{E_{xx}} & -\frac{\nu_{zx}}{E_{zz}} & 0 & 0 & 0 \\ -\frac{\nu_{xz}}{E_{xx}} & -\frac{\nu_{xz}}{E_{xx}} & \frac{1}{E_{zz}} & 0 & 0 & 0 \\ 0 & 0 & 0 & \frac{1}{2G_{xz}} & 0 & 0 \\ 0 & 0 & 0 & 0 & \frac{1}{2G_{xz}} & 0 \\ 0 & 0 & 0 & 0 & 0 & \frac{1}{2G_{xy}} \end{bmatrix} \begin{bmatrix} \sigma_{xx} \\ \sigma_{yy} \\ \sigma_{zz} \\ \sigma_{yz} \\ \sigma_{xz} \\ \sigma_{xy} \end{bmatrix}. \quad (1)$$

The E_{xx} and E_{zz} moduli can be found using standard tensile tests. The shear modulus in the plane of isotropy can be expressed as [14]

$$G_{xy} = \frac{E_{xx}}{2\left(1 + \nu_{xy}\right)}. \quad (2)$$

G_{xz} can be calculated using an off-axis tension test to determine E_{ϕ} on a sample built in the XZ plane with a rotation of $\phi = 45^{\circ}$ about the Y-axis. This E_{ϕ}, assuming negligible shear strain, can then be used to calculate G_{xz}, using the following equation [15]:

$$G_{12} = \frac{\sin^2\phi\cos^2\phi}{\left(1/E_{\phi}\right) - \left(\cos^4\phi/E_1\right) - \left(\sin^4\phi/E_2\right) + \left(\left(\nu_{21}/E_1\right) + \left(\nu_{12}/E_2\right)\right)\sin^2\phi\cos^2\phi}, \quad (3)$$

where the principal directions 1 and 2 correspond to Y and Z, respectively, and E_{ϕ} is the measured elastic modulus in the off-axis tension test. Note that E_{ϕ} can only be determined using the off-axis tension test if the following inequality is satisfied [15]:

$$G_{12} \geq \frac{E_1}{2\left(1 + \nu_{21}\right)}. \quad (4)$$

Another approach to determine the elastic constants is to implement the problem as an optimization problem by minimizing a function of several variables and constraints. A proposed solution for this is presented in Section 2.4.

1.2. Inconel 718. IN718 was developed in an effort to improve weldability and strain-aged cracking resistance. It is a hardenable alloy of Ni-Fe-Cr, which takes advantage of $Ni_3[Al,Ti]$ (γ' phase, fcc) and Ni_3Nb (γ'' phase, bct) precipitates to increase the resistance to cracking because of slow formation kinetics of these constituents. This alloy is known to have excellent corrosion and fatigue properties at service temperatures up to 760°C [16]. Typical IN718 compositions for wrought and powder material are given in Table 1.

The precipitates are mostly coherent with the austenite matrix that increases the mechanical strength of the material by co-straining with the matrix under different loading conditions. However, some precipitates can adversely affect the performance of these alloys, especially when they are exposed to high temperatures for a relatively long period of time [18]. Figure 1 shows the calculated and experimentally determined TTT diagrams for IN718 that will be used to determine the heat treatment cycle later in this work. The formation kinetics and thermodynamics of γ' and γ'' precipitates are analogous in these diagrams; γ' forms at a slightly higher temperature range, which is indicative of the importance of accurate temperature control in the ageing process. Despite that both the calculated and experimentally determined TTT curves are addressing the nominal IN718 material, slight variation in the chemical composition may result in discrepancy between the two approaches when exact formation kinetics of the γ' and γ'' are being investigated. In addition to solid solution strengthening, the strength can be obtained through formation of γ'' phase in this alloy. The γ' phase is brittle, and its formation is less preferred to increase the high and low temperature ductility. Other brittle phases such as δ may also form upon long exposure to elevated temperatures. Therefore, the service temperature of this alloy is limited by the formation of δ and σ phases.

1.3. Additive Manufacturing of Inconel 718. In laser powder bed fusion (LPBF), the starting point is a packed bed of powder material—usually prealloyed—that is fused locally using an adjustable laser heat source. Today's advanced machines spread the powder in layer thicknesses ranging from 30 to 200 μm [21, 22]. Selection of layer thickness is a trade-off between build rate and part quality, and the chosen layer thickness is a breakeven point for achieving the best built rate with satisfactory quality [23, 24].

There are several features that can cause anisotropy in AM materials. Porosity and pore shape [25], microstructure and texture [26–29], and surface finish [30, 31] of the AM material are among the most referred reasons for anisotropic behaviour. Previous studies have captured the anisotropic properties of various heat treatments of laser powder bed

Table 1: Chemical composition of a commercial wrought solution-treated IN718 [17] and the commercial IN718 powder used in this study (specifications from the supplier).

Element	Ni	Cr	Fe	Al	Mo	Nb (+Ta)	Ti
wt.%, commercial wrought	53	19	18.5	0.5	3.0	5.1	0.9
wt.%, commercial powder	50–55	17–21	Bal.	0.2–0.8	2.8–3.3	4.75–5.5	0.6–1.1

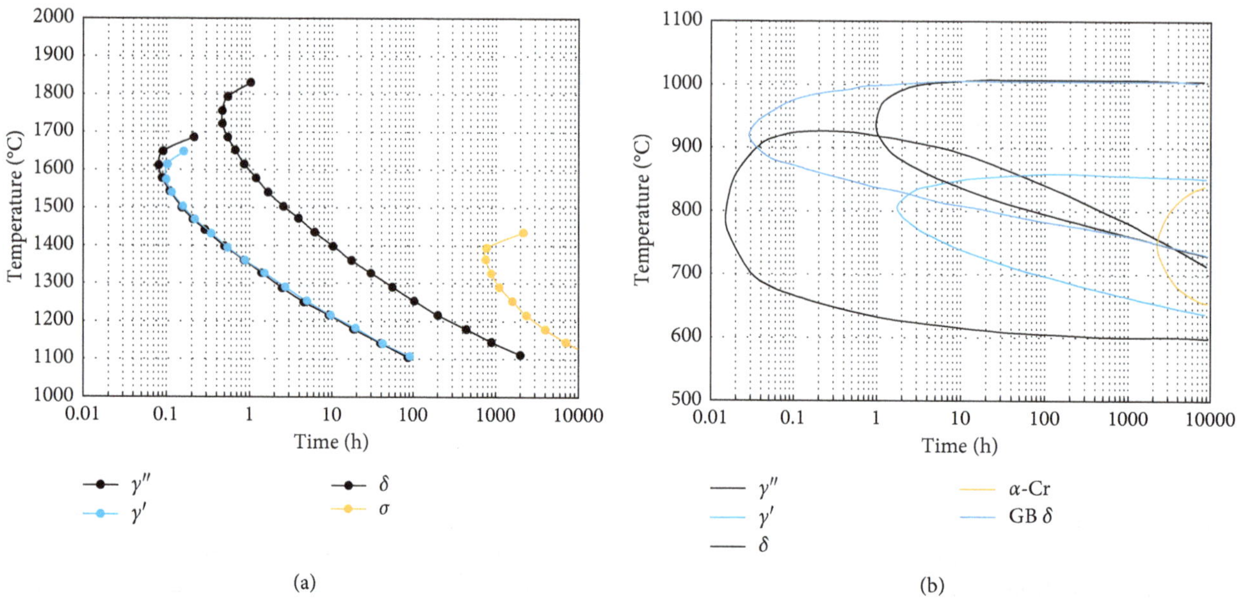

Figure 1: TTT diagram for IN718, calculated (a) and experimentally determined (b) [19, 20].

fusion Inconel 718 [11, 32–34]. The mentioned studies were not focused on determining anisotropic elastic properties, and thus excluded required orientations and mechanical properties (i.e., Poisson's ratio).

Since the powder bed surface is always set horizontally and parallel to the build plate, placement of the components at different angles will result in various grain growth orientations with reference to the local coordinate system of the component. Therefore, applying the same process parameters, including the laser scanning strategy, will cause evolution of disparate crystallographic textures depending on the build orientation [35]. Disregarding surface effects, this phenomenon is the major source of anisotropic behaviour in the same material that is processed at different build orientations. Figure 2 compares two possible grain growth directions induced by varied build parameters. These grains are assumed to undergo a biaxial loading in the presented direction annotated with *F*. It was also assumed that the grains are at one side of the melt pool and there is a symmetry in the growth shown by centreline annotation C_L. Moreover, the local coordinate system at each arbitrary location along the grain is indicated on the cross-cutting plane with <100> orientation as the normal vector. Figure 2 (a) schematically shows a slow cooling rate and Figure 2(b) presents a higher cooling rate. The latter is the dominant case in LPBF. After solidification, two differently oriented grains will be loaded along a certain crystal orientation. Therefore, the mechanical response of the material would apparently differ if identical parts are produced with various relative orientations to the build plate.

Choi et al. [36], Wang et al. [37], and Keshavarzkermani et al. [38] are among the researchers who investigated the microstructure and texture of IN718, processed by LPBF. In all these studies, the matrix mainly consists of γ grains. The grains were oriented nearly in the same direction, almost like a single crystal. According to their reported electron backscatter diffraction (EBSD) analysis, the γ matrix grains were oriented with <100> parallel to the build direction. Azar et al. [39] showed that the cubic crystal growth along <100> follows the maximum heat flow direction in welding, which is perpendicular to the powder layers in the LPBF case. According to Azar et al. [39], depending on the heat source dynamics and solidification kinetics, the crystals can follow a curvilinear path where the <100> is rotating (Figure 2).

Murr et al. [40, 41] reported that HIP treating the IN718 alloy increases the ductility of the material significantly, which is related to reduction in the residual stresses and hardness of the precipitates. Wang et al. [42] summarised the heat treatment cycles that were applied by previous researchers for IN718 processed by AM. The major driving force for optimising the heat treatment procedure is to maximise the evolution of advantageous phases while suppressing the formation of the detrimental ones.

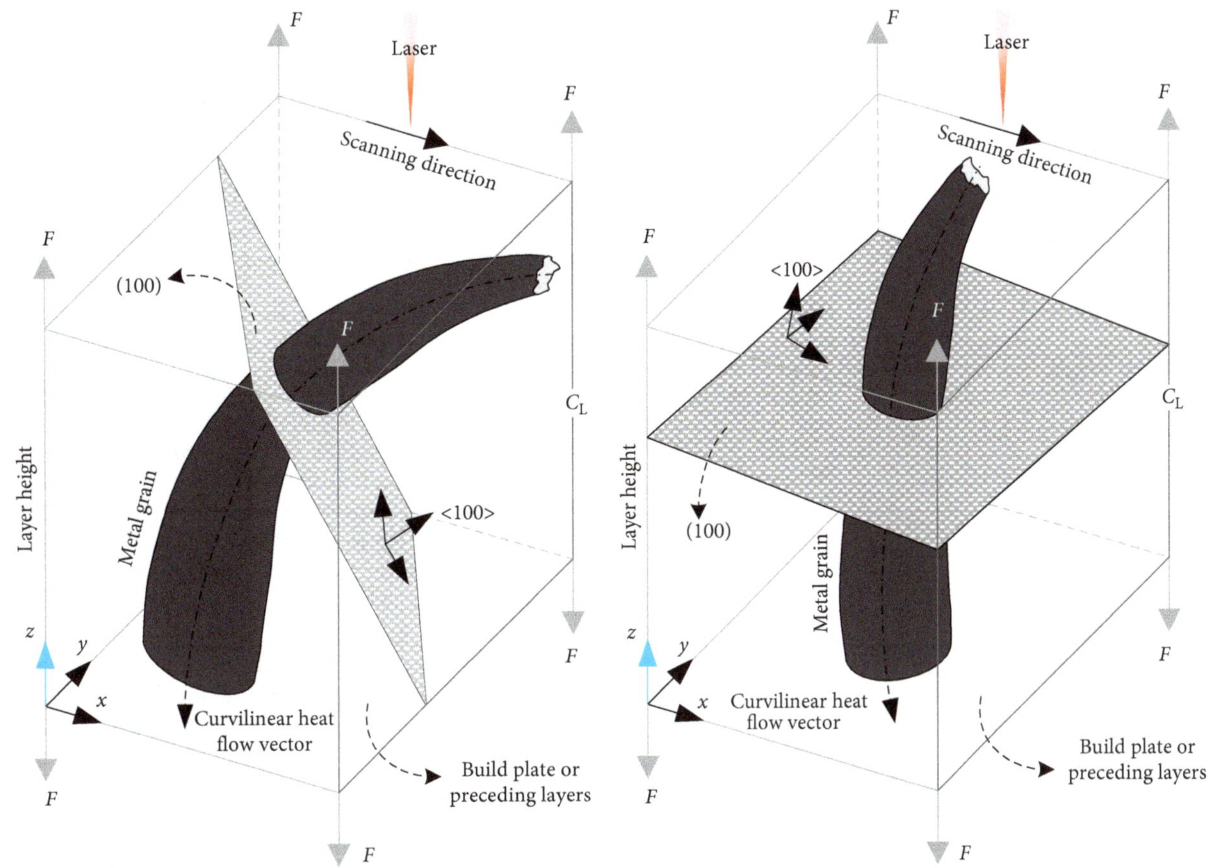

(a) (b)

FIGURE 2: Differences in growth kinetics of metal grains along maximum heat flow. This schematic illustration shows two different grain positions with respect to the future loading conditions. In (a), the loading is close to the crystallographic <111> while in (b), the loading is close to <100>.

2. Materials and Methods

In order to investigate the anisotropy of the additively manufactured IN718 material, the tensile specimens were manufactured at various orientations with respect to the horizontal build plate. In this section, the entire chain from powder to postprocessing, and testing will be summarised.

2.1. IN718 Powder. Prior to the LPBF processing, the powder morphology and chemistry were investigated. The powder used was from a recycled batch, which was sieved and dehumidified following standard practices. The morphology of the powder particles was determined using a Malvern G3 Morphology instrument in the dry dispersion mode. The powder was dispersed on a microscope slide and particles were captured for analysis one-by-one. About 33,000 particles were scanned in the analysis. The chemical composition of the powder was determined by energy-dispersive X-ray spectroscopy (EDS) mapping in a scanning electron microscope (SEM).

2.2. Laser Parameters and Sample Preparation. The tensile samples were built in an LPBF machine (SLM 280 HL from SLM Solutions, installed 2014), with process parameters for IN718 powder as supplied by SLM Solutions [43]. The main process parameters are given in Table 2.

The density of the specimen is related to the energy in the LPBF process [44], with the energy per area ($J \cdot m^{-2}$) defined as $E_A = P/vd$, where P is the laser power (W), v is the scanning velocity ($m \cdot s^{-1}$), and d is the hatch spacing (m). In this study, the energy density formula will be used to benchmark the build conditions against earlier published data. The relative density of the material was calculated on polished cross sections of the material using optical microscopy and an image-processing approach (ImageJ software).

Flat tensile specimens were produced in three sets, each consisting of 11 specimens built at different orientations. The sample orientation with respect to the build plate starts at horizontal (0°) increasing to vertical (90°) with 15° increments. Figure 3 shows the tensile specimen geometry, and Figure 4 shows the 11 specimen orientations for each set. The sample geometry and dimensions were initially designed for in situ tensile tests in an SEM, which is the subsequent study of this work. The colours correspond to the orientations in Table 3. The tensile specimens were not machined after LPBF, as the goal of the study is to assess

Table 2: Process parameters.

Laser power, P (W)	Layer thickness, t (μm)	Hatch spacing, d (μm)	Scan velocity, v (mm/s)
275	50	120	805

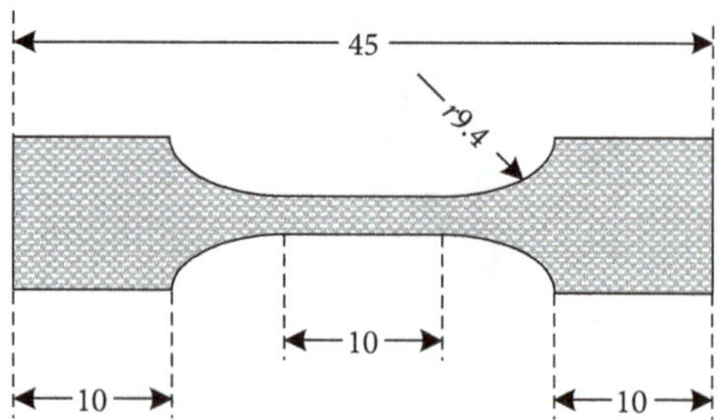

Figure 3: Geometry of the tensile specimen. Dimensions are in mm.

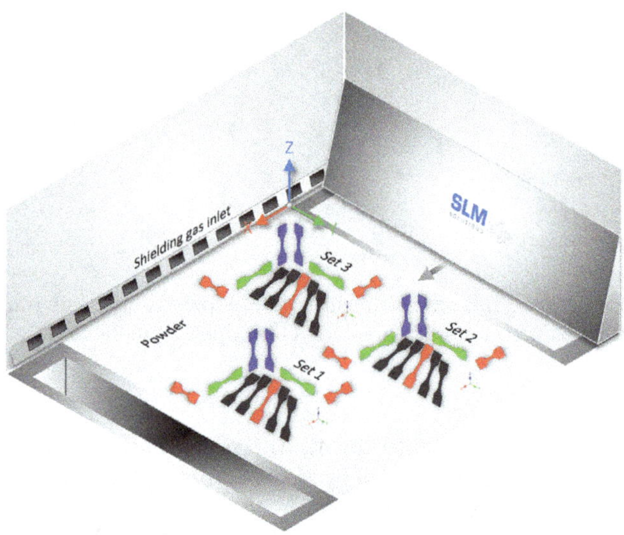

Figure 4: Orientation of the specimens and their configuration in the LPBF build platform. The colours indicate the samples that are used for determining the elastic constants in Section 2.3. Orientations are according to Table 3.

Table 3: Codes for test specimen orientation corresponding to Figure 4, with similar choice of colours. Orientation annotation is according to ISO/ASTM 52921:2013E.

ZX	ZY		
XZ	YZ		
$XZ-45$B	$XY+45$B	$YZ+45$A	
$XY+15$B	$XY+30$B	$XY+60$B	$XY+75$B

and classify the behaviour of LPBF components with as-built surfaces. In addition to the LPBF specimens, some reference specimens were machined from rolled and aged IN718.

Three different postprocessing routes (denoted as "sets") were designed. Each route was applied to a complete set of sample orientations. Tables 4–6 summarise the applied postprocessing parameters based on the instructions found in reference [45]. (S), (A), and (H) abbreviations were used to show the postprocessing steps of solution treatment, ageing, and HIP, respectively. The solution treatment was performed prior to the removal of the specimens from the build plate. HIP was performed by Bodycote AB.

2.3. Determination of Elastic Constants Using Digital Image Correlation. Using Digital Image Correlation (DIC) in combination with tensile testing [46] eliminates the need for strain gauges. It is sufficient to test at three orientations to determine the transversely isotropic compliance matrix. However, as shown in this study, it is necessary to increase the number of orientations to capture the mechanical response of the material. There are other methods available to capture anisotropy, such as magnetic resonance elastography [47] and nanoindentation [48]. However, these methods are dependent on the local properties of the materials and will only characterize the superficial surface layers.

A commercial DIC system (Vic3D) was used to determine the displacement fields during the tensile tests. The tensile tests were carried out using a Zwick tensile machine equipped with a 50-kN load cell. The cross-head speed was set to 1 mm/min. The cross-sectional area of each specimen was measured with a digital calliper and cross-checked with DIC to account for any deviation from the nominal dimensions. In addition, the surface roughness was measured with white light interferometry (WLI) to quantify the uncertainty of the measured cross-sectional area. Two cameras (2452 × 2052 pixels) equipped with Pentax 75 mm f/2.8 lenses were used. The cameras were mounted on a tripod such that two faces of the tensile sample were exposed to both cameras. This setup allows for the capture of out-of-plane displacements and in-plane displacements on the specimen surface. The correlation was performed with a subset of 29 pixels and a step size of 5 pixels. For a detailed description of the DIC setup, the reader is referred to reference [46].

The Lagrange strain was calculated using the Vic3D software, and the average principal and shear strains were exported for further processing in Matlab. For uniaxial loading of a specimen with gauge length L_0, the Lagrange strain can be written as

$$\varepsilon = \frac{\Delta L}{L_0} + \frac{1}{2}\left(\frac{\Delta L}{L_0}\right)^2. \tag{5}$$

Taking into account that the cross-sectional area varying as a load is applied, and introducing w as the current width, w_0 as the initial width, and similarly for the thickness t, the stress can be expressed as

Table 4: Set 1: instructions for solution treatment (S).

HT#	Description	Conditions	Holding temperature	Holding time	Cooling
1	Homogenization and solution treatment	Argon atmosphere. Fast heating to the target temperature	980°C	1 hr	Air cool

Table 5: Set 2: instructions for treatment (S + A).

HT#	Description	Conditions	Holding temperature	Holding time	Cooling
1	Homogenization and solution treatment	Argon atmosphere. Fast heating to the target temperature	980°C	1 hr	Air cool
2	Ageing	Vacuum/argon atmosphere. Fast heating to the target temperature	720°C	8 hr	Furnace cool to 620°C within 2 hours
3	Ageing	Vacuum/argon atmosphere. Fast heating to the target temperature	620°C	8 hr	Air cool

Table 6: Set 3: instructions for treatment (S + H + A).

HT#	Description	Conditions	Holding temperature	Holding time	Cooling
1	Homogenization and solution treatment	Argon atmosphere. Fast heating to the target temperature	980°C	1 hr	Air cool
2	HIP	100 MPa	1160°C	4 hr	Air cool
3	Ageing	Vacuum/argon atmosphere. Fast heating to the target temperature	720°C	8 hr	Furnace cool to 620°C within 2 hours
4	Ageing	Vacuum/argon atmosphere. Fast heating to the target temperature	620°C	8 hr	Air cool

$$\sigma = \frac{F}{A_0}\frac{w_0 t_0}{wt} = \frac{F}{A_0}\frac{1}{\sqrt{1+2\left[\left((w-w_0)/w_0\right)+\left((1/2)\left((w-w_0)/w_0\right)^2\right)\right]}\sqrt{1+2\left[\left((t-t_0)/t_0\right)+\left((1/2)\left((t-t_0)/t_0\right)^2\right)\right]}}, \tag{6}$$

or, in terms of strain, where subscript a denotes the axial direction:

$$\sigma_a = \frac{F}{A_0\left(\sqrt{1+2\bar{\varepsilon}_w}\,\sqrt{1+2\bar{\varepsilon}_t}\right)}. \tag{7}$$

Here, σ_a is the Cauchy stress, F is the current force, A_0 is the initial specimen cross-sectional area, and $\bar{\varepsilon}_w$ and $\bar{\varepsilon}_t$ are the average Lagrange strains [46] in the transverse directions for all points in the current cross section of the two AOIs, respectively. Thereafter, the Cauchy stress was used to find Young's moduli of the specimen in the load direction according to ASTM E111. Poisson's ratios were determined according to ASTM E132. With all the calculated terms, the compliance matrix can be constructed for a transversely isotropic material based on Equation (1) with colour codes corresponding to the samples in Figure 4.

2.4. Determination of Elastic Constants with an Optimization Algorithm. Determining anisotropic material data via direct measurements requires a large dataset to eliminate random errors in the measurements. To improve the significance of the measurements and reduce the number of test specimens, an optimization algorithm is implemented. This algorithm determines the elastic constants as a complementary solution to direct measurement using DIC. As it will be shown in this article, the benefit of determining the elastic constants by such optimization is that the data will be less vulnerable to noise in the strain measurements, since the algorithm weighs all the stress-strain pairs equally.

In this study, a set of measured stresses and corresponding strains is known for the tensile specimens with respect to the principal material direction. An optimization algorithm was implemented to optimize the compliance matrix.

Consider the transversely isotropic case for which Hooke's law can be expressed as Equation (1). Each pair of strain and stress is related by the compliance matrix and a transformation matrix $\mathbf{K}$, as derived by e.g., pp. 54–55 in Ting [14]. Hooke's law transforms to $' = \mathbf{S}'\boldsymbol{\sigma}'$, where $\boldsymbol{\sigma}' = \mathbf{K}\boldsymbol{\sigma}$, and $\mathbf{S}' = (\mathbf{K}^{-1})^T\mathbf{S}\mathbf{K}^{-1}$, which leads to $' = (\mathbf{K}^{-1})^T\mathbf{S}\mathbf{K}^{-1}\mathbf{K}\boldsymbol{\sigma} = (\mathbf{K}^{-1})^T\mathbf{S}\boldsymbol{\sigma}$. $\mathbf{S}'$ and $\mathbf{S}$ are the transformed and original compliance matrix.

In the current study the Lagrange strain tensor is measured directly using DIC. The shear stress and strain were found to be negligible, allowing the procedure described in Equation (3) to be used to calculate the shear moduli. When a relationship between stress and strain is established, the strain tensor can be expressed as

$$^{*} = \left(\mathbf{K}^{-1}\right)^{T} \mathbf{S}\mathbf{K}^{-1}\boldsymbol{\sigma}, \quad (8)$$

where $\boldsymbol{\sigma}$ is the Cauchy stress tensor from Equation (7). The objective of the optimization problem is to minimize a cost function $J(\theta)$. A common cost function for problems like this is the root-sum-of-squares function shown in the following equation:

$$J(\theta) = \sum_{i=1}^{m} \sqrt{^{2} - {}^{\prime 2}}, \quad (9)$$

where is the measured strain in the elastic region. An optimization problem like this can be implemented by e.g., the function fmincon in Matlab. The material is assumed to increase in volume and contract in the transverse directions when undergoing tensile stress. This in turn constrains the Poisson's ratio in any orientation to be bound by $0 < \nu_{ij} < 0.5$. As the elastic moduli in the 0°, 45° and 90° orientations are known, the problem is reduced to determining the Poisson's ratios. The elastic moduli in the compliance matrix (Equation (1)) are known and are used together with the constraints on the Poisson's ratios to determine ν_{xy}, ν_{xz}, and ν_{zx}. The shear moduli are determined based on Equations (2) and (3).

2.5. Electron Microscopy and Electron Backscattered Diffraction (EBSD) Analysis. In order to investigate the effect of postprocessing conditions on the grain structure using SEM, the samples were prepared by means of mechanical grinding and polishing followed by etching. Glyceregia solution was used for etching the samples. In the interest of studying crystallographic texture and preferred crystallographic orientations, EBSD analysis was employed. For this purpose, the 90° *ZY* specimens from set 1 and set 3 were selected. These samples were also prepared by mechanical grinding and polishing, though, instead of final etching, Oxide Polishing Suspension (OP-S, 0.04 μm) from Struers was applied. The EBSD sample was extracted from the side wall of this specimen. The backscattered patterns were collected in offline mode using FEI NOVA NANOSEM 650 ultralow vacuum field emission gun scanning electron microscope (SEM). An EBSD detector (Oxford Instruments NordlysMax3) was used to record approximately 1000 offline patterns per second. The working distance was set to 11 mm with a tilt angle of 70° under a high-current electron beam. The accelerating voltage of the gun was 20 kV. An area of 800 μm × 800 μm was scanned with step size 400 nm, accumulating crystallographic data for 4 million points for the selected area.

3. Results

3.1. Powder Characterization. Figure 5 shows the statistical values for the particle analysis. The figure contains information about circle equivalent (CE) diameter, high sensitivity (HS) circularity, aspect ratio and elongation for the analysed powder. Description of these entities is given in reference [49]. Moreover, selected individual particles in the vicinity of the peak of the distribution are shown in the figure. The vertical axis in the CE diameter figure is reported in arbitrary units (a.u.). The vertical and horizontal axes in the histogram figure are reflected in the upper and left figures in Figure 5.

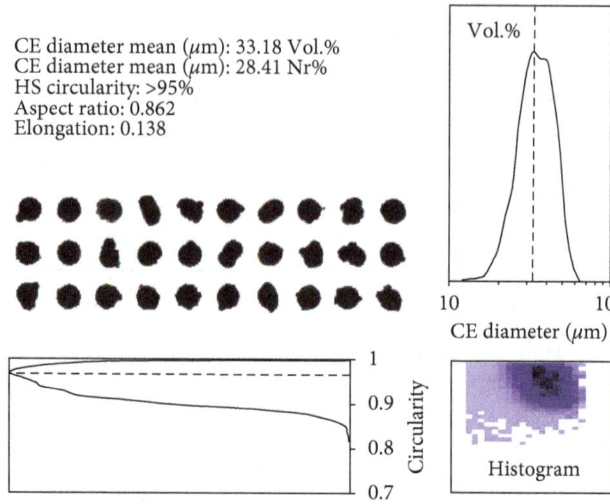

Figure 5: IN718 powder morphology analysis.

Figure 6 shows the chemical composition mapping of the particles. Some particles are rich in Al-Co-Ni content and depleted in other elements. Although there are local inhomogeneities in the chemical composition of the particles, the mass composition of the powder is within the specifications of IN718 datasheet.

3.2. Microstructure. Figure 7 shows SEM images of the *XY* + 45B tensile sample from set 1 at three locations. The observed steps are a result of the "stair-stepping" effect, and the distance between each step perpendicular to the page is the layer height. The edge steps were measured directly on the figure and the layer height was calculated based on the known nominal inclination (45°). The layer height was also checked and compared, both by focusing the scanning electron beam of the microscope on each two adjacent steps and measuring from the cross section. The results are tabulated in Table 7.

Figure 8 shows high-magnification SEM micrographs of a specimen from set 1. Figure 8(a) shows the fusion lines along the build direction, and Figure 8(b) shows the $\gamma + \gamma''$ lamellae structure at a higher magnification. White and blue arrows show some of the typical flaws in the matrix resulting from lack of fusion, and microsegregation in the interdendritic regions, respectively.

SEM micrographs of the aged specimens (SET 2) in Figure 9 show dendritic lamellae and grain boundary δ/η-phase evolution after ageing.

Figure 10(a) illustrates the acquisition conditions for the EBSD investigations. The build direction is along *Y*0 in the figure, leaving the LPBF layers perpendicular to *Y*0. Figure 10(b) presents the inverse pole figure *Y*0 (IPF-*Y*0) of the scanned area. In this frame, north of the figure is *X*0 and east

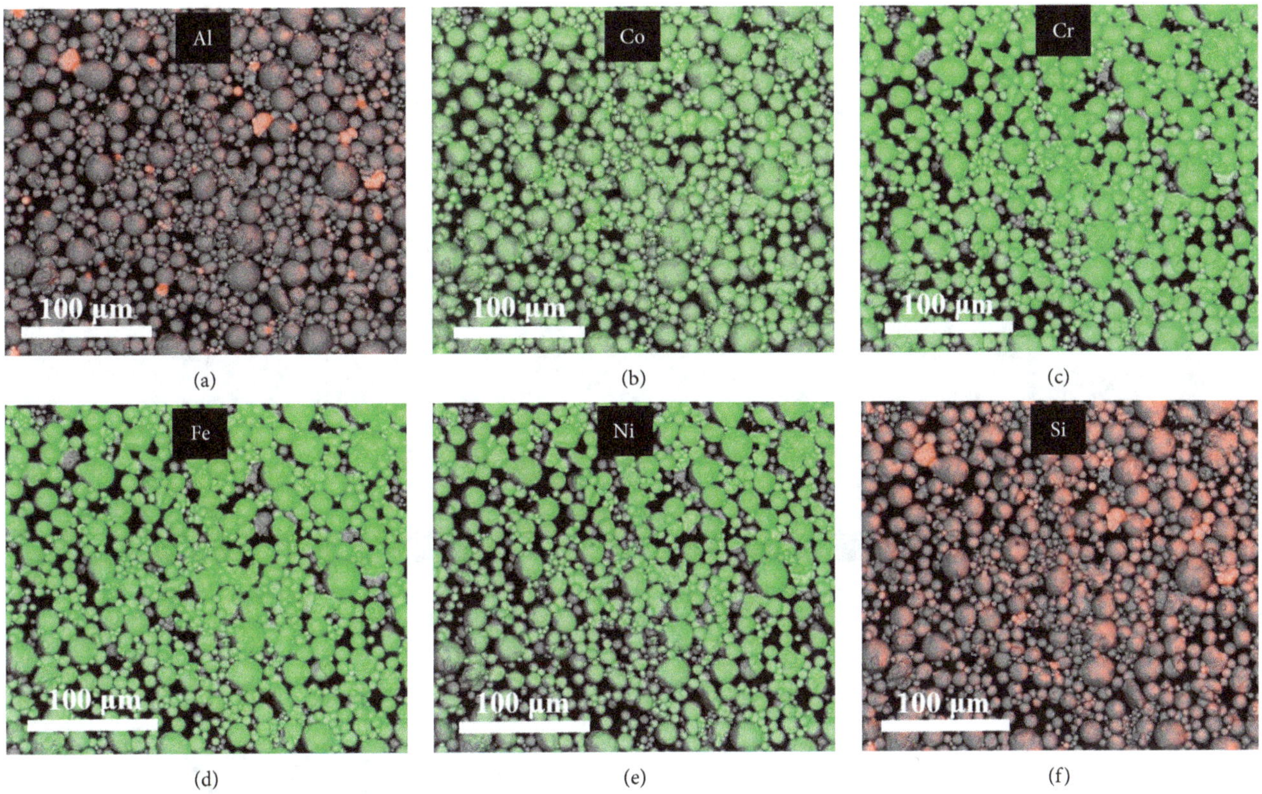

FIGURE 6: EDS chemical composition mapping of the used IN718 powder.

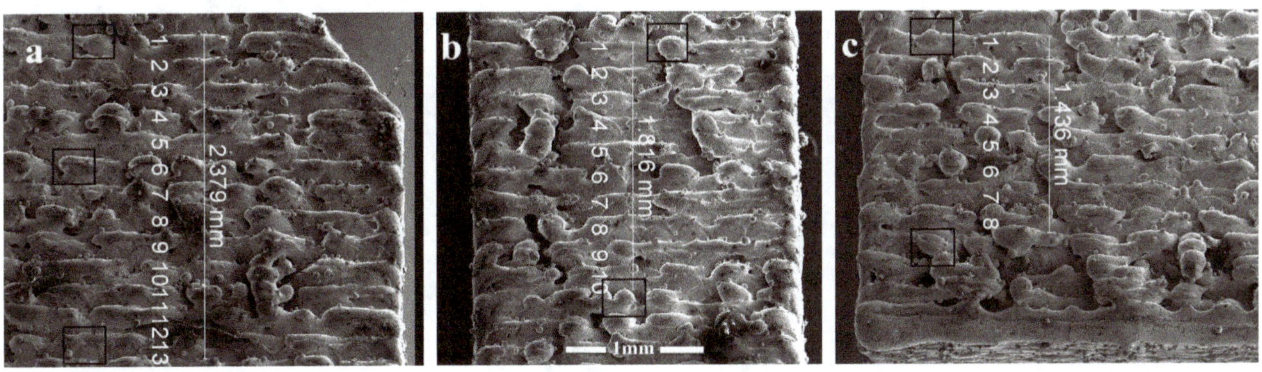

FIGURE 7: Surface topography of the upwards facing surface of the $XY + 45B$ specimen from set 1 at the (a) top; (b) middle; and (c) bottom of the tensile sample.

TABLE 7: Layer and edge step measurements.

Site	Layers	Total width (μm)	Width of single layer (μm)	Layer height (μm)
a	13	2379	183.0	49.0
b	10	1816	181.6	48.7
c	8	1436	179.5	48.1

of the frame is $Y0$. The dataset was rotated 90° counterclockwise to capture the texture of the material correctly.

Figures 11(a) and 11(b) show the inverse pole figures of the scanned area for samples from set 1 and set 3. It can be seen that the preferred orientation of {100}<100>||$Y0$ is dominating in set 1 as a result of the build conditions. In other words, the {100} family of planes is preferably oriented towards the build orientation if the sample is set at 90° with respect to the build platform. Traces of texture components in the samples from set 3 can still be seen. However, due to extensive recrystallization and ageing after postprocessing (HIPing + ageing), the texture is weakened. This can be observed by comparing the orientation clustering densities in the legend of both figures.

Figure 12 shows the orientation distribution function (ODF) space of the scanned area for the sample from set 1. The ideal texture component of <100>||$Y0$ was also plotted in each section. It is apparent that the hot spots on different sections of $\phi 2$ Euler angle coincide well with the presented ideal orientation, suggesting that there is a dominating texture component in the specimen.

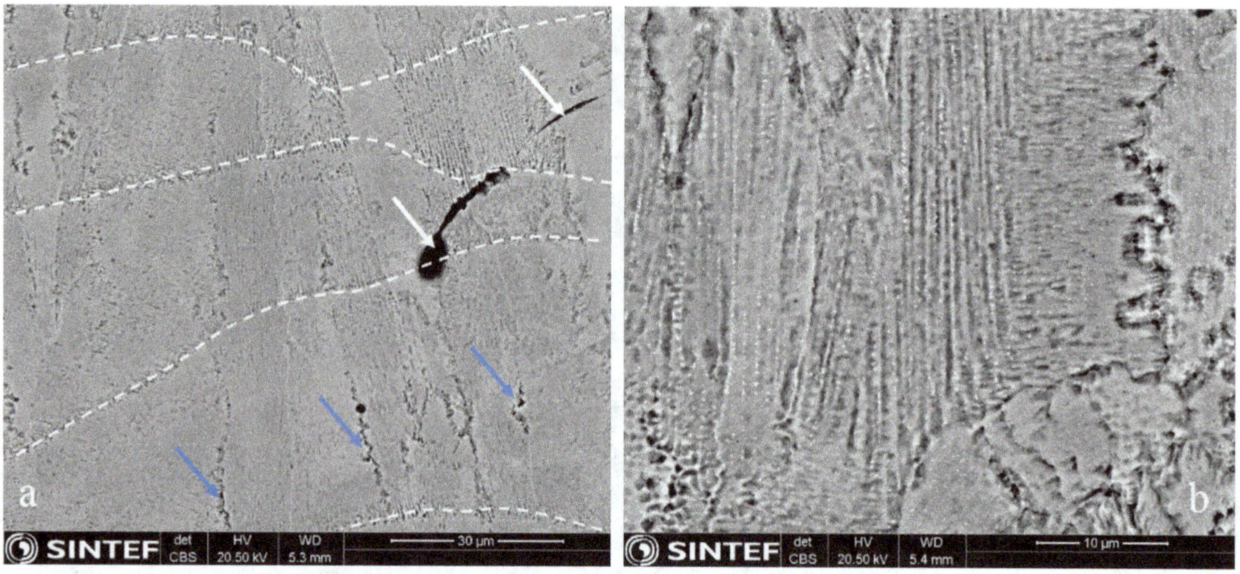

Figure 8: SEM micrographs from the solution-treated samples (Set 1) (a) showing the fusion lines, white arrows show the lack of fusion due to laser parameters and blue arrows show the inter-dendritic regions. (b) showing the dendritic $\gamma + \gamma''$ lamellae structure.

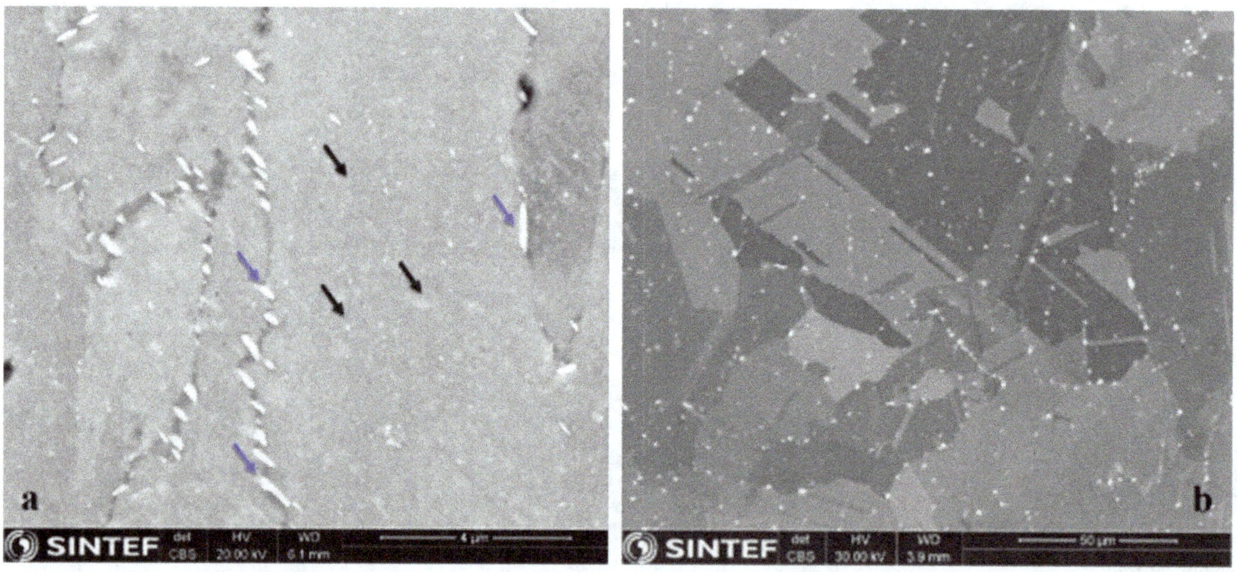

Figure 9: SEM micrograph of (a) aged IN718, and (b) HIPed and aged IN718. Black arrows show the interdendritic $\gamma + \gamma''$ lamellae, and blue arrows show the evolution of grain boundary δ/η-phase upon ageing. Extensive reconstructions of the grain structure, and grain boundary precipitates mostly on the prior boundaries of γ grains can be seen in (b).

3.3. Mechanical Properties. The relative density was measured to be 99.97% (set 1) and 99.98% (set 3). Figure 13 shows Cauchy stress versus Lagrange strain for set 1 (a), set 2 (b) and set 3 (c), with postprocessing as described in Tables 4–6, respectively.

Figure 14 shows the elastic modulus for set 1 (a), set 2 (b), and set 3 (c) for all build orientations. The error bars indicate the standard deviation of the strain field, as captured by DIC, in each specimen. Due to surface effects and porosity, the strain field on each specimen is not homogeneous. The elastic moduli are determined for the average strain (box), and for the average strain plus/minus the standard deviation (error bars).

Figure 15 shows the yield strength (a), ultimate tensile strength (b), and elongation at break (c) for all build orientations and heat treatment conditions. The average strain of each specimen is considered when determining the respective properties. The average value of the reference specimens is indicated with the green line.

The tensile properties of the IN718 specimens manufactured by LPBF, averaged over all the build orientations, are shown in Table 8. For comparison, the values for the

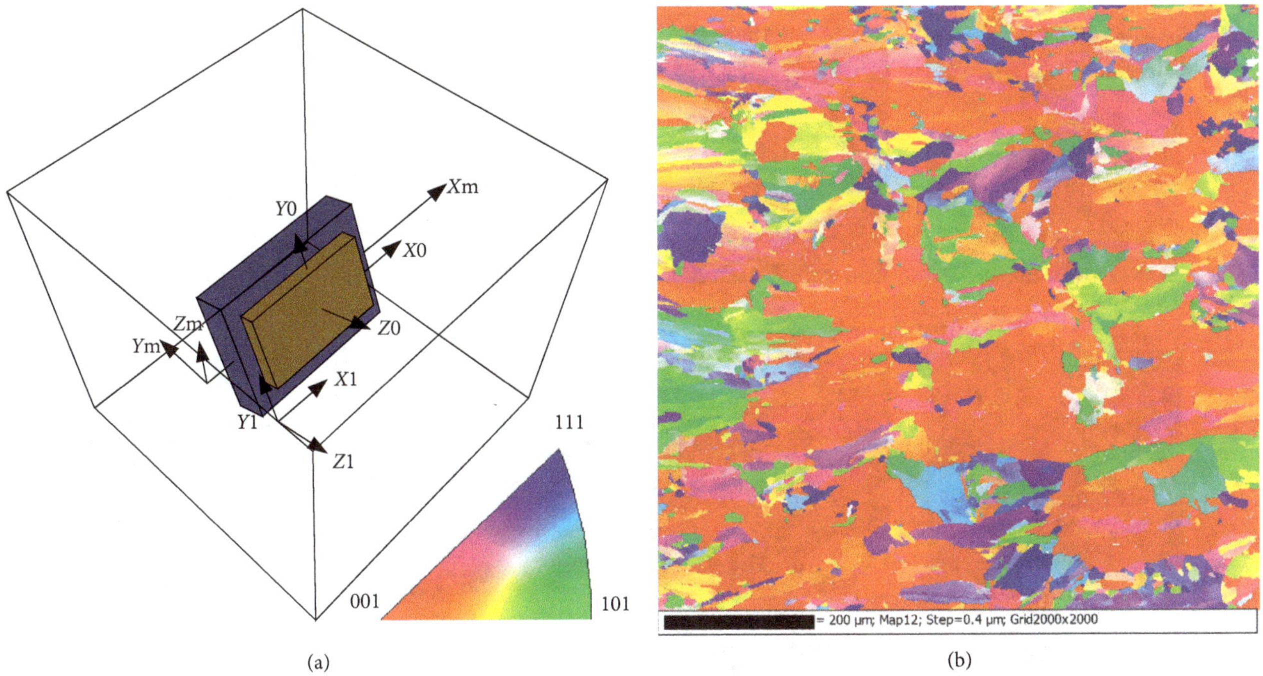

Figure 10: (a) Sample configuration and principal axes in the microscope chamber and (b) IPF-$Y0$ inverse pole figure grain map of the scanned area on the side wall of the sample oriented at 90° with respect to the build plate. The EBSD patterns were acquired with 0.4 μm step size.

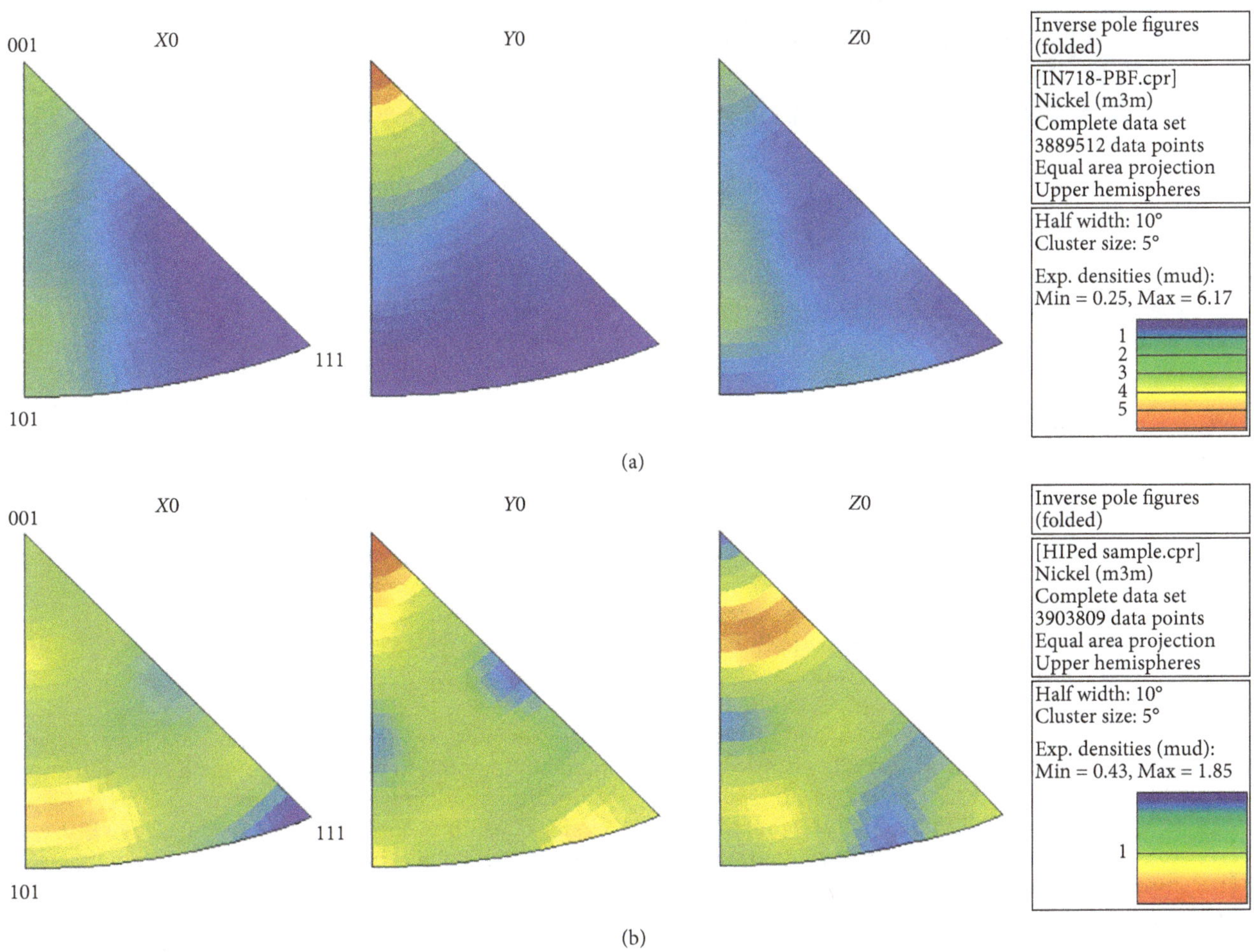

Figure 11: Inverse pole figure constructed based on 4 million data-points from the scanned area along the sample principal axes $X0$, $Y0$, and $Z0$ for (a) solution-treated and (b) HIP-treated sample.

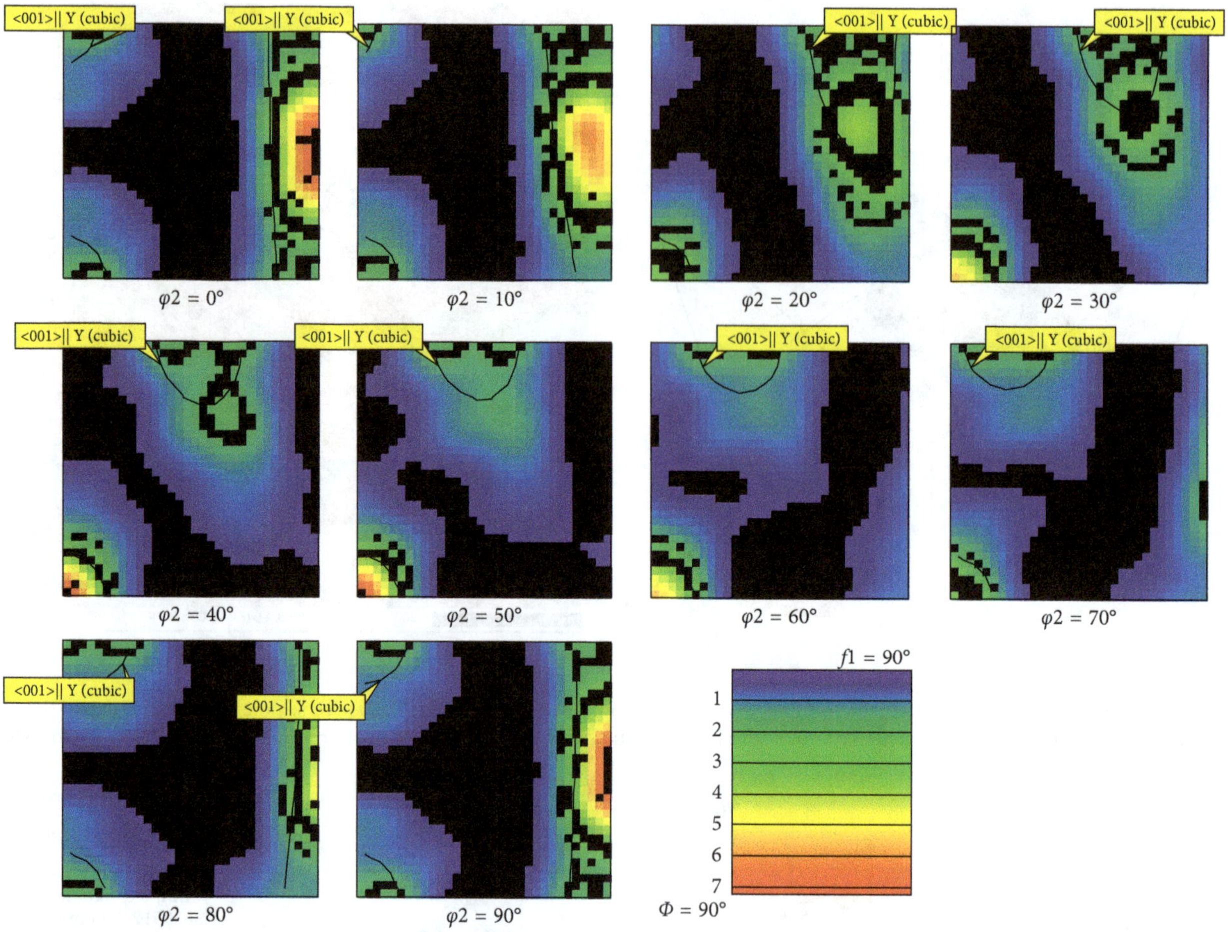

Figure 12: ODF sections of the scanned area. The boundaries of <001>||Y0 ideal texture components is shown with black solid line in different $\phi 2$ sections.

reference specimens (machined from rolled and aged IN718) are included in parenthesis.

3.4. Elastic Constants. The elastic constants, assuming transverse isotropy, acquired from the DIC strain fields are given in Table 9.

Table 10 shows the elastic constants obtained using the optimization algorithm described in Section 2.4 assuming transverse isotropy.

It is important to note that there are uncertainties associated with the calculated elastic constants. The elastic constants are determined from a calculated stress (Equation (7)), which is a function of the initial cross section area, the current transverse strain, and the applied force. The surface roughness is measured to be less than Ra = 1 μm for the 0° specimens, while peaking at the specimen oriented at 15° with Ra = 16 μm. The roughness then decreases to Ra = 5 μm as the specimens are raised towards 90°. HIP appears to reduce the surface roughness of the roughest surfaces by approximately 60%. The maximum uncertainty in the measurement of the initial cross section area due to the surface roughness is in the order of 80 μm^2, or 0.002% of the nominal cross section, introducing negligible uncertainty. The standard deviation of the transverse strain has a much larger contribution to the uncertainty. The measured force is assumed to be accurate followed by load cell calibration certificate. The error of the calculated stress can be quantified by introducing the uncertainty of the initial cross-sectional area and the standard deviation of the transverse strain to Equation (7). The average error of the calculated stress is 0.57%, 0.19%, and 0.19% for set 1, set 2, and set 3, respectively. Due to the low error, only average values have been used to determine the elastic constants.

4. Discussion

The presented anisotropic material data, and the general methodology to obtain these, provide a basis for improved numerical modelling (finite element analysis) of the mechanical response of components produced by LPBF without postproduction machining. Even if machining is required for tolerances and surface finish, complex geometries with large inaccessible surface areas will remain the weakest link, and this justifies the method for obtaining the mechanical properties of the unmachined specimens as proposed in this study.

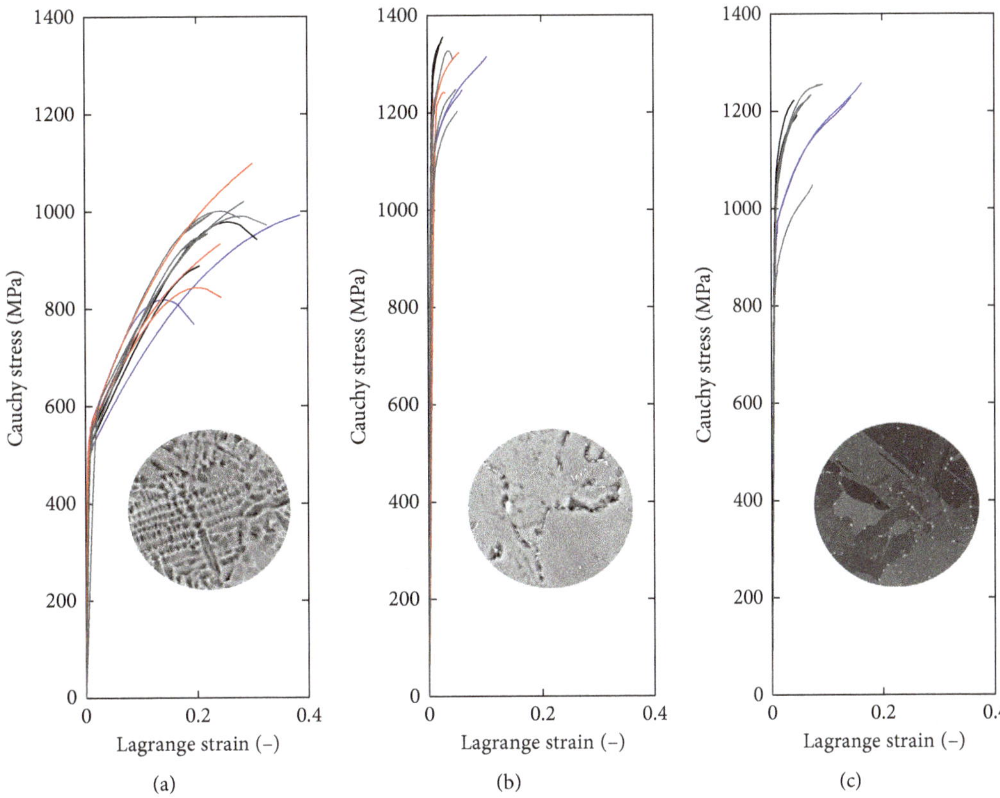

FIGURE 13: Cauchy stress plotted versus Lagrange strain for (a) S condition, (b) S + A condition, (c) S + H + A condition. Colours are according to Table 3.

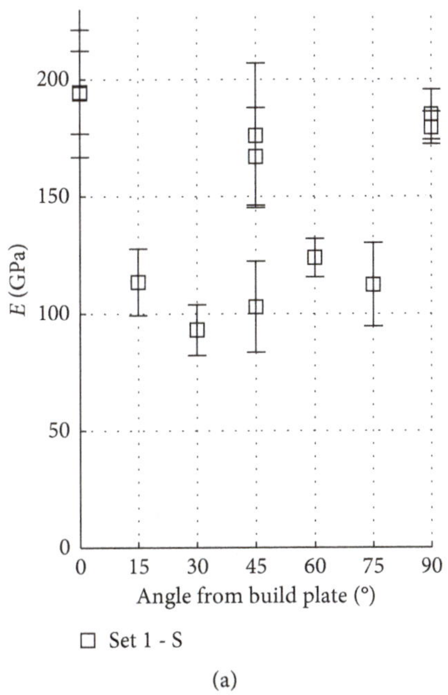

FIGURE 14: Mechanical properties of Set 1 (postprocessing "S"), Set 2 ("S + A"), and Set 3 ("S + H + A") with respect to orientation angle from build plate. Points indicate measured values (3 × 11 in each diagram), while the lines are either 2nd order polynomial fits (a), connecting lines (b and c) or least square fits (d), indicating the trend.

4.1. Powder, Laser, and Density Characteristics. Despite local inhomogeneities in the particle scale (Figure 6), the overall chemical composition of the powder is uniform. The presented defects in Figure 8(a), arising from metallurgical phenomena such as segregation or formation of eutectic phases, can possibly be related to local impurity enrichment

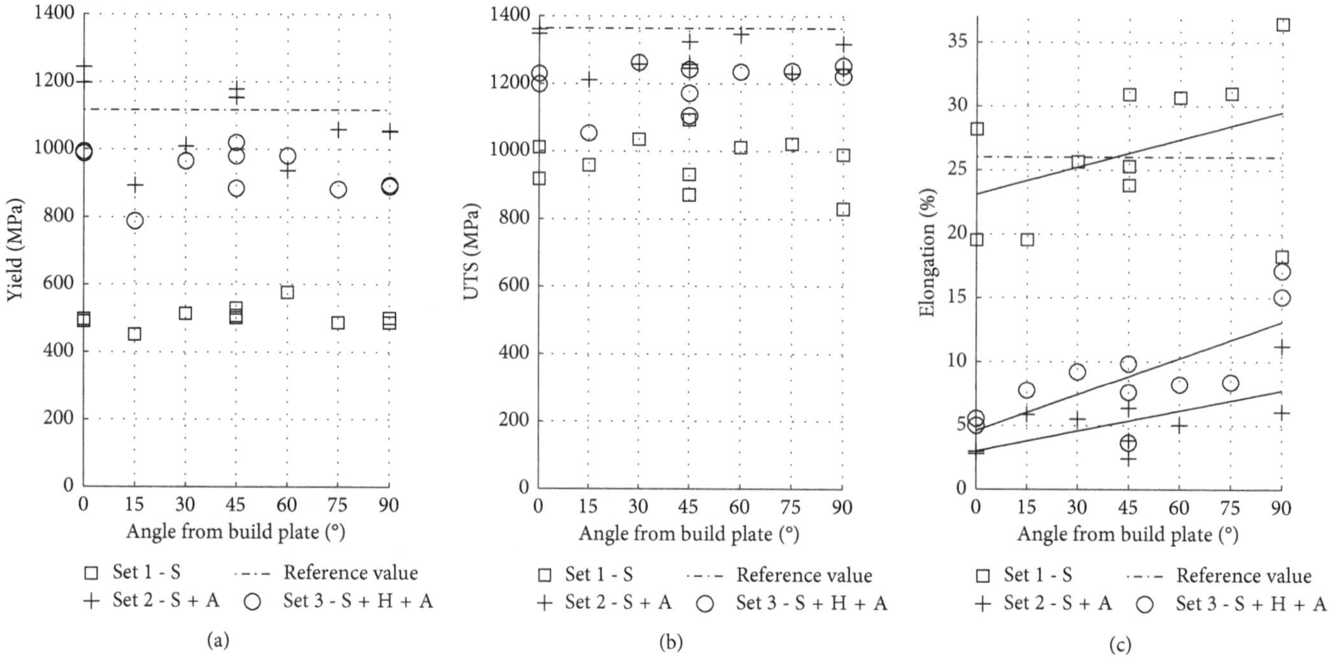

FIGURE 15: Yield strength (a), ultimate tensile strength (b), and elongation at break (c) of set 1, set 2, and set 3 with respect to orientation angle from build plate. Points indicate measured values while the lines in elongation indicate the least square fit.

TABLE 8: Average tensile properties for all eleven orientations. Values for the reference specimens (machined from rolled and aged IN718) are given in parenthesis.

	E (GPa)	UTS (MPa)	Yield stress (MPa)	Elongation at break (%)
SET 1	149 ± 40	970 ± 74	504 ± 30	26 ± 6
SET 2	181 ± 24 (193 ± 9)	1286 ± 52 (1364 ± 21)	1085 ± 106 (1116 ± 27)	5 ± 2 (26 ± 5)
SET 3	184 ± 30	1201 ± 63	933 ± 67	9 ± 4

TABLE 9: Elastic constants from DIC, for a transversely isotropic material model. This model is referred to as "DIC method" in figures below.

Elastic constants	Set 1 (S)	Set 2 (S + A)	Set 3 (S + H + A)
E_{xx} (GPa)	194	206	217
E_{zz} (GPa)	182	201	201
ν_{xy} (−)	0.4339	0.4748	0.4718
ν_{xz} (−)	0.4029	0.3650	0.4019
ν_{zx} (−)	0.3909	0.2045	0.2172
G_{xz} (GPa)	47	48	55
G_{xy} (GPa)	68	65	75

TABLE 10: Constants of optimized elastic model.

Elastic constants	Set 1 (S)	Set 2 (S + A)	Set 3 (S + H + A)
E_{xx} (GPa)	194	206	217
E_{zz} (GPa)	182	201	201
ν_{xy} (−)	0.3612	0.4994	0.3888
ν_{xz} (−)	0.3158	0.4999	0.4563
ν_{zx} (−)	0.3745	0.4877	0.4927
G_{xz} (GPa)	31	56	49
G_{xy} (GPa)	71	69	78

originating from non-IN718 powder particles. For instance, the presence of Al-rich particles (Figure 6) in the powder may result in anomalies and, therefore, proper cleaning of the build envelope, and powder-handling units play an important role in LPBF processing. Despite being recycled, the powder morphology and chemistry were well maintained (Figure 5), and the morphology measurements show no major deviations from the virgin powder specifications. The layer height after processing of the material decreased by 2–4% by comparing measured values in Table 7 with nominal values in Table 2. The source of such differences can be dependent on several parameters such as powder fusion and densification, or shrinkage and dilution between each two adjacent layers [50]. On the one hand, a small difference between the measured and nominal values implies a relatively high packing ratio of the powder particles prior to fusion. On the other hand, it may also imply limited dilution between two adjacent layers. The latter case can be detrimental if low dilution leads to lack of fusion between the layers and cause formation of cracks. Lack of fusion was rarely found in the large areas being investigated, although Figure 8(a) proves the presence of such anomalies. In these cases, subsequent HIP will be required. Full closure of the crack will depend on the local chemistry, location and morphology of the anomaly, type of trapped gas inside cavities (pores), and the HIP parameters for that specific material. It is worth mentioning that porosity in the powder particles may also play a central role in the evolution of postfusion porosity, since powder defects may retain throughout the entire process [51].

In this study, the energy per area (E_A) was 2.85 MJ/mm^2 (c.f. Table 2). Jia and Gu [52] achieved a relative density of 98.4% with an energy per area of 4.64 MJ/mm^2. Chlebus et al. [11] achieved a relative density in excess of 99.7% with an energy per area of 6.48 MJ/mm^2. According to the presented results in this work, high density can be achieved even with a low energy per area compared with the predecessor's applied parameters. Adjusting the process to build at high energy per area increases the baseline temperature and may cause the heat cycle to affect several layers below the last fusing surface. Depending on the peak temperature and the susceptibility of the material to form new phase constituents or anomalies, running the process at high temperature can be detrimental. Lower energy per area will confine the extent of the heat-affected zone (HAZ) in the previously fused layers.

4.2. Macro- and Microstructure. Beyond the effects of powder and part surface topography, the microstructure is also playing a significant role in determining the mechanical properties. Figures 8 and 9 show how the microstructure of a single alloy may change upon heat treatment. The grains are growing epitaxially throughout several layers before they are overgrown by other larger grains. The dendritic solidification pattern and $\gamma + \gamma''$ lamellae structure are visible in the microstructure. Observations of η and δ precipitates at the grain boundary were reported by several early researchers [42, 53, 54]. Their results unanimously propose that a small amount of these two phases may improve the notch sensitivity and resistance to intergranular crack propagation. The ageing of solution-treated IN718 is already showing these phases in the current study, and its indirect effect on the mechanical properties is shown in Figure 9(a). The benefit of HIP in eliminating anomalies was discussed earlier in Section 4.1. In addition to densification, the microstructure recrystallizes at high temperature during HIP, and the architecture of the grains and grain boundaries is largely affected [53–55]. According to Figure 9(b), the recrystallised grain boundaries transcend the primary γ grains, and subsequent ageing leads to the formation of precipitates not only on the recrystallised γ grains but also on the primary γ grains [26]. The presence of evenly distributed precipitates, even within the grains, seems to unitedly hamper the dislocation movements and causes a more uniform response of the material as can be seen when comparing set 1 with set 2 and set 3 in Figure 14(a).

In Figure 10(b), it was also illustrated that LPBF is a directional processing route that influences the grain structure, the grain morphology, and the crystallographic texture of the material. Preferred alignment of <100> along the build direction may weaken the material if it is built at an angle. The curvilinear heat flow vector argument is shown to be the dominant growth mechanism for high and low cooling kinetics [39]. The evolution of <100> along the build direction is primarily due to low layer thickness that embodies fewer grains within each layer, and high processing speed (i.e., laser scanning speed), which determines the range of grain growth kinetics. In such conditions, unlike welding, for example, the grains are not given sufficient time to bend behind the heat source (see e.g., Figure 2(a)).

4.3. Mechanical Properties. Consider the stress-strain curves presented in Figure 13. Ageing (Figures 13(b) and 13(c)) increases the yield stress and ultimate tensile strength at the expense of elongation, compared with the nonaged specimens in Figure 13(a). HIP before ageing will somewhat reduce the yield strength and ultimate tensile strength and increase the elongation (Figures 13(b) and 13(c)).

Figure 14 shows the elastic moduli for all three sets as function of the specimens' build angle relative to the build platform. For set 1, the elastic modulus is highest for the specimen built at 0° with respect to the build plate. It then drops as the specimens are raised towards 45°. The elastic modulus then increases as the angle is raised further up to 90°. The three measured values for the specimens at 45° are 167 ± 20.8 GPa, 103 ± 19.5 GPa, and 176 ± 30.1 GPa, for orientations *XZ*-45B, *XY* + 45B, and *YZ* + 45A, respectively. A possible explanation of the large variation in the 45° data is given in Section 4.4. As can be seen in Figure 14, ageing will increase the modulus of the material significantly. The average modulus is 149 ± 40 GPa, 181 ± 24 GPa, and 184 ± 30 GPa for set 1, set 2, and set 3, respectively.

Recall that the microstructure in set 1 primarily consists of FCC γ-phase with slip system $\{111\}\langle 1\bar{1}0\rangle$, and that the material exhibits a strong {100} fibre texture (Figures 11 and 12). Consider, for arguments sake, the material as single crystal FCC. The elastic modulus in any particular crystallographic direction can then be calculated using the following equation [56]:

$$\frac{1}{E_{[hkl]}} = \frac{1}{E_{<100>}} - 3\left(\frac{1}{E_{<100>}} - \frac{1}{E_{<111>}}\right)\left(\alpha^2\beta^2 + \alpha^2\gamma^2 + \beta^2\gamma^2\right), \tag{10}$$

where $E_{<100>}$ and $E_{<111>}$ are the moduli in the <100> and <111> directions, respectively, and α, β, and γ describe a unit vector $\mathbf{v} = [\,\alpha \;\; \beta \;\; \gamma\,]$ within the crystallographic plane of interest where α, β, and γ are the direction cosines with respect to [100], [010], and [001]. In the current study, we are interested in the modulus in the {100} plane under different loading angles. The inset in Figure 16 shows the {100} plane in a cubic system with the unit vector $\mathbf{v} = [0\cos(\theta)\sin(\theta)]$, and the blue line (Equation (10)) shows the moduli under different load angles with values for $E_{<100>}$ being greater than $E_{<111>}$ as measured with DIC for set 1. The calculated modulus in Figure 16 compares well with the observed behaviour of the material as seen in Figure 14(a).

Figure 15(a) shows the yield stress for all specimens as a function of build angle. For set 1, the yield stress is at its maximum at a 60° angle to the build plate. Although the sample size for set 1 is small, and thus not statistically significant, crystallographic theory supports the observed

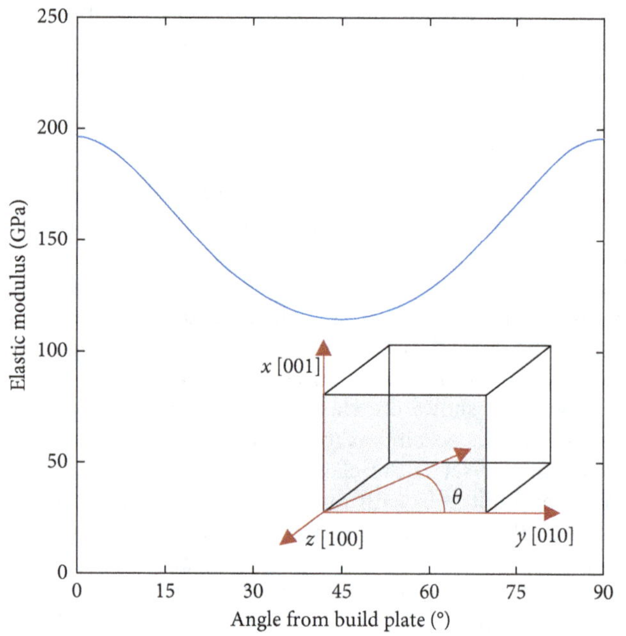

FIGURE 16: Elastic modulus is according to Equation (10) with a value for $E_{<100>} > E_{<111>}$ as measured with DIC for set 1. Inset: {100} plane in a FCC crystal with *xyz* and corresponding (hkl) coordinates.

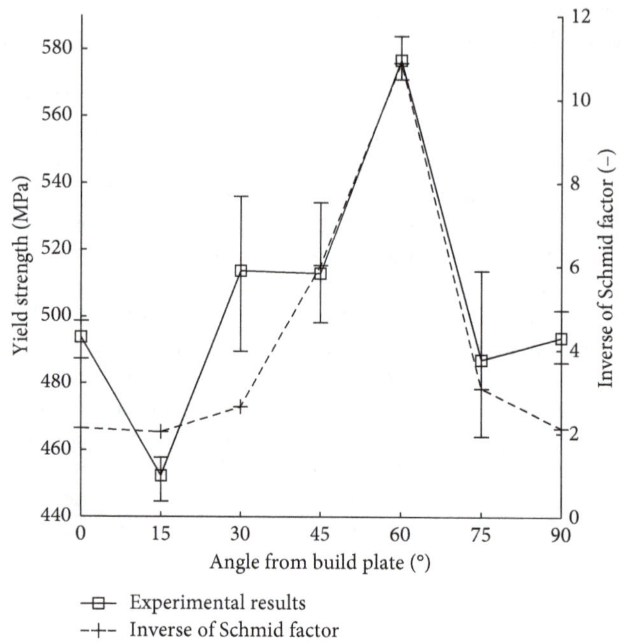

FIGURE 17: Yield stress for set 1 compared with the inverse Schmid factor for set 1.

maxima close to the theoretical maxima at 54°, as will be shown in the next paragraph. In the aged and HIP + aged condition, the scatter in the data is large and no clear anisotropic trends are observed. However, it is clear that the average yield stress increases as the material is aged, and HIP before ageing reduces the average yield stress somewhat.

Consider again the material as single-crystal FCC. The Schmid factor can then be employed to describe the anisotropic behaviour with respect to build angle, as previously done by Ni et al. [32], for the same material and process as in the present study. The yield stress can be expressed as

$$\sigma_{\text{yield}} \geq \frac{\tau_c}{\cos(\phi)\cos(\beta)}, \tag{11}$$

where τ_c is the critical resolved shear stress in slip systems, φ is the angle between the axial load and the normal direction of the slip plane, and β is the angle between the axial load and the slip direction. The term $\cos(\phi)\cos(\beta)$ is known as the Schmid factor. With the current slip system φ and β sum to 90° and the Schmid factor can be expressed as $(1/2)\sin(2\phi)$. For set 1, the yield stress and the inverse of the Schmid factor are plotted as function of build angle in Figure 17. The error bars represent the standard deviation of the strain field for each specimen. The general trend for the inverse Schmid factor and the yield stress is the same. If the material did not exhibit the strong texture observed in this study, Taylor's factor may be employed to estimate the deformation instead of the Schmid factor. However, with the texture exhibited in the present material the Taylor factor and the Schmid factor will be comparable.

The trend of the yield strength of set 1 corresponds with the Schmid factor, as shown in Figure 17, with a minimum in yield strength at a 15° angle from the build plate and a maximum yield strength at a 60° angle from the build plate. This trend is not observed for set 2 and set 3.

After ageing, and after HIP and ageing, the average yield stress increases to 1085 ± 106 MPa and 933 ± 67 MPa, respectively. The precipitation hardening that occurs during ageing will increase the yield stress due to the pinning of grain boundaries and the stopping of dislocation movement. The standard deviation of the yield stress is reduced when HIP is done prior to ageing, indicating a homogenization effect. The magnitude of the yield stress is also slightly reduced when HIP is done before ageing, most likely due to recrystallization and evolution of twin grain boundaries.

Figures 15 and 14(b) show the ultimate tensile strength for the three different heat treatment sets as function of build angle. Again, it is obvious that ageing increases the materials strength. The ultimate tensile strength in set 1 is significantly lower compared with the aged variants. It is observed that the ultimate tensile strength is lower when the material has undergone HIP processing, as was the case for yield stress. Further analysis of ultimate tensile strength is considered unnecessary for this material as its main application is in aerospace where plastic deformation would be considered as failure.

Figure 15(c) shows the elongation at break for the three different heat treatment sets as function of build angle. The average elongation at break for set 1 is 26 ± 5%, with increasing elongation at break as the build angle increases. The same trend is observed for the heat-treated samples, but the magnitude of the elongation at break is greatly reduced. The elongation at break for set 3 (S + H + A) is higher compared with the elongation at break for set 2 (S + A). This is attributed to a reduction in the residual stresses and hardness of the precipitates after HIP.

The angle between the axial load and the material texture will determine the failure mode. For the tensile specimens with main loading axis in the build plate plane (0°), the columnar grains are perpendicular to the loading direction (c.f. growth mechanisms in Section 1.3), promoting mode I type failure. For the specimens with main axis along the perpendicular to the build plate (90° build angle), the columnar grains are in general parallel to the load direction. This will limit the possibility of mode I failure. The lowest elongation at break should then be observed for the 0° specimens and increase gradually to a maximum for the 90° specimens. This hypothesis is supported by the data seen in Figure 15(c).

Consider the average mechanical properties for all orientations in Table 8. The results for the aged specimens (set 2) compare well with those of the reference specimens (machined from rolled and aged IN718). The specimens of set 2 have average modulus, yield stress, and UTS that are 94%, 94%, and 97% of the reference's values, respectively. However, the elongation at break is significantly higher for the reference specimens than for the set 2 specimens.

4.4. Elastic Constants. The elastic constants given in Tables 9 and 10 can be used in FE analysis (e.g., Abaqus) directly to predict the anisotropic elastic behaviour of a general three-dimensional part with a certain build orientation. Next, the model is discussed and verified by performing simulations with load conditions simulating the physical experiments. The first approach of determining the elastic constants, based on direct DIC measurements, is referred to as the "DIC method," while the approach employing the optimization algorithm is referred to as the "optimized method."

Figure 18 shows the elastic modulus numerically determined with the DIC method and the optimized method, compared with the experimental results for each heat treatment condition (sets 1, 2, and 3). When using the DIC method, the elastic modulus follows a 4th order polynomial as discussed in the previous section (Equation (10) with $E(0°) > E(90°) > E(45°)$). The same trend is observed with the optimized method, which follows the experimentally determined data more closely. This is a result of how the optimization algorithm treats each individual stress/strain pair equally, and thus the high values at 45° are given less weight.

The experimentally determined elastic moduli deviate from the calculated moduli, with higher modulus for the 45° angle, particularly for the two specimens with orientation *XZ*-45B and *YZ* + 45A, which are not simply rotated from a "flat" orientation such as *XY* + 45B (Figure 4 and Table 3). For the orientations *XZ*-45B and *YZ* + 45A, the projection of the specimen cross section along the *Z* axis (of the build chamber) is $w/\cos(45°)$, where w is the width of the sample (2.5 mm in the gauge section). On the other hand, for the orientation *XY* + 45B, the projection of the specimen cross section along the *Z* axis is $t/\cos(45°)$, where t is the thickness of the sample (1.5 mm). When loading specimens with 45° build angle, the grain boundaries will either align with the slip line and cause de-cohesion, or twin subgrains will form, which will introduce twinning-induced plasticity [57]. The experimental results shown in Figure 18(a) might indicate that for the orientations *XZ*-45B and *YZ* + 45A, subgrains are formed, while for *XY* + 45B, the grains align with the slip line and the plasticity is higher. Chlebus et al. [11] reported values for moduli in the solution treated only condition (set 1) with the trend $E(45)° > E(0°) > E(90°)$, which is in line with activation of slip lines under loading. Considering the magnitude of the moduli for specimens oriented at angles 15°, 30°, 60°, and 75° in the present study, the data suggest the worst-case scenario should be assumed, where the grains align with the slip line. This will in turn give a more conservative model. There are also significant differences between the study by Chlebus et al. and the present study when it comes to laser parameters, alloy composition, and heat treatment procedures. The spread in the $E(45°)$ data is far larger than the spread in the data for $E(0°)$ and $E(90°)$ for all three sets, which indicates that some competing effect—like the one described in this paragraph—might be the source of the presented disparity.

Using the DIC method, the modulus is calculated to be higher than the experimental values for 15°, 30°, 60°, and 75°, resulting in a possibility to overestimate the stiffness in components with walls, etc. oriented in this range of angles.

Both the DIC method and the optimized method predict moduli higher than the observed values for 15°, 30°, 60°, and 75°, with the exception of set 2 at 30° and set 3 at 30° and 60°. However, the optimized method is not as strongly influenced by the apparent outliers at 15°, 60°, and 75° compared with the DIC method. This in turn suggests that the optimized method is less vulnerable to noise in the data. For all heat treatment conditions, the optimized method gives a more conservative estimate of the moduli.

5. Conclusions

(i) A method has been demonstrated for obtaining anisotropic (transverse isotropic) elastic constants for material processed by laser powder bed fusion (LPBF). The method is based on building a set of tensile specimens, oriented in a systematic manner on the build plate, and using digital image correlation (DIC) during the tensile tests. The current study is focused on the response of nonmachined specimens, inspired by the fact that in complicated AM geometries, postmachining may be difficult or even impossible.

(ii) Extensive tensile testing of a single specimen geometry built with different orientations shows that the LPBF process induces mechanical anisotropy.

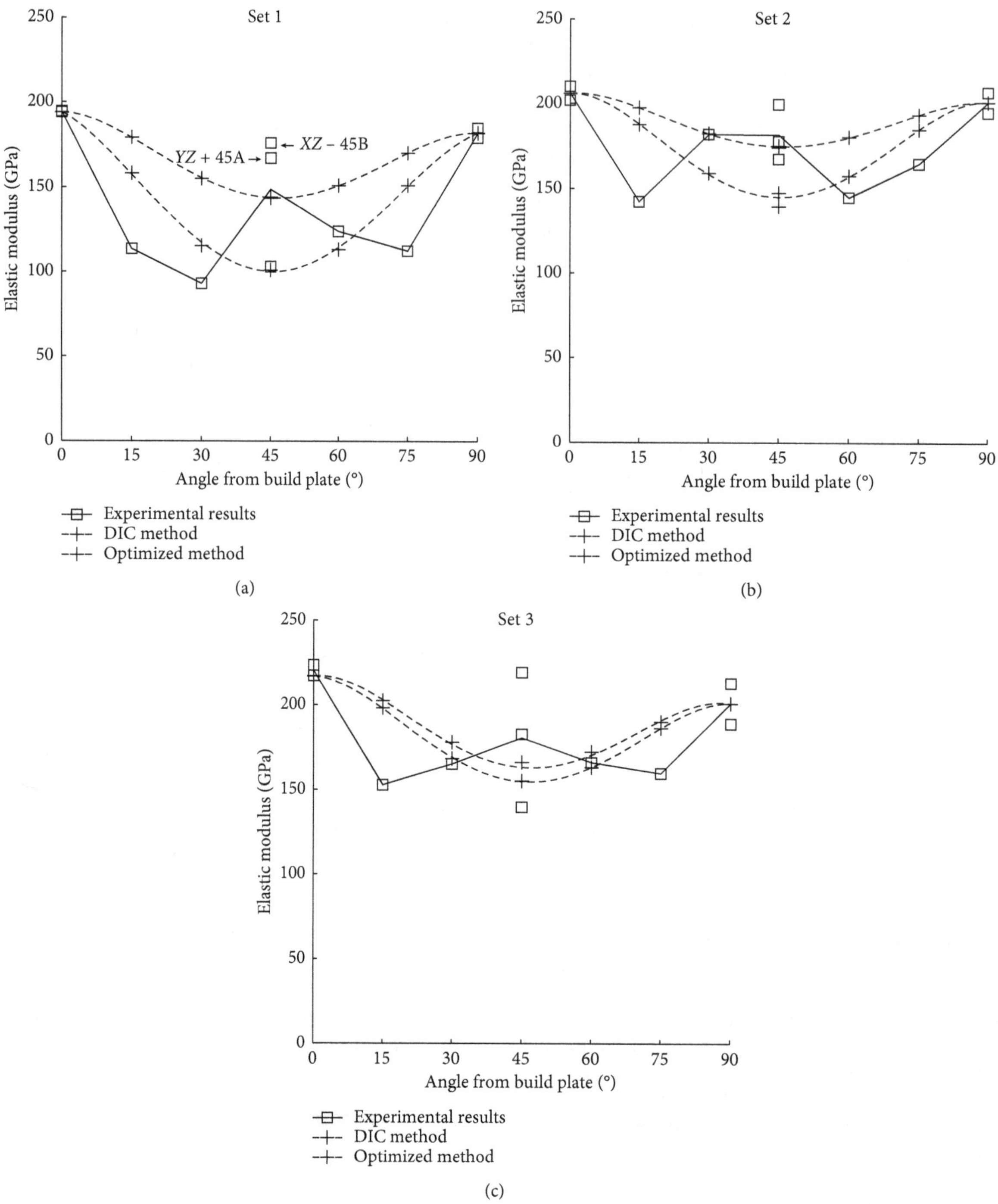

FIGURE 18: Elastic modulus in the load direction as measured directly, calculated based on the DIC measurements, and calculated based on the optimized elastic model for (a) as built condition, (b) aged condition, and (c) HIP + aged condition.

Therefore, care should be taken when orienting components in the build chamber to maximize the mechanical properties.

(iii) Two approaches for determining the anisotropic elastic constants are presented in this study. In the future, more experimental data will be included to consolidate the reliability and robustness of the determined elastic constants.

(iv) While all the tensile properties are influenced by the build angle of the test specimen relative to the build plate, the elastic modulus and the elongation at break show the highest dependence on build angle.

(v) Ageing has a large effect on the mechanical properties; it improves the elastic modulus, yield stress, and ultimate tensile strength, but reduces the elongation at break. Hot isostatic pressing (HIP) before ageing increases the elongation at break at the expense of a reduction in elastic modulus, yield stress, and ultimate tensile strength. Furthermore, HIP does not reduce the scatter in mechanical properties among the tested samples, and evidence of the initial structure and texture is still apparent.

(vi) This study confirms that LPBF processing of IN718 introduces texture, and the variation in mechanical properties can be explained by crystallographic theories such as Schmid's theory.

Conflicts of Interest

The authors declare that there are no conflicts of interest regarding the publication of this paper.

Authors' Contributions

The authors Even W. Hovig and Amin S. Azar contributed equally to this work.

Acknowledgments

The authors acknowledge funding from the Research Council of Norway (grant no. 248243/MKRAM project), GKN Aerospace Norway, Kongsberg Automotive Raufoss, Nammo Raufoss, OM BE Plast, and Sandvik Teeness. The additively manufactured specimens for this study were provided by PROMET AS. HIP and ageing heat treatments were performed by Bodycote AB in Sweden. Joachim Seland Graff, Martin Fleissner Sunding, and Alaa Mourad at SINTEF are greatly acknowledged for their technical support.

References

[1] B. E. Carroll, T. A. Palmer, and A. M. Beese, "Anisotropic tensile behavior of Ti-6Al-4V components fabricated with directed energy deposition additive manufacturing," *Acta Materialia*, vol. 87, pp. 309–320, 2015.

[2] W. E. Luecke and J. A. Slotwinski, "Mechanical properties of austenitic stainless steel made by additive manufacturing," *Journal of research of the National Institute of Standards and Technology*, vol. 119, p. 398, 2014.

[3] T. Wang, Y. Zhu, S. Zhang, H. Tang, and H. Wang, "Grain morphology evolution behavior of titanium alloy components during laser melting deposition additive manufacturing," *Journal of Alloys and Compounds*, vol. 632, pp. 505–513, 2015.

[4] J. J. Lewandowski and M. Seifi, "Metal additive manufacturing: a review of mechanical properties," *Annual Review of Materials Research*, vol. 46, no. 1, pp. 151–186, 2016.

[5] L. Thijs, M. L. M. Sistiaga, R. Wauthle, Q. Xie, J.-P. Kruth, and J. Van Humbeeck, "Strong morphological and crystallographic texture and resulting yield strength anisotropy in selective laser melted tantalum," *Acta Materialia*, vol. 61, no. 12, pp. 4657–4668, 2013.

[6] F. List, R. R. Dehoff, L. E. Lowe, and W. J. Sames, "Properties of Inconel 625 mesh structures grown by electron beam additive manufacturing," *Materials Science and Engineering: A*, vol. 615, pp. 191–197, 2014.

[7] N. Gardan and A. Schneider, "Topological optimization of internal patterns and support in additive manufacturing," *Journal of Manufacturing Systems*, vol. 37, pp. 417–425, 2015.

[8] C. C. Swan and I. Kosaka, "Voigt-Reuss topology optimization for structures with nonlinear material behaviors," *International Journal for Numerical Methods in Engineering*, vol. 40, no. 20, pp. 3785–3814, 1997.

[9] B. Vrancken, L. Thijs, J.-P. Kruth, and J. Van Humbeeck, "Heat treatment of Ti6Al4V produced by selective laser melting: microstructure and mechanical properties," *Journal of Alloys and Compounds*, vol. 541, pp. 177–185, 2012.

[10] T. Vilaro, C. Colin, and J.-D. Bartout, "As-fabricated and heat-treated microstructures of the Ti-6Al-4V alloy processed by selective laser melting," *Metallurgical and Materials Transactions A*, vol. 42, no. 10, pp. 3190–3199, 2011.

[11] E. Chlebus, K. Gruber, B. Kuźnicka, J. Kurzac, and T. Kurzynowski, "Effect of heat treatment on the microstructure and mechanical properties of Inconel 718 processed by selective laser melting," *Materials Science and Engineering: A*, vol. 639, pp. 647–655, 2015.

[12] T. Trosch, J. Strößner, R. Völkl, and U. Glatzel, "Microstructure and mechanical properties of selective laser melted Inconel 718 compared to forging and casting," *Materials Letters*, vol. 164, pp. 428–431, 2016.

[13] D. Tomus, Y. Tian, P. A. Rometsch, M. Heilmaier, and X. Wu, "Influence of post heat treatments on anisotropy of mechanical behaviour and microstructure of Hastelloy-X parts produced by selective laser melting," *Materials Science and Engineering: A*, vol. 667, pp. 42–53, 2016.

[14] T. C. T. Ting, *Anisotropic Elasticity: Theory and Applications*, Oxford University Press, New York, NY, USA, 1996.

[15] E. V. Morozov and V. V. Vasiliev, "Determination of the shear modulus of orthotropic materials from off-axis tension tests," *Composite Structures*, vol. 62, no. 3-4, pp. 379–382, 2003.

[16] J. C. Lippold, S. D. Kiser, and J. N. DuPont, *Welding Metallurgy and Weldability of Nickel-Base Alloys*, John Wiley and Sons, Hoboken, NJ, USA, 2011.

[17] R. Cozar and A. Pineau, "Morphology of y'and y" precipitates and thermal stability of inconel 718 type alloys," *Metallurgical Transactions*, vol. 4, no. 1, pp. 47–59, 1973.

[18] S. Singh Handa, J. Andersson, and M. Eynian, *Precipitation of carbides in a Ni-based superalloy*, Ph.D. thesis, Department of Engineering Science, University West, Trollhättan, Sweden, 2014.

[19] N. Saunders, Z. Guo, X. Li, A. Miodownik, and J.-P. Schille, "Modelling the material properties and behaviour of Ni-based superalloys," in *Proceedings of the 10th International Symposium on Superalloys*, pp. 849–858, Champion, PA, USA, September 2004.

[20] A. Oradei-Basile and J. F. Radavich, "A current TTT diagram for wrought alloy 718," in *Proceedings of Superalloys 718, 625 and Various Derivatives*, pp. 325–335, Pittsburgh, PA, USA, June 1991.

[21] A. Simchi, "Direct laser sintering of metal powders: mechanism, kinetics and microstructural features," *Materials Science and Engineering: A*, vol. 428, no. 1, pp. 148–158, 2006.

[22] M. E. Aydinöz, F. Brenne, M. Schaper et al., "On the microstructural and mechanical properties of post-treated additively manufactured Inconel 718 superalloy under quasi-static and cyclic loading," *Materials Science and Engineering A*, vol. 669, pp. 246–258, 2016.

[23] D. L. Bourell, H. L. Marcus, J. W. Barlow, and J. J. Beaman, "Selective laser sintering of metals and ceramics," *In-*

ternational Journal of Powder Metallurgy*, vol. 28, no. 4, pp. 369–381, 1992.

[24] M. Dressler, M. Röllig, M. Schmidt, A. Maturilli, and J. Helbert, "Temperature distribution in powder beds during 3D printing," *Rapid Prototyping Journal*, vol. 16, no. 5, pp. 328–336, 2010.

[25] D. H. Smith, J. Bicknell, L. Jorgensen et al., "Microstructure and mechanical behavior of direct metal laser sintered Inconel alloy 718," *Materials Characterization*, vol. 113, pp. 1–9, 2016.

[26] K. Amato, S. Gaytan, L. Murr et al., "Microstructures and mechanical behavior of Inconel 718 fabricated by selective laser melting," *Acta Materialia*, vol. 60, no. 5, pp. 2229–2239, 2012.

[27] X. Zhao, J. Chen, X. Lin, and W. Huang, "Study on microstructure and mechanical properties of laser rapid forming Inconel 718," *Materials Science and Engineering: A*, vol. 478, no. 1-2, pp. 119–124, 2008.

[28] F. Liu, X. Lin, C. Huang et al., "The effect of laser scanning path on microstructures and mechanical properties of laser solid formed nickel-base superalloy Inconel 718," *Journal of Alloys and Compounds*, vol. 509, no. 13, pp. 4505–4509, 2011.

[29] A. Simchi and H. Pohl, "Effects of laser sintering processing parameters on the microstructure and densification of iron powder," *Materials Science and Engineering: A*, vol. 359, no. 1-2, pp. 119–128, 2003.

[30] B. N. Turner and S. A. Gold, "A review of melt extrusion additive manufacturing processes: II. Materials, dimensional accuracy, and surface roughness," *Rapid Prototyping Journal*, vol. 21, no. 3, pp. 250–261, 2015.

[31] W. E. Frazier, "Metal additive manufacturing: a review," *Journal of Materials Engineering and Performance*, vol. 23, no. 6, pp. 1917–1928, 2014.

[32] M. Ni, C. Chen, X. Wang et al., "Anisotropic tensile behavior of in situ precipitation strengthened Inconel 718 fabricated by additive manufacturing," *Materials Science and Engineering: A*, vol. 701, pp. 344–351, 2017.

[33] O. Scott-Emuakpor, T. George, C. Holycross, J. Bruns, and A. Zearley, "Bending fatigue life comparison between DMLS and cold-rolled nickel alloy 718," in *Proceedings of the Prognostics and Health Management Solutions Conference (MFPT 2014)*, Virginia Beach, VA, USA, May 2014.

[34] M. M. Kirka, D. Greeley, C. Hawkins, and R. Dehoff, "Effect of anisotropy and texture on the low cycle fatigue behavior of Inconel 718 processed via electron beam melting," *International Journal of Fatigue*, vol. 105, pp. 235–243, 2017.

[35] V. Badiru Vhance and D. A. B. Liu, *Valencia, Additive Manufacturing Handbook: Product Development for the Defense Industry*, CRC Press Inc., Boca Raton, FL, USA, 2017.

[36] J.-P. Choi, G.-H. Shin, S. Yang et al., "Densification and microstructural investigation of Inconel 718 parts fabricated by selective laser melting," *Powder Technology*, vol. 310, pp. 60–66, 2017.

[37] X. Wang, T. Keya, and K. Chou, "Build height effect on the Inconel 718 parts fabricated by selective laser melting," *Procedia Manufacturing*, vol. 5, pp. 1006–1017, 2016.

[38] A. Keshavarzkermani, M. Sadowski, and L. Ladani, "Direct metal laser melting of Inconel 718: process impact on grain formation and orientation," *Journal of Alloys and Compounds*, vol. 736, pp. 297–305, 2018.

[39] A. S. Azar, E. Østby, and O. M. Akselsen, "Effect of hyperbaric chamber gas on transformation texture of the API-X70 pipeline weld metal," *Metallurgical and Materials Transactions A*, vol. 43, no. 9, pp. 3162–3178, 2012.

[40] L. Murr, S. Li, Y. Tian, K. Amato, E. Martinez, and F. Medina, "Open-cellular Co-base and Ni-base superalloys fabricated by electron beam melting," *Materials*, vol. 4, no. 4, pp. 782–790, 2011.

[41] L. E. Murr, E. Martinez, S. Gaytan et al., "Microstructural architecture, microstructures, and mechanical properties for a nickel-base superalloy fabricated by electron beam melting," *Metallurgical and Materials Transactions A*, vol. 42, no. 11, pp. 3491–3508, 2011.

[42] X. Wang, X. Gong, and K. Chou, "Review on powder-bed laser additive manufacturing of Inconel 718 parts," in *Proceedings of ASME 2015 International Manufacturing Science and Engineering Conference*, p. V001T02A063, American Society of Mechanical Engineers, Charlotte, NC, USA, 2015.

[43] SLM Solutions, "3D METALS, discover the variety of metal powders," October 2018, https://slm-solutions.com/sites/default/files/downloads/201en171023-01-001-powder_web.pdf.

[44] D. D. Gu, W. Meiners, K. Wissenbach, and R. Poprawe, "Laser additive manufacturing of metallic components: materials, processes and mechanisms," *International Materials Reviews*, vol. 57, no. 3, pp. 133–164, 2012.

[45] S. Ernst, "Postweld heat treatment of nonferrous high-temperature materials," in *ASM Handbook*, vol. 6, pp. 572–574, ASM International, Geauga County, OH, USA, 1993.

[46] F. Grytten, H. Daiyan, M. Polanco-Loria, and S. Dumoulin, "Use of digital image correlation to measure large-strain tensile properties of ductile thermoplastics," *Polymer Testing*, vol. 28, no. 6, pp. 653–660, 2009.

[47] R. Miller, A. Kolipaka, M. P. Nash, and A. A. Young, "Estimation of transversely isotropic material properties from magnetic resonance elastography using the optimised virtual fields method," *International Journal for Numerical Methods in Biomedical Engineering*, vol. 34, no. 6, article e2979, 2018.

[48] J. J. Vlassak and W. D. Nix, "Measuring the elastic properties of anisotropic materials by means of indentation experiments," *Journal of the Mechanics and Physics of Solids*, vol. 42, no. 8, pp. 1223–1245, 1994.

[49] A. Vogel, S. Diplas, A. Durant et al., "Reference dataset of volcanic ash physicochemical and optical properties," *Journal of Geophysical Research: Atmospheres*, vol. 122, no. 17, pp. 9485–9514, 2017.

[50] V. Gunenthiram, P. Peyre, M. Schneider, M. Dal, F. Coste, and R. Fabbro, "Analysis of laser–melt pool–powder bed interaction during the selective laser melting of a stainless steel," *Journal of Laser Applications*, vol. 29, no. 2, article 022303, 2017.

[51] M. Zhang, C.-N. Sun, X. Zhang et al., "Fatigue and fracture behaviour of laser powder bed fusion stainless steel 316L: influence of processing parameters," *Materials Science and Engineering: A*, vol. 703, pp. 251–261, 2017.

[52] Q. Jia and D. Gu, "Selective laser melting additive manufacturing of Inconel 718 superalloy parts: densification, microstructure and properties," *Journal of Alloys and Compounds*, vol. 585, pp. 713–721, 2014.

[53] A. Casanova, N. Martín-Piris, M. Hardy, and C. Rae, "Evolution of secondary phases in alloy ATI 718Plus® during processing," *MATEC Web of Conferences*, vol. 14, article 09003, 2014.

[54] E. McDevitt, S. Oppenheimer, R. Kearsey, and J. Tsang, *Crack Growth Behavior in ATI 718Plus® Alloy, Materials Science Forum*, Trans Tech Publications, Zürich, Switzerland, 2012.

[55] X. Wang, X. Gong, and K. Chou, "Review on powder-bed laser additive manufacturing of Inconel 718 parts," *Proceedings of*

the Institution of Mechanical Engineers, Part B: Journal of Engineering Manufacture, vol. 231, no. 11, pp. 1890–1903, 2017.

[56] K. M. Knowles and P. R. Howie, "The directional dependence of elastic stiffness and compliance shear coefficients and shear moduli in cubic materials," *Journal of Elasticity*, vol. 120, no. 1, pp. 87–108, 2014.

[57] J. Jung, J. I. Yoon, J. G. Kim, M. I. Latypov, J. Y. Kim, and H. S. Kim, "Continuum understanding of twin formation near grain boundaries of FCC metals with low stacking fault energy," *NPJ Computational Materials*, vol. 3, no. 1, p. 21, 2017.

20

Shear Parameters of Rammed Earth Material: Results from Different Approaches

R. El-Nabouch,[1] Q.-B. Bui,[2] P. Perrotin,[1] and O. Plé[1]

[1]*Univ. Grenoble Alpes, Univ. Savoie Mont Blanc, CNRS, LOCIE, 73000 Chambery, France*
[2]*Sustainable Developments in Civil Engineering Research Group, Faculty of Civil Engineering, Ton Duc Thang University, Ho Chi Minh City, Vietnam*

Correspondence should be addressed to Q.-B. Bui; buiquocbao@tdtu.edu.vn

Academic Editor: Zbyšek Pavlík

Rammed earth (RE) is a construction material which is manufactured by compacting the soil in a formwork, in different layers. Several recent studies have investigated this material. The seismic performance of RE buildings is an important topic which needs to be carefully investigated. The complex numerical model seems a performant approach to investigate the seismic performance of a whole building. To correctly establish the model, the shear parameters of the material, which are the cohesion and the friction angle, should be identified. This paper first presents experimental studies on the shear parameters of RE through the direct shear tests, at two different scales. The differences of the results at different scales are analyzed. Then, the obtained experimental values are used in a numerical model to simulate the shear behavior of RE walls which are loaded by a constant vertical stress and pushed horizontally on the top. From the obtained results, the values for numerical models are recommended.

1. Introduction

Rammed earth is an ancient technique of construction where the wall is built by compacting the soil in a formwork, in different layers. Each earthen layer has thicknesses of about 10–15 cm. More details about this technique of construction are presented in [1, 2]. This material is recently experiencing a renewed interest because of its low embodied energy [3], its recycling ability, and a positive hygrothermal behavior which offers an attractive living comfort [4]. Figure 1 illustrates an example of a recent school in France constructed by RE material.

A nonnegligible number of studies on RE material are observed during the last decade, on different topics: mechanical characteristics [5–9], thermal and hygrothermal properties [4, 10], and durability [11, 12]. For the earthquake performance of RE structures, although several studies have explored this topic [13–16], different aspects still need to be investigated more thoroughly. In order to study the earthquake behavior of an entire RE structure, robust numerical models are a necessary approach. For the numerical modelling of RE material, two important shear parameters are the friction angle φ and the cohesion c. Due to the superposition of successive earthen layers during the manufacture of RE walls, two kinds of characteristics should be distinguished: characteristics of the earthen layers ("intralayer") and those of the interfaces between earthen layers ("interlayer") [17]. The first scope of the present study is to identify the relevant values of the friction angle φ and the cohesion c for both intralayers and interlayers. The characterization of these parameters is the first important step to create performant numerical models for the seismic evaluation of RE buildings.

Several previous studies in the literature tried to identify these two parameters but important dispersions in the obtained results were observed. Cheah et al. [18] performed triaxial tests on cement-stabilized RE specimens and obtained friction angles from 45° to 56°. Bui et al. [19] used analytical calculations for RE specimens and indirectly obtained for intralayers a friction angle of 51° and a cohesion of 0.1 f_c, where f_c is the compressive strength. Nowamooz and Chazallon [20] used finite elements modelling with the

Figure 1: A recent school in France using RE material, architect: V. Rigassi.

Barcelona model on small compacted soil specimens and identified a friction angle and a cohesion of 41° and 13.4 kPa, respectively. Bui et al. [17] used discrete elements modelling to identify c and φ; the identified friction angle was of 45–50° and the identified cohesion was of 100–150 kPa. Miccoli et al. [21, 22] identified numerical values of 30–45° for intralayer friction angles.

The experimental approach is a feasible way which provides "direct" values, while for the numerical approach, a calibration of the results is needed and the identified values are "indirect". The difficulty with RE material is that the representativeness of the tested specimens should be insured because of a high inhomogeneity of the material [6]. El-Nabouch et al. [23] present an experimental study on a full-scale direct shear test where the friction angle and the cohesion of the intralayers were of 37.3° and 30 kPa, respectively. The present paper first presents the results about the cohesion and the friction angle of RE intralayers and interlayers and obtains the following different scales: on standard direct shear tests (10 cm × 10 cm × 4 cm) and on the full-scale direct shear tests (50 cm × 50 cm × 45 cm). Then, the experimental results are used in a numerical model to simulate the behavior of two RE walls which were tested under a constant vertical stress and a horizontal force on the top of the wall. The results and recommendations will be presented.

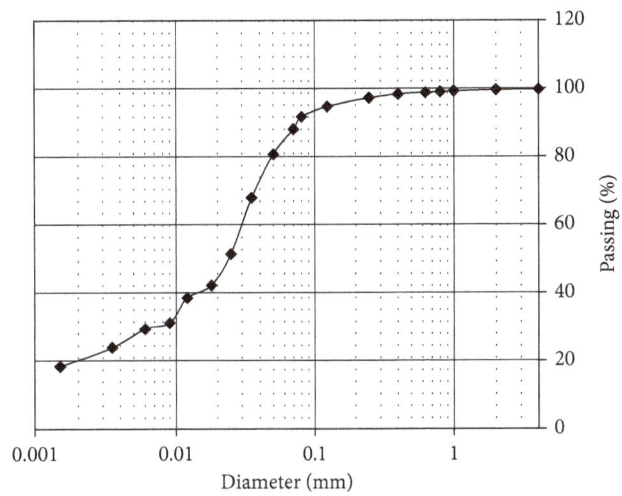

Figure 2: Grains size distribution of the earth used.

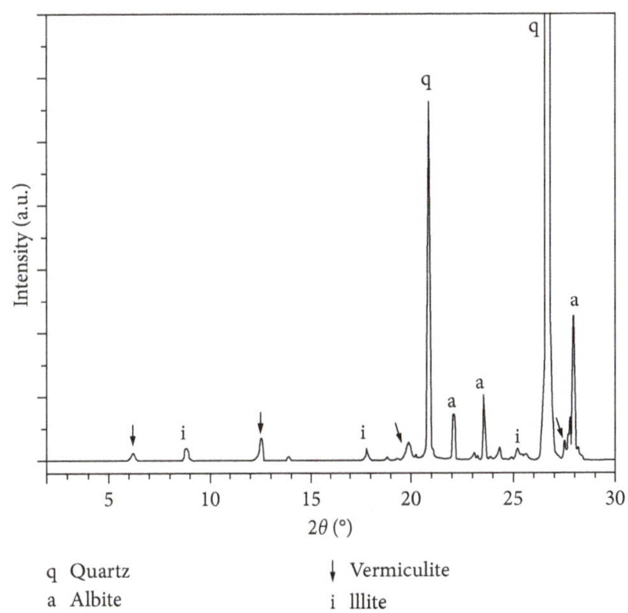

Figure 3: X-ray diffraction result of the used soil [24].

2. Experimental Studies

2.1. Material Used. The earth used was provided by a RE manufacturer company, in the Rhône-Alpes region (France). This soil had been used to build numerous ancient RE buildings in this region. The grain size distribution of the earth is presented in Figure 2. The composition of the soil was analyzed by X-ray powder diffraction where the data were recorded on small fractions of specimens that were crushed before the analysis. The diffraction patterns were recorded with a Bruker D8 diffractometer. The result (shown in Figure 3) reveals the presence of quartz (72.6%), albite (15.1%), illite (11%), and trace of vermiculite (1.3%) [24]. The Proctor optimum water content is of 12% (by weight), determined following the French standard [25].

2.2. Standard Direct Shear Tests

2.2.1. Test Method. Casagrande's shear test is a classical test in soil mechanics [26]. The soil specimen is placed in a shear box that consists of two independent half boxes; the contact between the two half boxes is at the midheight of the specimen. A vertical confining stress is applied to the specimen, and the upper half boxes are pulled laterally (speed of 1.5 mm/min) until the specimen fails. Three tests (at three different vertical stresses) are carried out to determine c and φ following Mohr–Coulomb's theory.

In the present study, the standard shear box (dimensions of 10 cm × 10 cm × 4 cm), which is usually used to identify the friction angle and the cohesion of sand or soil, was also used to test RE specimens (Figure 4).

In soil mechanics, specimens are directly fabricated in the shear box. However for RE material, the representativeness of the specimens manufactured in "small" moulds (compared to in situ RE walls) is usually problematic. Indeed, manufacturing a representative RE specimen by a pneumatic rammer in a small box (10 cm × 10 cm × 4 cm) is not easy. That was why the authors decided to take specimens directly from a 1.5 m width × 1.5 m height × 0.25 m thickness RE wall, which will be presented in detail in the next section. The wall was manufactured in a steel formwork with plywood plates. The earth was compacted by a pneumatic rammer, at 12% of manufacturing water content (by weight) and in 12 earthen layers (with 12 cm thick for each layer). Preliminary Proctor tests showed that this water content was the optimum value for the compaction energy of the pneumatic rammer used. The wall was unmolded from its formwork after the manufacturing and then cured at laboratory ambient conditions (20°C and 60% relative humidity, RH) during two months.

The specimens for direct shear tests were taken after two months and at a quasi-dry state (less than 3% of moisture content, by weight). Firstly, earthen blocks (about 35 cm × 35 cm × 15 cm) were taken from the wall (Figure 5(a)). Then, a table saw was used to cut specimens into accurate dimensions (10 cm × 10 cm × 3.5 cm), enabling to fit the shear box (Figures 5(b) and 5(c)). Since Bui et al. [6], it has been well known that the density in a RE layer decreased from the top to the bottom. That was why specimens for the standard shear tests were taken at different positions in an earthen layer: at 1 cm from the top (Figure 5(b)) where the compaction was the most important (dry density measured of $1864 \pm 51\ \mathrm{kg/m^3}$) and at the middle of the layer (Figure 5(c), dry density of $1703 \pm 45\ \mathrm{kg/m^3}$). The specimens at the bottom, close to the interface, could not be extracted because they were too brittle. It is worth noting that due to the fragility of unstabilised RE material, special attentions must be paid to correctly cut the specimens without visible damage (for example, [27]).

2.2.2. Test Results. For each direct shear test, the maximum shear stress was noted and the experimental results giving the relationship between the shear stress τ and the normal stress σ are summarized in Figure 6. Firstly, note that the tests are consistent with Mohr–Coulomb's theory since the experimental results are on the same straight line (a correlation coefficient greater than 97%). Furthermore, the graph shows that, at a same vertical stress, the specimens from the upper part had a shear resistance greater than that of the lower part. This result is not surprising because the upper part—which is better compacted during the manufacturing—has a greater density and therefore has a strength which is greater than that of the lower part.

Following Mohr–Coulomb's theory, the relationship between the maximum shear stress and the normal stress is determined. The result is presented in Figure 6 for the case of

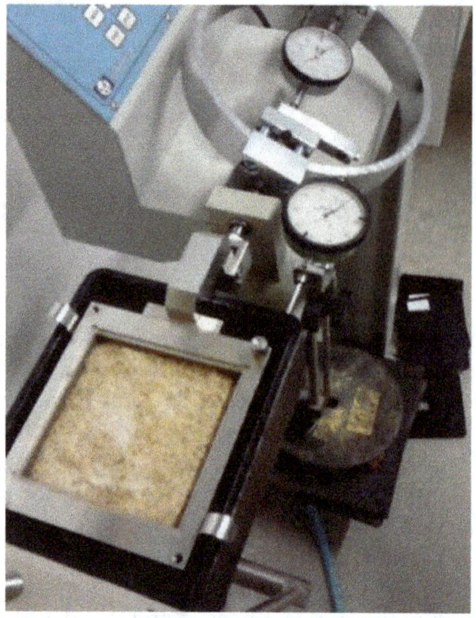

FIGURE 4: A standard direct shear test with a RE specimen inside the standard shear box.

the standard shear box. The apparent friction angle and the apparent cohesion were determined, giving, respectively, $\varphi = 44.1°$ and $c = 264$ kPa for the specimens taken from the layers' upper parts, whereas $\varphi = 44.4°$ and $c = 164$ kPa for the specimens taken for the layers' middle parts. It is interesting to note that the friction angles obtained were similar for the upper and middle parts of an earthen layer (about 44°). The main difference was the cohesion: the upper part had a more important cohesion (264 kPa), comparing to that of the middle part (164 kPa).

2.3. Full-Scale Direct Shear Tests. In a recent study, El-Nabouch et al. [23] have carried out an experimental study on a full-scale shear box (50 cm × 50 cm × 45 cm height) to take into account the scale effects (Figure 7). The soil used and the manufacturing process were the same as that of the present study. Two types of shearing tests were performed: specimens sheared at the middle of layer (intralayer) and specimens sheared at the interface between the layers (interlayer). Due to the large size of the specimens, there was a high variation of the moisture content inside of the specimens, which was from 3% to 6%, depending on positions (on the surface or at interior).

First, a vertical load is applied on the top of the specimen (by the vertical actuator VA), and then, a horizontal load is applied by the horizontal actuator VH. The results (Figure 8) showed that the friction angle and the cohesion of the intralayers were of 37.3° and 31 kPa, respectively, and that of the interlayers were of 34.8° and 24 kPa, respectively.

2.4. Discussion. The results obtained show that, for a same scale of test, the friction angle does not vary significantly within a RE wall: between the upper part and the middle part in the case of standard shear box tests (about 44°) and

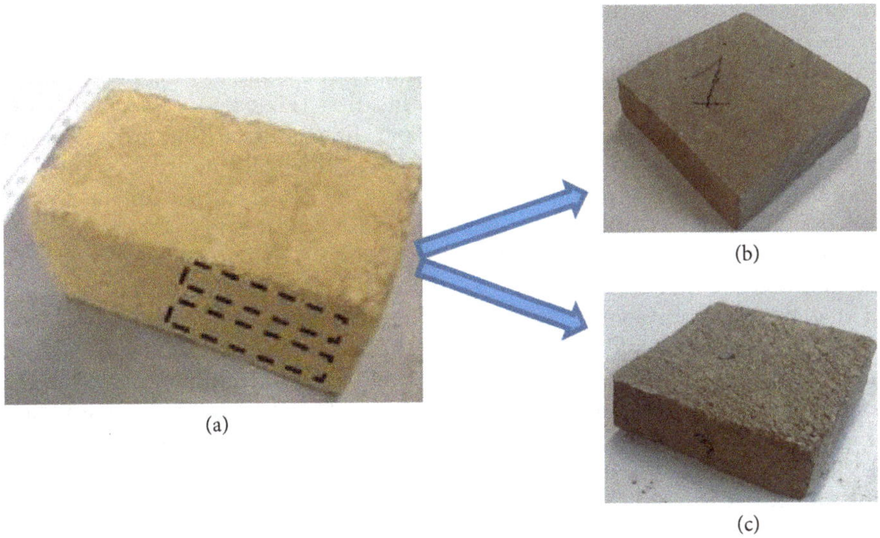

Figure 5: Location of the specimens inside one layer: upper (b) and middle (c) parts of the layer.

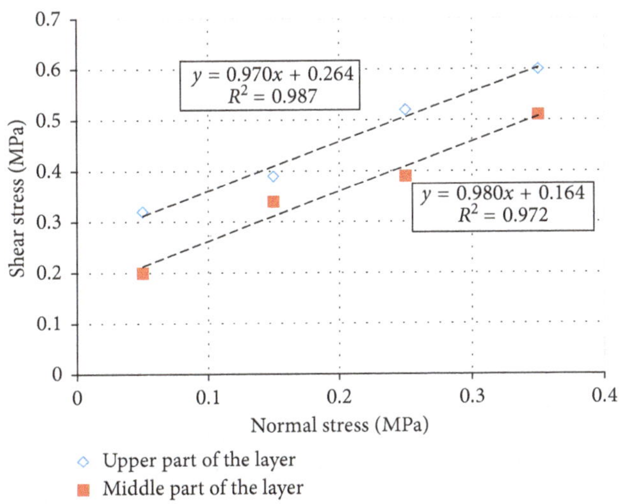

Figure 6: Results of standard direct shear tests.

Figure 7: The full-scale shear box installed on the loading frame.

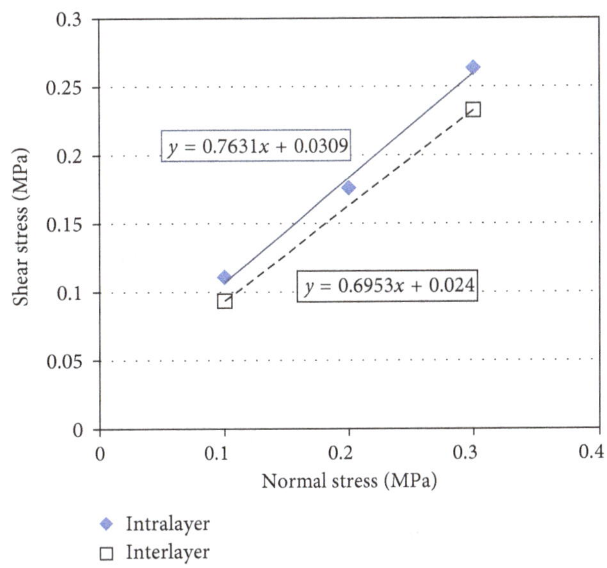

Figure 8: Results obtained from the full-scale shear box tests.

between the intralayer and the interlayer in the case of full-scale shear box tests (37.3° and 34.8°, respectively). It is suggested that the friction angle depends on the roughness of earth grains which is similar between the upper and middle parts of a layer. Besides the grain roughness, the density has also influence on the friction angle, which explains the slightly lower values of the interlayer friction angle, compared to that of the intralayer.

For the cohesion, the difference is clearer between the upper part and the middle part (264 and 164 kPa, respectively). It is suggested that, with a higher compactness, the dimensions of the micropores in the upper part are smaller than that of the middle part; the smaller micropores give a higher suction which provides higher cohesion and mechanical strength [7, 28]. This observation is also valid for the full-scale shear box tests where the interlayer cohesion is about 80% of the intralayer cohesion.

Another important observation is that the results obtained on the full-scale shear box are lower than that of the standard shear tests. Indeed, the shear tests at the midlayer on the standard box were conducted at a similar position with the intralayer shear tests on the full-scale box; for the standard box tests, the friction angle and the cohesion of the middle part are 44° and 164 kPa, respectively, while these parameters obtained on the full-scale box test are 37° and 31 kPa, respectively. The main differences between these two tests are the specimens' size and the moisture content. Indeed, it was observed that the moisture content was not homogeneous in the full-scale specimens and varied from 3% to 6%, depending on the positions, whereas the moisture content played a predominant role on the suction of earthen material which was the main source of the cohesion in earthen material [7, 8, 28, 29].

The influence of the moisture content is also probable because the cohesion obtained on intralayer full-scale specimens were relatively low (31 kPa). For comparison, following the formulas recommended in New Zealand Standard [30] or in [19], the cohesion would be 100–200 kPa for the RE in the present study; or following the values suggested in [22], the intralayer cohesion would be in the range from 132 to 264 kPa in the present study.

The low values of the cohesion and the friction angle obtained through the full-scale shear box are likely due to the effects of the moisture content, but also due to the size effects. Indeed, Flitti et al. [31] observed that when the moisture content increased (from 0 to 3%), the friction angle could decrease about 1° (for a friction angle of 40°). So, it can be suggested that the moisture may act as a lubricant which decreases the friction between the particles, but this phenomenon does not have enough effects to explain the low results obtained. This result confirms the nonnegligible role of the size effects.

From the results of intralayer and interlayer obtained on the full-scale box, the friction angle of the interlayer is 93% of the intralayer and the cohesion of the interlayer is 80% of the intralayer. This is an interesting result which serves for the numerical modelling.

3. Numerical Modelling of RE Shear Walls

For the numerical modelling, due to the simplicity, the earthen layers are usually assumed homogeneous and the interfaces are added between layers [17]. However, the above experimental results showed that there was a difference in the cohesions and the friction angles between the upper and middle parts of an earthen layer. So it is important to take the pertinent values for homogenized earthen layers (in numerical models) which represent the real inhomogeneous layers. On the contrary, it has already been observed that the most adapted values for a numerical model may sometimes be slightly different from the intrinsic characteristics of the material [32]. That was why a finite element (FE) model was built and simulated two horizontal loading tests in order to find the recommended values for the modelling of RE material.

3.1. Horizontal Loading Tests. Two walls (called wall A and wall B) were used for the numerical modelling of the present study. The walls having 1.5 m width × 1.5 m height × 0.5 m thickness (Figure 9) were manufactured with the same soil and the manufacturing process as the specimens presented above. The walls' moisture contents obtained after the test were 2.5–3% (i.e., the same as the specimens tested in the standard shear box).

The RE wall was directly built on a concrete beam. After the RE wall manufacturing, a second concrete beam was installed on the top of the RE wall, by applying a lime mortar. These walls were subjected to the pushover test which is a reliable approach to evaluate the walls' earthquake performance. First, vertical loads were applied on the top of the wall by means of two vertical actuators (VE1 and VE2 (Figure 9)). These loads corresponded to a vertical compressive stress of 0.3 MPa, which simulated the stress due to the vertical loads in a building (dead and live loads). Then, the horizontal pushing load was applied from a horizontal actuator (VH) with a displacement control (0.01 mm/s). The digital image correlation (DIC) technique was used to measure the displacements of the walls during the tests. In the DIC technique, the displacement fields are determined by comparing the images taken after and before loading (reference image). A professional camera with a resolution of 16 M pixels was used. The test was recorded by the camera, and the data processing was performed with the 7D software [33].

The bottom concrete beam was fixed to the steel frame by four steel brackets that could be mechanically adjusted (Figure 9). Another steel prop (*B* on Figure 9) was used as support to prevent the beam sliding when the top horizontal displacement is applied. The bottom concrete beam was also maintained by vertical tie rods (*T* on Figure 8) to avoid the beam rocking. Three displacement sensors M1 and M2 were installed to verify if there were any movement (vertical and horizontal) of the bottom concrete beam. The displacement sensor M3 was used to verify the accuracy of the results obtained from DIC.

Figure 10 shows the horizontal forces in function of the horizontal displacements on the top of both walls. These displacements were obtained from the DIC. Two walls tested exhibit similar initial stiffness and a nonbrittle behavior; the maximal horizontal load was 40–43 kN.

Figure 11 illustrates the crack propagation of wall A through the results of the major principal strains obtained from the DIC process. For the tested walls, three types of cracks were identified: first, a horizontal crack was observed at an interface between two earthen layers at the left-lower part of the wall. Second, inclined cracks which represent the diagonal compressive strut. These cracks appeared when the horizontal load reached about 80–85% of the maximal load. Third, the rocking of the walls at their base was also noted at the end of the tests (on the right side of the wall on Figure 11).

3.2. Numerical Modeling

3.2.1. Initial Considerations. The horizontal loading tests were modeled with Aster code [34] by using 3D solid

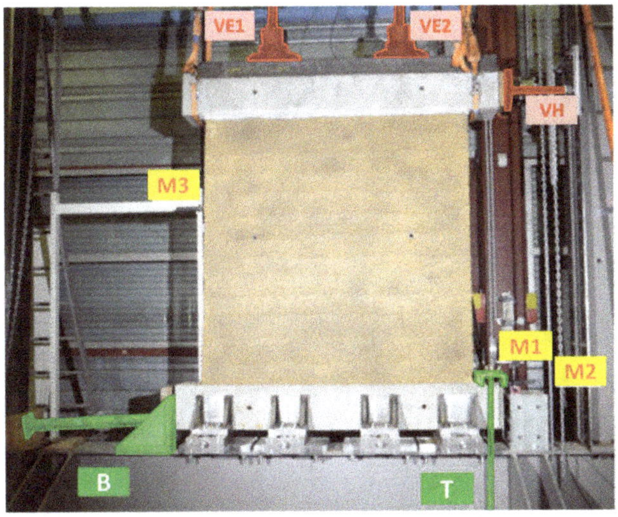

FIGURE 9: Test setup on a RE wall (1.5 m × 1.5 m × 0.25 m).

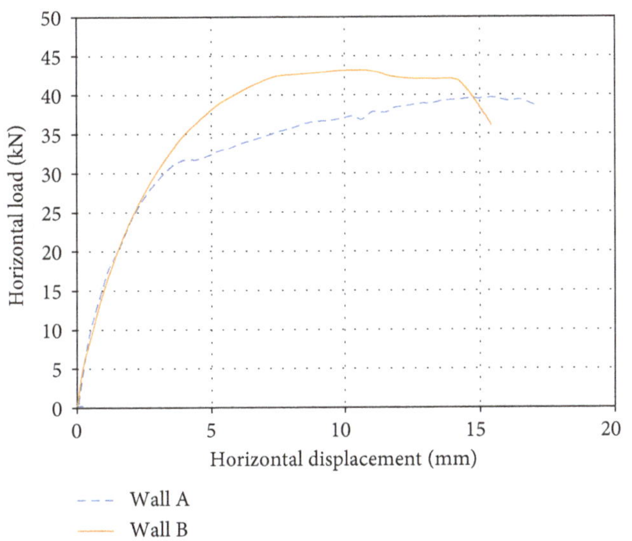

FIGURE 10: Horizontal forces on top of the wall in function of the top horizontal displacements.

elements. Aster is a FEM code, developed by Electricity of France. This code has different advanced options for the earthquake studies and was used to model different important and complex structures such as nuclear power plants. The initial aim of the authors of this paper is to first validate the numerical model at the wall scale and then apply for entire RE structures for further studies.

The well-known Drucker–Prager's model was chosen. In fact, Drucker–Prager's model does not directly use c and φ as parameters, but a conversion from Mohr–Coulomb' parameters is necessary. However, Drucker–Prager's model was used in the present study due to its robustness (working in 3D) and its simplicity by comparing to other advanced models. Drucker-Prager's yield criterion is written as follows:

$$F(\sigma, p) = \sigma_{eq} + \alpha I_1 - R(p) = 0, \tag{1}$$

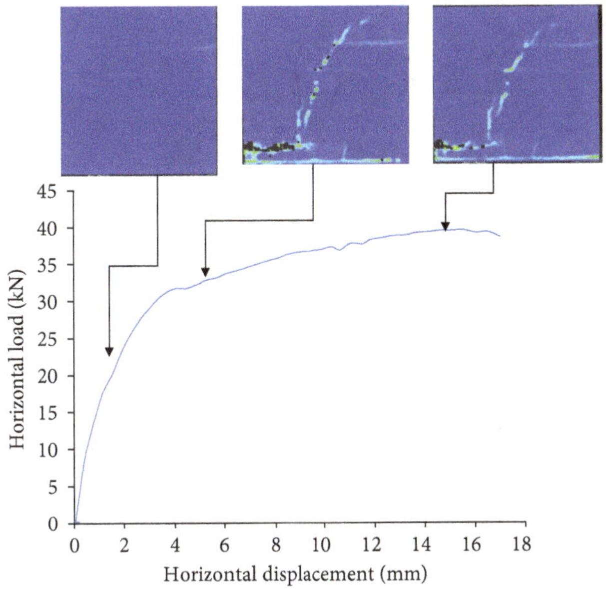

FIGURE 11: Evolution of cracking for the wall A in function of the horizontal displacement.

where $I_1 = \mathrm{tr}(\sigma)$ is the trace of the stress tensor, σ_{eq} is the second invariant of stress, and α is a function of the friction angle:

$$\alpha = \frac{2 \sin \varphi}{3 - \sin \varphi}. \tag{2}$$

$R(p)$ is the value taken by the criterion for $\mathrm{tr}(\sigma) = 0$, a function of the cumulated plastic strain p. The variation of R with the material plasticization presents the hardening. The parabolic hardening is chosen for this model.

The parameter σ_y is written in the function of φ and the cohesion c:

$$\sigma_y = \frac{6c \cos \varphi}{3 - \sin \varphi}. \tag{3}$$

Figure 12 illustrates the geometry and the meshing. The wall was modeled with 12 layers of 11.4 cm each (named "intralayer") and thin layers of 1 cm (named "interlayer") between the earthen layers. Indeed, the role of the interlayers may not be predominant in the case of vertical loads but important in the cases where the interfaces are more solicited [17]. In the present study where the walls are subjected to horizontal loads, the effects of interlayers on the model might have a significant impact on the results. In fact, the interlayers represent the bottom part of an earthen layer where the soil was less compacted than the upper part, and consequently, its mechanical properties are lower. The thickness of the interlayers was chosen following observations by DIC during compression tests where the thickness of the weak layers (excessive displacements compared to other parts) was about 1 cm. Modelling these two layers is a simple way which enables to consider each intralayer and each interlayer with different parameters. Each intralayer

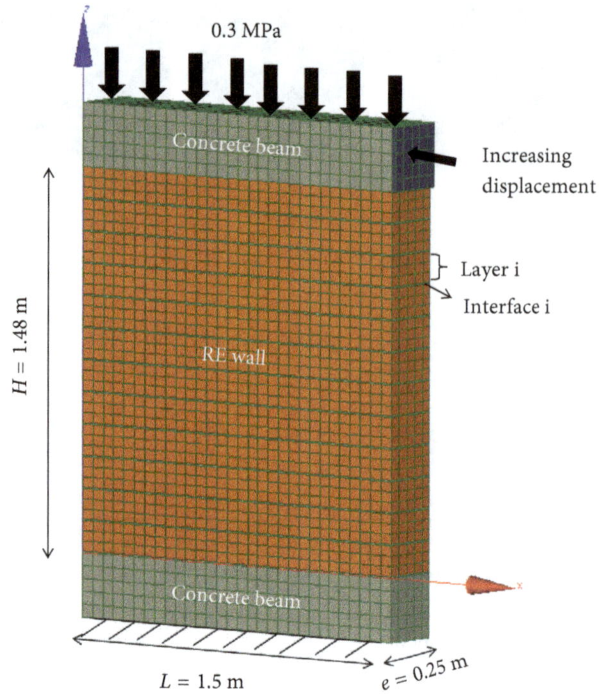

FIGURE 12: Geometry and boundary conditions of the numerical model.

and each interlayer are considered as homogeneous and isotropic. The bottom of the bottom concrete beam is considered as embedment.

3.2.2. Intralayer and Interlayer Characteristics. For the intralayers, Young's modulus was taken of 500 MPa that was measured by DIC during the vertical loading phase of the experiment, by using the vertical displacements at the walls' central part. Poisson's ratio was taken equal to 0.23, following the results of a previous study [7].

As mentioned above, the moisture content of the tested walls is similar to that of the specimens tested in the standard shear box, so the intralayer characteristics were taken following the results of the standard shear box. The interlayer characteristics were taken in percentage of the intralayer characteristics, as obtained on the full-scale shear box tests. Therefore, the intralayers were modeled with $\varphi_{intra} = 44^\circ$ and $c_{intra} = 164\,kPa$, which are the values of the middle part of a layer because the numerical model considered each earthen layer as a homogeneous material. The interlayer friction angle was taken as 94% of the intralayer, corresponding to $\varphi_{inter} = 40^\circ$; the interlayer cohesion was taken as $c_{inter} = 130\,kPa$ which represented 80% of the intralayer cohesion.

3.2.3. Numerical Results and Discussions. The numerical results are illustrated in Figure 13. From this figure, the numerical model could reproduce the maximum force of the experiments, the experimental initial stiffness, and the global elastoplastic behavior. The experimental second slope before the maximal load was less well reproduced;

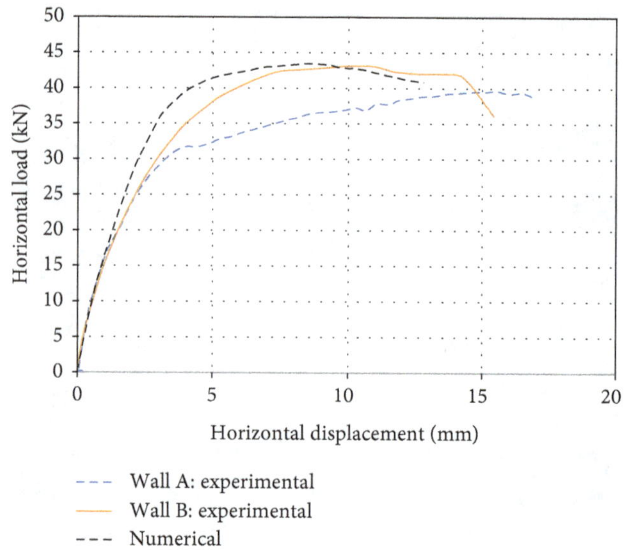

FIGURE 13: Numerical results for the cases of with and without interfaces.

this result is probably due to the limited number of parameters used in the Drucker–Prager model, compared to other more sophistical models which use more parameters [32].

Figure 14(a) illustrates the major principal strains for the model, which show the strain concentration at the left base of the RE wall on the diagonal line. The slight rocking at the right base is also observed. These strain concentrations correspond to the damages observed during experiments (Figure 14(b)).

If a pushover analysis following Eurocode 8 is carried out to assess the in-plane seismic performance of these walls by using the numerical curve of horizontal load-horizontal displacement (more details can be found in [13, 35]), the position of the numerical performance point will be close to the experimental performance point. This result means that the numerical model can provide satisfying results for the pushover analysis.

4. Conclusion

In the present paper, the shear parameters of RE material were first experimentally investigated by standard direct shear tests. For the standard direct shear tests, two types of specimens were tested: specimens taken at the upper parts of an earthen layer and specimens taken at the middle parts. The results obtained for the upper part were $\varphi_{layer} = 44^\circ$ and $c_{layer} = 264\,kPa$, and for the lower part, they were $\varphi_{layer} = 44^\circ$ and $c_{layer} = 164\,kPa$. The interesting finding was that the friction angles obtained for the upper and middle parts of an earthen layer were similar because of the similar roughness of earth grains and that the upper part had higher cohesion than the middle part due to a higher suction.

Then, the results obtained on a full-scale shear box were also presented to provide information about the interfaces between earthen layers. The results showed

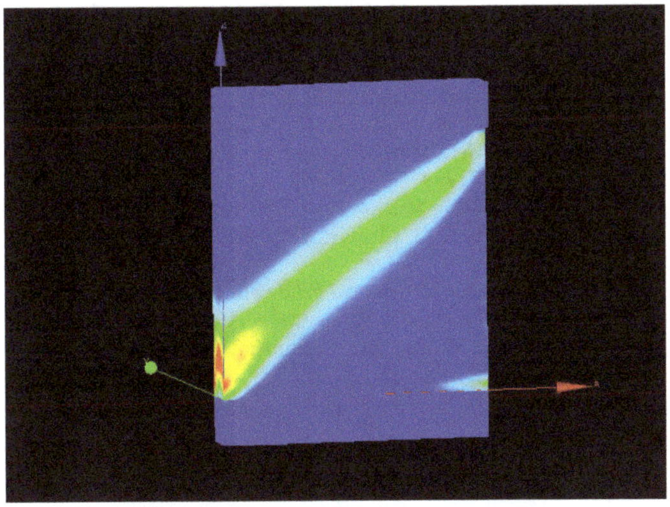

(a)

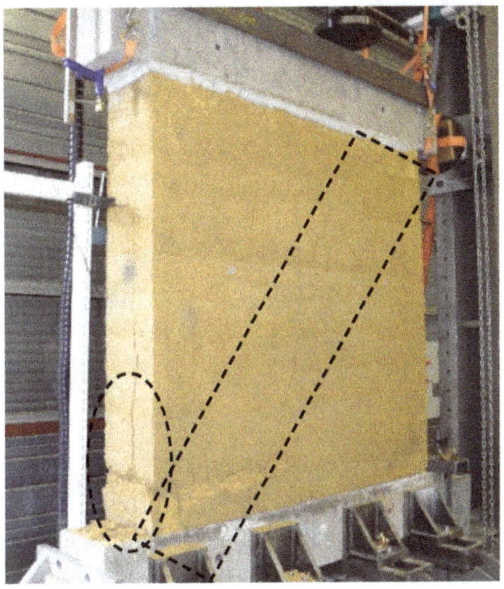

(b)

FIGURE 14: (a) Maximum principal strains in the numerical model; (b) wall damages.

that the interlayer friction angle was 93% of the intralayer and that the interlayer cohesion was 80% of the intralayer.

Finally, these experimental results were used for a numerical simulation of two RE walls subjected to horizontal loading tests. The numerical model could reproduce the maximum force of the experiments, the experimental initial stiffness, the global elastoplastic behavior, and the crack evolution during the experiments. These results showed the relevancy of the shear parameters used in the numerical model and that the model could be directly used for the pushover analysis on the seismic performance assessment of RE structures.

Conflicts of Interest

The authors declare that they have no conflicts of interest.

Acknowledgments

This research is funded by the Vietnam National Foundation for Science and Technology Development (NAFOSTED) under grant number 107.01-2017.22. The authors wish also to thank the French National Research Agency (ANR) for the PRIMATERRE project (ANR-12-Villes et Bâtiments Durables) which supports the Ph.D. thesis of the first author [35]. L. Teytu, M. Jaffré, and the technical staffs from LOCIE (USMB) are warmly thanked for their precious collaboration.

References

[1] P. Walker, R. Keable, J. Martin, and V. Maniatidis, *Rammed Earth: Design and Construction Guidelines*, BRE Bookshop, Watford, UK, 2005.

[2] M. Hall, R. Lindsay, and M. Krayenhoff, *Modern Earth Buildings-Materials, Engineering, Constructions and Applications*, Woodhead Publishing, Sawston, UK, 2012.

[3] J. C. Morel, A. Mesbah, M. Oggero, and P. Walker, "Building houses with local materials: means to drastically reduce the environmental impact of construction," *Building and Environment*, vol. 36, no. 10, pp. 1119–1126, 2001.

[4] L. Soudani, A. Fabbri, J. C. Morel et al., "Assessment of the validity of some common assumptions in hygrothermal modeling of earth based materials," *Energy and Buildings*, vol. 116, pp. 498–511, 2016.

[5] C.T. S. Beckett, C. E. Augarde, D. Easton, and T. Easton, "Strength characterisation of soil-based construction materials," *Géotechnique*, vol. 68, no. 5, pp. 400–409, 2017.

[6] Q. B. Bui, J. C. Morel, S. Hans, and N. Meunier, "Compression behaviour of non-industrial materials in civil engineering by three scale experiments: the case of rammed earth," *Materials and Structures*, vol. 42, no. 8, pp. 1101–1116, 2009.

[7] Q. B. Bui, J. C. Morel, S. Hans, and P. Walker, "Effect of moisture content on the mechanical characteristics of rammed earth," *Construction and Building Materials*, vol. 54, pp. 163–169, 2014.

[8] F. Champiré, A. Fabbri, J.-C. Morel, H. Wong, and F. McGregor, "Impact of relative humidity on the mechanical behavior of compacted earth as a building material," *Construction and Building Materials*, vol. 110, pp. 70–78, 2016.

[9] V. Maniatidis and P. Walker, "Structural capacity of rammed earth in compression," *Journal of Materials in Civil Engineering*, vol. 20, no. 3, pp. 230–238, 2008.

[10] P. Taylor, R. J. Fuller, and M. B. Luther, "Energy use and thermal comfort in a rammed earth office building," *Energy and Buildings*, vol. 40, no. 5, pp. 793–800, 2008.

[11] A. Arrigoni, C. Beckett, D. Ciancio, and G. Dotelli, "Life cycle analysis of environmental impact vs. durability of stabilised rammed earth," *Construction and Building Materials*, vol. 142, pp. 128–136, 2017.

[12] Q. B. Bui, J. C. Morel, B. V. V. Reddy, and W. Ghayad, "Durability of rammed earth walls exposed for 20 years to natural weathering," *Building and Environment*, vol. 44, no. 5, pp. 912–919, 2009.

[13] R. El-Nabouch, Q.-B. Bui, O. Plé, and P. Perrotin, "Assessing the in-plane seismic performance of rammed earth walls by using horizontal loading tests," *Engineering Structures*, vol. 145, pp. 153–161, 2017.

[14] M. I. Gomes, M. Lopes, and J. Brito, "Seismic resistance of earth construction in Portugal," *Engineering Structures*, vol. 33, no. 3, pp. 932–941, 2011.

[15] Q.-B. Bui, S. Hans, J.-C. Morel, and A.-P. Do, "First exploratory study on dynamic characteristics of rammed earth buildings," *Engineering Structures*, vol. 33, no. 12, pp. 3690–3695, 2011.

[16] Q.-B. Bui, T.-T. Bui, and A. Limam, "Assessing the seismic performance of rammed earth walls by using discrete elements," *Cogent Engineering*, vol. 3, no. 1, article 1200835, 2016.

[17] T.-T. Bui, Q.-B. Bui, A. Limam, and J.-C. Morel, "Modeling rammed earth wall using discrete element method," *Continuum Mechanics and Thermodynamics*, vol. 28, no. 1–2, pp. 523–538, 2015.

[18] J. S. J. Cheah, P. Walker, A. Heath, and T. K. K. B. Morgan, "Evaluating shear test methods for stabilised rammed earth," *Proceedings of the Institution of Civil Engineers-Construction Materials*, vol. 165, no. 6, pp. 325–334, 2012.

[19] T.-T. Bui, Q.-B. Bui, A. Limam, and S. Maximilien, "Failure of rammed earth walls: from observations to quantifications," *Construction and Building Materials*, vol. 51, pp. 295–302, 2014.

[20] H. Nowamooz and C. Chazallon, "Finite element modelling of a rammed earth wall," *Construction and Building Materials*, vol. 25, no. 4, pp. 2112–2121, 2011.

[21] L. Miccoli, A. Drougkas, and U. Müller, "In-plane behaviour of rammed earth under cyclic loading: experimental testing and finite element modelling," *Engineering Structures*, vol. 125, pp. 144–152, 2016.

[22] L. Miccoli, D. V. Oliveira, R. A. Silva, U. Müller, and L. Schueremans, "Static behaviour of rammed earth: experimental testing and finite element modelling," *Materials and Structures*, vol. 48, no. 10, pp. 3443–3456, 2015.

[23] R. El-Nabouch, Q. B. Bui, O. Plé, and P. Perrotin, "Characterizing the shear parameters of rammed earth material by using a full-scale direct shear box," *Construction and Building Materials*, vol. 171, pp. 414–420, 2018.

[24] A. Arrigoni, A. C. Grillet, R. Pelosato et al., "Reduction of rammed earth's hygroscopic performance under stabilisation: an experimental investigation," *Building and Environment*, vol. 115, pp. 358–367, 2017.

[25] NF P94-093, Soils: reconnaissance et essais—Détermination des références de compactage d'un matériau—Essai Proctor Normal—Essai Proctor modifié, 2014..

[26] NF P 94-071-1. Sols: Reconnaissance et Essais—Essai de cisaillement rectiligne à la boîte—Partie 1: Cisaillement direct," AFNOR 08/1994..

[27] Q. B. Bui, S. Hans, and J. C. Morel, "The compressive strength and pseudo elastic modulus of rammed earth," in *Proceedings of the International Symposium on Earthen Structures*, pp. 217–223, Interline Publishers, Bangalore, India, August 2007.

[28] P. A. Jaquin, C. E. Augarde, D. Gallipoli, and D. G. Toll, "The strength of unstabilised rammed earth materials," *Géotechnique*, vol. 59, no. 5, pp. 487–490, 2009.

[29] P. Gerard, M. Mahdad, A. R. McCormack, and B. François, "A unified failure criterion for unstabilized rammed earth materials upon varying relative humidity conditions," *Construction and Building Materials*, vol. 95, pp. 437–447, 2015.

[30] NZS 4297, *Engineering Design of Earth Buildings*, New Zealand Standard, Auckland, New Zealand, 1998.

[31] A. Flitti, N. Della, and R. D. V. Flores, "Experimental study of the shear resistance of granular material: influence of initial state," *Journal of Theoretical and Applied Mechanics*, vol. 55, no. 2, pp. 523–533, 2017.

[32] R. Borja, *Plasticity-Modeling and Computation*, Verlag Berlin Heidelberg: Springer, Heidelberg, Germany, 2013.

[33] P. Vacher, S. Dumoulin, F. Morestin, and S. Mguil-Touchal, "Bidimensional strain measurement using digital images," *Proceedings of the Institution of Mechanical Engineers, Part C: Journal of Mechanical Engineering Science*, vol. 213, no. 8, pp. 811–817, 1999.

[34] Aster, *Analysis of Structures and Thermomechanics for Studies and Research*, EDF, Clamart, France, 2011, http://www.code-aster.org.

[35] R. El-Nabouch, *Mechanical behaviour of rammed earth walls under pushover tests*, Ph.D. thesis, Univ. Grenoble Alpes-Univ. Savoie Mont Blanc, Chambéry, France, 2017.

21

In Vitro Bioactivity of Binary Nepheline-Fluorapatite Glass/Polymethyl-Methacrylate Composite

Dalia Y. Zaki[1] **and Esmat M. A. Hamzawy**[2]

[1]*Restorative and Dental Materials Department, National Research Center, Cairo, Dokki 12622, Egypt*
[2]*Glass Research Department, National Research Center, Cairo, Dokki 12622, Egypt*

Correspondence should be addressed to Esmat M. A. Hamzawy; ehamzawy9@gmail.com

Academic Editor: Michele Iafisco

In vitro bioactivity of stoichiometric nepheline-fluorapatite glass/polymethyl-methacrylate (PMMA $(C_5O_2H_8)_n$) composite was evaluated. Four glasses of nepheline/fluorapatite with different ratios (75/25, 80/20, 85/15, and 90/10 mole%) were added in 20 and 40 wt.% to PMMA. The composite samples were soaked in simulated body fluid (SBF) for 25 days. A scanning electron microscope with energy dispersive X-ray microanalysis (SEM/EDX) and thin film X-ray diffraction analysis was used to evaluate the composite materials after immersion in SBF. Inductively coupled plasma (ICP) and pH changes were used in the determination of the ions released and the alkalinity, respectively, after the immersion in Tris-buffered solution. Effect of glass filler loading on compressive strength of the cements was also evaluated. The four binary nepheline-fluorapatite glasses/PMMA cements composites are good potential bioactive materials. The compressive strength was between 70.36 ± 6.47 and 97.30 ± 3.90 MPa. In general, decreases of the compressive strength follow the increase of the glass ratio (i.e., 40 wt.%), but all meet that specified by the ASTM F-451.

1. Introduction

Some compositions of glasses, namely, bioactive glass composed mainly of silica, sodium oxide, calcium oxide, and phosphates, proved to have the ability to bond chemically with bone when implanted into living tissues. Such bonding is a result of a series of reactions in the glass and its surface. An exchange of monovalent cations from the glass, with H_3O^+ from surrounding body fluid, takes place resulting in an increase in the pH of the surrounding fluids. A slight alkaline medium is a favorable medium for osteoblasts responsible for bone formation. Such increase in pH should be within a limit to prevent the inhibition of osteoblast activity and to prevent cell necrosis or apoptosis [1, 2].

Bioactive glass has the ability to develop an adherent interface with tissues through the formation of a biologically active hydroxyl carbonate apatite (HCA) layer that resists significant mechanical forces. The strength of the developed adherent interface may be equivalent or greater than the cohesive strength of the implant material in some cases. The rapid reaction at the surface is important to provide a fast bonding with the tissues [1].

Several bioactive glasses, ceramics, glass ceramics, and composites with different compositions were discovered. Their behavior toward tissues was found to be dependent on their composition. Slight changes in the composition were found to greatly affect the properties and hence their application. Bioactive glasses have been used as a reconstruction material in treatment of the bone, and certain compositions have the ability to bond to soft tissues. The controlled rate of degradation and bonding to the tissue could be made through providing a special composition of the prepared glass [3].

The hydrolysis time of glass is a significant property when intended for use in the human body as an implant material. Phosphate glasses with high solubility rates are commonly used as suture thread and as drug delivery carriers. For glass-based materials designed for long-term application in the human body, decreasing solubility as much as possible without losing the bioactive property is essential [4]. Previous studies indicated that the addition of

Al_2O_3 decreased the solubility of phosphate glass to an acceptable limit, thus increasing the long-term stability of the glass materials which is essential for bone defect repairing [4, 5]. Decreasing solubility through the addition of Al_2O_3 allows the HCA layer to live for longer time on the glass surface. A study done by Mikhailenko et al. [6] revealed that aluminum oxide increases the glass stability in water media, while silicon oxide strongly decreases the resistance of glasses rendering them hydrolytically unstable.

In a recent study done by Hamzawy et al. [7], in Na_2O–CaO–Al_2O_3–SiO_2–P_2O_5–F system, the stoichiometric binary nepheline-fluorapatite phases were prepared and investigated. Transparent glasses were obtained in a high stoichiometric ratio of nepheline, that is, 75, 80, 85, and 90%, whereas the low ones, that is, 25, 50, and 70%, gave devitrified glass samples. In vitro bioactivity testing of the prepared glass batches showed good bioactive behavior of the prepared glass, where Ca and P ion release was detected after immersion in Tris-buffered solution for various time intervals.

Despite the advantages of bioactive glass toward tissues, its high modulus of elasticity and brittleness limits its applications; therefore, it has been used in combination with polymethyl-methacrylate (PMMA) to form bioactive bone cement and with metal implants as a coating to form a calcium-deficient carbonated calcium phosphate layer [8].

Bone cements have different applications in dentistry and orthopedic surgery. In dentistry, bone cements are used for sinus floor augmentation, retrograde filling, and in certain cases for anchorage of dental implants. It is widely used in anchoring artificial prostheses to bone. Polymethylmethacrylate (PMMA) is an amorphous self-curing polymer, which is supplied as solid and liquid phases. The solid phase is composed of PMMA, together with benzoyl peroxide as an initiator and barium sulphate as a radiopaque element. The liquid phase is composed of MMA monomer and *N*,*N'*-dimethyl-*p*-toluidine as an activator. Despite its high mechanical properties, reaching its full strength rapidly, thus providing immediate support after setting, its poor bioactivity has been reported as a cause of debonding at the cement-bone interface [9, 10].

This work aims at studying the bioactivity of commercially available PMMA as bone cement through adding the stoichiometric ratio of nepheline-based glasses, that is, 75, 80, 85, and 90 in mole%. The effect of 20 and 40 wt.% glass loading on the in vitro bioactivity of the cements and on the compressive strength was evaluated.

2. Materials and Methods

2.1. Preparation of PMMA/Glass Composite Cement Specimens. Preprepared four stoichiometric nepheline/apatite glasses (75/25, 80/20, 85/15, and 90/10 mole%) were used and mixed with commercially available PMMA (polymethylmethacrylate, $(C_5O_2H_8)_n$, Cemex® Isoplastic, Tecres, Verona, Italy) in two ratios (20 and 40 wt.%) to form composite samples (Table 1). The chemical composition of the prepared glass batches is listed in Table 2. The glass powder was between 0.212 and 0.150 mm in grain size. The composite samples were prepared by addition of the glass powder to the monomer liquid in a sonicator, and then, the PMMA powder was manually mixed with the glass-filled monomer liquid at a powder : liquid ratio of 3 : 1 ($g{\cdot}mL^{-1}$) to obtain a homogeneous paste. At the dough stage, the mixture was poured into molds for bioactivity and compressive strength. Five samples for each mixture and the control were evaluated for each test.

Table 1: Chemical composition of the tested composite samples.

Sample	PMMA (weight%)	Glass	
		Weight%	NeAp
NAP75	80	20	NeAp 75
	60	40	NeAp 75
NAP80	80	20	NeAp 80
	60	40	NeAp 80
NAP85	80	20	NeAp 85
	60	40	NeAp 85
NAP90	80	20	NeAp 90
	60	40	NeAp 90

PMMA: polymethyl-methacrylate.

2.2. In Vitro Bioactivity Assessment. For bioactivity testing, molds to obtain disc specimens with a diameter of 19 mm and a height of 5 mm were used. After setting, the specimens were ground on SiC papers ranging from 600 to 2400 grit and finally polished with 1 and 3 μm diamond pastes [10, 11]. Specimens were immersed in a simulated body fluid (SBF) prepared according to the procedure described by Kokubo [12]. Specimens were individually suspended in 150 ml of freshly prepared SBF for 25 days at physiological conditions of pH and temperature.

Environmental SEM for high-resolution imaging and elemental analysis system EDX (inspect S with accelerating voltage 30 kV, magnification 13x up to 1000.000, and resolution 3 nm, FEI Company, Netherlands), using EDX Genesis software (Version 3.6), were used to analyze the specimens' surfaces before and after immersion in SBF.

For identification of the crystalline phases that developed on the specimens' surfaces after immersion in SBF, thin film X-ray diffraction analysis (XRD) (Philips X-ray diffractometer, Amsterdam, Netherlands) was performed. Cu-Kx radiation ($\lambda = 1.5405$) was used as the X-ray source.

2.3. Assessment of Ion Release. The initial calcium (Ca) and phosphorus (P) ion concentration release behavior up to 4 hours from the prepared specimens for each mixture was determined. Composite discs were individually suspended in a tightly sealed container filled with 150 ml of freshly prepared Tris-buffer solution at physiological conditions of pH and temperature (pH ~ 7.13 at 37°C) using a nylon thread to ensure complete solution coverage [13, 14]. Immersion was done for 10, 125, and 240 minutes. Ten milliliter of the immersion solution was withdrawn for ion concentrations measurement using inductively coupled plasma atomic emission spectroscopy (ICP-OES) (Ultima 2 ICP, Horiba, USA). The concentration of Al ions released in SBF after specimen immersion for 25 days was also evaluated.

TABLE 2: Chemical composition of the glass batches.

Sample number	Molar ratio		Composition/mass-%						Product
	Ne	Ap	SiO_2	Al_2O_3	CaO	Na_2O	CaF_2	P_2O_5	
NeAp 75	75	25	31.72	26.92	12.83	16.36	1.35	10.83	Transparent glass
NeAp 80	80	20	33.83	28.71	10.26	17.46	1.08	9.06	Transparent glass
NeAp 85	85	15	35.95	30.51	7.70	18.55	0.81	6.79	Transparent glass
NeAp 90	90	10	38.06	32.30	5.13	19.64	0.54	4.53	Transparent glass

Ne: nepheline ($NaAlSiO_4$); Ap: fluorapatite ($Ca_5(PO_4)_3F$).

The pH changes of Tris-buffer solution after specimens' immersion were monitored every 10 minutes up to 195 minutes using a calibrated digital pH meter (Jenway 3510 bench pH meter, UK).

2.4. Compressive Strength Testing. The compressive strength testing was done according to the standard specification for acrylic bone cement specified in the ASTM F-451-99 standard [15]. Five cylindrical specimens with a diameter of 12 mm and a height of 6 mm for each composition were prepared and then placed in a desiccator before testing. Testing was done using a computer-controlled universal testing machine (LRX-plus, Lloyd Instruments Universal Test Machine, Fareham, UK) by applying a compressive load at a crosshead speed of 0.5 mm/min. Data were recorded using Nexygen-MT software (Version 4.2). The average and standard deviation were calculated.

2.5. Statistical Analysis. The mean and standard deviation values were calculated for each group in each test. Data were explored for normality using Kolmogorov–Smirnov and Shapiro–Wilk tests. Repeated measure ANOVA was used to compare between dependent samples for more than two groups. One-way ANOVA was used to compare between independent samples for more than two groups. The significance level was set at $P \leq 0.05$. Statistical analysis was performed with IBM® SPSS® Statistics Version 20 in Windows.

3. Results and Discussion

A fast and simple method for determination of the in vitro bioactive behavior of the material is performed through studying the pH changes and ion release in the solution. Additionally, the ability of a material to form a calcium phosphate layer in vitro after partial leaching and dissolution is often taken as a measure of its bioactivity in vivo [2].

Figures 1 and 2 show SEM images and the corresponding EDX spectra of the control and the glass-filled specimens after immersion in SBF for 25 days. No obvious changes were evident on the surface of the control specimens; it shows only the polymer beads were surrounded by in situ PMMA. Such finding is confirmed by the EDX spectrum, which shows the absence of Ca and P precipitates. Barium in the form of barium sulphate (9.00% w/w) added by the bone cement manufacturer as radio-opacifier (i.e., rendering the bone cement visible with X-ray imaging of the patient after treatment) appeared on the surface of the control and the tested specimens. On the contrary, SEM images revealed irregular precipitates on the surfaces of the glass-filled specimens of all compositions. The morphology and thickness of the formed layer varied with the glass composition and the weight percentage predominantly on 40% glass concentration. EDX spectrum shows the presence of Ca- and P-rich agglomerates on their surfaces. The Ca/P atomic ratio of the layers formed on the tested specimens was calculated from the EDX results and is listed in Table 3. These values are close to the Ca/P ratio of hydroxyapatite (1.67), fluorapatite (1.67), tetracalcium phosphate (2.00), and octacalcium phosphate (1.33) [16].

The thin film XRD patterns of the cement composite surfaces after immersion in SBF for 25 days are shown in Figures 3 and 4. The results confirmed the formation of apatite phase [$Ca_5(PO_4)_3(OH)$], whereas patterns of barite ($BaSO_4$) and low quartz (SiO_2) were also detected on the surface of the tested specimens.

The process of formation of the calcium phosphate layer on a substrate surface starts, provided that the critical supersaturation of the surrounding solution is reached. Seeded crystal growth on a substrate containing calcium phosphate takes place at considerably lower supersaturations of aqueous solutions. Epitaxy is one of the reasons why certain materials can function as seeding nuclei for other materials. The growth of one crystal on another takes place when atomic dimensions of one or more commonly occurring faces of each are similar. Therefore, a number of phases may form in the order of decreasing solubility: amorphous calcium phosphate (ACP), dicalcium phosphate dihydrate (DCPD) $CaHPO_4{\cdot}2H_2O$, anhydrous calcium phosphate (DCPA) $CaHPO_4$, octacalcium phosphate $Ca_8H_2(PO_4)$ $6.5H_2O$ (OCP), β-tricalcium phosphate (β-TCP), $Ca_3(PO_4)_2$, and hydroxyapatite ($Ca_{10}(PO_4)_6(OH)_2$; HAP) [16]. Such a finding explains the formation of the calcium phosphate phase [$Ca_5(PO_4)_3(OH)$] precipitated on the specimen's surfaces with different glass compositions and loading, which are revealed in this study.

The effect of incorporation of Al_2O_3 in glass composition and its effect on ion release and bioactivity were investigated in previous studies. The results of the present investigation is in agreement with a study done by El-Kheshen [4] who reported that the bioactivity increased by gradual addition of Al_2O_3, which was indicated by the acceleration in the formation of the hydroxyl apatite layer on the glass surface after immersion in simulated body fluid. However, in a study done by Melchers et al. [5], it was postulated that the incorporation of Al_2O_3 did not significantly affect the bioactivity and may slightly improve it; however, increasing its concentration adversely affected the bioactivity. They

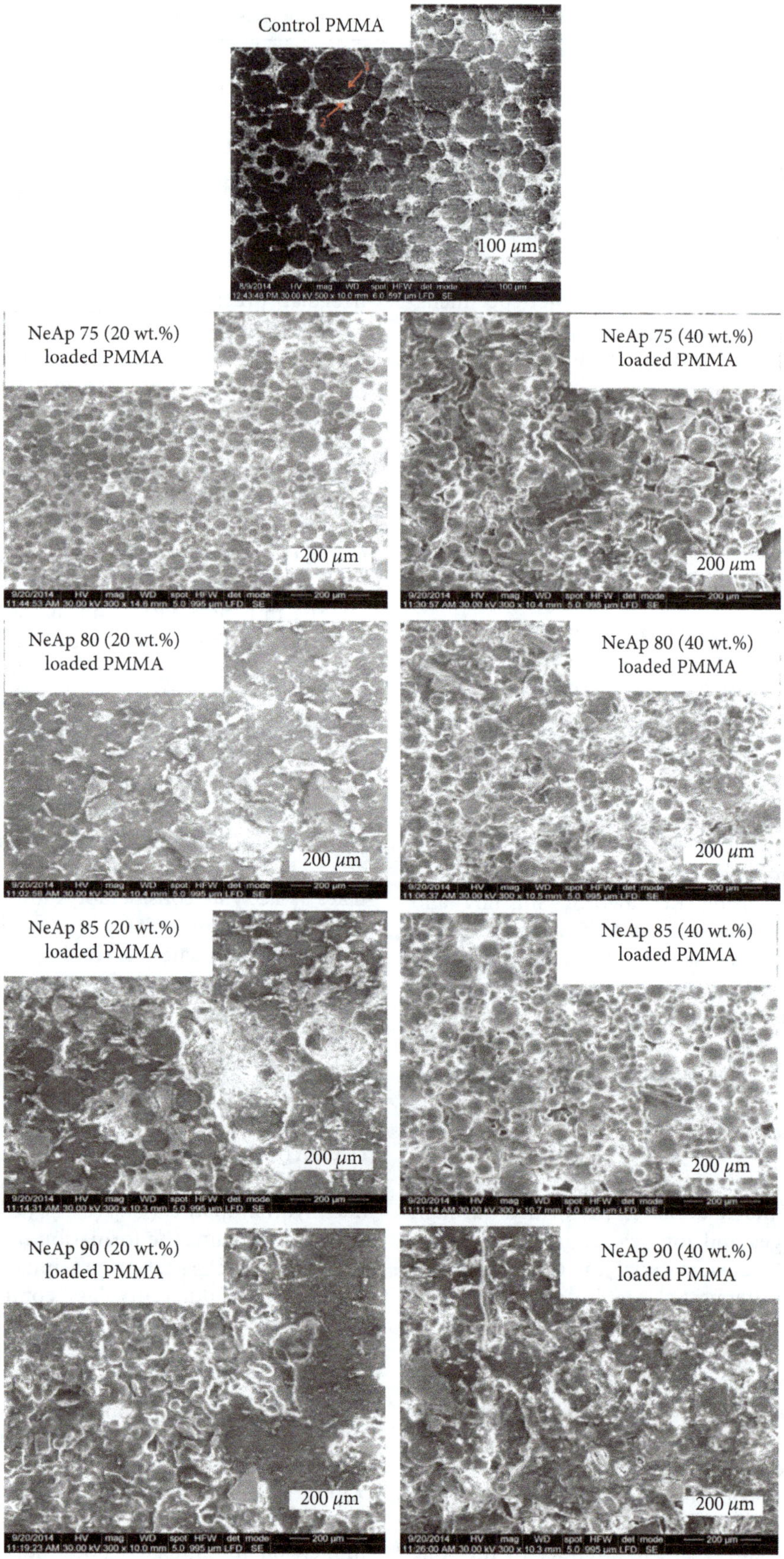

Figure 1: SEM images of the control and the glass-filled specimens after immersion in SBF for 25 days.

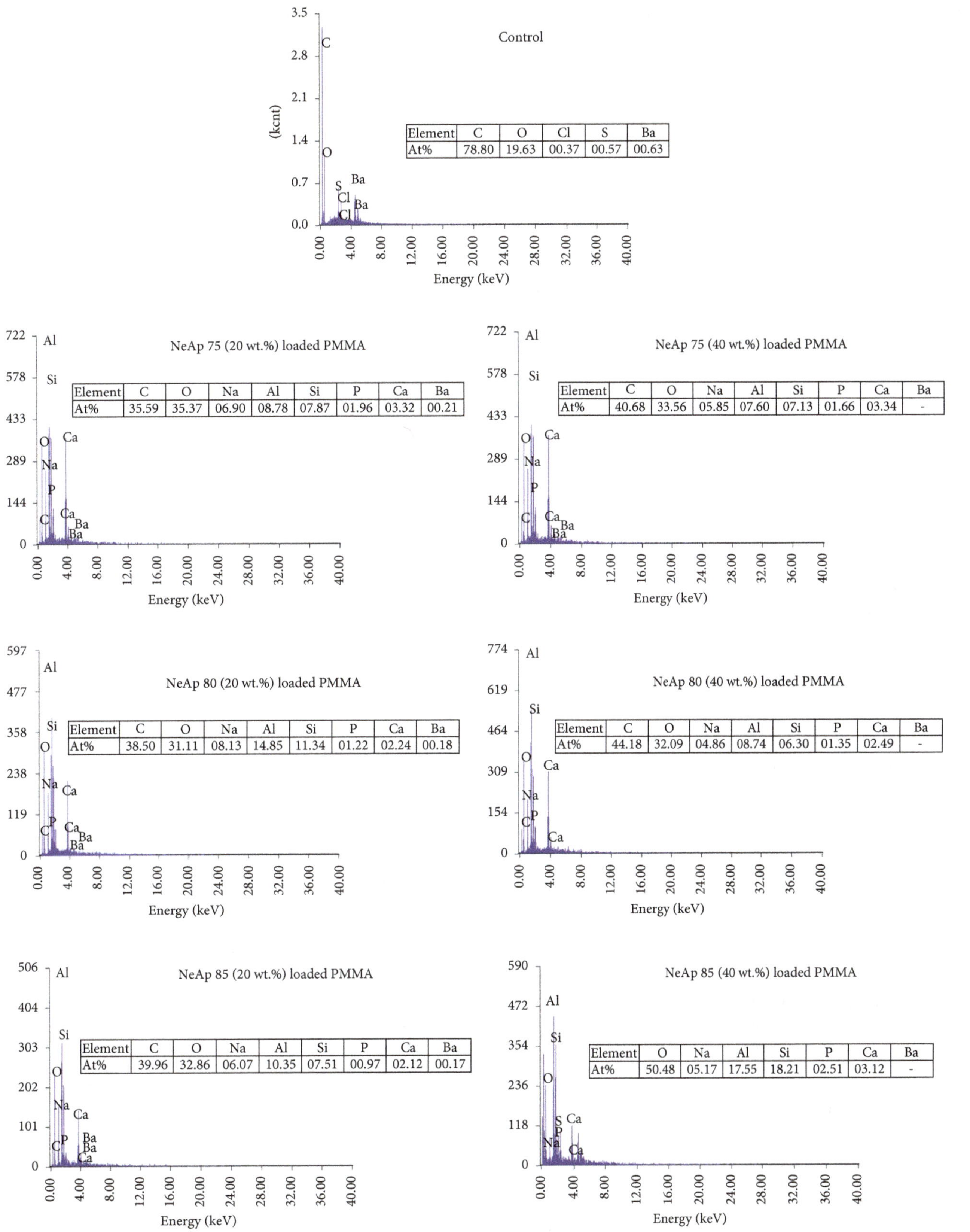

FIGURE 2: Continued.

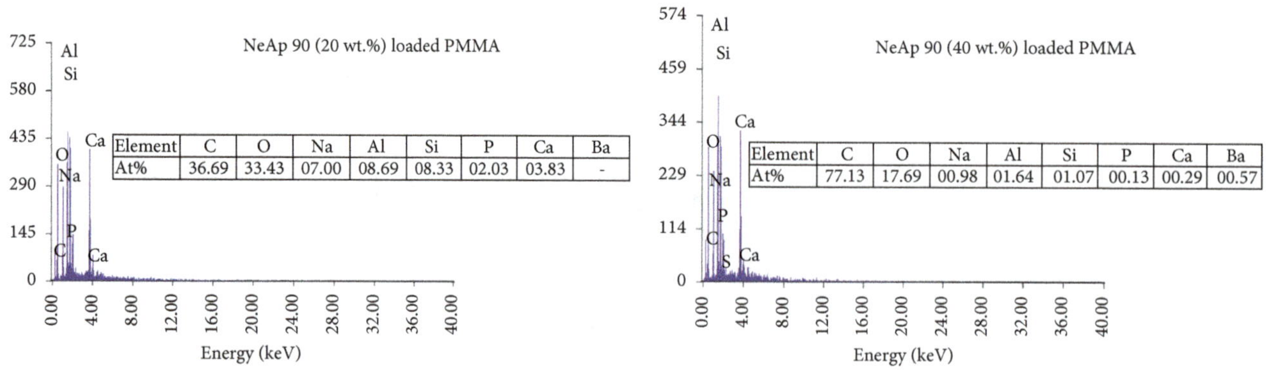

FIGURE 2: EDX spectra of the control and the glass-filled specimens after immersion in SBF for 25 days.

TABLE 3: Calcium/phosphate molar ratios as revealed by EDX analysis of the tested specimens surfaces.

Specimens	NeAp 75 (20%)	NeAp 75 (40%)	NeAp 80 (20%)	NeAp 80 (40%)	NeAp 85 (20%)	NeAp 85 (40%)	NeAp 90 (20%)	NeAp 90 (40%)
Ca/P	1.69	2.12	1.83	1.84	2.2	1.2	1.8	1.6

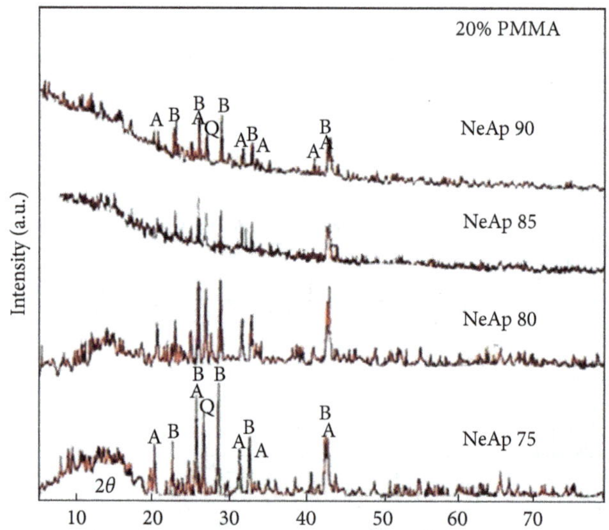

FIGURE 3: Thin film X-ray diffraction patterns of the specimens' surface with 20% glass loading.

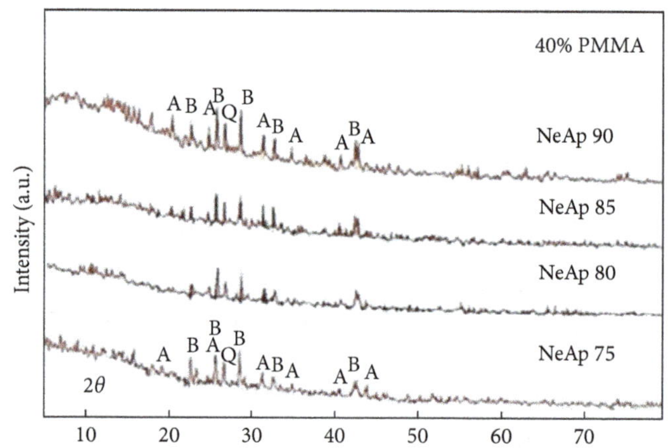

FIGURE 4: Thin film X-ray diffraction patterns of the specimens' surface with 40% glass loading.

postulated that at higher Al_2O_3 concentrations the required charge compensation produced through the interaction of Al^{3+} and PO_4^{3-} and trapping of Ca^{2+} may obstruct the release of Ca, P, and Si ions required for the formation of hydroxycarbonate apatite. Another study [17] suggested that the presence of both mullite ($Al_6Si_5O_{13}$) structural units and fluorapatite structural units in glass might act as bonding agents between the bone mineral and the glass.

The early release of Ca and P is confirmed by the results of ICP analysis, and the results are shown in Figures 5 and 6. For all compositions and concentrations, the high statistically significant mean value of cation concentration was found after 240 min of immersion ($p \leq 0.001$), indicating a gradual increase in Ca ion release with time. Specimens loaded with NeAp 75 glass showed the highest mean value of calcium ion release probably due to the higher fluorapatite content and the least Al_2O_3 percent. On the contrary, the lowest statistically significant mean value of P ion concentration was found after 240 mins, except that of NeAp 75 (20%) where the highest significant mean value of P ion concentration was found after 240 min of immersion ($p \leq 0.001$). Such finding is an indication of early release of P ion with a gradual decrease with time due to the consumption of P in apatite layer formation. Regarding NeAp 75 (20%), results indicate that the rapid release starts after 125 min of immersion.

Due to concerns regarding the biocompatibility of Al, the concentration of Al ion in SBF after specimen's

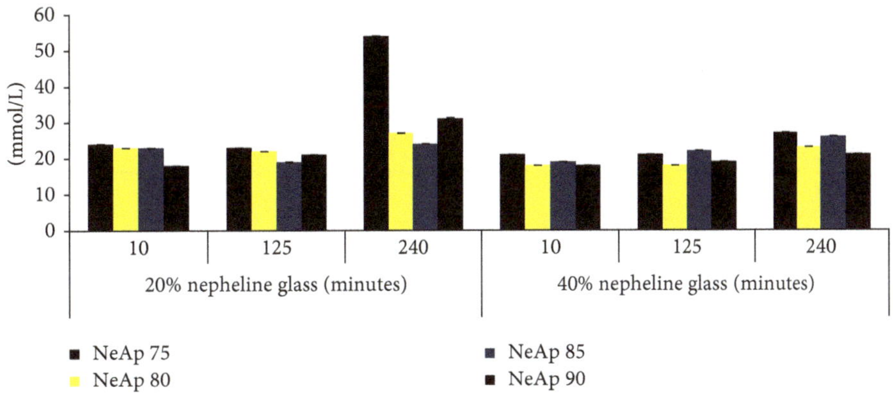

Figure 5: Change in Ca concentrations (mmol/L) in Tris-buffer solution as a function of soaking time.

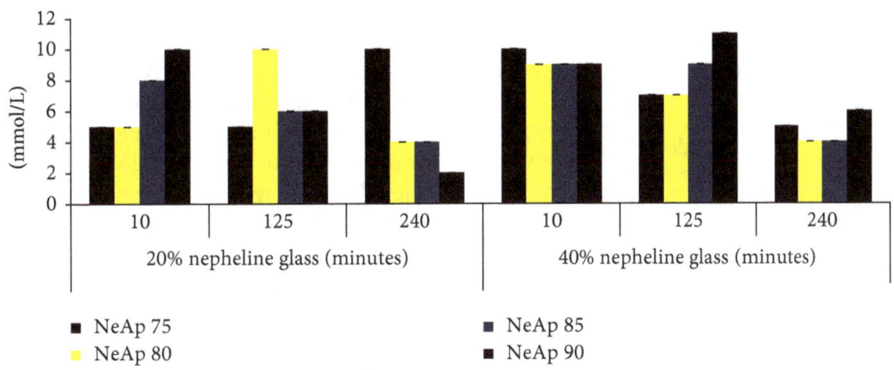

Figure 6: Change in P concentrations (mmol/L) in Tris-buffer solution as a function of soaking time.

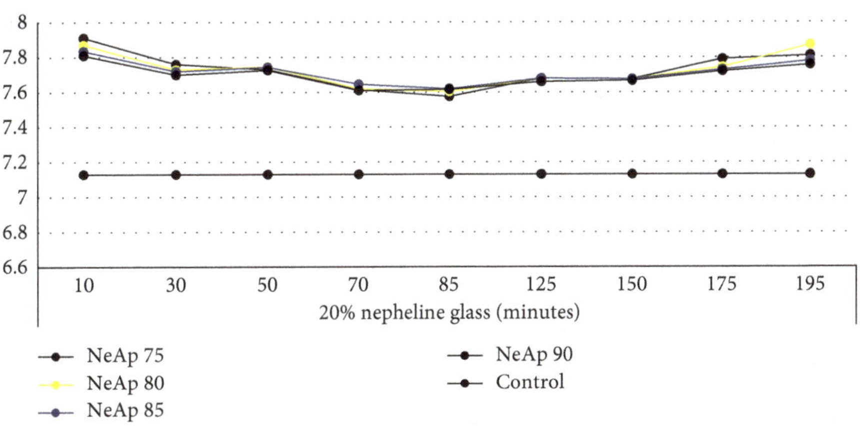

Figure 7: pH changes of Tris-buffer solution along different time periods for each 20% NeAp glass type.

immersion for 25 days was evaluated. Results revealed that the concentration of Al ions was less than 0.01 mg/l for all the tested materials. Meyera et al. [18], evaluated biologically glass ionomer bone cement (GIC) by osteoblast cell culture methods. GIC is a calcium aluminosilicate glass containing fluoride, to be mixed with a homopolymer or copolymer of alkenoic acids. The average composition (wt. %) of GIC bone cement is SiO_2 (35%), Al_2O_3 (30%), CaO (15%), fluorine (10%), Na_2O (3%), and P_2O_5 (7%). It was reported that although accumulation of aluminum was noticed in osteoblasts cultivated in vitro in the presence of glass ionomer bone cement, the cells revealed normal physiological activity with no signs of toxicity as determined by light and scanning electron microscopy. However, further investigations regarding the biocompatibility and cytotoxicity for the prepared PMMA/nepheline-fluorapatite glass composites are required.

Results of the changes in pH measured in this study showed a rapid increase in pH after 10 minutes and during the testing period (Figures 7 and 8). The rapid increase in pH

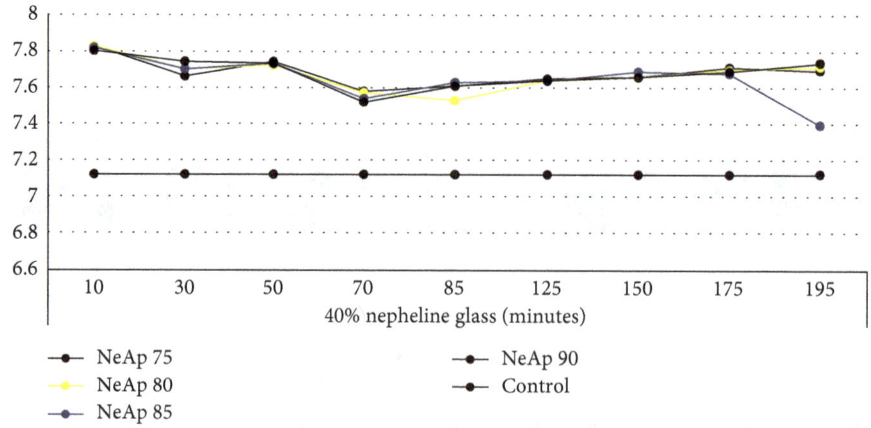

FIGURE 8: pH changes of Tris-buffer solution along different time periods for each 40% NeAp glass type.

TABLE 4: The mean and standard deviation (SD) values of compressive strength (MPa) in each group.

Variables	20% glass (mean ± SD)	40% glass (mean ± SD)	p value
Control	97.30 ± 3.90^{aA}	97.30 ± 3.90^{aA}	1, ns
NeAp 75	87.14 ± 6.65^{aA}	70.83 ± 3.31^{acA}	0.309, ns
NeAp 80	91.13 ± 6.26^{aB}	76.34 ± 5.57^{acA}	0.038*
NeAp 85	92.98 ± 5.93^{aA}	70.36 ± 6.47^{bcB}	0.011*
NeAp 90	96.90 ± 2.94^{aB}	85.17 ± 6.82^{aA}	0.019*
p value	≤0.001*	0.085, ns	

Means with different lower case letters in the same column indicate statistically significant difference, while means with different upper case letters in the same row indicate statistically significant difference; *significant ($p < 0.05$); ns: nonsignificant ($p > 0.05$).

is the result of rapid release of alkali ions into the solution where the particles started to leach and dissolve immediately when in contact with Tris-buffer solution [19]. Such finding is in good agreement with observations of immersion of glass 45S5 particles in Tris-buffered solutions by Cerruti et al. and Greenspan et al. [19, 20] in which a rapid increase in pH took place in the solution during the first 2–6 h. The highest mean pH value was found after 10 min of immersion in all groups. 20% NeAp 75 specimens showed the highest mean pH value (pH = 7.910) after 10 min of immersion, which falls within the physiological limit of human tissues.

Compressive strength testing was done to evaluate the proper glass weight percent that is able to bioactivate the PMMA polymer without adversely affecting the compressive strength of bone cement. Table 4 shows the mean and standard deviation (SD) values of compressive strength in each group. A statistically significant decrease in compressive strength was observed as the filler content was increased in all groups except for NeAp 75 where no statistical significant difference was observed ($p > 0.05$). No statistically significant difference was observed between all groups with different glass percentages and the control group except for NeAp 85 (40%) where a statistical significant decrease in compressive strength (70.36 MPa) was noticed. This may be attributed to the fact that the glass particles act as stress concentration areas. According to the ASTM F-451, a minimum value of 70 MPa is essential for bone cements [15]; thus, the tested groups meet this mechanical requirement. The decrease in compressive strength of PMMA bone cement with increasing the filler loading is in agreement with the results obtained by Rentería-Zamarrón et al. [11], where the compressive strength of the cement was decreased with increasing of wollastonite ($CaSiO_3$) content.

4. Conclusions

Bioactive PMMA bone cements could be obtained by adding 20 and 40 wt.% in all the composite samples containing all glass ratios of stoichiometric nepheline, that is, 75, 80, 85, and 90%. On the contrary, SEM/EDX spectrum and thin film XRD confirmed the formation of the apatite phase on the surfaces of all the composite specimens. The increase in the glass ratio (i.e., 40 wt.%) decreases the compressive strength values, but all meet that specified by the ASTM F-451.

Conflicts of Interest

The authors declare that they have no conflicts of interest.

References

[1] C. Vichery and J.-M. Nedelec, "Bioactive glass nanoparticles: from synthesis to materials design for biomedical applications," *Materials*, vol. 9, no. 4, pp. 288–296, 2016.

[2] D. Zhang, M. Hupa, and L. Hupa, "In situ pH within particle beds of bioactive glasses," *Acta Biomaterialia*, vol. 4, pp. 1498–1505, 2008.

[3] P. Lopes, M. Corbellini, B. L. Ferreira et al., "New PMMA-co-EHA glass-filled composites for biomedical applications: mechanical properties and bioactivity," *Acta Biomaterialia*, vol. 5, no. 1, pp. 356–362, 2009.

[4] A. A. El-Kheshen, F. A. Khaliafa, E. A. Saad, and R. L. Elwan, "Effect of Al_2O_3 addition on bioactivity, thermal and mechanical properties of some bioactive glasses," *Ceramics International*, vol. 34, no. 7, pp. 1667–1673, 2008.

[5] S. Melchers, T. Uesbeck, O. Winter, H. Eckert, and D. Eder, "Effect of aluminium ion incorporation on the bioactivity and structure in mesoporous bioactive glasses," *Chemistry of Materials*, vol. 28, no. 10, pp. 3254–3264, 2016.
[6] N. Y. Mikhailenko, E. E. Stroganova, and N. V. Buchilin, "Solubility of calcium phosphate glasses and glass ceramic materials in water and physiological media," *Glass and Ceramics*, vol. 70, no. 3-4, pp. 3-4, 2013.
[7] E. M. A. Hamzawy, O. A. Alharbi, and D. Y. Zaki, "Characterization and bioactivity in high cristobalite-nepheline-apatite glass and glass ceramics," *Interceram*, vol. 65, pp. 32–36, 2016.
[8] J. Slane, J. Vivanco, J. Meyer, H.-L. Ploeg, and M. Squire, "Modification of acrylic bone cement with mesoporous silica nanoparticles: effects on mechanical, fatigue and absorption properties," *Journal of the Mechanical Behavior of Biomedical Materials*, vol. 29, pp. 451–461, 2014.
[9] R. Vaishya, M. Chauhan, and A. Vaish, "Bone cement," *Journal of Clinical Orthopaedics and Trauma*, vol. 4, no. 4, pp. 157–163, 2013.
[10] A. L. Fernandes da Silva, A. M. Borba, N. R. Simão, F. L. M. Pedro, A. H. Borges, and M. Miloro, "Customized polymethyl methacrylate implants for their construction of craniofacial osseous defects," *Case Reports in Surgery*, vol. 2014, Article ID 358569, 8 pages, 2014.
[11] D. Rentería-Zamarrón, D. A. Cortés-Hernández, L. Bretado-Aragón, and W. Ortega-Lara, "Mechanical properties and apatite-forming ability of PMMA bone cements," *Materials and Design*, vol. 30, no. 8, pp. 3318–3324, 2009.
[12] T. Kokubo and H. Takadama, "How useful is SBF in predicting in vivo bone bioactivity?," *Biomaterials*, vol. 27, no. 15, pp. 2907–2915, 2006.
[13] M. Marques, "Simulated biological fluids with possible application in dissolution testing," *Dissolution Technologies*, vol. 18, no. 3, pp. 15–28, 2011.
[14] L. Chen, D. Zhai, Z. Huan et al., "Silicate bioceramic/PMMA composite bone cement with distinctive physicochemical and bioactive properties," *RSC Advances*, vol. 5, no. 47, pp. 37314–37322, 2015.
[15] *ASTM F451-08 Standard Specification for Acrylic Bone Cement*, 2003, http://www.astm.org/Standards/F451.htm.
[16] Z. Amjad, *Calcium Phosphates in Biological and Industrial Systems*, p. 515, Kluwer Academic Publishers, New York, NY, USA, 1998.
[17] D. Wood and R. Hill, "Glass ceramic approach to controlling the properties of a glass-ionomer bone cement," *Biomaterials*, vol. 12, no. 2, pp. 164–170, 1991.
[18] U. Meyera, D. H. Szulczewskia, R. H. Barckhaus, M. Atkinson, and D. B. Jonesa, "Biological evaluation of an ionomeric bone cement by osteoblast cell culture methods," *Biomaterials*, vol. 14, no. 12, pp. 917–924, 1993.
[19] M. G. Cerruti, D. Greenspan, and K. Powers, "An analytical model for the dissolution of different particle size samples of bioglass in Tris buffered solution," *Biomaterials*, vol. 26, no. 24, pp. 4903–4911, 2005.
[20] D. C. Greenspan, I. P. Zhong, and G. P. La Torre, "Effect of surface area to volume ratio on in vitro surface reactions of bioactive glass particulates," *Bioceramics*, vol. 7, pp. 55–60, 1994.

Permissions

All chapters in this book were first published in AMSE, by Hindawi Publishing Corporation; hereby published with permission under the Creative Commons Attribution License or equivalent. Every chapter published in this book has been scrutinized by our experts. Their significance has been extensively debated. The topics covered herein carry significant findings which will fuel the growth of the discipline. They may even be implemented as practical applications or may be referred to as a beginning point for another development.

The contributors of this book come from diverse backgrounds, making this book a truly international effort. This book will bring forth new frontiers with its revolutionizing research information and detailed analysis of the nascent developments around the world.

We would like to thank all the contributing authors for lending their expertise to make the book truly unique. They have played a crucial role in the development of this book. Without their invaluable contributions this book wouldn't have been possible. They have made vital efforts to compile up to date information on the varied aspects of this subject to make this book a valuable addition to the collection of many professionals and students.

This book was conceptualized with the vision of imparting up-to-date information and advanced data in this field. To ensure the same, a matchless editorial board was set up. Every individual on the board went through rigorous rounds of assessment to prove their worth. After which they invested a large part of their time researching and compiling the most relevant data for our readers.

The editorial board has been involved in producing this book since its inception. They have spent rigorous hours researching and exploring the diverse topics which have resulted in the successful publishing of this book. They have passed on their knowledge of decades through this book. To expedite this challenging task, the publisher supported the team at every step. A small team of assistant editors was also appointed to further simplify the editing procedure and attain best results for the readers.

Apart from the editorial board, the designing team has also invested a significant amount of their time in understanding the subject and creating the most relevant covers. They scrutinized every image to scout for the most suitable representation of the subject and create an appropriate cover for the book.

The publishing team has been an ardent support to the editorial, designing and production team. Their endless efforts to recruit the best for this project, has resulted in the accomplishment of this book. They are a veteran in the field of academics and their pool of knowledge is as vast as their experience in printing. Their expertise and guidance has proved useful at every step. Their uncompromising quality standards have made this book an exceptional effort. Their encouragement from time to time has been an inspiration for everyone.

The publisher and the editorial board hope that this book will prove to be a valuable piece of knowledge for researchers, students, practitioners and scholars across the globe.

List of Contributors

Shiyue Pan, Fei Chen, Qiang Shen and Lianmeng Zhang
State Key Laboratory of Advanced Technology for Materials Synthesis and Processing, Wuhan University of Technology, Wuhan 430070, China

Meijuan Li
School of Chemistry, Chemical Engineering and Life Sciences, Wuhan University of Technology, Wuhan 430070, China

Zhixiong Huang
School of Materials Science and Engineering, Wuhan University of Technology, Wuhan 430070, China

Miguel Couto
Earth Sciences Department, School of Sciences, University of Minho, 4710-057 Braga, Portugal

Jorge Sanjurjo-Sánchez
Instituto Universitario de Xeoloxía "Isidro Parga Pondal", University of A Coruña, 915001 A Coruña, Spain

Carlos Alves
Earth Sciences Department, LandS/Lab2PT-Landscape, Heritage and Territory Laboratory (FCT UID/AUR/04509/2013, FEDER COMPETE POCI-01-0145-FEDER-007528), School of Sciences, University of Minho, 4710-057 Braga, Portugal

Haiting Xia, Rongxin Guo and Feng Yan
Faculty of Civil Engineering and Mechanics, Kunming University of Science and Technology, Kunming, Yunnan 650500, China
Key Laboratory of Yunnan Province for Disaster Reduction in Civil Engineering, Kunming University of Science and Technology, Kunming, Yunnan 650500, China

Heming Cheng
Faculty of Civil Engineering and Mechanics, Kunming University of Science and Technology, Kunming, Yunnan 650500, China

Jia-xiang Wu, Xiang Fang, Zhen-ru Gao, Huai-xi Wang, Jun-yi Huang, Shuang-zhang Wu and Yu-chun Li
College of Field Engineering, PLA Army Engineering University, Hou Biaoying Road No. 88, Nanjing, Jiangsu 210007, China

Jun-yi Huang, Xiang Fang, Yu-chun Li, Bin Feng, Huai-xi Wang and Kai Du
College of Field Engineering, PLA University of Science and Technology, Hou Biaoying Road No. 88, Nanjing, Jiangsu 210007, China

José Manuel Gómez-Soberón
Barcelona School of Building Construction (EPSEB), Polytechnic University of Catalonia, Barcelona 08028, Spain

Francisca Guadalupe Cabrera-Covarrubias and Jorge Luis Almaral-Sánchez
Faculty of Engineering Mochis, Autonomous University of Sinaloa, Los Mochis 81210, Mexico

María Consolación Gómez-Soberón
Civil Engineering School, Metropolitan Autonomous University, Mexico City, Mexico

Lang Bai, Yan Li, Mingshun Yang, Zimeng Yao and Zhiyuan Yao
School of Mechanical and Precision Instrument Engineering, Xi'an University of Technology, Xi'an, Shaanxi 710048, China

Tianbo Hua, Xingguo Yang, Qiang Yao and Hongtao Li
State Key Laboratory of Hydraulics and Mountain River Engineering, Sichuan University, Chengdu, Sichuan 610065, China
College of Water Resource and Hydropower, Sichuan University, Chengdu, Sichuan 610065, China

Hyung-Jun Kim
Hazard Mitigation Evaluation Technology Center, Korea Conformity Laboratories, Cheongju 28115, Republic of Korea

Won-Jun Park
Department of Building System Engineering, Kangwon National University, Samcheok 25913, Republic of Korea

J. Jakubowicz, M. Sopata, G. Adamek and T. Kachlicki
Institute of Materials Science and Engineering, Poznan University of Technology, Jana Pawla II 24, 61-138 Poznan, Poland

P. Siwak
Institute of Mechanical Technology, Poznan University of Technology, Piotrowo 3, 60-965 Poznan, Poland

Ziyang Zhao and Jun Zhang
Key Laboratory of Transportation Tunnel Engineering, Ministry of Education, Chengdu 610031, China
Key Laboratory of Highway Construction and Maintenance Technology in Loess Region, Shanxi Transportation Research Institute, Taiyuan 030006, China

Yuandong Wang
College of Geography and Planning, Gannan Normal University, Ganzhou 341000, China
Key Laboratory of Digital Earth Science, Institute of Remote Sensing and Digital Earth, Chinese Academy of Sciences, Beijing 100094, China

Xibin You, Lianfang Yu, Lihong Meng, Xiangming Xu and Guangxu Liu
College of Geography and Planning, Gannan Normal University, Ganzhou 341000, China

Jingqin Wang, Ying Zhang and Huiling Kang
State Key Laboratory of Reliability and Intelligence of Electrical Equipment, Hebei University of Technology, Tianjin 300130, China

Byung Jae Lee
R&D Center, JNTINC Co. Ltd., 9 Hyundaikia-ro, 830 Beon-gil, Bibong-Myeon, Hwaseong, Gyeonggi-do 18284, Republic of Korea

Seong-Hoon Kee
Department of Architectural Engineering, Dong-A University, 37 Nakdong-Daero, 550 Beon-gil, Saha-gu, Busan 49315, Republic of Korea

Taekeun Oh
Department of Safety Engineering, Incheon National University, 119 Academy-ro, Yeonsu-gu, Incheon 22012, Republic of Korea

Yun-Yong Kim
Department of Civil Engineering, Chungnam National University, 99 Daehak-ro, Yuseong-gu, Daejeon 34134, Republic of Korea

Roberto Baca Arroyo
Department of Electronics, National Polytechnic Institute, 07738 Mexico City, Mexico

Zijun Wang and Boming Zhao
School of Civil Engineering, Beijing Jiaotong University, Beijing, China

A. C. D. Royal
School of Civil Engineering, University of Birmingham, Edgbaston, Birmingham, UK

Yuichiro Tashiro, Satoshi Komasa, Hiroshi Nishizaki and Joji Okazaki
Department of Removable Prosthodontics and Occlusion, Osaka Dental University, 8-1 Kuzuhahanazono-cho, Hirakata-shi, Osaka, Japan

Akiko Miyake
Department of Oral Health Engineering, Faculty of Health Sciences, Osaka Dental University, 1-4-4 Makino-honmachi, Hirakata-shi, Osaka, Japan

Jing-Qin Wang, Hui-Ling Kang and Ying Zhang
Province-Ministry Joint State Key Laboratory of Reliability and Intelligence of Electrical Equipment, Hebei University of Technology, Tianjin 300130, China

Even W. Hovig, Amin S. Azar, Frode Grytten and Knut Sørby
Norwegian University of Science and Technology (NTNU), Department of Production and Quality Engineering, Trondheim, Norway

Erik Andreassen
SINTEF Industry, Oslo, Norway

R. El-Nabouch, P. Perrotin and O. Plé
Univ. Grenoble Alpes, Univ. Savoie Mont Blanc, CNRS, LOCIE, 73000 Chambery, France

Q.-B. Bui
Sustainable Developments in Civil Engineering Research Group, Faculty of Civil Engineering, Ton Duchang University, Ho Chi Minh City, Vietnam

Dalia Y. Zaki
Restorative and Dental Materials Department, National Research Center, Cairo, Dokki 12622, Egypt

Esmat M. A. Hamzawy
Glass Research Department, National Research Center, Cairo, Dokki 12622, Egypt

Index

www.ingramcontent.com/pod-product-compliance
Lightning Source LLC
LaVergne TN
LVHW060603210926
848485LV00005B/1364